Ninth International Conference on Environmental Science and Technology (ICEST 2018)

Houston, Texas, USA
25-29 June 2018

Editors:

George A. Sorial **Jihua Hong**

ISBN: 978-1-5108-8079-5

Printed from e-media with permission by:

Curran Associates, Inc.
57 Morehouse Lane
Red Hook, NY 12571

Some format issues inherent in the e-media version may also appear in this print version.

Printed by Curran Associates, Inc. (2019)

For permission requests, please contact American Academy of Sciences
at the address below.

American Academy of Sciences
9720 Town Park Drive, Suite 18
Houston TX 77036
USA

Phone: 504-486-6332

info@AASci.org

Additional copies of this publication are available from:

Curran Associates, Inc.
57 Morehouse Lane
Red Hook, NY 12571 USA
Phone: 845-758-0400
Fax: 845-758-2633
Email: curran@proceedings.com
Web: www.proceedings.com

Environmental Science and Technology

(2018)

Edited by

George A. Sorial
Jihua Hong

Library of Congress Cataloging-in-Publication Data

Environmental Science and Technology 2018

Proceedings from the 8th International Conference on Environmental Science and Technology, held on June 25-29, 2018 in Houston, Texas, USA

Includes bibliographical references

I. Sorial, George A.
II. Hong, Jihua
III. International Conference on Environmental Science and Technology
(9th : 2018 : Houston : Texas)

Email: press@AASci.org
Website: www.AASci.org/conference/env

Environmental Science and Technology
(2018)

Edited by

George A. Sorial
Jihua Hong

American Science Press, Houston, USA

TABLE OF CONTENTS

Drinking Water

Water Quality Assessment/Management

Nitrogen-Phosphorus Wastewater Treatment

Sludge Treatment

Municipal Wastewater Biotreatment

Industrial Wastewater Biotreatment

Adsorption/Desorption for Wastewater Treatment

Physico-chemical Wastewater Treatment

Reactions and Degradation of Wastewater Contaminants

Nanotechnology Applications

AIR POLLUTION AND AIR QUALITY CONTROL

Aerosol / Air Quality Assessment

Transport of Pollutants

Waste Gas Control Techniques

Air Pollutant Monitoring

Hazardous Gas Biofiltration / Catalysts for Reducing Emission

Catalysts for Reducing Emission

LAND (SOIL, SOLID WASTE) POLLUTION AND REMEDIATION

Contaminants in the Subsurface / Natural Attenuation of Contaminants

Solid Waste Management

Waste Site Remediation / Waste Recycling

Human Exposure

Bio-response and Ecotoxicology

Bioavailability and Bio-accumulation

Microbiology and Microbial Degradation

WETLANDS AND SEDIMENTS

06-1 Wetland Conservation

Assessment of Sediments

Remediation of Contaminated Sediments

GLOBAL CHANGE GLOBAL WARMING AND CLIMATE CHANGE

Global Warming and its Impacts

Carbon Discharge Reduction

METALS

Metal Distribution

Metal Removal and Remediation

Metal Processing

MODELING

Environmental Simulation

Water Quality Modeling

GIS, DATABASE, AND REMOTE SENSING

GIS for Environmental Assessment

12-2 Data Management and Statistics

ENVIRONMENTAL ANALYSIS AND MEASUREMENTS

Environmental Analysis

Field Measurment Technologies

New Method Applications

Environmental Monitoring

SOCIETY AND THE ENVIRONMENT

Society and the Environment / Environmental Ethics and Laws

SUBSTAINABLE DEVELOPMENT AND ENVIRONMENTAL MANAGEMENT

Environmental Quality and Planning

Substainable Development

INTRODUCTION

The Ninth International Conference on Environmental Science and Technology 2018 was held on June 25-29, 2018 in Houston, Texas, USA. The Program included 15 sections, containing 60 sessions with approximately 400 platform and poster presentations. This conference series strives to provide a platform for an extremely diverse group of environmental topics for engineers and scientists from around the world.

Authors of the presentations accepted for the program were invited to submit their papers to the Conference Organizing Committee. These papers were received and then reviewed by the editors, session chairs, and the members of the Scientific/Technical Committee of the conference. The papers and abstracts accepted for publication were assembled into this volume.

Sections are arranged basically according to their order listed in the original program. The conference also consisted of having a plenary session with three speakers from different universities at the United States.

Environmental Science and Technology 2018 contains the following sections:

- Plenary Presentations
- Water Pollution and Water Quality Control
- Air Pollution and Air Quality Control
- Land (Soil, Waste Solid) Pollution and Remediation
- Ecosystem Assessment and Restoration
- Bio-Assessment and Toxicology
- Wetlands and Sediments
- Global Warming and Climate Change
- Metals (Distribution, Removal, Remediation, Speciation, and Phytoremediation)
- Modeling
- GIS, Database, and Remote Sensing Applications
- Environmental Analysis and Measurements
- Society and the Environment
- Substainable Development and Environmental Management

• Renewable Energy Development

We would like to especially thank the session chairs who were instrumental in the success of the conference. The Conference was sponsored and organized by the American Academy of Sciences, with financial contributions from the co-sponsors and supporting organizations.

The papers in these proceedings represent the authors' results and opinions. No sponsors, cosponsors, participating organizations or editors should be construed as endorsing any specific contents or conclusions in the proceedings.

George A. Sorial, Ph.D.

University of Cincinnati

Jihua Hong, Ph. D.

American Academy of Sciences

PLENARY PRESENTATIONS

Engineering Applications in Food, Energy, and Water Nexus

Bahram Asiabanpour
(Ingram School of Engineering, Texas State University, San Marcos, TX 78666, USA;
ba13@txstate.edu)

This presentation will look at the applications of the engineering methods in design, development, and optimization of the processes in the Food, Energy, and Water nexus. Specifically the development of the Ever***Green*** project, an off grid hydroponic system, including its subsystems in soil-free agriculture, atmospheric water generation, natural light and renewable energy collection and delivery, smart technology systems, and their effects on the development of an Agriculture-STEM education pipeline will be demonstrated.

To See or Not To See the Shortcomings of Priority Pollutants - The Case of the 16 EPA PAHs

Jan T. Andersson
(University of Muenster, 48149 Muenster, Germany; anderss@uni-muenster.de)

In 1976 the Environmental Protection Agency (EPA) put together a list of 16 polycyclic aromatic hydrocarbons (PAH) as "priority pollutants" for water quality studies. It consists of unsubstituted hydrocarbons, from the two-ring compound naphthalene up to two six-ring aromatics. At that time this was a reasonable selection and fulfilled the expectations demanded of it. Notwithstanding the fact that water quality studies were the aim of the list, over the years it has been diverted from its intended use and employed for very different other purposes. The suitability of these compounds for many of these purposes - but also for its original intent - is now dubious.

In the over 40 years since then, environmental research has grown dramatically. It is clear that today this limited set of compounds is unsatisfactory for most purposes for which it is currently used but this is rarely acknowledged. Therefore the question arises how to proceed in such a situation. Should a discussion be initiated to investigate if not other sets of pollutants would meet the demands of researchers and regulators better?

To advance such a discussion, the benefits of finding a replacement that reflects the demands of environmental studies in this century need to be highlighted. Possibly the outcome will be that one set of pollutants will not be enough to cover such a wide field of research and monitoring adequately but several, each designed with a particular purpose in mind. Is that a loss or a gain?

On the other hand, we need to look at what makes this list so attractive and widely used and therefore what would be possible (and plausible) arguments for retaining it, despite its obvious shortcomings.

These thought-provoking questions will be discussed and illustrated with examples from the current literature and legislation.

Production, Evaluation and Application of Commercial Grade Biochar and Activated Carbon for Contaminant Adsorption

Hafiz Ahmad
(Florida State University, Panama City, FL 32405, USA; hahmad@pc.fsu.edu)

Carbonaceous materials, such as biochar, activated carbon, have been the subject of increased interest due to their unique physico-chemical characteristics (e.g., large surface area, high micro-porosity, and superb sorption capacities) and potential value in capturing pollutants, soil amendment. This presentation will review our recent efforts on commercial production, application and evaluation of these value-added carbon products.

In the first part, the two general methods of interest involving biomass-pyrolysis and char-activation will be reviewed. Then application of these two principles/methodologies in our state of art pilot plants that are capable of producing various grades of biochar, bio-oil, biogas, and activated carbon will be shown. The second part of this talk will summarize the extensive lab-work to determine the feasibility of utilizing our two major products (biochar and activated carbon) for commercial application (contaminant adsorption and soil amendment). Specifically, the summary of some of our experimental batch adsorption test results to remove pollutants such as phenol, COD from wastewater will be presented. Finally, the third part of this presentation will show our pilot-scale study of in-line application of Powdered Activated Carbon (PAC) in a municipal wastewater treatment plant using air-diffusion process and quick removal of the PAC from the waste-stream using Cloth Media Filtration (CMF) system. The results will argue that this PAC application and quick removal process can reduce the challenges of the treatment facilities where the existing hydraulic profile do not allow the use of large tanks that are normally associated with long retention time for PAC removal.

WATER POLLUTION
AND
WATER QUALITY CONTROL

Hydrochemistry of the Hyporheic Zone in Coastal California Watersheds Dominated by Monterey-Modelo Formation Substrate

Marissa De Hoyos and Barry Hibbs
(California State University, Los Angeles, CA, USA)

ABSTRACT: Hyporheic zone is the interstitial environment in streambeds and streambanks below and adjacent to streams, where active fluid flux and biogeochemical cycling occurs during surface/subsurface interactions. We investigated hyporheic zone processes in the Malibu Creek Watershed which feeds into the marine environment at the Pacific coastline. Earlier studies by researchers suggested reaction with geologic strata by percolating groundwater may be an important source of nutrients (P and N) during dry weather flow conditions (2 weeks after any rain event). Miocene marine shales comprised of the Monterey-Modelo Formation is the dominant substrate thought to contribute nutrients to the watershed. Our studies of hyporheic zone in one part of Malibu Creek Watershed, along a site in Las Virgenes Creek, indicates urban runoff of both tap and recycled water is an important contributing source of nutrient loading, although some geologic input is also evident. We addressed the issue by studying hyporheic zone processes at the site in Las Virgenes Creek receiving tap water runoff and recycled water runoff, along with normal baseflow. Where recycled water is used the data indicate nutrient loading and important biogeochemical processes take place in the hyporheic zone, including denitrification and/or dissimilatory nitrate reduction (producing ammonium) and gradual accumulation of phosphate by evapotranspiration and sorption/desorption processes. Selenium concentrations vary but reveal a strong correlation between nitrate and selenium concentrations in the hyporheic zone. A model of selenium oxidation and mobilization from reduced forms in marine strata is proposed with nitrate acting as an oxidizing agent. Stable water isotopes indicate that stream water at the study site consists mostly of locally derived water from recharge of local precipitation, with some imported water from urban water amounting to about 15 to 20% of the total streamflow. Hyporheic zone samples contain 5 to 20% imported water. Two hyporheic samples are isotopically distinct from other samples. The two distinct samples have very different nutrient content compared to other hyporheic zone samples and stream water, implying a unique nutrient source.

INTRODUCTION

Treated wastewater is brought from municipal wastewater treatment plants for spreading on urban landscapes. Due to drought and imported water cutbacks, treated water use has risen in California as a substitute for use of potable water. Treated wastewater can make its way into local watersheds and shallow aquifers via runoff after irrigation watering and from infiltration and leakage from pipes carrying recycled water to urban landscapes. Treated wastewater has higher concentrations of nitrate than most potable waters, often up to 10 mg/L NO3-N, while tap water from state project sources runs about 1 mg/L NO3-N (Horns and Kim, 2007). Nitrate has been proven to be an oxidizing agent that can release toxic elements such as selenium contained naturally in geologic strata (Wright, 1999). The Monterey-Modelo Formation in California is known to be a selenium bearing stratum (Harrison, 2016). Water quality studies in the Malibu Watershed indicate this formation is one of the principle stratum found in the watershed (Horns and Kim, 2007). Our studies test three processes and properties: 1) is recycled water and tap irrigation creating higher loads of nutrients in urban watersheds, specifically Las Virgenes Creek? 2) what are the biogeochemical processes occurring in the hyporheic zone? and 3) is selenium being oxidized by nitrate in groundwater near or in the hyporheic zone?

Previous studies showed moderate to high selenium concentrations (20 to 75 ug/L) in creeks where Monterey-Modelo formation is the principal outcrop, and where nitrate from unknown sources is elevated above 2 mg/L NO3-N (Hibbs et al., 2009; Harrison, 2016). This project may illuminate the effects of recycled treated water on the environment. As the demand for water rises with the increase of population throughout California and the United States, treated wastewater is flowing through rivers and creeks and draining into coastal receiving waters. By studying the nutrient and trace element content of these water systems, it will create understanding of the effects of recycled water on the environment. Since recycled water application and use is expanding immensely in coastal California, the research will examine possible effects due to leaching of geologic strata that is known to contain selenium, uranium, and other oxidizable elements (Horns and Kim, 2007). Looking at recycled water use and nutrient and trace element loading from groundwater seepage will help in understanding these agents and how they react with local geology.

At the scientific level, the work will expand understanding of geologic role in trace element loading when oxidizing agents such as nitrate from recycled water are applied to urban and other landscapes. The uncertainties evaluated include what concentrations of nitrate and what concentrations of dissolved oxygen will release toxic trace elements into groundwater, which is loaded into watersheds by groundwater seepage during the lengthy dry season in California (Hibbs et. al., 2012).

STUDY AREA

The Las Virgenes Creek is a major tributary stream in Malibu Creek Watershed with outfall leading directly into the Pacific Coast (Figure 1). Las Virgenes Creek is the largest tributary to Malibu Creek, the second-largest stream flowing into Santa Monica Bay (CDM, 2006). Several water quality monitoring programs have reported that water quality is impacted as it flows through the City of Calabasas, and the creek is listed as impaired under the Clean Water Act Section 303(d). Impairments include nitrate, selenium, and organic enrichment. This creates speculation that these impairments may be due to urban runoff (National Parks Conservation Association, 2008).

Urbanization of Las Virgenes Creek Watershed has occurred in two stages. Most of the urban area above the 101 Freeway was already developed to present levels by 1976 (Figure 1). By 1994, most of the areas below the 101 Freeway were built out to present density (Owens, 2001). Additional development has taken place sporadically in a few areas since 1994. The area studied in Las Virgenes Creek extends for 10 km from the natural wildlands of Ahmanson Ranch, through urbanized Calabasas, to the natural open space of White Oak Farm. Ahmanson Ranch and White Oak Farm are now government owned and state park properties used for outdoor recreation.

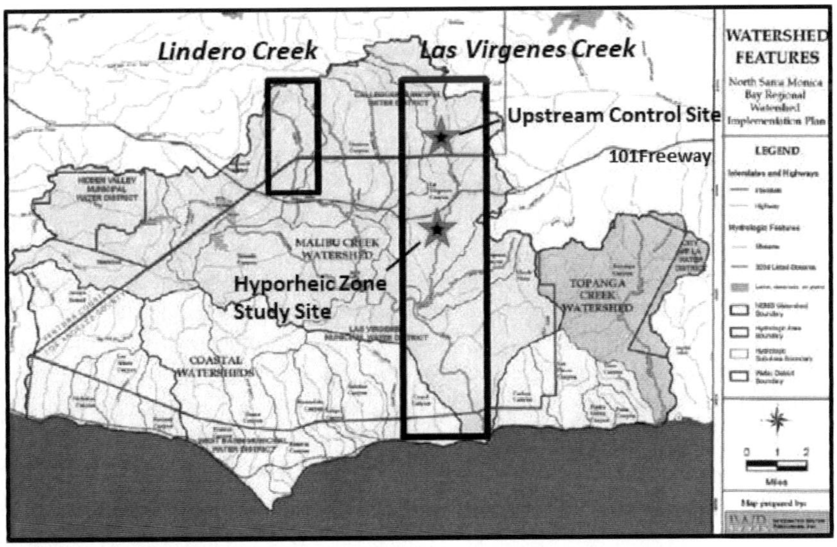

FIGURE 1. The Santa Monica Mountains is divided into three watersheds areas: the Malibu Creek Watershed, the Coastal Watersheds and the Topanga Creek Watershed where numerous creeks and tributaries empty into the Pacific Ocean. The study area takes place within the Malibu Creek Watershed near where the Las Virgenes Creek intersects with the 101 Freeway (modified from Integrated Water Resources Inc., 2006).

Along the first part of this stream reach, Las Virgenes Creek flows across the Monterey-Modelo Formation and its eroded alluvial valley fill products. The Monterey-Modelo Formation is a white punky diatomaceous rich mudstone/turbidite complex of Miocene age (Rumelhart and Ingersoll, 1997). Further downstream, Las Virgenes Creek flows across outcrop of the Tertiary volcanic and intrusive rocks of the Conejo Formation. The Conejo Formation forms the core of the Santa Monica Mountains and underlies a large part of the Malibu Creek Watershed (CDM, 2006). Overlying the bedrock formations are a series of recent alluvial units formed from the weathering of the regional formations. These include stream deposits, alluvial fan and floodplain deposits, dissected and older alluvial deposits, landslides and colluviums (CDM, 2006). Stream deposits and floodplain deposits formed by Las Virgenes Creek are of interest to this study because they are in direct hydraulic contact with the creek. The study areas extends through open space, through urbanized section of stream in the City of Calabasas, and reenters open space area below Calabasas.

METHODS

The hyporheic zone is the area within the surface and/or substrate below ground where groundwater and surface water mix (Figure 2). Within the study area seven sampling sites were designated for testing of the hyporheic zone, including a surface water sampling site and six hyporheic zone samples in both in-channel and adjacent-channel locations (Figure 3). Sample stations 2, 5, and 6 are adjacent-channel locations and samples stations 1, 3, and 4 are in-channel sample locations. The downstream surface water sampling station was taken near stations 1 and 2. These stations are below an urban reach of the City of Calabasas. In addition, an upstream surface water control site was sampled in open space in an undeveloped area above the City of Calabasas (Figure 1). The upstream site measures ambient nutrient concentrations in the stream.

Surface water samples were collected using designated high-density polyethylene plastic bottles at designated stations. The bottles were labeled with the date, time and location prior to collecting sample. Sample bottles were rinsed with sample three times prior to filling to avoid possible contamination. All samples were filled to the rim and capped without leaving room for air bubbles to form to avoid any possible evaporation. Samples were placed in an ice chest after being collected.

For hyporehic zone studies, a mini piezometer and peristaltic pump were used to collect sample. The pump tubing was inserted into the mini piezometer (Figure 4) where hyporheic water was extracted. The peristaltic pump was set at a flow rate of about 0.25 to 0.40 liters per minute to avoid induced flow away from the piezometer screen. After purge time of about 10 minutes, a clean plastic bottle was used to collect sample after it was rinsed.

Prior to collecting samples, specific conductance, temperature, pH, and dissolved oxygen were measured for each station using calibrated YSI field meters. Laboratory analysis was performed in accordance with Standard and EPA Methods. Nutrient analysis was performed by spectrophotometry in the Hydrogeology Laboratory at California State University-Los Angeles using EPA approved or equivalent procedures. Tests were done for nitrate, orthophosphate, and ammonium-nitrogen. Methods included laboratory procedures equivalent to EPA procedures EPA 350.1, EPA 351.1, EPA 351.2, EPA 365.1, and 40 CFR 141. Selenium was tested by ICP-MS by Weck Laboratories, following standard methods. Stable isotope measurements of water were made at the Laboratory of Isotope Geochemistry at the University of Arizona. The hydrogen and oxygen isotopic composition of water were determined using a Finnigan Delta-S Isotope Ratio Mass Spectrometer (IRMS) following reduction with Cr and CO2 equilibration. Results were expressed as δ^2H and $\delta^{18}O$ in per mil (‰) relative to the standard VSMOW with analytical precisions of 0.9 ‰ and 0.08 ‰, respectively.

Sampling of streamflow and hyporheic zone was done in January and August 2018. Some additional data on nutrient and stable water isotopes in tap water and treated wastewater were taken from Hibbs et al., (2012) and Harrison (2016). Water isotope data were plotted with reference to the global meteoric water line. Other data were plotted as bivariate scatter plots. Mixing lines and evaporation lines were characterized in the standard plotting procedures.

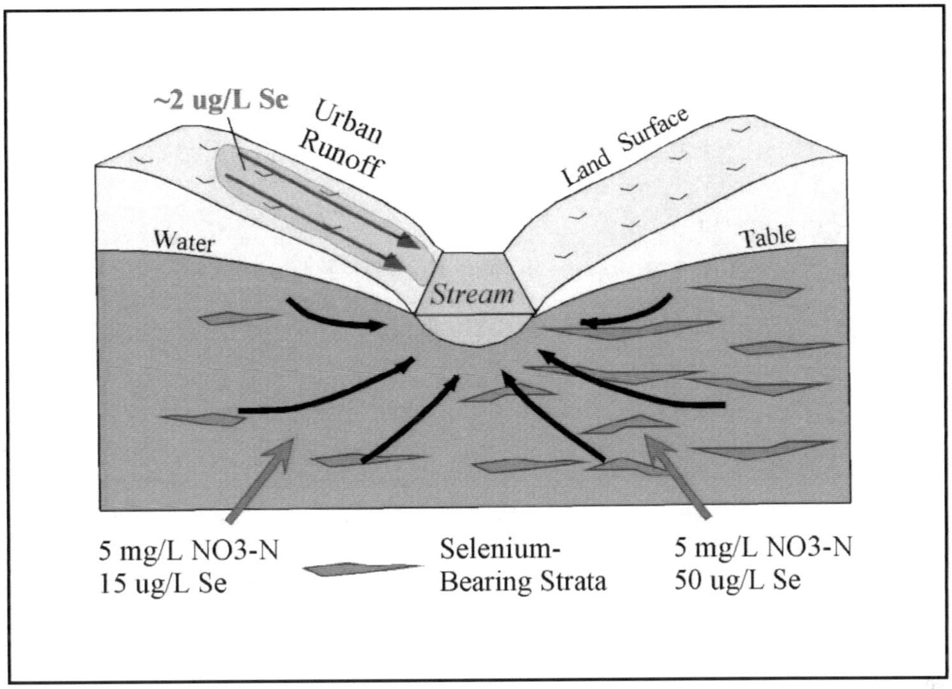

FIGURE 2. The hyporheic zone is illustrated. The mixing of groundwater and surface water in the hyporheic zone allows the leaching of selenium into local groundwater where selenium bearing strata exists and where nitrate is enriched in groundwater.

FIGURE 3. The Las Virgenes Creek hyporheic zone study area. The hyporheic zone sampling sites are indicated by the red symbols. The bold arrow indicates the direction of creek flow.

FIGURE 4. A peristaltic pump (right photo) is used to draw water samples into a mini-piezometer that is installed temporarily in both in-channel and off-channel hyporheic zone sites (left photo).

RESULTS

Data results from sampling indicates that treated wastewater has nitrate-nitrogen and orthophosphate concentrations above 6.0 mg/L and variable ammonium concentrations less than 1.0 mg/L (Table 1). Tap water has nitrate-nitrogen less than 0.7 mg/L NO3-N, orthophosphate below detection limits (<0.15 mg/L PO4) and ammonium of 0.52 mg/L NH3.

Hyporheic zone water has variable concentrations of nutrients. Important common characteristics are noted for hyporheic zone samples 1, 2, 3, and 6 (Table 1). These samples contain considerable phosphate (1.0 to 3.5 mg/L PO4), moderately low to moderate nitrate (<0.23 to 1.32 mg/L NO3-N), and low to moderate ammonium (<0.015 to 0.45 mg/L NH3). Alternatively, hyporheic zone samples 4 and 5 have low orthophosphate concentrations (0.4 to 0.6 mg/L PO4), high nitrate concentrations (6.4 to 11.2 mg/L NO3-N), and low ammonium concentrations (ranging from 0.030 to 0.04 mg/L NH3). Stream water at the upstream site contains low nutrient concentrations of <0.20 mg/L NO3-N, 0.22 mg/L PO4, and 0.021 mg/L NH3. The upstream surface water data represent ambient nutrient concentrations at the natural site upstream. Stream water at the downstream hyporheic zone site contains nutrient concentrations similar to hyporheic zone sets 1, 2, 3 and 6, except nitrate is higher in the stream (1.10 to 1.45 mg/L NO3-N). Ammonium is very low (<0.077 mg/L NH3) and orthophosphate varies significantly in the stream water during the two sampling events, from 0.66 to 8.93 mg/L PO4 (Table 1).

Selenium concentrations are also variable; at or below detection limits of 0.40 ug/L Se in treated wastewater and tap water (Table 1), while varying from 2.4 to 28 ug/L Se in hyporheic zone samples. Stream selenium concentrations at the downstream site are intermediate, and register at 12.0 ug/L Se (Table 1). There is no selenium data available at the upstream Las Virgenes Creek site at the present time.

Stable water isotope data collected from Las Virgenes Creek and from hyporheic zone are plotted with tap water and treated wastewater samples from this study, and from studies published by Hibbs et al., (2012) and Harrison (2016) (Figure 5). Studies have identified stable water isotopic signatures of California State Project waters imported to the Los Angeles Basin from Northern California (USGS, 2003). By the time waters arrive at the Los Angeles Basin, State Project waters range somewhat narrowly about a mean of -9.5 per mil $\delta^{18}O$ and -75 per mil $\delta^{2}H$ (USGS, 2003). With minor variability, State Project waters sampled in Las Virgenes Creek Watershed plot within the range of isotopic signatures shown previously by the USGS, 2003 (Figure 5).

Recycled water delivered by Las Virgenes Municipal Water District ranged from -8.0 to -9.0 per mil $\delta^{18}O$ and from -63 to -67 per mil $\delta^{2}H$ (Figures 5). Local waters from several reference springs at or near the upstream of Las Virgenes Creek had constant signatures of about -5.7 per mil $\delta^{18}O$ and -43 per mi $\delta^{2}H$

(ambient local groundwater, Figure 5). Local isotopic values are consistent with mean isotopic values recorded for precipitation in the Los Angeles Basin. Las Virgenes Municipal Water District tap water, recycled water, and local groundwater are reliable end members representing defined water types in the study (Figures 5).

When plotted, the isotopic shift of treated wastewater trends along a mixing line between tap water and local reference (native local) water and is interpreted to result from the infiltration of shallow groundwater into sewer pipes through cracks in pipes which occur where sewer pipes penetrate the water table (Hibbs et al., 2012). A small amount of evaporation may have occurred at the sewage treatment plant, accounting for a smaller part of the isotopic shift between tap water and recycled water (Figure 5). Hyporheic zone and stream samples collected at or below the developed reach of Las Virgenes Creek plot along the mixing line between imported waters (both tap water and treated wastewater) and local reference groundwater (Figure 5). Slight evaporation is shown in hyporheic zone samples 1, 2, and 3 and in downstream surface water. The data indicate downstream surface water and hyphoreic zone water are mostly composed of locally derived water, with smaller amounts of imported water.

Table 1. Study Results for Nutrients and Selenium Collected from Upstream and Downstream Las Virgenes Creek Sites, Hyporheic Zone Samples, and Treated Wastewater and Tap Water Samples.

Sample	Date	PO4 (mg/l)	NO3-N (mg/l)	NH3 (mg/l)	Se (ug/l)
Las Virgenes (upstream)	06/12/2017	0.22	<0.20	0.021	n/a
Las Virgenes (downstream)	01/22/2018	0.66	1.19	0.077	12.0
	07/11/2018	8.93	1.45	0.065	n/a
Hyporheic #1	01/22/2018	1.07	<0.20	0.433	2.7
	07/11/2018	3.12	0.27	0.634	n/a
Hyporheic #2	01/22/2018	2.83	0.42	0.453	2.4
	07/11/2018	1.59	0.25	<0.015	n/a
Hyporheic #3	01/22/2018	1.00	0.72	0.203	6.3
	07/11/2018	1.18	1.32	0.086	n/a
Hyporheic #4	01/22/2018	0.41	11.20	<0.015	28.0
	07/11/2018	0.67	10.60	<0.015	n/a
Hyporheic #5	01/22/2018	0.60	6.41	0.040	13.0
Hyporheic #6	07/11/2018	3.32	0.35	0.030	n/a
Treated Wastewater	09/29/2017	7.47	6.84	0.046	<0.40
	12/18/2017	14.70	8.18	0.987	0.40
Tap Water	12/18/2017	<0.15	0.63	0.520	<0.40

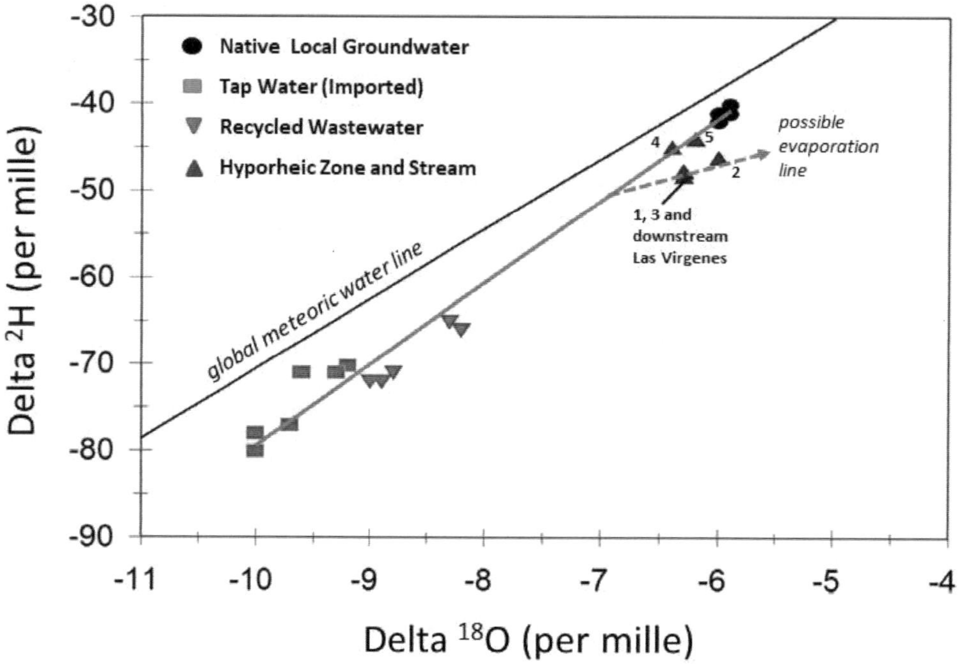

FIGURE 5. Stable water isotopes plotted for tap water, recycled (treated) wastewater, native local groundwater, hyporheic zone samples, and stream water collected at the downstream hyporheic zone site. Stable water isotopes indicate the data collected at the downstream site consists mostly of local (ambient native) groundwater.

DISCUSSION

The data collected in this study indicates that nutrients are more concentrated in downstream surface sites of Las Virgenes Creek due to dry weather urban runoff (Table 1 and Figure 6). Nutrients are very low at the upstream control site in Las Virgenes Creek, indicating an agent of nutrient addition to the creek at the downstream site (Table 1). Almost certainly these higher nutrient concentrations at the downstream site are due to dry weather urban runoff. Stable water isotopes suggest that nutrients enter the hyporheic zone in Las Virgenes Creek at samples sites 1, 2 and 3 where nitrate is probably removed from the hyporheic zone by denitrification and dissimilatory nitrate reduction, according to the respective reactions:

$$\mathbf{4NO_3^- + 5CH_2O = 2N_2 + 5HCO_3^- + H^+ + 2H_2O} \qquad \textbf{(equation 1)}$$

$$\mathbf{NO_3^- + H_2O + 2CH_2O = NH_4^+ + 2HCO_3^-} \qquad \textbf{(equation 2)}$$

Reaction 2 accounts for higher ammonium in hyporheic zone at sites 1, 2 and 3 relative to surface water (Table 1).

Stable water isotopes indicate that hyporheic zone samples 1, 2 and 3 and downstream surface water in Las Virgenes Creek are genetically linked and follow a common evaporation trend from a mixture that includes about 15 to 20% imported water (mixture of tap and recycled wastewater, Figure 5). The evaporation line extends back to the mixing line between native local groundwater and imported water. The evaporation line originates at around 15 to 20% imported water and 80 to 85% native local groundwater.

Hyporheic zone water samples 4 and 5 are not evaporated and are a different water type consisting mostly of native groundwater and about 5% imported water (Figure 5). In addition, samples 4 and 5 have the highest nitrate and selenium concentrations (Table 1 and Figure 7). The correlation between nitrate and selenium observed in Figure 7 is probably due to oxidation of selenium bound in reduced forms in sedimentary stratum and residuum, according to the equation:

$$5FeSe_2 + 14NO_3^- + 4H^+ \rightarrow 7N_2 + 10SeO_4^{2-} + 5Fe^{2+} + 2H_2O \quad \textbf{(equation 3)}$$

Nitrate is a known oxidizing agent that will raise oxidation potential in groundwater where nitrate is loaded in an aquifer. Miocene marine shales and siltstones in the Monterey-Modelo Formation may be the original sources of selenium at the study site. Selenium probably leaches into groundwater from low-permeability marine strata, along with standard inorganic constituents, primarily sulfate (Hibbs et al., 2012). Elevated selenium concentrations develop in shallow groundwater, and baseflow carries selenium into urban streams. Groundwater baseflow accounts for most of the selenium loading to streams.

Positive correlations are observed between nitrate and selenium in both groundwater and surface water in the hyporheic zone site we investigated (Figure 7). Previous theoretical calculations showed favorable Gibbs free energies for oxidation of selenium in marine shales by dissolved nitrate (Wright, 1999). Miocene marine shales are exposed in vast areas of Southern California. We are undertaking regional studies to determine if other Southern California Watersheds are affected by elevated nitrate and selenium concentrations, and where selenium may threaten habitat.

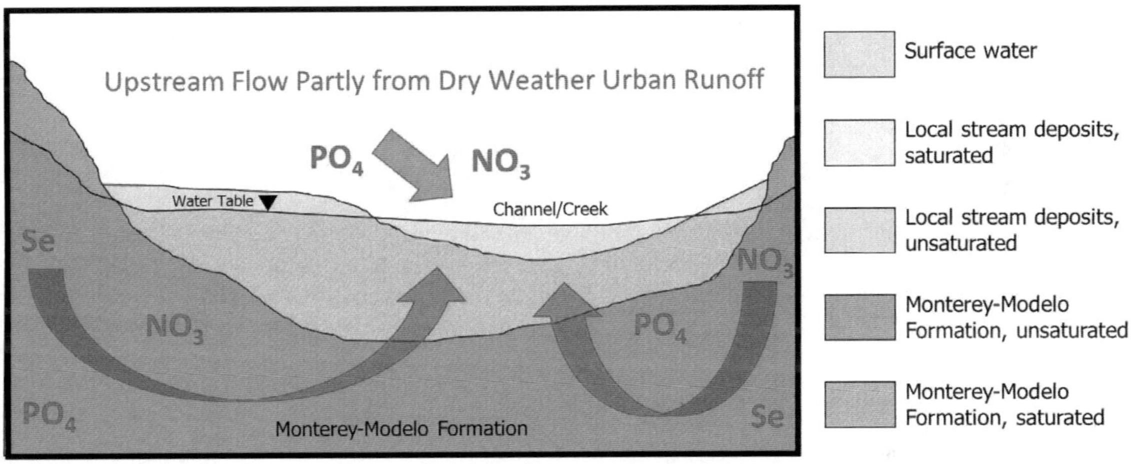

FIGURE 6. Model for hyporheic zone and surface water mixing showing nitrate and orthophosphate loading to surface streams due to dry weather urban runoff, along with oxidation of selenium contained in Monterey-Modelo strata by dissolved nitrate (modified from Harrison, 2016).

CONCLUSIONS

Study results indicate a complex set of biogeochemical processes in the hyporheic zone affecting nutrient and trace element availability and possible solute loading to Las Virgenes Creek. Stable water isotopes indicate most of the water in the hyporheic zone and stream water at the downstream site is locally derived, with imported water (tap and recycled) making up 5 to 20% of the water content. Hyporheic zone samples in the lower part of the study site (samples 4 and 5) were collected at the bottom of a riffle where a unique water type was determined. The unique water type showed no evaporation, about 5% imported water, high nitrate and selenium concentrations, and low orthophosphate concentrations. This water comes from groundwater baseflow and is not genetically related to surface water and other hyporheic zone water at the same site. Surface water and hyporheic zone water at the upstream portion of the study site are slightly evaporated and are similar water type. Complex processes that are probably occurring in the upstream hyporheic zone include denitrification and/or dissimilatory nitrate reduction. These chemical processes remove nitrate. Dissimilatory nitrate reduction also adds ammonium in the hyporheic zone. Orthophosphate is mostly from urban runoff and it becomes enriched due to evaporative concentration in unsaturated soils above the hyporheic zone. Selenium concentrations are correlated to nitrate concentrations and probably result from nitrate-facilitated oxidation of selenium that is bound in marine strata. Anthropogenic nitrate, a known oxidizing agent, raises oxidation potential allowing reduced forms of selenium in the strata to speciate into the soluble aqueous phase.

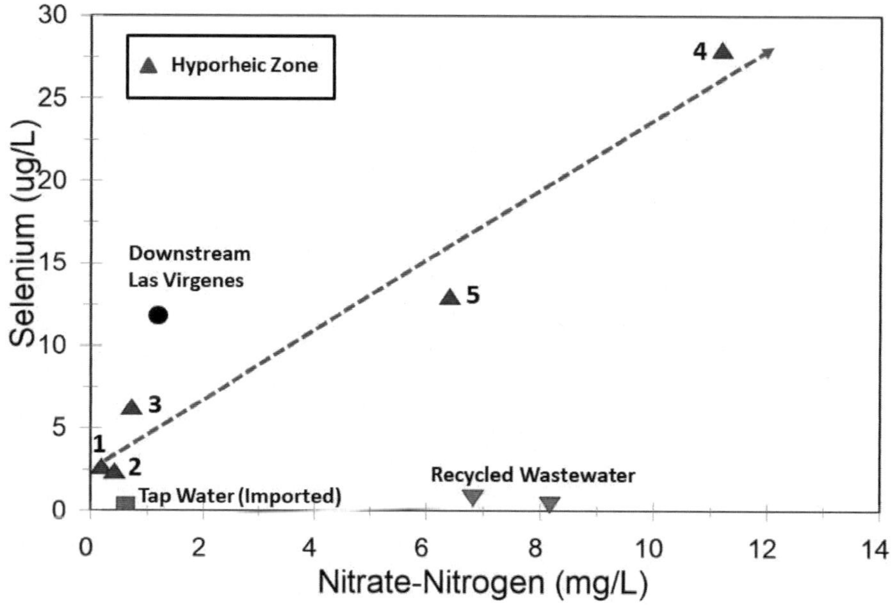

FIGURE 7. Selenium versus nitrate scatter plot for sample sites in Figure 3 and Table 1. A clear correlation exists in the hyporheic zone between nitrate and selenium. High nitrate acts as an oxidizing agent releasing selenium that is in reduced forms in marine strata and in the erosional products from weathering of marine strata near Las Virgenes Creek.

ACKNOWLEDGEMENTS

Support for this research provided by a student grant from the Geological Society of America and from the CSULA LaKretz Endowment.

REFERENCES

CDM. 2006. Technical Memorandum Task 3.2A, *Hydrogeology and Aquifer Characteristics, North Santa Monica Bay Watersheds Regional Watershed Implementation Plan and Malibu Creek Bacteria TMDL*: Prepared for County of Los Angeles Watershed Division: 14 p.

Harrison, M. 2016. *Isotopic & hydrochemical identification of source waters & pollutants in the Malibu watershed, California.* M.S. Thesis. Department of Geosciences and Environment, California State University, Los Angeles: 69 p.

Hibbs, B., M. Merino, R. Andrus, W. Hu, and A. Doro-on. 2009. *Groundwater baseflow sourced from Miocene rocks and residuals carries elevated selenium into southern California streams.* in Starrett, S., ed., Great Rivers, American Society of Civil Engineers, Reston, VA: 1919-1928.

Hibbs, B., W. Hu, and R. Ridgway. 2012. "Origin of stream flows at the wildlands-urban interface, Santa Monica Mountains, California." Environmental and Engineering Geoscience. XVIII (1): 51-64.

Horns, M. and K. Kim. 2007. *Las Virgenes Creek pollution source investigation.* City of Calabasas Environmental Services Division: 22 p.

National Parks Conservation Association. 2008. *State of the Parks, Southern California's Mediterranean Biome Parks, a Resource Assessment*: 120 p.

Owens, B. 2001. *A Protection and Revitalization Plan for Las Virgenes Creek.* M.S. Thesis. Graduate Program in Landscape Architecture, California State Polytechnic University, Pomona. various pagination.

Rumelhart, P., and R. Ingersol. 1997. "Provenance of the upper Miocene Modelo Formation and subsidence analysis of the Los Angeles Basin, Southern California, implications for paleotectonic and paleogeographic reconstructions." Geol. Soc. Am. Bull. 109 (7): 885-899.

USGS. 2003. *Geohydrology, geochemistry, and ground-water simulation optimization of the Central and West Coast Basins, Los Angeles County, California*. USGS Water Resources Investigations Report 03-4065: 196 p.

Wright, W. 1999. "Oxidation and mobilization of selenium by nitrate in irrigation drainage." Journal of Environmental Quality. 28 (4): 1182–1187.

Application of Engineered Cockle Shells and Zeolites for Phosphorus Removal in Controlling Algae Bloom in Eutrophic Water

Richard Zhang and ***Steven Wu*** (Clear Lake High School, Houston, Texas, USA)
Rio Lirag and Chunlong Zhang (University of Houston-Clear Lake, Houston, Texas, USA)

ABSTRACT: Eutrophication is a significant environmental issue caused by excessive nutrients that lead to algae growth and fish kill. The objective of this project was to develop an *in-situ* deployable product to remove phosphorus (P) for algae bloom control in eutrophic water. This study showed that the naturally abundant cockle shells and zeolites as the raw materials yielded promising P removal efficiencies (30.4% for cockle shell; 99.5% for zeolites), and the effectiveness increased as the size of engineered cockle shells decreased. Fourier transform infrared-spectrum (FTIR) indicated cockle shells and zeolites have a characteristic $CaCO_3$ peak (850 cm^{-1}) and silicate peak (1000 cm^{-1}), suggesting that precipitation and sorption are the respective mechanism. Adsorption to zeolite was found to fit Freundlich multi-layer adsorption isotherm. When applied to two local water bodies in Houston, engineering feasibility calculations on deployment dose, settling velocity using Stokes's law, and cost compared to the conventional dredging technique revealed that these novel materials are cost-effective, sustainable, and safe to use in controlling algae bloom in eutrophic water.

INTRODUCTION

Eutrophication is a widespread issue that occurs in many water bodies around the world. The excessive amount of nutrients from various sources causes an algae bloom. A notable example of large-scale disasters resulting from eutrophication is the 8,000-square mile hypoxic dead zone in the Gulf of Mexico downstream from the Mississippi River delta. The river carries nitrate- and phosphate-rich materials from all point and non-point sources in the region to the Gulf of Mexico. Algae bloom makes survival extremely difficult for aquatic life. It not only affects aquatic life, but also causes economic and recreational loss (Quigg and Thronson, 2008). Phosphorus has been identified as the limiting nutrient in algae blooms. Unlike nitrogen, phosphorus cannot be volatilized from water, making it harder to remove. Existing methods such as adsorption, precipitation, and enhanced biological removal are not effective in surface water applications. For example, adsorption is very effective in wastewater treatment plants, since sorbing materials can be easily regenerated or disposed of in a small setting. Chemical precipitation using calcium-containing materials have been effective in removing phosphorous from water (Goh et at., 2015; Klapper, 1992); however, in open water, settled phosphorus in sediment can potentially release back into water column when water conditions change. Biological treatments render phosphorous to remain in suspended bacterial cells, which can be easily removed in treatment plants as sludge (Buda et al., 2012). However, this is problematic in surface water.

The use of synthetic chemicals in open water may pose adverse effects on surface water quality. Therefore, several laboratory studies have explored the use of natural materials. Cockle shell, a calcium-containing natural material, has been used previously for phosphorous removal (Goh et al., 2015; Hariharan et al., 2014). Its enhancement was studied by the introduction of sulfate. Bioretention media enhanced by cockle shells removed up to 94 – 96% phosphorus in a laboratory experiment simulating storm water runoff (Goh et al., 2015). Zeolites, minerals that can be obtained in volcanic deposits, range from ineffective to very effective phosphorous removal depending on the type (Kapapinar, 2009). The use of such laboratory results from batch reactor studies, however, can be inadequate to extend to *in-situ* deployment and applications. Key considerations for *in-situ* uses include whether these materials are environmentally friendly and how they can be reused or regenerated.

The objectives of this study were (1) to search for a material with promise of cost-effective phosphorous removal from surface water, (2) to reveal the physicochemical mechanisms through solubility product (K_{sp}), adsorption isotherm and Fourier transform infrared-spectrum (FTIR) spectrum, and (3) to conduct engineering calculations for material feasibility such as dose and potential adverse impact on water quality, as well as cost analysis in field-scale applications using chemical and hydrological data of two local water bodies in Houston.

MATERIALS AND METHODS

Chemicals and Reagents. Calcium acetate ($Ca(CH_3COO)_2$), calcium carbonate ($CaCO_3$), and calcium chloride ($CaCl_2$) were purchased from Bakers, VWR, and Sigma respectively. Stock phosphorous solution containing 1,000 mg P/L was obtained from Merck. De-ionized (DI) water was produced by a Milli-DI Water Purification system (Millipore Co.). Zeolite was generously provided from a vendor in New Zealand and used without further preparation. Cockle shell was purchased from a local source in Houston, and was ground using a mechanical grounder (Model HM-375 Grinder, Houghton Manufacturing Co.). USA standard testing sieves No. 18 and No.10 (SoilTest, Inc.) were used to ground cockle shell into four different sizes (raw material size of 1 cm, >2mm, 1-2mm, <1mm).

ICP-OES Measurement of Phosphorous and QA/QC. Phosphorous concentrations were analyzed using direct injection of aqueous supernatant at an emission wavelength of 214.914 nm using inductively coupled plasma-optical emission spectrometry (ICP-OES) with Win32 software (PerkinElmer). A standard quality assurance/quality control (QA/QC) protocol was followed to ensure data quality (Rice et al., 2017). Triplicate measurements were conducted for every sample to minimize analytical variations. Calibration curve ($R^2 > 0.999$) was made daily using calibration solutions containing 0.1, 1.0, 10, 25, and 50 mg P/L. The method detection limit was 0.15 mg/L.

Determination of Characteristic Atoms / Functional Groups by FTIR. Cockle shell, zeolite, and pure chemicals were ground and pressured into a thin film for Fourier Transform Infrared (FTIR) analysis using Frontier IR (PerkinElmer). The FTIR spectrum of two natural materials were compared with that of pure chemicals.

Batch Studies on Phosphorous Removal. The first batch study was to compare the effectiveness of phosphorous removal by several candidate materials, including two natural materials. Phosphorous concentrations were measured in triplicate after 2 and 30 days to evaluate short- and long-term removal. The removal was calculated using the initial and final concentration in mg P/L. The influence of cockle shell particle sizes on phosphorous removal was conducted similarly in a set of 50 mL test tubes at an initial concentration of 5 mg P/L in DI water and 0.53 g of cockle shell with four different sizes. Tubes were manually shaken three times a day and the remaining phosphorus level was measured with the ICP-OES (Rice et al., 2017; Xie et al., 2014). The third batch study was to introduce sulfate that could solubilize calcium and improve phosphorous removal. Potassium sulfate was added to each vial in six different sulfur-to-phosphorus (S:P) molar ratios at 0, 1, 2, 5, 10, and 50.

Adsorption Isotherm Study. Adsorption studies were conducted at 22^0C in 100 mL tubes. All sorption studies were done with duplicate analysis and supplemented with a P-free control. The amount of sorbed phosphorous after five days was calculated by the difference in phosphorus before and after adsorption experiment (Chen, 2015). Sorption data was fit to three adsorption isotherm models including Linear model, Langmuir model for saturated sorption, and Freundlich model for multilayer sorption (Eq. 1):

$$Q = K \times C_e^{1/n} \quad (1)$$

where Q = sorbed phase concentration (mg/kg), C_e = aqueous phase concentration (mg/L), K = partition (adsorption) coefficient, n = constant related to the intensity of adsorption.

Case Application – Required Dose and Settling Velocity in Lake Houston and Lake Conroe. The amount of cockle shell and zeolite required to remove phosphorous (kg/day) is estimated from phosphorous loading (kg/day) divided by the sorbed phosphorous concentration in cockle shell and zeolite (mg/kg). The best fit Freundlich isotherm model parameters were used to calculate the sorbed phase concertation. Phosphorus loading and hydrological data for Lake Houston and Lake Conroe were obtained from USGS (2005) and the Texas Water Development Board (2013). Settling velocity of cockle shell and zeolite particles in the two Houston lakes were calculated using the Stokes's law, assuming laminar flow (no wind and other turbulences).

$$v\left(\frac{m}{s}\right) = \frac{g \times d^2 \times (\rho_p - \rho_w)}{18\,\mu} \qquad (2)$$

where v = settling velocity (m/s), g = gravitational constant (9.8 m/s^2), d = particle diameter (m), ρ_p = particle density (g/cm^3; 2.5 g/cm^3 for cockle shell and 1.1 g/cm^3 for zeolite), ρ_p = density of water (1.0 g/cm^3), μ = water dynamic viscosity (0.00891 g/cm-s). Following Eq. 2, the time to settle in the water column of depth (h) was calculated based on the water depth (Lake Houston: median water depth = 3.7 m, maximum h = 15 m; Lake Conroe: median water depth = 6.2 m, maximum depth = 23 m).

RESULTS

Screening of Materials for Maximal Phosphorus Removal. An initial screening test using ($Ca(CH_3COO)_2$, $CaCO_3$, $CaCl_2$) and one natural product (cockle shell) indicates that $Ca(CH_3COO)_2$ has higher phosphorous removal than cockle shell following a two-day treatment. This was extended to 30 days to examine if a longer time will improve the removal efficiency. Interestingly, about 30% increase of phosphorous removal after 30 days was observed in $Ca(CH_3COO)_2$, and a slight increase of about 5% in cockle shell. An additional screening test was conducted by including a second natural product (zeolite designated as Aqua-P) for the comparison of phosphorous removal (Figure 1). It clearly shows the promising results of two natural products; in particular, zeolite has the highest phosphorous removal, followed by calcium acetate and cockle shells after 30 days. Hence, subsequent studies will focus on the application of these two natural materials.

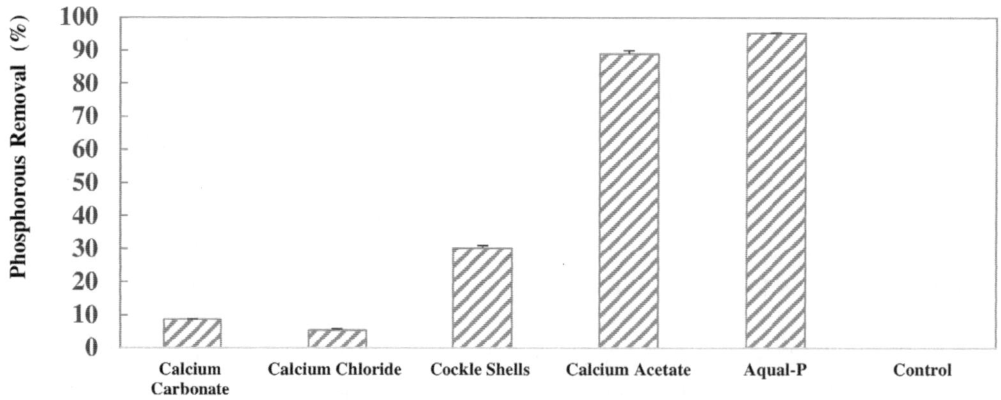

FIGURE 1. Comparison of Phosphorous Removal (%) by Various Materials

Effects of Particle Size on Phosphorus Removal. The effects of particle size were conducted in duplicate using cockle shell exposed to 5 mg P/L. The three largest size categories, all greater than 1 mm in diameter, display dismal rates of removal between 0% and 10%. For the smallest size category (< 1 mm diameter), the percent removal significantly increased to 50%. Therefore, these results demonstrate a promising correlation between decreasing particle size and improving phosphorus removal (**Figure 2**).

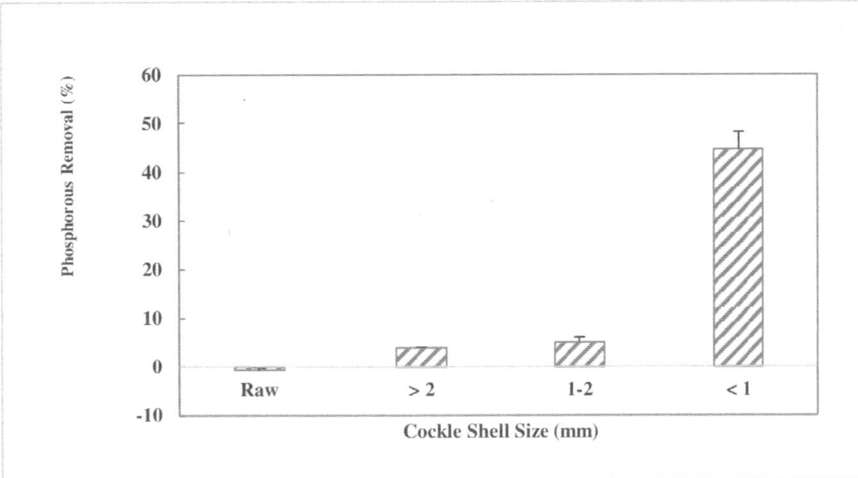

FIGURE 2. Effects of Cockle Shell Particle Size on Phosphorous Removal

Improvement of Phosphorus Removal with Sulfate. Sulfate is supposed to enhance phosphorous removal by dissolving more calcium into the solution. The removal peaked when sulfate was absent and when the S:P ratio was 2:1. The amount of phosphorous removal by only cockle shells was approximately 16.32%, slightly lower than the 25% removal in the previous study. However, when S:P ratio was 2:1, phosphorous removal actually improved slightly to 18.75%, larger than phosphorous removal by purely cockle shells. Unlike the previous study using calcite with pH adjustment, the addition of sulfate largely seems to have hindered the cockle shells' ability to precipitate with phosphorus.

Characteristic Atoms / Functional Groups Responsible for Removal. FTIR spectrum show that cockle shell shares similar function groups with calcium carbonate with the characteristic FTIR peak of carbonate (850 cm^{-1}), strongly suggesting removal through precipitation with calcium. Zeolite matches the FTIR spectrum of silicate with a silicate peak at 1500 cm^{-1}, indicating phosphorous is removed through the adsorption of the silicate.

Best Fit Adsorption Isotherm Model Describing Phosphorus Adsorption to Zeolite. None of the isotherm data fits the linear model. The data fits Freundlich model well with an R^2 of 0.7665 using the linearized form of Eq. 3:

$$\ln Q = ln\ K + \frac{1}{n} lnC_e \qquad (3)$$

The K and 1/n value estimated from the intercept and slope were 207.46 and 6.8, respectively. These model parameters are essential to design the dose required for adsorption-based materials, which will be described in the case studies in the two Houston lakes.

Feasibility Study on Cockle Shell and Zeolite Deployment in the Houston Lakes. This case study considers two major lakes in the Houston area (Lake Houston and Lake Conroe). **Figure 3** depicts the dose required as a function of equilibrium phosphorous concentration and flow conditions in Lake Houston. The required dose is small at the high concentration range, but in order to maintain very low (e.g., below 0.1 mg/L) phosphorous concentration in the lake, the required dose increases exponentially as the required phosphorous concentration decreases. A log-log relationship exists between the daily dose and the desired equilibrium phosphorous concentration in the lake.

The safety of deployed zeolite or cockle shell in both lakes is evaluated based on the suspended solid (SS) concentrations which were calculated using the dose and the lake volume. Under the worst-case scenario, SS concentrations increase exponentially when the applied dose reaches certain high doses. If 10 mg/L of SS is assumed to be the safe level for lake turbidity, the corresponding safe dose at the low flow is 208 tons/day. Under the worst-case scenario in Lake Houston, a negligible turbidity increases by 0.24 mg/L following aerial spraying at 40 ton/day. These results indicate that both natural materials are safe to use without adverse impact on water quality.

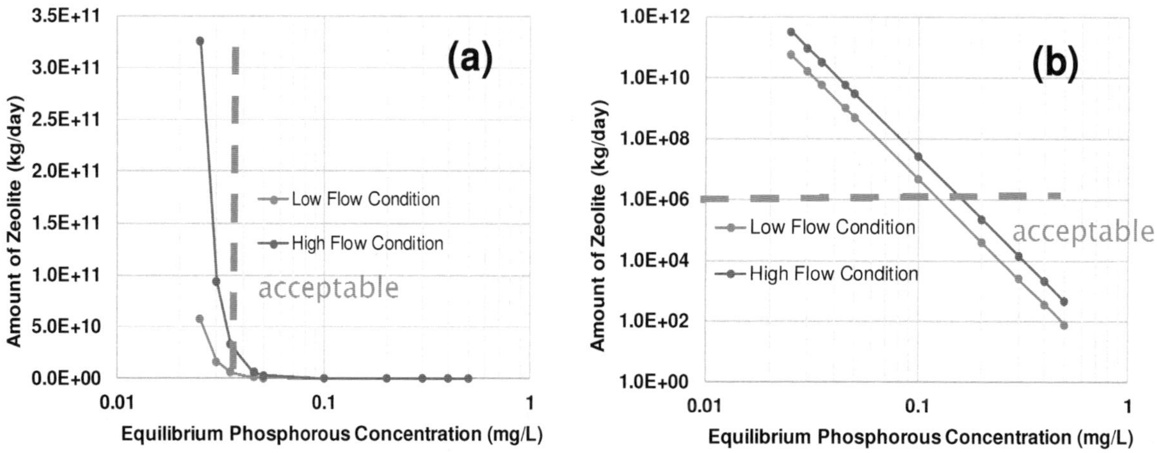

FIGURE 3. Zeolite dose required for the removal of phosphorous from Lake Houston

The third question relevant to *in-situ* deployment regards the settling time of these particles. While small particle size enhances phosphorous removal, smaller particles tend to settle more slowly and create turbidity. According to Stokes' law, cockle shells at 1 mm diameter settles in water column within 4 – 17 sec at the median to maximum water depth, and zeolite particles with a much smaller size (0.1 mm) settle within 100 min – 6.93 hrs at the median to maximum water depth in Lake Houston. Results clearly indicate that ground cockle shell will settle in the water column within a day, and zeolite particles will settle within a day or more under the median water depth of 3.7 m. This implies that ground cockle shell should not interfere with water turbidity, whereas the deployment of zeolite may encounter temporary water turbidity issues only when a large dose is deployed into a deep portion of lake water.

DISCUSSION

Precipitation and Adsorption Mechanism with the Novel Materials. Cockle shells and other sea shells are abundantly available in the coastal U.S., which are typically disposed in landfills as wastes. Zeolites are common in volcanic deposits and coastal plains in certain areas in the U.S. Their beneficial uses have never been comprehensively studied for their potential phosphorous removal capabilities in eutrophic waters. This study demonstrated that two novel natural materials have distinct mechanisms in removing phosphorus. The very small solubility product of $Ca_3(PO_4)_2$ attributes to the phosphorous removal through precipitation. Zeolite does not have any element that is ionically attracted to phosphorous, but it is known to be an aluminosilicate material with a porous crystalline structure as an adsorbent. Additionally, the best fit Freundlish adsorption isotherm further reveals that the adsorption onto zeolite is multilayer sorption.

New Deployment Methods Suitable for Surface Water Use. This study has explored size reduction and chemical modification using sulfate. It is expected that synthesized nanoparticles from cockle shell (Hariharan et al., 2014) to further reduce size and enhance porosity could produce even higher phosphorous removal. With these engineered cockle shell and zeolite from their raw materials, several deployment methods can be adopted for *in-situ* surface water uses, including aerial spraying, floating nets, bagged shells, and porous zeolites. Aerial spraying is preferred for emergency response of algae bloom at high phosphorous levels. The floating bags encapsulating the active ingredients through porous materials can be deployed directly in surface water for recycling and regenerating if necessary.

Cost Analysis of Natural Materials vs. Traditional Methods. Cost data shows that the aerial spraying of natural materials costs at $168 k vs. sediment dredging at $1.075 billion. Clearly, the estimated dose of zeolite in Lake Houston is economically feasible, and it presents a significant cost reduction compared to dredging. It should be noted that cockle shells also meet financial constraints due to existing shell recycling programs in the Gulf Coast area, which can provide a steady and abundant supply of shells virtually for free.

CONCLUSIONS

Two natural materials were found to effectively remove phosphorous. Further improvement was made by reducing particle size and adding sulfate at 2:1 S:P molar ratio. The predominant mechanism for cockle shell (precipitation through $(Ca)_3(PO_4)_2$) and zeolite (multilayer adsorption through silicate) was elucidated. Applying to two local water bodies in Houston, engineering feasibility calculations with deployment dose, settling velocity, and cost revealed that these novel materials are cost-effective and safe to use in the control of algae bloom in eutrophic water. The promising results warrant further studies to develop these natural materials into commercial products in eutrophication control. Significant economic and environmental benefits are expected from the extension of this study.

ACKNOWLEDGMENTS

We are grateful to the support from the University of Houston - Clear Lake for materials and instrumentations. Thanks are also due to Blue Pacific Minerals for providing the Aqual-P sample.

REFERENCES

Buda, A., G.F. Koopmans, R.B. Bryant, and W.J. Chardon. 2012. "Emerging technologies for removing nonpoint phosphorus from surface water and groundwater". *J. Environ. Qual.* 41(3), 621-627.

Chen, X.J. 2015. "Modeling experimental adsorption isotherm data". *Information* 6, 14-20.

Goh, H., N. Zakaria, C.K. Chang, L.T. Liang, and K.Y. Foo. 2015. "Influence of hydraulic conductivity and organic matter content in different bioretention media on nutrient removal". *Appl. Mech Mat.* 802, 448-453.

Hariharan, M, N. Varghese, A.B. Cherian, P.V. Sreenivasan, J. Paul, and A. Antony. 2014. "Synthesis and characterization of $CaCO_3$ nano particles from cockle shells using chitosan as precursor". *Int. J. Sci. Res. Pub.* 4(10), 1-10.

Karapınar, N. 2009. "Application of natural zeolite for phosphorus and ammonium removal from aqueous solutions". *J. Haz. Mat.* 170(2-3), 1186-1191.

Klapper, H. 1992. "Calcite covering of sediments as a possible way of curbing blue-green algae". In Eutrophication: Research and Application to Water Supply (ed. DW Sutcliffe and JG Jones), pp. 107-111. Freshwater Biological Association, London.

Rice, E.W., R.B. Baird, A.D. Eaton. 2017. *Standard Methods for the Examination of Water and Wastewater,* 23rd Edition, American Public Health Association, American Water Works Association, Water Environment Federation.

Xie, J., X. Zhang, Z.W. Xu, G.F. Yuan, Z. Tang, X. Sun, and D. J. Ballantine. 2014. "Total phosphorus concertation in surface water of typical agro- and forest ecosystems in China: 2004-2010". *Front. Environ. Sci. Eng.* 8(4), 561-569.

Distribution and Evaluating the Fate of Cyclic Volatile Methyl Siloxanes in a High Altitude Lake

Junyu Guo and Jianbo Zhang
(Peking University, Beijing, China)

Cyclic volatile methyl siloxanes (cVMS) used in personal care products are emitted to aquatic environments with wastewater effluents. Persistent, toxic, bioaccumulated, and potential to ecological harm of cVMS have been identified. In this study, the environmental behavior of octamethylcyclotetrasiloxane (D_4), decamethylcyclopentasiloxane (D_5) and dodecamethylcyclohexasiloxane (D_6)) were evaluated in a high altitude lake (Dian Lake, 24°58′N 102°39′E). We collected and determined three cVMS compounds in air, water and sediment samples in both winter (January) and summer (July) 2017. The mean atmospheric concentrations of D_4, D_5 and D_6 in winter were 10.44±4.91, 18.40±4.99, 2.56±1.12 $ng \cdot m^{-3}$, respectively, and 7.44±3.50, 9.67±3.99, 1.5±0.66 in summer, respectively. Seasonal variability of the concentrations of D_4, D_5 and D_6 was explained by the seasonality in the OH radical concentrations. The mean concentrations of D_4, D_5 and D_6 in water were 0.83±0.45, 18.08±4.03, 20.78±5.77 $ng \cdot L^{-1}$ in winter, respectively, and 0.23±0.13, 15.52±3.64, 20.42±5.79 $ng \cdot L^{-1}$ in summer, respectively. However, mean concentrations of D_4, D_5 and D_6 in sediments were 9.64±2.45, 183.75±22.62, 280.61±45.83 $ng \cdot g^{-1}$ dw in winter, respectively, and 1.57±0.29, 91.87±19.36, 270.27±31.27 $ng \cdot g^{-1}$ dw in summer, respectively. To compare with measurements, a fugacity-based mass balance chemical fate model for lakes (QWASI) was used for Dian Lake, and explored the behaviour of cVMS in lake. Modeling result showed that most of cVMS distributed in sediment, and persistence in Dian Lake was influenced by sediment sorption and burial.

Modeling and Assessing Uncertainty Factors in Salt Balances in Groundwater Basins

Barry J. Hibbs
(California State University-Los Angeles, Los Angeles, CA, USA)

ABSTRACT: This paper discusses criteria and methodology for developing salt balances in groundwater basins. The project describes a basic approach to account for salt balance terms with factors responsible for the variation of the respective loading coefficients. The ability to predict salt loading from land use activities, such as expanded use of recycled water as an irrigation source, is subject to uncertainty. To evaluate uncertainty, we suggest testing of models of salinity loading to aquifers factoring in limits of uncertainty to consider ranges of possible loading under multiple possible loading scenarios. A demonstration model is developed to account for salt loading when recycled water is a loading factor added to the salt balance.

INTRODUCTION

The State of California now requires Salt and Nutrient Management Plans (SNMP) to be developed for groundwater basins. The state policy requires plans to be developed for every groundwater basin in California. Development of SNMPs will lead to a more comprehensive approach to basin water quality management. Acceptable loading of salts, nutrients, and other constituents that may adversely affect groundwater quality requires development of accurate water balance terms, knowledge of aquifer properties and processes, and understanding of environmental conditions that modify water balances in groundwater basins. Similarly, loading of natural and anthropogenic sources of salts and nutrients require an understanding of the water balance. Water balance inputs (e.g., interbasin groundwater flow; natural recharge) can enrich or dilute nutrients and salts in groundwater flow systems depending on the specific type of input, the input rate, and the salt or nutrient load contained in the input. The ambient groundwater quality will determine if an input either degrades or improves the water quality in an aquifer, and if the degradation is within acceptable tolerance limits and assimilative capacity.

Water balance estimates in a given basin range from well constrained and highly accurate to very suspect estimates that have been developed with anecdotal information. Salt and nutrient assessments must integrate uncertainty to provide a margin of safety based on the reliability of information used to develop accordant water balance and mass transfer assessments in the salt balance. This paper discusses technical approaches to estimating the extent of buffering that should be provided for good quality groundwater basins in light of the potential impact of, for example, increased recycled water and stormwater use on groundwater quality. While the issues described in this paper are germane to recent policy developed in California, USA, the guidelines and issues discussed in this paper can be extended to any basin where salt management issues are relevant.

THE SALT BALANCE

The salt balance for any given land area or soil unit can be expressed by the following equation (Carter, 1975):

$$Sp + Si + Sr + Sd + Sf = Sc + Sppt + Sdw$$

where:

Sp = salt in natural precipitation falling upon the area
Si = salt in the irrigation water used in an area
Sr = residual salts in the soil
Sd = salt dissolved into solution from soil minerals
Sf = salt in applied fertilizers or soil amendments

Sc = salt taken up by or retained on the plant and removed
Sppt = salt chemically precipitated in the soil
Sdw = salt in the drainage water from the area

With respect to salt inputs, the salt in natural precipitation, Sp, is usually very small (<5 mg/L TD), except near coastal areas where the values can reach at least mentionable values of 10 to 20 mg/L TDS (Junge and Werby, 1958; Whitehead and Feth, 1964; Davis and Deweist, 1966). Si is a function of the salt concentration in the irrigation water which may be moderate (e.g., 400 mg/L TDS) when irrigation source comes from various imported or domestic sources, or may be moderately high (e.g., 1000 mg/L TDS) when recycled water is used (U.S. Department of Agriculture, 1954). Sr can be large or small depending on climate, basin hydrological conditions, and land use. Sr is often significant in natural arid landscapes that have seen substantial accumulation of salts in shallow soils, usually due to high evapotranspiration rates combined with little rainfall. Sd is highly variable, depending on type of minerals present in vadose zone and aquifer strata. The presence of soluble marine rocks and detritus in groundwater basins often creates high salt loading to soils and aquifers due to in situ mineral dissolution. Where sediments in aquifers are derived from erosion of igneous or metamorphic rocks in highlands, loading factors due to Sd are usually low. Sf is usually low but can be significant when soils are chemically amended with gypsum and other soil conditioners used in agricultural landscapes.

Regarding salt outputs, Sc can be important when light watering is done with sprinklers in aesthetic and recreational landscapes (golf courses, parks, and yards). This leaves evaporative salt dust on plant tissues which may be removed by mowing, runoff, or wind deflation. Sppt can be important where light watering is done in urban and recreational landscapes, but is usually managed in agricultural areas as a result of controlled leaching of croplands to protect plant health. Sdw drains into surface water or groundwater. Sdw can be reduced through effective land management strategies (U.S. Department of Agriculture, 1954; Garcia-Fresca and Sharp, 2005). Several of these salt balance components must be evaluated for their importance and uncertainty.

The salt loading from in situ mineral dissolution, Sd is a key evaluation parameter due to its importance in many areas, particularly where sedimentary rock aquifers and detritus are present in formations. Sd is highly variable, depending on type of minerals present in vadose zone and aquifer strata. The presence of soluble marine rocks and detritus in groundwater basins often creates high salt loading to soils and aquifers due to in situ mineral dissolution. Sd and Sr are grouped together as the principal factors in salt loading in California groundwater basins. In arid regions, significant salt concentration effects occur as plant roots osmotically exclude salts in soil water, leaving behind salt enriched soils. When salts are later leached in soils by precipitation recharge, the leachate eventually reaching groundwater may contain several hundred to a few thousand milligrams per of dissolved salts (Malecki and Matyjasik, 2002). This is especially common in dry areas where the recharge amounts to only a few millimeters per year. Where wastewater is applied to urban landscapes, the salts are already enriched in the feed water and evapoconcentration effects are amplified.

Within groundwater systems, some of the natural salinization develops in aquifers by retention/pickup of ions from the atmosphere during recharge of precipitation and runoff. Usually the more important natural source of dissolved salts is from weathering of the common rock minerals contained in aquifers. The salt loading from in situ mineral dissolution is a key evaluation parameter due to its importance in many areas, particularly where sedimentary rock aquifers and detritus are present in water bearing formations. Silt and clay bearing rocks (mudstones and shales) often form in anaerobic environments due to high organic matter content (Hem, 1992). Reduced mineral species such as pyrite often forms in these fine textured rocks. When the rocks are eventually exposed by tectonic uplift, conditions can turn from anaerobic to aerobic and pyrite and other reduced minerals can be oxidized. Pyrite oxidation releases sulfate and other ions into subsurface waters, sometimes in rather high concentrations (Hem, 1992).

While most land application projects predict accumulation of salts in soil profiles during land application of wastewater and irrigation water, the tacit assumption is that mass balance eventually occurs (salt mass applied to soil profiles equals salt mass lost to runoff, plus leaching fraction lost to deep percolation to groundwater). This is not a strictly valid assumption because precipitation of less soluble ions contained in soil solutions (Ca, HCO_3) may lead to permanent loss of soil solutes (Sanz et al., 2011).

These mineral precipitation losses are usually minimal however, accounting for no more than a very few percent of the dissolved mineral content in soil water. If evaporative salts form at land surface, such as the surface salt crust thernardite, the salts may be stripped from the land surface by heavy windstorms. This process might represent a measurable percentage loss of dissolved solids in soils.

UNCERTAINTY RELATED TO SALT LOADING FACTORS

Where in situ mineral dissolution processes, Sd, does not factor as an important source of salt loading to a groundwater basin, other factors dominate salt loading processes. Factors such as Si - salt in the irrigation water used in an area take on greater importance. Related factors such as drought and water conservation and how they affect salt loading factors such as Si must be considered. How use and management of groundwater basins affects Si, and ultimately Sdw - salt in the drainage water from an area, are often uncertain.

Leaching of soluble salts from agricultural lands and reducing salinity in the crop root zone is essential for good crop production, especially in arid lands. Systematic procedures have been developed for maintaining desirable salinity in irrigable croplands and the methodology has been used for leaching excess salts from agricultural lands for over 100 years (United States Department of Agriculture, 1954). Leaching is most often done by applying excess irrigation water on irrigable landscapes. Empirical water balance calculations specify anticipated leaching losses based on soil characteristics, soil salt concentration, irrigation water concentration, and irrigation quantities (United States Department of Agriculture, 1954).

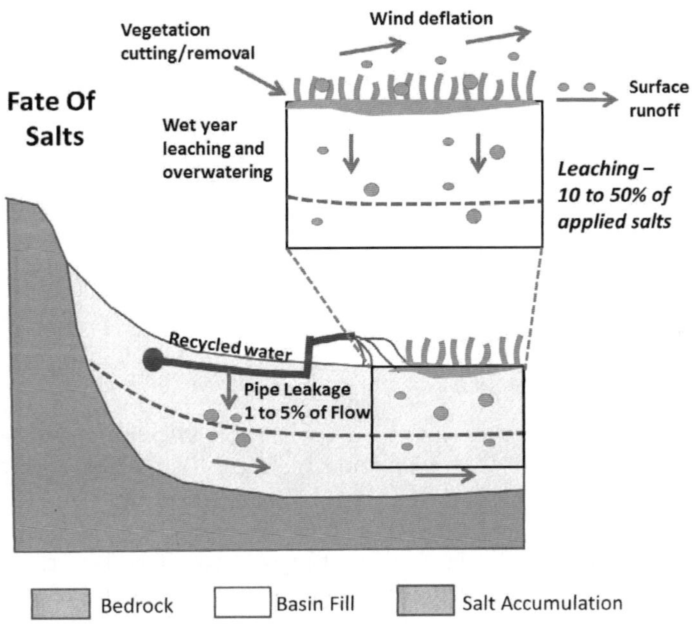

Figure 1. Generic Model 1 - Fate of salts due to application of recycled water in irrigated landscapes. Pipe leakage is considered to be minimal due to repair procedures, but may reach 1 to 5% of flows in systems. Some of the recycled water salts accumulating in soils may be lost by surface runoff, removal of vegetation that accumulated small amounts of mineral salts in tissues, accumulated salts as dusty evaporative crusts on tissues, and wind deflation of surface crusts of salts. A percentage is lost and the remainder is leachable below the plant routing zone.

Unfortunately irrigation of urban landscapes with recycled water does not generally include very sophisticated engineered calculations and the level to which salts accumulate following application with recycled water is understood anecdotally. The ability to predict salt loading from use of recycled water as an irrigation source is subject to uncertainty. Therefore we suggest that testing of models of salinity loading to aquifers should factor in limits of uncertainty to consider ranges of possible loading under multiple possible loading scenarios. This discussion provide understanding of the limitations based on ranges of salt loading that may be expected from use of recycled water, ranging from lower limits to upper limits, with recommendations for use of what are considered to be the more reliable estimates (Figures 1 to 4).

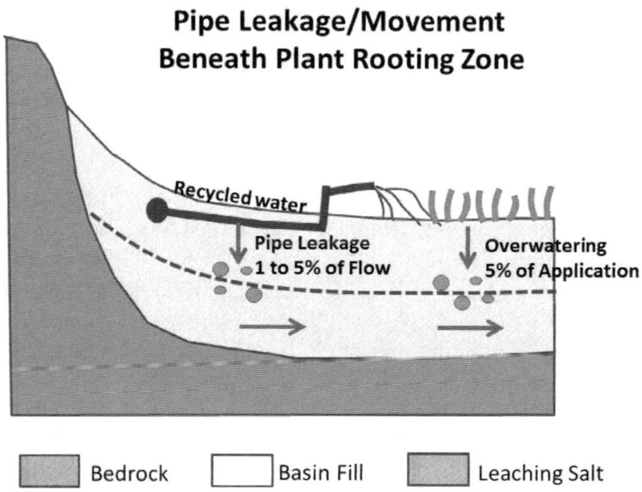

Figure 2. Generic Model 2 - Fate of salts due to application of recycled water in irrigated landscapes under a conservative loading scenario. Pipe leakage is 1 to 5% of flows and overwatering by 5% creates additional leakage. In this scenario, >90% of salts are lost from surface removal mechanisms and <10% of applied recycled salts reach the water table. In other test scenarios the values can be changed to reflect other infiltration and loading rates, for example 15% of pipe leakage and 25% infiltration by overwatering, where 40% of the recycled water salts reach groundwater.

Land surface irrigation with recycled water requires special consideration. Because of environmental concerns related to minimizing surface runoff, efforts are usually made to keep recycled water pipelines and other infrastructure in good repair. Some pipe leakage is inevitable due to local land subsidence, soil compaction, and slope instability that will stress and fracture recycled water pipes. After many years recycled water pipe leakage might reach 1 to 5% of flows (Figure 1), while water pipes and sewer pipe exfiltration can easily approach 5 to 10% in older systems (Norin et al., 1999; Farley and Trow, 2003; Garcia-Fresca, 2004). Some of the recycled water salts accumulating in soils may be lost by surface runoff, removal of vegetation that accumulated some small amounts of mineral salts in tissues, accumulated salts as dusty evaporative crusts on tissues, and wind deflation of surface crusts of salts. A percentage is lost and the remainder is leachable below the plant rooting zone (Figure 1). Especially in arid lands, salts may accumulate in soils due to evaporation. Concentrated salts may then leach into the aquifer at a fixed rate at higher concentration due to equilibration between soil water evaporation and soil water leaching. The leachate enrichment factors often vary from 1.5 to 8 relative to application concentration (Figure 3). Salts may also accumulate in soils for many years of normal precipitation, and then leach substantially during a year of heavy precipitation (Figure 4). The scenario of occasional leach cycles during wet years probably reflects the most common condition in Southern California (Figure 4).

A method is needed to test conceptual models of salt accumulation, loss, and leaching from recycled water and other water applications, as depicted in Figures 1 to 4. The method should test lower to upper limits of salt loading due to recycled and other water use, while providing suggestions on how uncertainty can be factored into the analysis of salt loading.

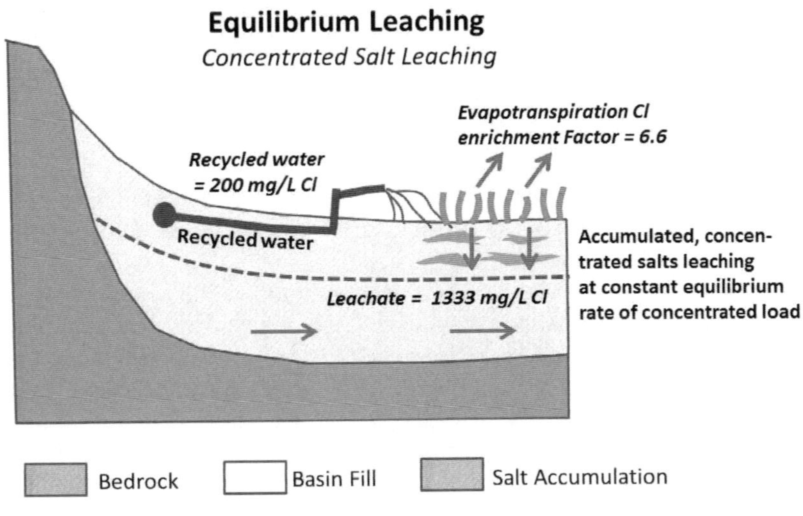

Figure 3. Generic Model 3 - Fate of salts due to application of recycled water in irrigated landscapes. In this generic model, salts accumulate in soils due to evaporation and then leach into the aquifer at a fixed rate at higher concentration due to equilibration between soil water evaporation and soil water leaching. The enrichment factor presented here is 6.67 between application concentration and leachate concentration.

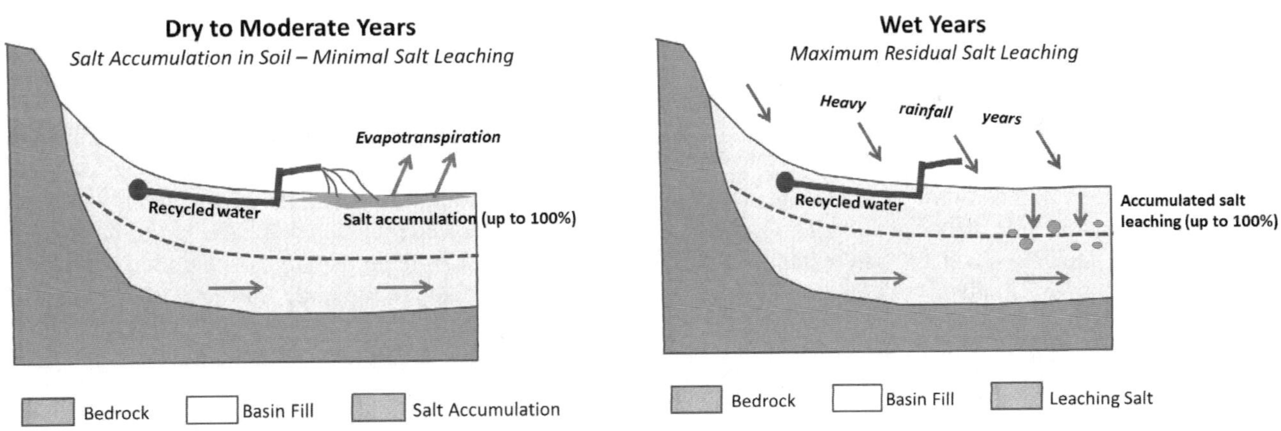

Figure 4. Generic Model 4 - Fate of salts due to application of recycled water in irrigated landscapes. In this generic model salts accumulate in soils for many years of normal precipitation and then leach substantially during a year of heavy precipitation. This type of meteorological conditions is characteristic in Southern California.

GENERIC MODEL TESTING OF UNCERTAINTY IN SALT LOADING

To evaluate uncertainty in salinity loading factors related to Si, and how drought, water conservation, and land use change affect Si and other salt loading factors, a model is developed to calculate salinity in a generic groundwater basin. Testing is done with the model on variable controls due to storage variation in the basin, loading of recycled water constrained by uncertainty in salinity loading, drought testing scenarios, and other control inputs. The model is developed to perform salt balance in any groundwater basin as desired. The model requires an input hydrological budget for an aquifer or aquifer zone, as well as input salinity values for each flow term in the budget. An initial input salinity for each aquifer or zone is also required. The model prescribes source inputs (e.g., mountain front recharge, delivered water return recharge, recycled water return recharge, etc.) and outputs (e.g., pumping, underflow to another aquifer zone or subbasin, river leakage, etc.).

The model calculates salinity changes for individual ions as well as for bulk TDS. The model calculation mixes influent flow volumes and salinities to develop output values showing changes in storage, changes in salinity, and pounds of solute. In a series of steps, the calculation process is repeated with the calculated output providing time series calculations.

The model is constructed to evaluate uncertainty in salinity loading factors related to Si and Sf in a generic groundwater basin that is recharged in predevelopment times by two sources; precipitation recharge at the valley floor and mountain front recharge. Discharge in predevelopment times is due to river seepage (baseflow) and underflow to an adjacent basin. The mountain front recharge in predevelopment times was 3000 acre-ft/year with another 7000 acre-ft/year coming from precipitation recharge on the valley floor. River leakage and underflow to an adjacent basin combine for an average 10,000 acre-ft/year with no change in storage.

Chloride is the salinity loading parameter tested in the generic salinity loading model. Any ion or TDS could be used in place of chloride in the demonstration calculations. It is assumed that the basin fill consists mostly of granitic detritus eroded from surrounding mountains. Most of the surrounding mountains consist of igneous rocks with some sedimentary rocks exposed on one side of the basin. Chloride loading in the generic groundwater basin includes 40 mg/L (geologic and rain water) in precipitation recharge, and 55 mg/L in mountain front recharge. Based on the flows and loads, predevelopment chloride concentration in the generic groundwater basin is computed:

[(3000 acre-ft/year x 40 mg/L) + (7000 acre-ft/year x 55 mg/L Cl)] / 10,000 acre-ft/year

= 50.5 mg/L Cl

The generic groundwater basin is assumed to develop around 1935. In most recent decades the average pumping rate in the basin is 80,000 acre-ft/year. Underflow and baseflow still occur but their combined output has decreased to 3000 acre-ft/year. Mountain front recharge and precipitation recharge at the valley floor, and their salinities are unchanged. Imported water is delivered to the basin. A combined delivered water return recharge of 50,000 acre-ft/year results from recharge of imported water and recharge of domestic groundwater supply. Spreading basins collect and recharge 23,000 acre-ft/year of surface runoff. The chloride for spread water is the same as that of precipitation recharge. Delivered water return recharge includes chloride in tap water plus input of geological chloride, giving a combined concentration of 90 mg/L for recharge of delivered water (Table 1).

In the "storage 1" test scenario used in calculations, the groundwater basin contains 5 million acre-ft of storage. In the "storage 2" test scenario used in calculations, the groundwater basin contains 1 million acre-ft of storage. The test scenarios have inflow and outflow rates and respective salinities specified the same in both storage 1 and storage 2 calculations. This specification provides a comparison of sensitivity of inputs in small to moderate sized basins. After decades of groundwater basin use, the chloride concentration in the groundwater aquifer has re-equilibrated to a new chloride concentration in response to delivered water return recharge:

[(3000 acre-ft/year x 55 mg/L) + (7000 acre-ft/year x 40 mg/L Cl) +

(50,000 acre-ft/year x 90 mg/L Cl) + (23,000 acre-ft/year x 40 mg/L Cl)] / 83,000 acre-ft/year

= 70.66 mg/L Cl

Table 1. Hydrological budget and loading concentrations for chloride in a generic groundwater basin used for testing scenarios. A ten-year average hydrological budget is specified prior to testing scenarios. Two storage scenarios are used in all calculations of loads (storage 1 and storage 2).

Inflow:	**Acre-Ft** (annual)	**Cl** (mg/L)
Hill and Mountain Front Recharge (excluding spreading)	3000	55
Precipitation Recharge (areal recharge basin floor)	7000	40
Delivered Water Return (tap)	50000	90
Spreading Basins	23,000	40
Outflow:	**Acre-Ft** (annual)	**Cl** (mg/L)
Pumping	80,000	70.66
Underflow to Adjacent Basin	2,000	70.66
Baseflow, River Leakage	1,000	70.66
Storage = 5 million acre-ft (storage 1) and 1 million acre-ft (storage 2).		

In generic testing scenarios, it is assumed that all inflows, outflows, and concentrations remain stable from 2000 to 2010. New management strategies are devised to generically test uncertainty in salinity loading factors in a "changing hypothetical basin" through 2040. Five scenarios were used to test in recharge and loading due to surface application of recycled water for irrigation (Hibbs, 2017). The amount of recycled water that is applied to surface landscapes is 20,000 acre-ft/year and a variable amount of the recycled water and salt reach groundwater. In this short paper, only one testing scenarios is tested, as follows:

• Recharge of 10,000 acre-ft/year recycled water salts with 1000 mg/L enriched chloride due to a wet year that occurs every 1 in 5 years, leaching the accumulated recycled water salts, for 30 years

In this test scenario, 50% of the applied salts in recycled water are assumed to accumulate completely during average years with no salt leaching to the aquifer during the years of average rainfall. Half of the salts are assumed to be lost due to runoff, collection of vegetation clippings that were slightly encrusted with evaporative salt, and wind deflation. Once every five years a wet year occurs and the residual salts (100% of the total accumulated during average years) is assumed to leach completely into the aquifer, with 10,000 acre-ft of additional precipitation recharge. The remaining evaporative salts (50%) are completely leached with the additional input of precipitation recharge. This hypothetical scenario matches generic conceptual models 1 and 4 (Figures 1 and 4), and represents realistic limits of practical salt loading due to the addition of 20,000 acre-ft/year of reclaimed water to the landscape, that has become enriched due to evaporation, is diminished 50% by removal processes, and is completely leached by rain every five years during a wet cycle.

Table 2. Hydrological budget and loading concentrations for chloride in a generic groundwater basin. This test scenario follows generic conceptual models 1 and 4 (Figures 1 and 4), and represents moderate limits of practical salt loading. In this test scenario, 50% of the applied salts in recycled water are assumed to accumulate completely during average years with no salt leaching to the aquifer during the years of average rainfall. Half of the salts are assumed to be lost due to runoff, collection of vegetation clippings that were slightly encrusted with evaporative salt, and wind deflation. Once every five years a wet year occurs and the residual salts (100% of the total accumulated during average years) is assumed to leach completely into the aquifer, with 10,000 acre-ft of additional precipitation recharge. The remaining evaporative salts (50%) are completely leached with the additional input of precipitation recharge (Figures 1 and 4).

Inflow:	**Acre-Ft** (annual)	**Cl** (mg/L)
Hill and Mountain Front Recharge (excluding spreading)	3000	55
[1]Precipitation Recharge (areal recharge basin floor component)	7000	40
Delivered Water Return (tap)	50000	90
Spreading Basins	23,000	40
Recycled Water (avg. year, 4 of 5 years)	0	0
[1]Recycled Water Salts Leached by Rainwater (wet year, 1 of 5 years)	10,000	1000
Outflow:	**Acre-Ft** (annual)	
Pumping	80,000 – Cl varies with time, Figure 4.9	
Underflow to Adjacent Basin (avg. year, 4 of 5 years)	3,000 – Cl varies with time, Figure 4.9	
Underflow to Adjacent Basin (wet year,1 of 5 years)	6,000 – Cl varies with time, Figure 4.9	
Baseflow, River Leakage (avg. year, 4 of 5 years)	None	
Baseflow, River Leakage (wet year, 1 of 5 years)	7,000 – Cl varies with time, Figure 4.9	
Storage = 5 million acre-ft (storage 1) and 1 million acre-ft (storage 2). Water budget includes no change in storage (inflows = outflows) over 30 year hypothetical analysis.		

[1]During a wet year that occurs every five years, there is an extra 10,000 of precipitation recharge that is lumped under "recycled water salts leached by rainwater". This is in addition to the 7,000 acre-ft of precipitation recharge that occurs every year.

The water budget in the analysis is presented in Table 2. The gain of 10,000 acre-ft/year of recharge during the wet year is balanced by an assumed loss from the system of 6000 acre-ft/year in river leakage and 4000 acre-ft/year in underflow to adjacent basin, with no change in storage. The 30 year result indicates moderate change in chloride concentration in the basin for both 5 million and 1 million acre-ft of storage

(Figure 5). Pulses of precipitation recharge carrying accumulated salts to the aquifer are more pronounced in the 1 million acre-ft storage basin, with higher chloride concentration reached at the end of 30 years. The additional storage in the 5 million acre-ft basin dampens the concentration of the salt loading that leaches once every five years. The changes after 30 years are moderate in this scenario, testing reasonable limits of expected loading due to application of recycled water of 20,000 acre-ft/year that accumulates 50% of the recycled water salts in soils that is then 100% leached during a wet year at 5 year interval (Figure 5). Results indicate storage is an important control parameter, and salinization of the hypothetical aquifer is mild to moderate, depending on storage. Similar analysis were performed for multiple test scenarios to account completely for uncertainty (Hibbs, 2017).

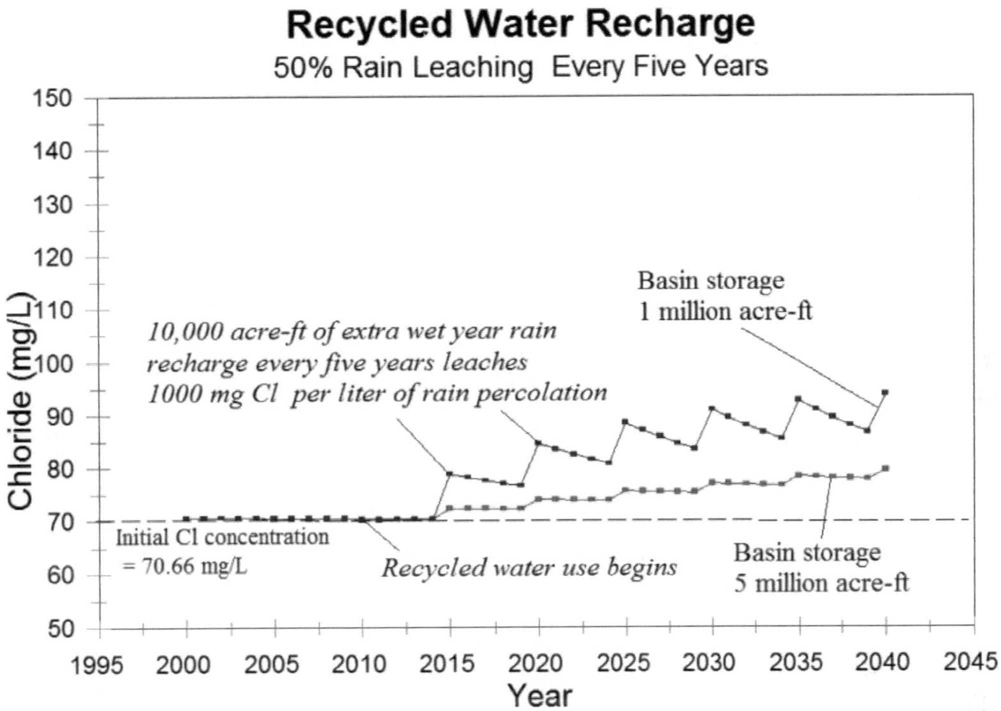

Figure 5. Fate of salts due to application of recycled water in irrigated landscapes from recharge of 10,000 acre-ft/year recycled water salts with 1000 mg/L enriched chloride due to a wet year that occurs once every 5 years, leaching the accumulated recycled water salts, for 30 years. The 30 year result indicates moderate change in chloride concentration in the basin for both 5 million and 1 million acre-ft of storage. Pulses of precipitation recharge carrying accumulated salts to the aquifer are more pronounced in the 1 million acre-ft storage basin, with higher chloride concentration reached at the end of 30 years. The additional storage in the 5 million acre-ft basin dampens the concentration of the salt loading that leaches once every five years. The changes after 30 years are moderate in this scenario, testing reasonable limits of expected loading due to application of recycled water of 20,000 acre-ft/year that accumulates 50% of the recycled water salts in soils that is then 100% leached during a wet year at 5 year interval.

To further describe a means of evaluating uncertainty of parameters of salt balance, a model sensitivity analysis should be done. The purpose of sensitivity analysis is to vary model input parameters within plausible limits to see how variation affects model output. Variation of model input values is done within limits of uncertainty; usually high/low values that bracket realistic ranges of values for a parameter or process. Frequently a single input parameter is varied uniquely during sensitivity analysis while other input parameters are held constant. The sensitivity calculation is done with the input variable to determine the corresponding change from a baseline calculation. If plausible variation of the model input causes a significant change in model output, then the input parameter should be reevaluated to develop a less uncertain value. This process may require additional modeling steps or collection of new field data acquisition to better characterize an uncertain parameter or process (e.g., the salt concentration in delivered water return flow). If variation of model input causes minor changes in model output, then model input parameter does not have to be rigorously re-evaluated or estimated.

Sensitivity analysis of an input value should be performed independently by changing a concentration, or flow, within plausible limits based on the uncertainty of the input parameter value. The sensitivity to an input values will depend on the flow rate and concentration of the input relative to the groundwater storage and initial chloride in a given zone. A complete sensitivity analysis should be done on the full suite of parameters and variables, for every zone or subdomain in the whole groundwater basin.

CONCLUSIONS

The salt balance of the soil water zone of a groundwater basin includes multiple components, including inputs and outputs. The inputs include salt in natural precipitation falling upon the area, salt in the irrigation water used in an area, residual salts in the soil, and salt dissolved into solution from soil minerals. The outputs include salt in applied fertilizers or soil amendments, salt taken up by or retained on the plant and removed, salt chemically precipitated in the soil, and salt in the drainage water from the area. These inputs and outputs operate in the groundwater zone and in the soil water zone. The soil water zone is often the main source of salt loading source to groundwater. Outputs of salts from groundwater include salt mass lost from pumping, river leakage, evapotranspiration, underflow, and small amounts of salts lost by precipitation as mineral cements.

Several of these salt balance components have been evaluated in this report for their importance and for their uncertainty. The salt loading from in situ mineral dissolution was discussed first. Other factors were next considered in cases when in situ mineral dissolution is not so important. Evaluations were made of salt in the irrigation water used in an area and salt in applied fertilizers or soil amendments. Factors such as drought and water conservation and how they affect salt loading factors should be included in other analysis. The analysis considered how use and management of groundwater basins affects salt in the drainage water from an area.

To improve estimates in the face of uncertainty, empirical "field" studies should be carried out to determine what are the anticipated loading rates of salts and water due to irrigation with recycled water. Without additional studies, the evaluations must consider average as well as high/low loading scenarios, as performed in this investigation. Uncertainty must factor in how much difference is represented in the high/low calculations. If the amount of recycled water used produces limited changes to the salt loading in a basin, even in "high percentage" loading scenarios, then the impact of loading of salts due to recycled water use might not be evaluated further. If the "high-low percentage" loading scenarios produce significant differences, then additional work should focus on improving estimates in a groundwater basin.

REFERENCES

Carter, D.L. (1975). Problems of salinity in agriculture, in Poljakoff-Mayber, A. Gale, J., and Carter, D.L. eds., Plants in saline environments: Springer, p. 25-35.

Davis, S.N., and DeWiest, R.J. (1966). Hydrogeology: John Wiley and Sons, Inc., NewYork, USA 463 p.

Farley, M., and Trow, S. (2003). Losses in Water Distribution Networks: IWA Publishing House: London, Alliance House, 282p.

Garcia-Fresca, B., and Sharp, J.M., Jr. (2005). Hydrogeologic considerations of urban development – Urban-induced recharge. In Ehlen, J., Haneberg, W.C., and Larson, R.A (editors), Humans as Geologic Agents: Geol. Soc. America, Reviews in Engineering Geology, vol. XVI, p. 123-136.

Hem, J.D. (1992). Study and interpretation of chemical characteristics of natural water (3rd ed.): USGS Water-Supply Paper 2254, 263 p.

Hibbs, B. (2017). Technical approach to estimating the appropriate extent of assimilative capacity use for groundwater basins in the Los Angeles Region: Technical contract report prepared for the Santa Ana Regional Water Quality Control Board, 206 p.

Junge, C.E., and Werby, D. (1958). The concentration of chloride, sodium, potassium, calcium, and sulfate in rainwater over the United States: J Meteorology, vol 15, p. 417-425.

Malecki, J. and Matyjasik, M. (2002). Vadose zone – challenges in hydrochemistry: Acta Geologica Polonica, vol. 52, no 4, p. 449-458.

Norin, M., Hulten, A.-M., and Svensson, C. (1999). Groundwater studies conducted in Goteborg, Sweden,. In Chilton, J. (Editor), Groundwater in the Urban Environment - Selected City Profiles: Rotterdam, Balkema, p. 209-216.

Sanz, E., Ayora, C. and Carrera, J. (2011). Calcite dissolution by mixing waters - geochemical modeling and flow-through experiments: Geologica acta, vol. 9, no. 1, p. 67-77.

U.S. Department of Agriculture. (1954). Diagnosis and improvement of saline and alkali soils: Agricultural Handbook No. 60.

Whitehead, H.C., and Feth, J.H. (1964). Chemical composition of rain, dry fallout, and bulk precipitation at Menlo Park, California, Journal of Geophysical Research, vol. 699, p 3319-3333.

Assessment of the Water Supply Alternative for a Mining Prospect in the Western Region of Saudi Arabia: A Case Study

Bassam. Tawabini and Abdulaziz Shaibani
(Geosciences Department, King Fahd University of Petroleum and Minerals (KFUPM), P.O. Box 189,Dhahran 31261, Saudi Arabia; E-mail: bassamst@kfupm.edu.sa)

Extending over 2 million square km, the Kingdom of Saudi Arabia is the largest country by landmass in the Middle East and the 13th largest in the world. The Kingdom's geology gives it an abundance of natural resources and raw materials. There are more than 48 minerals that have been identified in the Kingdom, including gold, zinc, aluminum, copper, phosphate and feldspar. Extraction of these metals is taking place at a rapid pace driven by the huge needs of these metals in both local and international markets and the Kingdom's vision2030 which identifies mining as a pillar for economic growth. The Kingdom has identified over two thousand mining sites of precious and non-precious metals. The scarcity of fresh water needed for domestic and process uses has always been a great challenge to the mining industry in the Kingdom. One example was a copper mining project at Jabal Sayid located approximately 150 km southeast of Al-Madinah City in the western region of Saudi Arabia. A study was conducted to assess the water supply alternatives near and around the project site. Due to the absence of any surface water resources exist in the region, the alternatives water resources that were assessed for this mining project included: local groundwater aquifers, desalinated seawater and treated wastewater. A detailed survey of these water resources around the project site was carried out through literature review, field survey and site visits. Local groundwater resources were mapped and the main uses of groundwater in the area were identified. The study revealed that the estimated maximum daily needed supply of potable water will be about 21 m^3 based on a consumption rate of 70 liter per capita per day for 300 workers. Such quantity of water can be obtained from nearby local wells. Although, this is a small quantity, it still pauses some stress on the extremely limited shallow renewable groundwater in the region. The estimated total amount of process water for copper extraction needed over 15 years of project life was around 50 million m^3. Such huge quantity of water cannot be met by local groundwater resources, which are barely sufficient for local domestic, agricultural and livestock raising uses. On the other hand, calculation estimated that transporting 50 million m^3 of seawater would cost from 14 to 25 million US dollars and the pipes would be susceptible for corrosion problem due to the high salinity. The study concluded that the treated wastewater purchased from Al-Madinah tertiary-treated municipal wastewater treatment plant (WWTP) would provide the required quantity of a fairly better quality than seawater at an estimated cost of around US$ 9.0 million. The cost is basically needed for transporting the treated wastewater from Al-Madinah WWTP to the project site. Based on this study recommendation, the mining company signed an agreement with the Al-Madinah municipality to purchase and transport a daily amount of 20,000 m^3 from the city WWTP to the mine through a 160-km of 300mm-diameter pipeline. This amount can be increased based on mining needs. Detailed assessment methods along with the description of both geological and hydrogeological characteristics of the project site will also be covered in the paper.

Redox Transformation of Arsenic and Selenium Along a Groundwater Flowpath Beneath a Drained Terrestrial Marsh

Barry J. Hibbs, Rachel Ridgway, and James Walker
(California State University, Los Angeles, CA, USA)

ABSTRACT: An evolving groundwater flowpath was studied beneath a historic swamp. Located in San Diego Creek Watershed of Orange County California, the Historic "Swamp of the Frogs Marsh" was drained in the late 1800s. Today, groundwater beneath the historic swamp is suboxic to oxic. Groundwater was sampled at nine points along the groundwater flowpath. Along the first half of the 0.6 mile flowpath, groundwater moves through the outer fringes of the historic swamp. There, groundwater salinity does not change and remains slightly saline, while arsenic increases from 13 ug/L to 90 ug/L As. Selenium increases from 50 to 228 ug/L Se in this region. Over 98% of the arsenic and selenium along the first half of the flowpath is in the form of arsenate and selenate, the most oxidized forms of these elements. Iron and manganese are not detectable, and almost all of the nitrogen is in the form of nitrate. Along the second half of the flowpath, groundwater moves into and through the interior of the historic swamp. In this region, salinity doubles and a correlative increase is observed between chloride, bromide, sulfate, and arsenic, the latter reaching 196 ug/L As. Selenium decreases to about 60 ug/L Se in the lower flowpath region. Over 98% of the arsenic is arsenate, but only 85% of the selenium is selenate in the lower flowpath region, the rest is present as selenite (an intermediate redox form). No organic forms of arsenic or selenium are present along the full flowpath. There is a small amount of manganese (<70 ug/L) and trace amounts of ammonium in the lower flowpath region, but no detectable iron. Based on these data, we propose a model of selenium oxidation and arsenic desorption (but not reductive dissolution) from iron and aluminum oxides and clays within the upper part of the flowpath, and leaching of evaporative salts, containing arsenate, in the lower flowpath region. Geomorphic and geochemical analysis indicates that the lower flowpath region was once a phreatic salt playa.

INTRODUCTION

The research undertaken and presented in this paper expands on work previously completed by others (Hibbs and Lee, 2000; Lee, 2001; Meixner and others 2004; Sjolin, 2004; and Hibbs et al., 2017). The study purpose is to focus extensively on the sources, spatial distribution, and concentrations of selenium, arsenic, and associated hydrochemical and isotopic parameters in shallow groundwater along groundwater flowpaths. Local areas were studied in the vicinity of open and underground channels and drains. This work provides supplemental and newer information on selenium, arsenic, and associated contaminants and assesses the controls affecting their mobility. This information is expected to help policy-makers and managers to mitigate the effects of these trace elements in the surface waters and groundwater of the watershed.

The San Diego Creek Watershed covers approximately 112 square miles in Orange County, California and occupies a major portion of the historic San Joaquin Ranch (Figure 1). Regional streams drain headwaters from the Santa Ana Mountains in the north and northeast and the San Joaquin Hills in the southeast. These waters drain ultimately into Upper Newport Bay, one of the last remaining coastal estuaries of southern California (Hibbs and Lee, 2000). Before 1890, waters draining from the highlands collected in the central low-lying part of the watershed creating a wetland environment known as the "Swamp of the Frogs" Marsh (Figure 2.0.1) (Trimble, 2003). Wetlands existed where the water table was at ground surface, creating an area where water-logged soils prevented land development. In the late part of the nineteenth century, part of this marsh was drained to make way for the expansion of ranching and agriculture (Mendenhall, 1905; Trimble, 2003). By dredging channels to drain the marsh, the water table was effectively lowered by several feet so that today a shallow groundwater system exists above deeper regional aquifers.

Land use history and agriculture play important roles in controlling groundwater quality in the shallow aquifer. During the late 1860s, the San Joaquin Ranch supported at least 40,000 sheep (Conard, 1990). By the early 1880s, the primary emphasis had shifted to cattle. Crop farming took on primary importance in the 1890s, and approximately 31,000 acres were planted with barley, with other lands planted with beans, wheat, corn, and potatoes (Conard, 1990). By 1911, lima beans had replaced barley as the principal crop grown on San Joaquin Ranch (Conard, 1990).

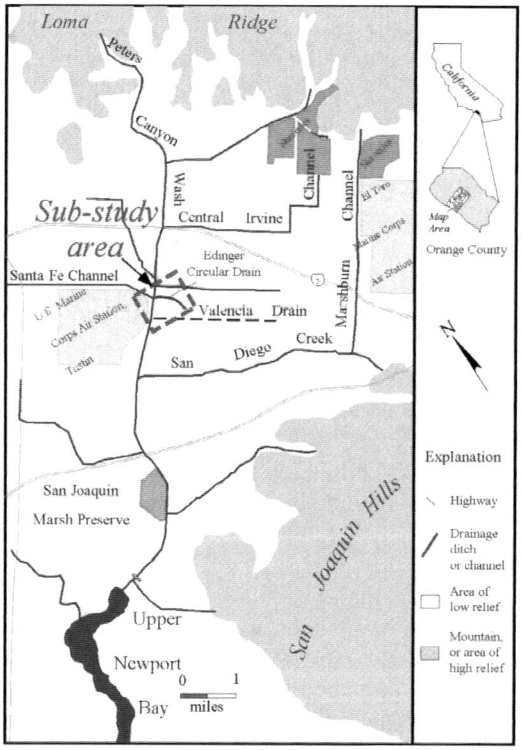

Figure 1. San Diego Creek Watershed and location of Edinger Drain sub-study area.

The Swamp of the Frogs Marsh had not been completely drained by 1919, especially the lower lying areas of the marsh. Some limited improvements of the lands once occupied by the marsh had been effected by this time for the purposes of growing alkali resistant crops (Eckmann et al., 1919). Eckmann et al., (1919) saw little potential for planting orchard crops in the area once occupied by Swamp of the Frogs Marsh, noting that "the deep-rooted orchard crops are not grown under average alkali conditions, as much because of adverse drainage conditions as the presence of alkali." Historic topographic maps produced by the USGS in 1948 show numerous orchards that encompass the boundaries of the former marsh, but there was almost no historical orchard development within the historic marsh boundary.

Considerable effort was subsequently made to improve the poorly drained, alkali lands and increase their productiveness. Additional lowering of the water table did much toward reducing salt concentration, while irrigation water, natural rainfall, and in some instances overflow waters assisted in leaching away the alkali (Eckmann et al., 1919). Since the late 1960s, the area evolved from mostly agriculture to combined residential and commercial development.

High selenium (>30 ug/L) and in some areas concentrated arsenic (<10 ug/L) is found in shallow groundwater and in surface flows in San Diego Creek Watershed. Selenium is an essential micro-nutrient without which livestock can develop muscular dystrophy and humans can suffer from heart disease (Thornton, 1983; Frankenberger, 1998). Selenium can also be a toxic trace element. Above a certain threshold, selenium can induce birth defects in fish and waterfowl. Extensive deformities documented in wildlife at the selenium-impacted Kesterson Reservoir led the US EPA to adopt an interim chronic water quality criterion of 5 µg/L selenium for surface water bodies.

Selenium can exist in one of four oxidation states. Under oxidizing conditions selenium exists dominantly in the +VI oxidation state as the highly soluble anion selenate (Se VI). Under intermediate redox conditions selenate (Se VI) can be reduced to selenite (Se IV) in the +4 oxidation state, which tends to exist as a sorbed ion. Under more reducing conditions, selenium can be further reduced to elemental selenium (Se 0) or selenide (Se -II). Reduced selenide can bind with hydrogen to produce a volatile gas. Alternatively, it can also react with ferrous iron and other metals and precipitate as a mineral phase. Redox transformations of selenium tend to be biologically-mediated in natural systems (Dowdle and Oremland, 1998; Losi and Frankenberger, 1998; Zawislanski and Zavarin, 1996).

Abundant organic matter found in marshlands tends to create anoxic reducing conditions. Such conditions would have acted as a sink for selenium. Hibbs and Lee (2000) proposed a scenario by which selenium eroded from upland marine shales was oxidized to soluble forms allowing it to advect with surface and groundwater flows until reaching the reducing waters of the Swamp of the Frogs Marsh. There, aqueous selenate would have been reduced and immobilized as solids like iron selenide (FeSe2) and perhaps elemental selenium and selenite. Drainage of the marsh permitted the infiltration of oxygen-rich waters that act to oxidize and re-mobilize the selenium sequestered in the marsh. Thus groundwater acts as a non-point source of selenium contamination to the watershed.

Arsenic can exist in numerous organic and inorganic forms in the environment. It is a known toxin that can induce various types of cancer as well as adverse neurological and cardiovascular effects in humans (Le, 2002). For this reason the EPA reduced the Maximum Contaminant Level for drinking water from 50 µg/L to 10 µg/L. Presently, there is no criterion for arsenic concentrations in surface waters. In the mineral phase, arsenic is found to be most abundant as arsenopyrite (FeAsS). Arsenic is often found in association with pyrite deposits, which are common secondary minerals in wetland and watershed sediments (Frankenberger, 2001). The predominant forms of arsenic found in natural waters tend to be in one of two oxidation states; in oxic waters arsenate (As V) dominates, whereas in suboxic or more reducing waters arsenite (As III) becomes more abundant. Under normal environmental oxygen levels, arsenate is thermodynamically more stable (Ferguson and Gavis, 1972, Robertson, 1989; Smith, 1998). However, it also tends to be less mobile than arsenite due to the tendency of arsenate to adsorb strongly to oxy-hydroxides of iron (Drever, 1997).

OBJECTIVES AND METHODS

The overall objective of this study is to examine the regional sources and concentrations of selenium and arsenic in the shallow groundwater in the San Diego Creek Watershed. The primary objective of this paper is to focus extensively on the sources, spatial distribution, and concentrations of selenium, arsenic, and associated hydrochemical parameters along a shallow groundwater flowpath in the watershed. The focal point is the flowpath along Edinger Drain (Figure 1), where several springs and groundwater flows were identified and sampled (Figures 1 and 2).

All groundwater samples were collected in triple-rinsed HDPE or LDPE bottles. Aliquots were separated and preserved within twenty-four hours of collection for various chemical analyses. Filtration was performed using 0.45 µm polycarbonate filter paper. Samples collected for anion analysis were filtered and kept refrigerated below a temperature of 4°C. Samples to be analyzed for dissolved selenium, dissolved arsenic, and selenium speciation were also preserved in this way. Samples to be analyzed for cations and other metals (iron, manganese, molybdenum and vanadium) were filtered and acidified to less than pH 2 with nitric acid (HNO3) then kept refrigerated (< 4°C). Samples collected for arsenic speciation were preserved with 2-mL buffered EDTA ascetic acid/acetate solution.

Anion analysis was conducted by ion chromatography in the Hydrogeology Laboratory at California State University-Los Angeles. Analysis of total dissolved selenium and selenium speciation was conducted by hydride generation atomic absorption (HG-AA) following the method described by Zhang and others (1999) at the University of California-Riverside. Cation analysis was conducted by hydride gas generation-atomic absorption (HGG-AA) at the Water Quality Center Lab, University of Arizona. Other metals analysis (iron, manganese, molybdenum and vanadium) were conducted by inductively coupled argon plasma mass spectrometry (ICAP-MS) by MWH Laboratories (Monrovia, CA). Arsenic speciation was performed by ion chromatography inductively coupled plasma dynamic reaction cell mass spectrometry (IC-ICP-DRC-MS) by Applied Speciation (Tukwila, WA). Only standard anions, total arsenic and selenium, and arsenic and selenium speciation data are reported in this paper. Further details are found in Hibbs et al., (2008).

RESULTS

During channel reconnaissance, the research team discovered 14 groundwater seeps and springs. The groundwater sample collected at the furthest upstream point in Edinger Drain (marked 1854 ft in Figure 2) contained about 50 ug/L total Se (Table 1 and Figure 3). Further downstream, the selenium values increased gradually to 228 ug/L total Se at the boundary between the "Transitional Swamp Region" and the "Historic Salt Flats" (Figure 3). At the sample point closest to the outlet, the selenium values decrease to about 65 ug/L total Se (Table 1 and Figure 3). Almost all of the selenium is in the (+VI) oxidation within the Transitional Swamp Region, while about 11% of the selenium is in the (+IV) oxidation state within the Historic Salt Flats Region (Table 1). Rarely is shallow groundwater detected in San Diego Creek Watershed with more than 4% Se(IV), except beneath San Joaquin Marsh and beneath this Historic Salt Flats (Hibbs et al., 2008).

The groundwater sample collected at the furthest upstream point (1854 ft in Figure 2) contained 13.5 ug/L total As (Table 2 and Figure 3). Arsenic increased steadily to a concentration of 90 ug/L at the boundary between the "Transitional Swamp Region" and the "Historic Salt Flats". Below the boundary, arsenic continues to increase to as much as 196 ug/L (Table 2 and Figure 3). This is, as far as we know, the highest arsenic concentration ever measured in shallow groundwater in San Diego Creek Watershed. Nearly all of the arsenic is in the (+V) oxidation state (Table 2).

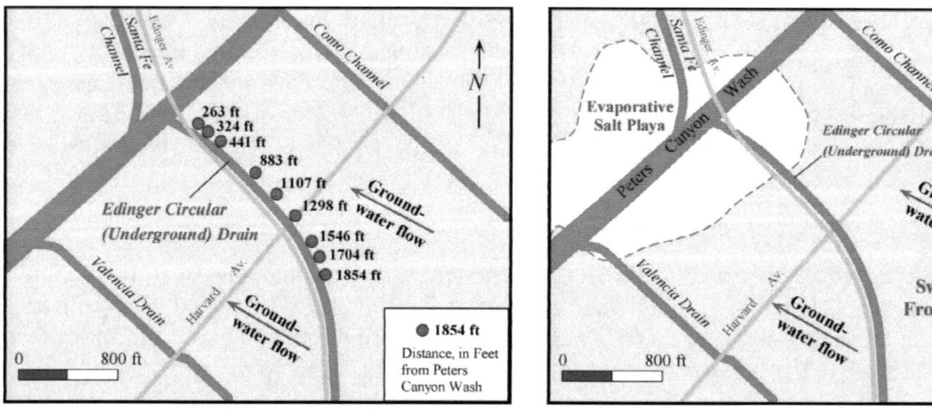

Figure 2. Groundwater sampling sites in Edinger Drain during March, 2007 sampling run. Edinger drain crosses a newly interpreted "Historic Salt Flats" (i.e., evaporative salt playa; or phreatic playa) that is thought to be an important source of high arsenic in the area.

Visual comparison of parameters is obtained by creating a "reference point" that is adjusted so that the concentration scales match with respect to the parameter values at the reference point (see Figure 3 and find "reference point"). The parameter values overlap directly with one another at the "reference point". This provides a comparison of how parameters evolve downgradient of the "reference point" (Figure 3). Correlative results are observed between sulfate, bromide, chloride, and arsenic, but *not* with selenium. All except selenium increase in a proportional manner within the Historic Salt Flats (Figures 3). The concentration of selenium decreases in groundwater in the Historic Salt Flats (Figures 2 and 3). This implies unique hydrochemical processes operate along Edinger Drain Flowpath.

Ion correlation trends between arsenic, chloride, bromide, and sulfate are calculated with an enrichment coefficient developed at the "reference point" (Figure 3) and other points downgradient of the reference point. For example, the concentration of sulfate at the reference point at 883 ft is 1246 mg/L. At the next sample point at 441 ft, sulfate has increased to 2354 mg/L. Thus, an enrichment factor of 1.89 (2364/1246) is calculated. Similar enrichment factors for chloride, bromide, and arsenic are strongly correlated at the various points downgradient from the reference point (Figure 3). Enrichment factors for sulfate, arsenic, chloride, and bromide increase progressively within the Historic Salt Flats. Evaporative concentration of these soluble salts and arsenic (V) in evaporative salt crusts (in predevelopment times) and subsequent leaching into shallow groundwater is the probable cause.

Table 1. Selenium Speciation Data for Edinger Drain -
Groundwater Seeps, March 2007

Sample ID	Se VI (ug/L)	Se IV (ug/L)	Organic Se (ug/L)	Total soluble Se (ug/L)	% Se VI
Edinger Drain 1854 ft	49.57	0.86	<0.2	50.43	98.3%
Edinger Drain 1704 ft	57.55	0.89	<0.2	58.44	98.5%
Edinger Drain 1546 ft	72.04	0.98	<0.2	73.02	98.6%
Edinger Drain 1298 ft	91.86	1.08	<0.2	92.94	98.8%
Edinger Drain 1107 ft	126.97	1.25	<0.2	128.22	99.0%
Edinger Drain 883 ft	226.68	1.83	<0.2	228.51	99.2%
TRANSITION POINT					
Edinger Drain 441 ft	63.19	7.70	<0.2	70.89	89.14%
Edinger Drain 324 ft	61.32	7.47	<0.2	68.79	89.14%
Edinger Drain 263 ft	65.31	7.32	<0.2	72.63	89.9%
PETERS CANYON WASH – 0 FT					

Table 2. Arsenic Speciation Data for Edinger Drain
Groundwater Seeps, March 2007

Sample ID	As V (ug/L)	As III (ug/L)	DMAs (ug/L)	MMAs (ug/L)	% As V
Edinger Drain 1854 ft	13.5	<0.08	<0.08	<0.08	~100
Edinger Drain 1704 ft	20.8	<0.08	<0.08	<0.08	~100
Edinger Drain 1546 ft	23.3	<0.08	<0.08	<0.08	~100
Edinger Drain 1298 ft	29.2	--	--	--	--
Edinger Drain 1107 ft	33.6	<0.08	<0.08	<0.08	~100
Edinger Drain 883 ft	90.3	<0.08	<0.08	<0.08	~100
TRANSITION POINT					
Edinger Drain 441 ft	170.0	<0.08	<0.08	<0.08	~100
Edinger Drain 324 ft	176.0	<0.08	<0.08	<0.08	~100
Edinger Drain 263 ft	196.0	<0.08	<0.08	<0.08	~100
PETERS CANYON WASH – 0 FT					

DMAs – dimethylated arsenic
MMAs – monomethylated arsenic

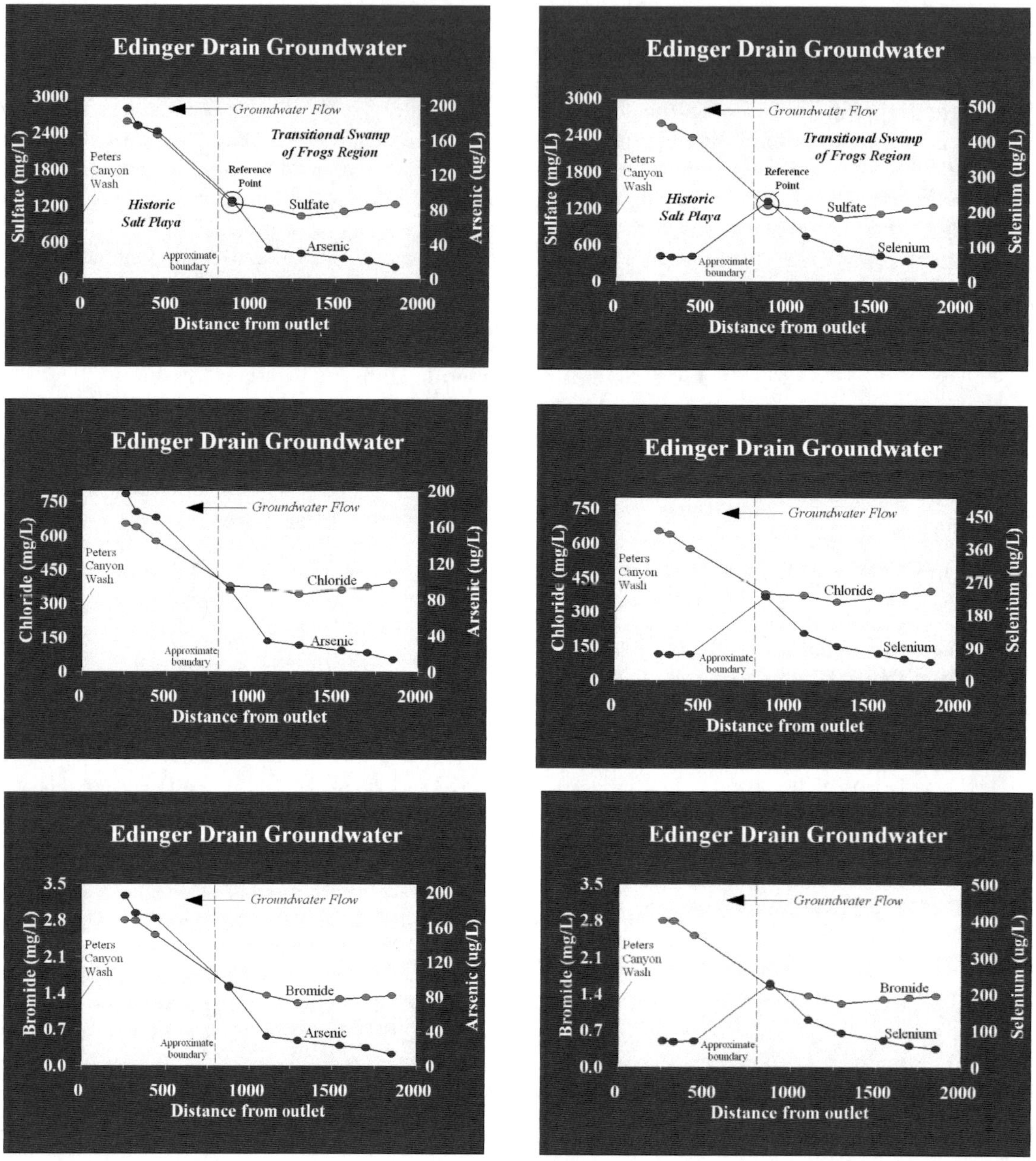

Figure 3. Graphs showing concentrations of sulfate, chloride, and bromide plotted against arsenic and selenium concentrations in groundwater along Edinger Drain. The reference point is calibrated on the axis scale so the points overlap directly for the ions on the relative axes right above the salt flats (dashed line). Sulfate, chloride, and bromide don't vary much in the transitional region (right of dashed line), where selenium and arsenic both increase. As groundwater moves into the alkali region (left of dashed line), sulfate, chloride, bromide and arsenic increase in concentration in a proportional manner, whereas selenium decreases sharply in concentration. Evaporative enrichment of arsenic, sulfate, chloride, and bromide is implied in the alkali salt flats region.

DISCUSSION

Results of sampling indicate local shallow groundwater contains high concentrations of selenium and arsenic near Edinger Drain. Our findings indicate that most of the dry weather flows of Edinger Drain are fed by percolating groundwater, not by urban runoff. Further down the hydraulic gradient, close to Peters Canyon Wash, the concentration of selenium decreases once groundwater enters the Historic Salt Flats area (Figures 2 and 3). Selenium is even lower in groundwater right by Peters Canyon Wash. Arsenic, on the other hand, increases continuously in samples down the hydraulic gradient (Table 2 and Figure 3). Concentrations of bromide, chloride, and sulfate do not change much in the Transitional Swamp Region (Figure 3), but increase (much like arsenic) in the Historic Salt Flats Region (Figures 3). These hydrochemical facies are associated closely with land use change over time. Details on the *predevelopment* and *postdevelopment* models are provided below. The conditions outlined in these models reflect only those along Edinger Drain groundwater flowpath.

Predevelopment Model. Arsenic and selenium bearing runoff, and groundwater containing fairly dilute concentrations of selenium and moderate concentration of arsenic entered the Transitional Swamp Region. There the entrant waters encountered an intermediate redox environment. This region, along the current path of Edinger Drain, was not highly reducing because of fluctuating water tables that were associated with dry seasons and drought years (Figures 4 and 5) (Meixner et al., 2004; Sjolin, 2005). Evidence of intermediate oxidation conditions in the Transitional Swamp Region are shown in soil profiles near Edinger Drain and Harvard Avenue. These profiles contain alternating layers of phreatic caliche (stable, shallow water tables) and layers of iron-stained soils which indicate semi-oxidizing conditions during dry periods when water tables were deeper. Semi-oxidizing conditions frequently transitioned to intermediate redox to slightly reducing conditions during periods when water tables were high and when wetlands were stable (Figures 4 and 5). The overall average was an intermediate redox condition in the Transitional Swamp Region. This aspect of the model, within the Transitional Swamp Region, corresponds closely to interpretations provided by Meixner et al., (2004) and Sjolin (2005).

Intermediate oxidation states dominated in the Transitional Swamp Region during predevelopment times (Figures 4 and 5). Iron oxide coatings indicate that soils were not highly reducing and that water tables probably fluctuated. During dry times when water tables were deeper, arsenic probably wicked up into the more oxidizing vadose zone and was oxidized to arsenic (V). Arsenic (V) sorbed onto iron and aluminum oxide surfaces, organics, and clays. During wet times when water tables were shallow, redox conditions were more reducing and arsenic was probably reduced from arsenic (V) to arsenic (III). The reduced form of arsenic is more soluble than arsenic (V). Soluble arsenic moved with groundwater into the Historic Salt Flats region (Figures 4 and 5).

Selenium (VI) was probably reduced to selenium (IV) in the intermediate redox zone discussed above. Selenium (IV) probably sorbed onto organic particles and clay minerals in the Transitional Swamp Region. Selenium was not reduced sufficiently in this region to produce iron selenides (-II) or elemental selenium (0). Sulfate was not reduced to a great degree because this area was not sufficiently reducing (Figures 4 and 5).

Soluble ions that are not redox sensitive in intermediate redox zones (bromide, chloride, and sulfate) and arsenic (III) moved with subsurface water into the Historic Salt Flats (Figures 4 and 5). Salt precipitates were not persistent in the Transitional Swamp Region because soluble ions were flushed by episodes of surface runoff and/or periodic high water tables. Instead, soluble salts accumulated in the Historic Salt Flats. Some residual selenium (IV) in groundwater and selenium (VI) in surface runoff also moved into the Salt Flats, but the supply of selenium may have been limited because much of the mass of selenium (as Se-IV) had been sequestered in the Transitional Swamp Region. In the Salt Flats Region where topography was low-lying, the water table was perennially shallow. There the waters evaporated strongly to produce crusts of salts consisting of sulfate, bromide, and chloride, along with considerable amounts of arsenic (V), but with little selenium. Most of the selenium had already been removed in the Transitional Swamp Region, and/or within the shallow groundwater of the hypersaline, strongly alkaline Historic Salt Flats. These saturated environments beneath salt flats are known for their strongly anoxic conditions below the water table (Figures 4 and 5) (Boyd and Kreitler, 1986; Ryu et al., 2003). Likewise some of the sulfate was reduced beneath the Historic Salt Flats to form iron-sulfide compounds.

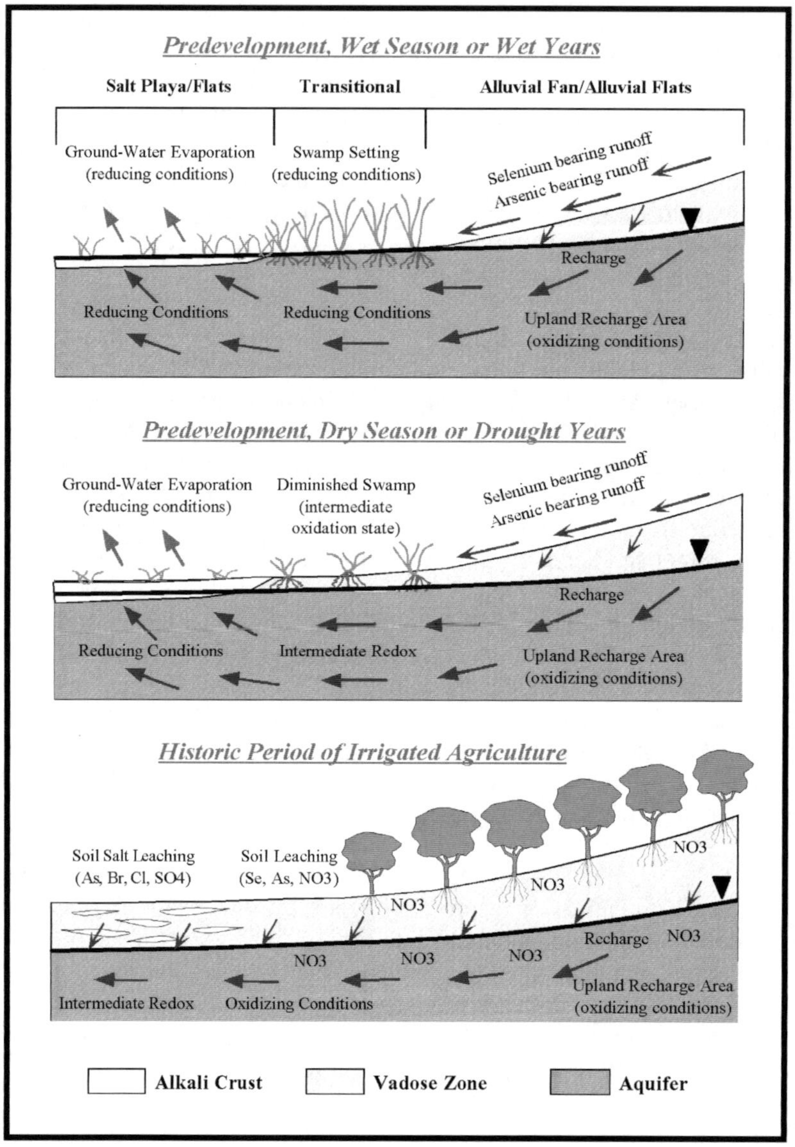

Figure 4. Time series conceptual models for water table and land surface conditions along and above Edinger Drain covering wet and dry conditions during predevelopment times and after the area was developed for irrigated agriculture.

Gao et al., (2006) identified oxidizing conditions in the exposed salt crusts of phreatic salt flats, due to exposure of crusts to the atmosphere. Arsenic and sulfur in these salt crusts are stable as arsenic (V) and evaporite sulfate (VI) salts. Gao et al., (2006) found that about 98% of arsenic in the crusts of the salt flats was arsenic (V). Beneath salt crusts, shallow groundwater is usually highly anoxic (Boyd and Kreitler, 1986; Ryu et al., 2003). In these saturated anoxic zones, sulfur is reduced to iron-sulfide while selenium is reduced to elemental selenium and iron-selenide. Thus, much of the sulfur and arsenic in the Historic Salt Flats remained in oxidized forms in the salt crusts until and after periods of swamp drainage and eventual leaching of residual salts. Furthermore, the oxidized form of arsenic (As-V) deposited in the salt crusts continues to leach out of the residuum of the crusts into shallow groundwater beneath the Historic Salt Flats (Table 2).

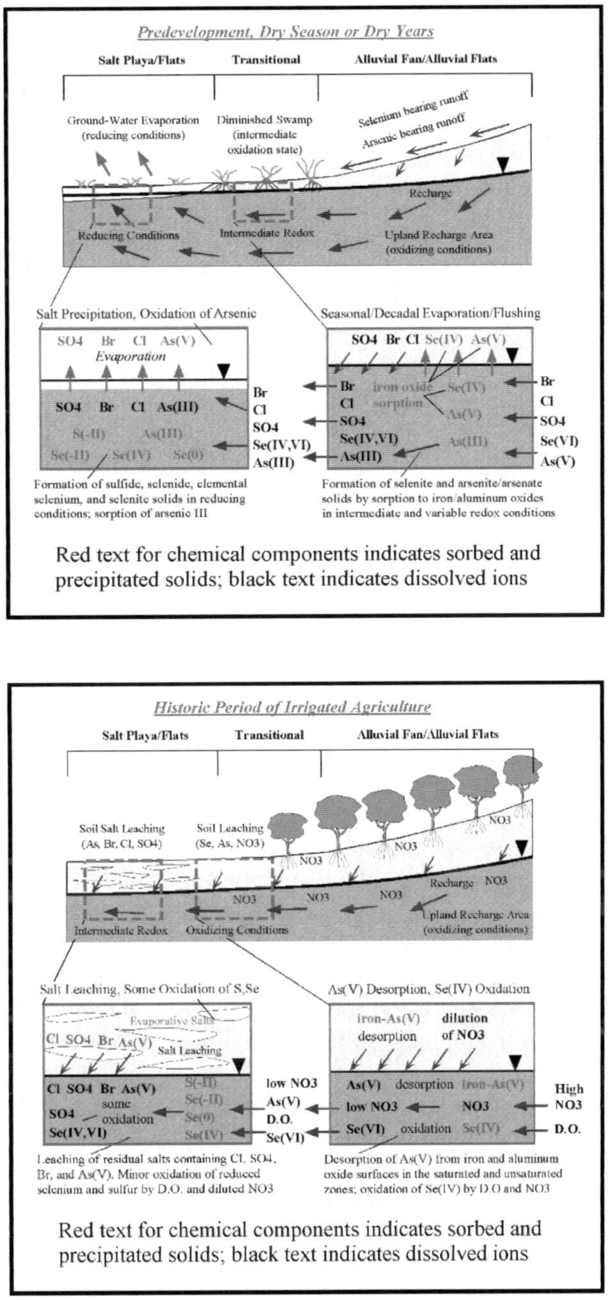

Figure 5. Predevelopment and postdevelopment conceptual models for water table and land surface conditions along and above Edinger Drain. Inset regions show hypothesized and interpreted details for hydrochemical processes, and migration and attenuation of solutes in the Transitional Swamp Region and Historic Alkali Salt Flats Region.

This summary predevelopment model includes a strong sink for selenium (IV) and arsenic (V) via sorption onto clay minerals, Fe and Al-oxides, and organic material within the intermediate redox sites of the Transitional Swamp Region. There was little persistent enrichment of soluble ions (Cl, SO4, and Br) by evapoconcentration or sulfate reduction in this transitional zone. Arsenic (III) and soluble anions moved downgradient into the Historic Salt Flats Region. Further, the model proposes strong concentration of large amounts of Cl, SO4, Br, and arsenic (V) in the evaporative salts crusts in the Historic Salt Flats Region. In the groundwater beneath the Salt Flats, where soils were anoxic, a smaller mass of sulfate was probably reduced to iron-sulfides. A limited pool of selenium was probably reduced to iron-selenide and elemental selenium in the anoxic groundwater beneath the Historic Salt Flats (Figures 4 and 5).

Postdevelopment Model. After drainage of the Swamp of Frogs Marsh, the area was developed for irrigated agriculture (Figures 4 and 5). Arsenic and selenium concentrations in groundwater moving through the transitional region increase along the flowpath, while chloride, bromide, and sulfate do not change very much. Contemporary groundwater is highly oxidizing in the Transitional Swamp Region, as indicated by the absence of iron, manganese, and ammonium (Hibbs et al., 2008); and abundance of oxidized forms of arsenic (V) and selenium (VI) (Tables 1 and 2).

Residual salts remaining in unsaturated soils are now leaching into groundwater within the Historic Salt Flats. This carries large masses of chloride, bromide, sulfate, and arsenic (V) into shallow groundwater. Bound selenium is probably present in limited amounts in the soils of the Historic Salt Flats Region. Selenium was sequestered in the Transitional Swamp Region before masses of Se ever made it to the Historic Salt Flats. The slightly suboxic (intermediate to slightly oxidizing) redox states that exist in contemporary groundwater in the Historic Salt Flats is indicated by small amounts of manganese and relative abundance of selenium (IV) (Table 2 and Figures 4 and 5) (Hibbs et al., 2008). Groundwater is more oxidizing in the Transitional Swamp Region where no manganese is present and where selenium (IV) is meager (Hibbs et al., 2008). Suboxic redox state does not allow selenium to be oxidized in the same amount as in the Transitional Swamp Region. Most likely, selenium was never enriched in soils of the Historic Salt Flats to begin with.

CONCLUSIONS

The geochemical patterns of selenium, arsenic and anion salts exhibit spatial variability within the shallow aquifer of San Diego Creek Watershed. Groundwater within the boundary of the historic Swamp of the Frogs Marsh show a different geochemical signature than those found outside this boundary. An arsenic "hotspot", where arsenic concentrations are anomalously high, was identified inside the boundary of the historic swamp (Meixner et al., 2004). New information presented in this investigation suggests that the highest concentration of arsenic arises by evapoconcentration of groundwater that was sourced from a local geologic terrain in the headwaters of San Diego Creek Watershed.

Spatial relationships suggest that the important sources of selenium and arsenic came from mountain terrains that drain into different regions in San Diego Creek Watershed (Meixner et al., 2004; Hibbs et al., 2008). These spatial relationships seem to suggest that the important arsenic sources came from mountain terrains that drain into the San Diego Creek region. The arsenic may very well be dissolved in-situ from mineralogical or sediment deposits sourced from the mountains that were eroded and deposited in the low lying Tustin Plain. In this possible scenario, arsenic bearing minerals in the upper San Diego Creek drainage area (above Swamp of the Frogs Marsh) could be releasing arsenic directly into groundwater at concentrations up to 12 ug/L. Alternatively, the arsenic may come from solution of minerals *within* the mountains followed by drainage of arsenic bearing runoff onto the low lying plain. Once groundwater and surface water moved into the Swamp of the Frogs Marsh region, other processes enriched arsenic to the abnormally high concentrations observed along Edinger Drain.

REFERENCES

Boyd, F.M., and C.W. Kreitler. 1986. *Hydrology of a Gypsum Playa, Northern Salt Basin, Texas*. University of Texas Austin, Bureau of Economic Geology Report of Investigations 158. 37 p.

Conard, R. 1990. *Historic American Buildings Survey, Irvine Ranch Agricultural Headquarters, Boyd Tenant House*. Report by PHR Associates, Santa Barbara, California.

Dowdle, P.R., and R.S. Oremland. 1998. "Microbial Oxidation of Elemental Selenium in Soil Slurries and Bacterial Cultures." *Environmental Science and Technol*ogy. 32 (23): 3749–3755.

Drever, J.I. 1997. *The Geochemistry of Natural Waters: Surface and Groundwater Environments*. Third Ed. Prentice Hall, New Jersey.

Eckmann, E.C., A.T. Strahorn, L.C. Holmes, and J.E. Guernsey. 1919. *Soil Survey of the Anaheim Area, California*. U.S. Department of Agriculture, Advance Sheets-Field Operations of the Bureau of Soils. 79 p.

Ferguson, J.F. and J. Gavis. 1972. "A Review of the Arsenic Cycle in Natural Waters." *Water Research*. 6: 1259-1274.

Frankenberger, W.T. and R.A. Engberg.1998. *Environmental Chemistry of Selenium*. Marcel Dekker, Inc., New York.

Frankenberger, W.T. 2001. *Environmental Chemistry of Arsenic*. CRC Press.

Gao, S., S.R. Goldberg, A.T. Chalmers, M.J. Herbel, R. Fujii, and K.K. Tanji. 2006. "Sorption Processes Affecting Arsenic solubility in Tulare Lake Bed Sediments, California." *Chemical Geology*, 228: 33-43.

Hibbs, B., and M. Lee. 2000. *Sources of Selenium in the San Diego Creek Watershed, Orange County, California*. Defend the Bay & California Urban Environmental Research and Education Center. 43 p.

Hibbs, B., M. Merino, M., R. Andrus. 2008. *Sources and Distribution of Selenium and Arsenic in Shallow Groundwater: San Diego Creek Watershed*. In Selenium, Nitrate, and Other Constituents in San Diego Creek and Newport Bay Watersheds, 2004 - 2008: California State University, Los Angeles. Technical contract report prepared for the Santa Ana Regional Water Quality Control Board. 300 p.

Hibbs, B.J., M. Lee, and R. Ridgway. 2017. "Land Use Modification and Changing Redox Conditions Releases Selenium and Sulfur from Historic Marsh Sediments." *Journal of Contemporary Water Research and Education*. 161: 48-65.

Le, X.C. 2002. Arsenic Speciation in the Environment and Humans. In *Environmental Chemistry of Arsenic*. W.T. Frankenberger (ed.): 95-116.

Lee, M. 2001. *Sources, Distribution, and Mobility of Selenium in San Diego Creek Watershed: Master's Thesis*, California State University- Los Angeles.

Losi, M.E., W.T. Frankenberger. 1998. "Microbial Oxidation and Solubilization of Precipitated Elemental Selenium in Soil." *Journal of Environmental Quality*. 27(4): 836-843.

Meixner, T., B. Hibbs, J. Sjolin, and J. Walker. 2004. *Sources of Selenium, Arsenic, and Nutrients in the Newport Bay Watershed*. Contract report prepared for the Santa Ana Regional Water Quality Control Board. 129 p.

Mendenhall, W.C. 1905. *Development of Underground Waters in the Eastern Coastal Plain of Southern California*. USGS Water Supply Paper 137. 140 p.

Robertson, F.N. 1989. Arsenic in Groundwater Under Oxidizing Conditions, Southwest United States. *Environmental Geochemistry and Health*. 11(3/4): 171-186.

Ryu, J., R. A. Zierenberg, R.A. Dahlgren, and S Gao. 2003. "Sulfur Biogeochemistry and Isotopic Fractionation in Shallow groundwater of Owens Dry Lake, California." AGU Abstracts with Programs, Fall Meeting. Abs. B31D-0342.

Sjolin, J.A. 2004. *Process-level Controls on the Concentration of Arsenic and Selenium in the Groundwater of an Urbanizing Catchment*. Master's thesis, University of California-Riverside.

Smith, E., R. Naidu, A.M. Alston. 1998. "Arsenic in the Soil Environment: A Review." *Advances in Agronomy*. 64: 149-195.

Thornton, I. 1983. *Applied Environmental Geochemistry*. New York. Academic Press.

Trimble, S.W. 2003. "Historical Hydrographic and Hydrologic Changes in the San Diego Creek Watershed, Newport Bay, California." *Journal of Historical Geology*. 29: 422-444.

USGS. 1948. *Topographic Map of the Tustin California Quadrangle*.

Zawislanski, P.T., M. Zavarin. 1996. "Nature and Rates of Selenium Transformations: a Laboratory Study of Kesterson Reservoir Soils." *Soil Science Society of America Journal*. 60: 791-800.

Zhang, Y.Q., J.N. Moore, and W.T. Frankenberger, Jr. 1999. "Speciation of Soluble Selenium in Agricultural Drainage Waters and Aqueous Soil–Sediment Extracts Using Hydride Generation Atomic Absorption Spectrometry." *Environmental Science and Technol*ogy. 33: 1652–1656.

Gibberellic Acid Surface Complexation and Desorption Retention on the Ferrihydrite/Water Interface

Li Zhang and Fei Liu (China University of Geosciences (Beijing), Beijing, China)
Liang Chen (Tianjin University, Tianjin, China)

Gibberellic acid (GA_3) is a wildly used plant growth regulator and environmental toxin especially in China, but few studies have focused on GA_3 interactions with minerals/soils. In this study, GA_3 surface complexation and desorption retention mechanism on ferrihydrite was investigated by combining sorption–desorption batch studies with Fourier transform infrared (FTIR) spectroscopy and moving-window two-dimensional (MW2D) correlation spectroscopy. The Fh-GA_3 surface complexes and desorption retention depended strongly on pH. For pH>3.1, electrostatic interaction played an important role in GA_3 sorption on Fh via two ways. One is forming outer-sphere complex directly by electrostatic attraction, which is in the minority. Another is acting as driving force to facilitate the formation of surface hydration-shared ion pair (mainly formed at pH<5.7) and solvent-surface hydration-separated ion pair (mainly formed at pH>5.7). Those three outer-sphere complexes are partly reversible according to the higher total desorption percentage of GA_3 (58–77%). For pH<3.1, the generated monodentate complex was observed and increased with decreased pH, but was more desorption retention than the outer-sphere complexes according to the lower total desorption percentage of GA_3 (37%). At typical soil and groundwater pH levels (4.5–9), the outer-sphere complexes is dominated, but maybe out-competed by nitrate and other anions and then easily desorbed from Fh. Besides, desorption retention of GA_3 on Fh could inhibite the (bio-)availability of GA_3 in soil and groundwater , which could lengthen its half-life period.

Protection of the Columbia River from Contamination of the Hanford Site Nuclear Facility, Washington State

Dibakar (Dib) Goswami
(Nuclear Waste Program, Washington State Department of Ecology, 3100 Port of Benton Blvd., Richland, WA 99354, USA; Email:dgos461@ecy.wa.gov)

The Hanford nuclear operations and chemical separations processes supported the generation of plutonium for the Cold War and left world's most complicated "problem areas" of contaminated soil and groundwater with chemicals and radionuclides. At present, about 160 square kilometers (62 square miles) of groundwater beneath the Hanford Site are contaminated above the state and federal drinking standard. In 1989, the Washington State Department of Ecology (Ecology), Environmental Protection Agency and US Department of Energy (USDOE) formed an agreement to clean up the Hanford Site and developed a comprehensive site wide groundwater remediation strategy with a vision to address various contaminated plumes of hazardous and radioactive waste to protect the Columbia River. Accordingly, the strategy addressed plumes found in the reactor areas adjacent to the Columbia River to protect the river from major contaminants of chromium, strontium-90, and uranium. It also includes containment of major plumes of chemicals and radionuclides found in the central plateau region so that these plumes never reaches the Columbia River. To meet these goals, pump and treat remediation systems were placed along with the deployment of innovative technologies such as apatite sequestration barrier, uranium sequestration for radionuclides to address complex groundwater problems.

Quality Assessment of Groundwater by Graphical Models in Rural Provinces of Korea

Jong Yeon Hwang, Sunhwa Park, Moon-Su Kim, Hun-Je Jo, Ji-In Kim, Gyeong–Mi Lee, InKyu Shin, Sang Ho Jeon, Da Hee Song, Deok-Hyun Kim, Hyen Mi Chung, Hyun-Koo Kim
(NIER, Hwangyong-ro 42, Incheon, 22689, Korea)

The chemical composition of groundwater is controlled by many factors that included composition of precipitation, mineralogy of the watershed and aquifers, climate, and topography. One of the major functions for the validations is to link the relationship between distribution of ionic contents in the groundwater and graphical models for tracing important factors. As being known, a variety graphical and statistical techniques have been devised since the early 1900s in order to facilitate the classification of waters, with the ultimate goal of dividing a group of sample into homogeneous groups such as hydrochemical facies. On the presumption, we focused on the quality assessment as a agricultural and industrial usage and the tracing of ion sources in groundwater in the rural provinces of Korea. For the major goals of this study we categorized the approaching steps of this study as three categories such as followings; 1. Assessment of groundwater qualities as agricultural and industrial usages, 2. Classifications of groundwater by using of Piper, Gibbs and Chadah diagram methods, 3. Tracking of the ionic sources by using of rock-dominance types and relative ionic distribution patterns. First, for the assessments of groundwater suitability, we tried to evaluate the groundwater qualities samples from 2015 to 2017(n=2,153 samples) as agricultural and industrial usages by applying such as following equations and diagrams; SAR, Na (%), RSC, PI, SSP, MH, PS, and Kelly's ratio, Wilcox diagram, U.S. salinity Laboratory's Diagram, Langelier Saturation Index (LSI), Corrosivity Ratio (CR) diagram. Most of the results from the assessment were well satisfied with the guidelines of each criteria. And second, the classifications of groundwater by Piper diagram showed that the groundwater types were classified as the Ca^{2+}-(Cl^{-}-NO_3^{-}) and Ca^{2+}-HCO_3^{-} types, and these types are generally known as the general characteristics of groundwater in Korea. And, by the applications of the CAI (Chloro Alkaline Index) diagrams for groundwater samples it was known that two mechanisms of ion exchange (direct exchange and reverse exchange) are coexisted in soil-groundwater environment. Finally, in the applications of graphical models for the determining of weathering types there were many various mechanisms in weathering processes and most of the weathering types were grouped as dolomite and gypsum type. And, through Gibbs diagrams graphed by TDS via the concentration of ions it was shown that most of the ionic sources in groundwater have a rock/soil-dominance origin. And also, through the results of ion sources tracking it was also known that there are mutual interactions between the alkaline earths and alkali metals originated from soil or rock interactions with weak acidic anions and strong acidic anions in groundwater.

Risk Reduction / Cost Savings by Effective Oil Spill Response and Spill Countermeasure Plans

Steven Pedigo (OSEI Corporation, Dallas, TX, USA); Diane Wagenbrenner, Cody Kaduce (Lawrence Anthony Earth Organization U.S. (LAEO)); Paul W. Sammarco (Louisiana Universities Marine Consortium (LUMCON), 8124 Highway 56, Chauvin, Louisiana 70344)

In this study, the *Oil Spill Comparative Analysis* provides details and documentation on the conclusions summarized herein and supports that high costs associated with oil spill cleanup would be substantially reduced while also bringing better environmental protections, clean up effectiveness, and mitigated liabilities.

It is a cost comparison of each available cleanup method. The comparative model illustrates how much oil is actually removed from the environment giving the stakeholder a true cost picture on a per gallon basis. The comparisons analyze data from recent spills making a compelling case for an alternative response solution that mitigates both cost and the risk of further environmental damage.

In many instances oil/fuel emergency response contractors (Oil Spill Response Organizations [OSROs]) will employ the absolute maximum amount of labor and material they are able to invoice within their response in order to maximize profits. For decades, the responsible parties and their insurers have assumed the high cost of inadequate mechanical and chemical response methods.

Mechanical clean-up is generally the first-choice response solution during a spill. According to US Coast Guard reports, mechanical clean up methods will only remove 2 to 8 percent of the oil in calm seas, leaving 92 to 98 percent of the oil remaining. The lingering oil leaves the insurer and the responsible party culpable for significant damages that could have been avoided if a more comprehensive response solution had been utilized. Mechanical skimmer cleans up typically produces more collected seawater than oil; astonishingly, skimmed water is collected and disposed of at a rate of more than 9 to 1 over collected oil using mechanical oil spill response methods.

Chemical clean-up has long term adverse effects so global leaders and oil spill response professionals continue to take action to reduce use of chemical dispersants that are inherently toxic. Not only do chemical dispersants fail to effectively cleanup spills, they exacerbate the problem by introducing yet another harmful substance to the contaminated environment.

A time for change that benefits everyone. OSE II is a tried and true technology that has been effective in oil spill clean-up for more than 25 years turning reducing costs, mitigating risk, and having no adverse environmental impact. OSE II can reduce the risk of environmental natural resource destruction while reducing associated clean up costs up to 77%, while producing no adverse impact on the environment, which has been carried out over 36,000 times to date, in almost every conceivable scenario where hydrocarbons or hydrocarbon based material can be released or spilled.

Challenges of Municipal Waste Management and Disposal Criteria

Eyilachew Yitayew Abate
(Ole Brumms Vei-24, 0979 Oslo, Norway)

The rapid increase in population and unplanned urban development with limited resource; the city of exposed for informal settlement, environmental hazards, uncollected and untreated municipal wastes that often pollute water resources and breading diseases. The study aimed at analyzing the challenges of municipal waste management and setting certain criteria for disposal site to the city of Addis in particular and to the country in general. Conceptualising and analysing of the city structural plan and the physical environment are the main areas to understand the urbanization issue and to set certain criteria. In order to select waste disposal sites certain criteria and evaluation techniques are indispensable for the city development.

Environmental problems in Addis are clearly due to a number of factors, unplanned urban development is the main issue including the absence of proper waste disposal site and management system. Addis Ababa has faced many challenges due to unbalance urban population growth and limited infrastructure facilities. When the landfill was established, the only factors that were considered in selecting the site were hauling distance, availability of land and the distance from the city centre. There were no applied any other criteria. As a result, the nearby surface water changed its color and odour due to leachates of different mixed wastes disposed in the area. Generally municipal waste collection and disposal in the city of Addis Ababa is in a state of deplorable condition.

There are still opportunities to handle municipal wastes even though a lot of challenges that threaten Addis if the city administration utilize properly; private sector is one of the potential asset to contribute a part under conducive environment. The city council is responsible to implement the proposed disposal sites and making conducive policy in order to accommodate all interested stakeholders. Sitting certain criteria and selecting disposal sites are not the only solution to handle the municipal waste challenges in the absence of systematic collection and management system.

Formation of Disinfection Byproducts in Desalinated Drinking Water: An Assessment and Control Strategy

Shakhawat Chowdhury
(King Fahd University of Petroleum and Minerals, Dhahran, Saudi Arabia)

ABSTRACT: Desalinated seawater is the major source of drinking water in many regions including the Middle East, Mediterranean Basin, Australia and USA. During desalination, several activities, such as, pre-treatment of source water, desalination through distillation or membrane separation, stabilization, mixing, storage and distribution are performed. Few disinfectants (e.g., chlorine, chloramines, chlorine dioxide) are used during these activities to control the biofouling agents and microbiological regrowth. The unintended reactions between the disinfectants and natural organic matter (NOM), bromide and iodide result in the formation of disinfection byproducts (DBPs) in the product water. The product water is stabilized and blended with treated freshwater (e.g., groundwater) to meet the domestic water demands. The DBPs in desalinated and blended water is an issue due to their possible cancer and non-cancer risks to humans. In this paper, formation and distribution of DBPs in different steps of desalination and in water distribution system prior to reaching the consumer tap were investigated. The variability of DBPs among different sources and desalination processes was explained. The toxicity of various DBPs were compared and the strategies to control DBPs in desalinated and blended water were proposed. Several research directions were identified to achieve the comprehensive control on DBPs in desalinated and blended water, which are likely to protect the humans from the adverse consequences of DBPs.

INTRODUCTION

During desalination, feed water is collected through different types of intakes. Chlorine ($HOCl/OCl^-$) is typically applied during pretreatment to prevent bacterial growth and biofouling, and to enhance the performance of filters. Pretreatment is performed using continuous or intermittent chlorination with initial doses of 0.5 – 2.0 mg/L to achieve the target residuals of 0.25 – 0.5 mg/L. Chlorine is also applied in different stages of desalination and in final disinfection to inactivate microorganisms. The disinfectants (e.g., chlorine) react with natural organic matter (NOM) and various inorganic compounds to form disinfection byproducts (DBPs), many of which are known to have cancer and non-cancer risks to humans (USEPA, 2016). Past studies have reported that seawater could contain bromide and iodide in the ranges of 50,000 – 80,000 and 21 – 60 µg/L respectively (Agus et al., 2009; Duranceau, 2010). The thermal desalination processes remove bromide and iodide to non-detectable levels in the feed water. Although the RO processes remove more than 99% bromide, the RO permeate contains considerable amounts of bromide (250-600 µg/L) and iodide (< 4 – 16 µg/L), which can increase concentrations of brominated and iodinated DBPs in desalinated and blended water (Le Roux et al., 2015; Kim et al., 2015). The brominated and iodinated DBPs are more cytotoxic and genotoxic (Kim et al., 2015).

To date, over 100 DBPs have been reported in drinking water while approximately 600 DBPs are likely to be present in supply water (Richardson et al., 2007). The most investigated DBPs in desalinated and blended water include trihalomethanes (THMs) and haloacetic acids (HAAs) (Kim et al., 2015). Many DBPs are of human health concern due to their possible cancer and non-cancer risks. The regulatory agencies, such as, USEPA has the regulatory limits of 80 and 60 µg/L for THMs and HAA_5 respectively (USEPA, 2016). Despite the public health relevance, not much is known about DBPs in desalinated and blended water, and the consequent exposure and risk. Formation of DBPs in desalinated water can vary depending on the sources and the type of desalination processes. Further, concentrations of DBPs in different stages of desalination needs better understanding to control DBPs in desalinated and blended water. In this study, the research on DBPs in desalinated and blended water were reviewed. Formation of DBPs in different stages of desalination processes were investigated. The variability of DBPs depending on the type of desalination process and sources were characterized. The effects of bromide on the formation and distribution of DBPs in desalinated water were investigated. The cytotoxicity and genotoxicity of different DBPs on the mammalian cells were compared. The strategies were identified to better control DBPs in tap water.

DBPs FORMATION

Pretreatment of Feed Water. In chlorinated seawater, concentrations of THMs and HAAs were in the ranges of 0 – 680 and 0 – 175 µg/L respectively (Le Roux et al., 2015) while several emerging DBPs were also reported. In Saudi Arabia, THMs, HAAs, HANs and I-THMs in chlorinated seawater were in the ranges of 3.1 – 27.9, ND – 7.0, 0.43 - 0.76 and 1.9 – 2.57 µg/L respectively. In UAE, THMs and HAAs were < 25 – 95 and ND – <14.5 µg/L respectively while in Qatar, THMs were in the range of 21.7 – 61.4 µg/L. The highest concentrations of THMs (490 - 680 µg/L) and HAAs (69 – 175 µg/L) were reported for a pilot desalination plant in Tampa Bay, Florida. The total organic carbon and chlorine doses were in the ranges of 4.3 – 10.9 and 2.5 – 5.0 mg/L respectively (Agus et al., 2009). In UAE, concentrations of THMs and HAAs were below 100 µg/L in most cases; however, THMs were reported up to 200 µg/L following an oil spill near the intake (El Din et al., 1991).

The lab studies showed various DBPs when chlorine (Cl_2), chloramines (NH_2Cl), chlorine dioxide (ClO_2) and ozone (O_3) were used as disinfectants. In chlorinated samples, THMs, HAAs, HANs and haloacetamides (HAcAms) were in the ranges of ND – 206, 0 – 122, 4.0 - 155 and 110 – 240 µg/L respectively. In the NH_2Cl and ClO_2 treated samples, concentrations of DBPs were much lower while the O_3 treated samples showed higher brominated THMs (B-THMs), HAAs, bromate and HNMs. Overall, THMs in the lab studies from Saudi Arabia, Qatar, Spain, USA and Canada were reported to be 35 - 180, 60-165, 0.34 - 154, 43-206 and 22 - 33 µg/L respectively. In Halifax (Canada), bromate were in the range of 500 - 4500 µg/L (Brookman et al., 2011). In another ozone treated plant, TBM, TBAA and HNMs were reported to be 180, 61 and 16.5 – 34.8 µg/L respectively. The higher concentrations of B-THMs, bromate and total DBPs in O_3 treated water could be a public health concern (USEPA, 2018). Overall, THMs were higher than HAAs in the feed water and lab studies. Approximately 99.9% and 83.7% of THMs and HAAs, respectively, were brominated (Kim et al., 2015). The DBAA and TBAA contributed approximately 41% of total HAAs.

Desalination Plants Prior to Distribution. The pretreated feed water is passed through the thermal or membrane separation processes. In thermal plants, THMs in the MSF processes were reduced by 94% during distillation (El Din et al., 1991; Le Roux et al., 2015). In the East Coast of Saudi Arabia, THMs in feed water were in the range of 3.1 – 27.9 µg/L, which were reduced to 0.09 – 3.48 µg/L in the thermal distillate (Kutty et al., 1989). In thermal distillate, BDCM, DBCM and CHBr3 were 39.5%, 31.6% and 28.9% respectively. However, CHBr3 were 94.7% in the feed water, indicating higher fraction of $CHBr_3$ removal during this process (Kutty et al., 1989). In the feed water from Red Sea (Saudi Arabia), HAAs were in the range of 5.35 – 6.86 µg/L while these were not detected in thermal distillate (Le Roux et al., 2015).

In contrast, the RO plants often function with one, two or four RO passes based on the initial and final water quality and design parameters (Kim et al., 2015). In the pretreated feed water of Tampa Bay RO plant, THMs and HAAs were in the ranges of 490 – 860 µg/L and 69 – 175 µg/L respectively (Agus et al., 2009). In the RO permeate, these were in the ranges of 2.3 – 6.4 µg/L and 1.0 – 2.5 µg/L respectively, indicating more than 99% removal of THMs and HAAs from feed water (Agus et al., 2009). However, Le roux et al. (2015) reported THMs in a RO plant with cellulose triacetate membranes in the range of 39 – 67 µg/L while THMs were 14.5 – 18.4 µg/L in feed water. THMs in RO permeate was increased due to the reaction of residual chlorine with the adsorbed and accumulated reactive organic matter in the fouling layer Le roux et al. (2015). In some RO plants, chloramines are used to limit biofouling and to reduce the membrane damage (Kim et al., 2015). However, chloraminated RO permeate had I-THMs in the range of 1.7 – 2.6 µg/. In addition to the regulated DBPs, several emerging DBPs were reported in desalinated water. Le Roux et al. (2015) reported I-THMs in desalinated water from two plants in Jeddah, Saudi Arabia. The averages of DCIM, BCIM, DBIM, CDIM, BDIM and IF were ND, ND, 1.73, ND, 0.6 and ND respectively. It is to be noted that I-THMs were reported in many surface and ground water systems (Ioannou et al., 2016). Overall, DBPs in the thermal distillate were much lower than the RO permeate. The thermal distillation needs higher energy, which primarily use fossil fuels leading to higher costs for desalinated water. In addition, thermal plants are likely to release higher levels of CO_2 into the environment.

Distribution Systems within Desalination Plants. The DOC in the RO permeates were in the range of <0.1 – 0.6 mg/L (Agus et al., 2009), which are much lower than the typical freshwater sources. However, bromide (Br^-) and iodide (I^-) were reported to be in the ranges of 250 – 600 μg/L and <4 – 16 μg/L respectively, which were much higher than the freshwater sources. The higher levels of bromide and iodide increase the brominated and iodinated DBPs (Ioannou et al., 2016). Typically, the toxicity follows the order of C-THMs < B-THMs < I-THMs.

In a RO plant in Jeddah, THMs in the distribution systems were in the range of 13.5 – 25.7 μg/L. In another RO plant in the Eastern region, Saudi Arabia, $CHCl_3$, BDCM, DBCM and $CHBr_3$ were 0.94%, 1.02%, 2.14% and 95.9% respectively while in the MSF plant in Doha, THMs in the distribution systems were in the range of 0.37 – 8.15 μg/L. In Al-Khobar, THMs in the distribution system of a MSF plant were in the range of 2.4-12.1 μg/L (Chowdhury 2013, 2016). Several other MSF plants showed similar low concentrations of THMs in the distribution systems. In all plants, B-THMs were dominant. The higher concentrations of THMs in the distribution systems of the RO plants might be due to high initial concentrations of THMs and higher levels of bromide in the RO effluents than the thermal effluents.

Blended Water. The desalinated water generally has lower pH than the neutral range, which is adjusted through stabilization by blending with the treated brackish/ground water or untreated seawater prior to supplying to the water distribution network. The RO permeates with low DOC are blended with treated groundwater (e.g., high DOC) resulting in higher concentrations of DOC in the blended water. The high concentrations of DOC and higher levels of bromide and iodide in blended water are likely to increase the concentrations of B-DBPs, I-DBPs and total DBPs upon chlorination.

In Jeddah plant, concentrations of THMs in blended water were in the range of 9.25 – 30.12 μg/L, in which bromoform contributed 42 – 96%. In Al-Khobar, THMs were in the range of 2.1 – 52.4 μg/L, in which B-THMs were 60 – 86% (Chowdhury et al., 2018). In Doha (Qatar), THMs in blended water were reported to be in the range of 10.5 – 36.3 μg/L (Table 4). In the Eastern coast of Saudi Arabia, concentrations of THMs in blended water were in the range of 3.1 – 12.8 μg/L while bromoform contributed 61 – 86% of total THMs. In a lab study, averages of THMs were reported to be 29, 40.5 and 42.6 μg/L for desalinated and treated groundwater blending ratios of 1:10, 1:5 and 1:2 respectively (Tawabini et al., 2011). Overall, THMs in the blended water were significantly higher than the THMs in desalinated water (Chowdhury et al., 2018).

Water Distribution System. Past studies have reported higher concentrations of DBPs in the WDS than the water treatment plants (WTP) (Chowdhury et al., 2008; MOE, 2018). In Ontario, the data from 162 MWS showed the ranges of THMs in WTP and WDS of 0.5 – 273 and 0.5 - 289 μg/L respectively [averages = 31.6 and 40.2 μg/L respectively] (Chowdhury et al., 2008). Fayed (1993) reported THMs in the WDS from eight major cities (Dammam, Riyadh, Makkah, Madinah, Jeddah, Abha, Hail, Buraidah) in the range of 0.03 – 41.7 μg/L with the averages of 0.83 – 18.2 μg/L. In other areas including Dhahran, Jubail, Qatif and Khafji, average of THMs were in the range of 4.05 – 11.02 μg/L. In these areas, concentrations of $CHCl_3$, BDCM, DBCM and $CHBr_3$ were in the ranges of ND – 9.3, ND – 3.4, ND – 1.5 and 1.6 – 12.0 μg/L respectively. In all locations, B-THMs were in the range of 70 – 95% of the total THMs. Alsohaimi et al. (2012) reported the concentrations of bromate in Saudi Arabia in the range of 3.4 – 75.0 μg/L.

In Kuwait, THMs were reported in the ranges of 8.4–92.4 μg/L (Latif, 1991). In four locations of Kuwait, averages of $CHCl_3$, BDCM, DBCM and $CHBr_3$ were in the ranges of ND – 5.9, 1.3 – 7.9, 2.8 – 13.7 and 3.4 – 77.4 μg/L respectively with the averages of 0.8 – 1.0, 2.2 – 2.9, 5.6 – 9.2 and 24.3 – 47.3 μg/L respectively. In Qatar, THMs were in the ranges of 1.5– 89.3 μg/L (Al-Otoum, 2014). The concentrations of $CHCl_3$, BDCM, DBCM and $CHBr_3$ were in the ranges of ND – 4.6, 0.1 – 26.2, 0.1 – 7.2 and 1.4 – 56.0 μg/L respectively with the averages of 0.52 - 0.82, 2.0 - 2.59, 1.04 – 4.61 and 9.6 - 14.6 μg/L respectively (Table 5). In Egypt, THMs were reported in the range of 7.5 – 87.4 μg/L. In Bahrain, averages of $CHCl_3$, BDCM, DBCM and $CHBr_3$ were in the ranges of 0.5 – 1.2, 0.32 - 0.73, 0.27 – 2.9 and 1.28 – 6.4

μg/L respectively (Al-Saleh and Al-Haddad, 1994). In most locations, concentrations of brominated DBPs were much higher than the chlorinated DBPs.

Plumbing Premise: Past studies have investigated the increase of DBPs in the plumbing pipes (PP) and hot water tanks (HWT) (Chowdhury, 2016). In three locations from Quebec, Canada, THMs in PP were 136-181% the THMs in the WDS. In HWT, THMs were 191-269% the THMs in the WDS and 132-159% the THMs in the PP. Concentrations of HAAs in PP and HWT were 23-224% and 53-261% of HAAs in the WDS, respectively. The reduction of HAAs from WDS to the PP and HWT might be due to increased microbiological activity in PP because of the consumption of free chlorine residuals and the destruction of HAAs by microorganisms (possible presence of a biofilm). In the blend (mixture of desalinated and treated groundwater) in Dhahran, Saudi Arabia, THMs in PP and HWT were 1.1 – 2.4 and 1.6 – 3.0 times the THMs in the WDS respectively. Concentrations of HAAs were 0.9 – 1.8 and 1.2 – 1.9 times the HAAs in the WDS respectively (Chowdhury, 2016). The lifetime excess cancer risks based on the samples from PP and HWT were predicted to be 1.46 (0.40 – 4.3) and 1.68 (0.35 – 5.1) times the cancer risks from WDS (Chowdhury, 2016).

The populations are exposed to the water from taps in houses, while the regulatory and monitoring agencies recommend measuring DBPs at various points within the WDS. Depending on the size of the plumbing premise, water may stay in the PP and HWT for a considerable amount of time before it reaches taps in the house (Weinberg et al., 2006; Dion-Fortier et al., 2009). This stagnation may be even longer during off-peak hours (e.g., midnight to morning; late morning to early evening), which allows extended reactions between the residual organics and free residual chlorine, and increase DBPs in PP and HWT

HUMAN HEALTH RISK AND TOXICITY

Several DBPs can pose cancer and non-cancer risks to humans (Table 1). The effects include colorectal/bladder cancers, neural tube defects, pre-term deliveries, miscarriages and low birth weight babies (King and Marrett, 1996). Exposure to DBPs can occur throughout lifetime via multiple pathways, such as, ingestion with drinking water, and inhalation and dermal absorption during showering, bathing, cooking, house cleaning and swimming. Past studies have reported increased cancer and non-cancer risks to humans from exposure to DBPs in supply water (Chowdhury, 2016; Chowdhury et al., 2018). Among the DBPs, brominated, nitrogenated and iodinated DBPs were reported to be more toxic than the chlorinated DBPs (Plewa et al., 2002; 2008; Wagner and Plewa, 2017; Zhang et al., 2012). The summary of few DBPs, their cytotoxicity, genotoxicity and risk factor data are presented in Table 1. Among THMs, iodoform (TIM) induced cytotoxicity to the Chinese hamster ovary (CHO) cell at the lowest concentration of 3937 μg/L while bromodichloromethane (BDCM) induced genotoxicity to human derived hepatoma line (HepG2) cell at the lowest concentration of 164 μg/L (Table 1). In HAAs, the lowest values for cytotoxicity and genotoxicity were 93 and 930 μg/L respectively, for the iodoacetic acid (IAA) (Table 1). In HANs, iodoacetonitrile (IAN) induced cytotoxicity to the CHO cell at the lowest concentration of 17 μg/L while bromoacetonitrile (BAN) induced genotoxicity at the lowest concentration of 4798 μg/L (Table 1). Among the cyanogen halides, cyanogen bromide induced cytotoxicity at the lowest concentration of 106 μg/L (Table 1). The iodinated, nitrogenated and brominated DBPs were more cytotoxic and genotoxic than the chlorinated DBPs (Table 1). In drinking water, concentrations of DBPs to date were much lower (e.g., USEPA, 2012; MOE, 2018; NLDOE, 2016) than the lowest concentrations needed to induce cytotoxicity and genotoxicity to the CHO and/or HepG2 cells (Wagner and Plewa, 2017; Li et al., 2012). However, the slope factor (SF) and drinking water unit risk (DWUR) demonstrated the possibility of cancer risks from several DBPs through oral pathway (USEPA, 2018). Although the concentrations of individual DBPs in supply water were much lower than the lowest concentrations for inducing cytotoxicity and genotoxicity to the CHO cells, the effects of the mixture of DBPs need better understanding. In general, DBPs in drinking water are present in the form of mixture. Consequently, the synergistic effects might play an important role toward inducing the cytotoxic and genotoxic effects. Future research in needed to explain the effects of the mixture of DBPs.

Table 1. Health Effects and Toxicity Data for some DBPs

DBPs	Name	Symbol	Formula	CAS No.	MW	LCC (μM)	LGC (μM)	LCC [μg/L]	LGC [μg/L]	SF (mg/kg-day)$^{-1}$	RfD (mg/kg-day)	DWUR (per μg/L)
THMs	Chloroform	TCM	$CHCl_3$	67-66-3	119.37	NA	10000	NA	1193690	RA	0.01	
	Bromodichloromethane	BDCM	$CHBrCl_2$	75-27-4	163.82		1		**164**	0.062	0.02	1.8×10^{-6}
	Dibromochloromethane	DBCM	$CHBr_2Cl$	124-48-1	208.28		10		2083	0.084	0.02	2.4×10^{-6}
	Bromoform	TBM	$CHBr_3$	75-25-2	252.73	NA	1000	NA	252731	0.0079	0.02	2.3×10^{-7}
	Bromodiiodomethane	BDIM	$CHBrI_2$	557-95-9	346.73	1500	NS	520098	NS			
	Chlorodiiodomethane	CDIM	$CHClI_2$	638-73-3	302.28	100	2000	30228	604556			
	Bromochloroiodomethane	BCIM	$CHBrClI$	34970-00-8	255.28	2200	NS	561609	NS			
	Dichloroiodomethane	DCIM	$CHCl_2I$	594-04-7	210.83	2000	NS	421658	NS			
	Iodoform	TIM	CHI_3	75-47-8	393.73	10	NS	**3937**	NS			
HAAs	Bromochloroacetic Acid	BCAA	$C_2H_2BrClO_2$	5589-96-3	173.39	300	3000	52017	520170			
	Bromodichloroacetic Acid	BDCAA	$C_2HBr_2ClO_2$	7113-314-7	207.84	500	NS	103920	NS			
	Chlorodibromoacetic Acid	CDBAA	$C_2H_2Br_2O_2$	5278-95-5	252.3	100	13000	25230	3279900			
	Dichloroacetic Acid	DCAA	$C_2H_2Cl_2O_2$	79-43-6	128.94	2000	NS	257880	NS	0.05	0.004	1.4×10^{-6}
	Dibromoacetic Acid	DBAA	$C_2H_2Br_2O_2$	631-64-1	217.86	200	750	43572	163395			
	Bromoacetic Acid	BAA	$C_2H_3BrO_2$	79-08-3	138.95	2	13	278	1806			
	Chloroacetic Acid	CAA	$C_2H_3ClO_2$	79-11-8	94.5	250	300	23625	28350			
	Tribromoacetic Acid	TBAA	$C_2HBr_3O_2$	75-96-7	296.76	5	3000	1484	890280			
	Trichloroacetic Acid	TCAA	$C_2HCl_3O_2$	76-03-9	163.39	400	NS	65356	NS	0.07	0.02	2.0×10^{-6}
	Iodoacetic acid	IAA	$C_2H_3IO_2$	64-69-7	185.95	0.5	5	**93.0**	**930**			
	Diiodoacetic acids	DIAA	$C2H_2I2O_2$	598-89-0	311.84	100	1000	31184	311840			
	Bromoiodoacetic acids	BIAA	$C_2H_2BrIO_2$	71815-43-5	264.84	250	2500	66210	662100			
	Chloroiodoacetic Acid	CIAA	$C_2H_2ClIO_2$	53715-09-6	220.39	170	NA	37466.3	NA			
HANs	Bromochloroacetonitrile	BCAN	$C_2HBrClN$	83463-62-1	154.39	7	250	1081	38598			
	Dibromoacetonitrile	DBAN	C_2HBr_2N	3252-43-5	198.85	1	30	199	5965			
	Dichloroacetonitrile	DCAN	C_2HCl_2N	3018-12-0	109.94	10	2400	1099	263849			
	Trichloroacetonitrile	TCAN	C_2Cl_3N	545-06-2	144.38	25	1000	3609	144379			
	Bromoacetonitrile	BAN	C_2H_2BrN	590-17-0	119.95	1	40	120	**4798**			
	Chloroacetonitrile	CAN	C_2H_2ClN	107-14-2	75.50	50	250	3775	18874			
	Iodoacetonitrile	IAN	C_2H_2IN	624-75-9	166.95	0.1	30	**17**	5008.5			

LCC: Lowest concentration to induce cytotoxicity in the Chinese hamster ovary (CHO) cell; LGC: Lowest concentration to induce genotoxicity in the CHO cell: SF: Slope factor (mg/kg-day)$^{-1}$; R_fD: Reference dose (mg/kg-day); DWUR: Drinking water unit risk (per μg/L); NS: not significant; NA: not available.

CONTROLLING DBPs

The DBPs are formed in many stages from the source to the consumer's tap in the house. The formation and control of DBPs are summarized in Figure 1. The final step prior to water consumption (Post-Stage VI) may include filtration of water through the granular activated carbon (GAC) filters. Past studies reported the decrease of THMs and HAAs using the GAC filters (Chowdhury et al., 2010; Weinberg et al. 2006). Filtration through the point of use (PoU) GAC filters can be an option while the cost for the filters and handling of contaminated GAC laden waste can be the other issues, which need better attention. The application of alternative disinfectants might reduce the formation of DBPs. However, application of alternative disinfectants needs to be carefully evaluated in context to emerging DBPs, health risks, cost and feasibility.

CONCLUSIONS

This study presented a review on the formation of DBPs in different stages of desalinated and blended water production and supplies from source to tap, and their toxic effects to the mammalian cells. The high concentrations of bromide in seawater result in higher fractions of brominated and total DBPs while the presence of iodide and iodinated DBPs have been reported in the recent years.

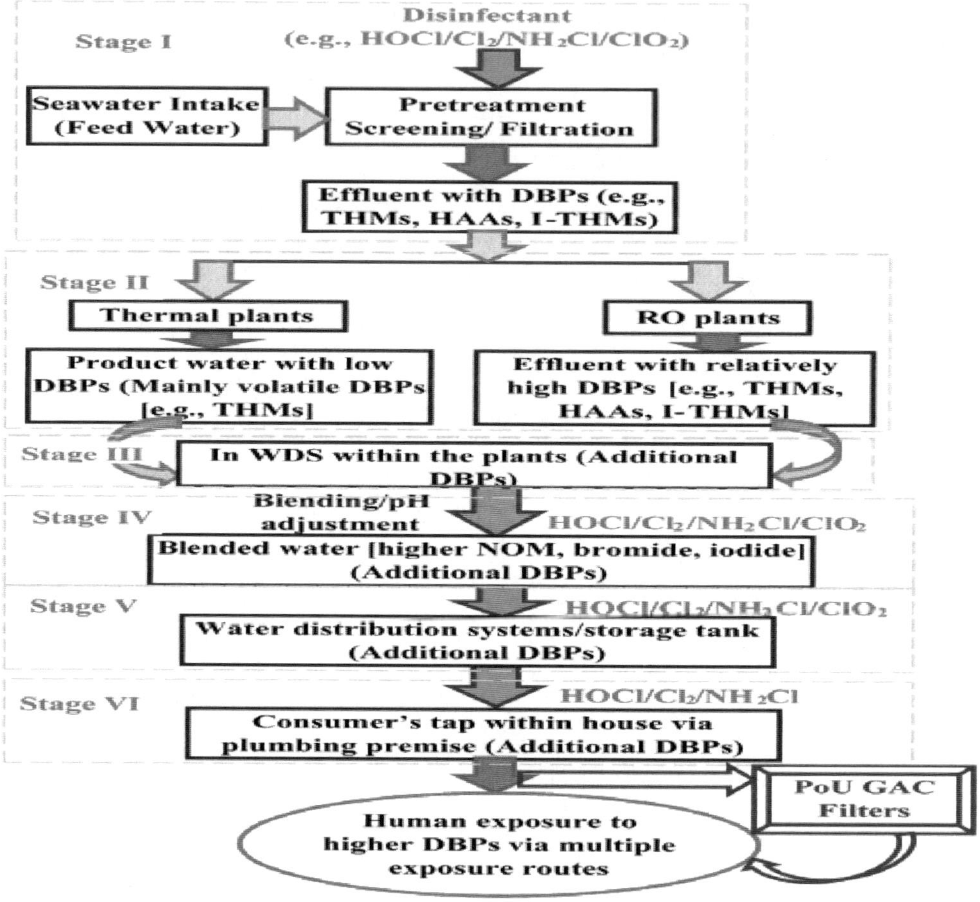

Figure 1. Formation of DBPs in different stages of supplying desalinated water

The brominated and iodinated DBPs are more cyto and geno-toxic than the regulated DBPs. In addition, new generation DBPs are being identified in the supply water. There is a need to better characterize DBPs, their formation kinetics, effects to humans and control strategies. The coastal regions on the earth are the homes of approximately 2 billion populations while the groundwater aquifers in these regions might be polluted due to seawater intrusion. These populations as well as the populations supplied with the desalinated and blended water are likely to be affected by the high concentrations of bromide and iodide in seawater. There is no single approach to provide the solution to the entire spectrum of the problems. A multi-barrier approach for different stages of desalinated and blended water supplies might assist in controlling DBPs at the exposure points, and thus better protect the humans.

ACKNOWLEDGEMENT

The author acknowledges the support of research group funding for the Water Research Group at the King Fahd University of Petroleum and Minerals (KFUPM), Dhahran 31261, Saudi Arabia.

REFERENCES

Agus, E., N. Voutchkov, D.L. Sedlak, Disinfection by-products and their potential impact on the quality of water produced by desalination systems: A literature review, Desalination. 237 (2009) 214–237.

Al-Otoum, FK, Disinfection by-products of chlorine dioxide (chlorite, chlorate, and trihalomethanes): Occurrence in drinking water in Qatar, QATAR UNIVERSITY, 2014.

Al-Saleh, FS, A.S. Al-Haddad, Levels of trihalomethanes in the blended water of Bahrain, Sci. Total Environ. 152 (1994) 221–225.

Alsohaimi, IH, Z.A. Alothman, M.R. Khan, M.A. Abdalla, R. Busquets, A.K. Alomary, Determination of bromate in drinking water by ultraperformance liquid chromatography-tandem mass spectrometry, J. Sep. Sci. 35 (2012) 2538–2543.

Brookman, RM, R. Lamsal, G. Gagnon, Comparing the Formation of Bromate and Bromoform Due to Ozonation and UV-TiO2 Oxidation in Seawater, Adv. Oxid. Technol. 14 (2011) 23–30.

Chowdhury S., 2013. Exposure assessment for trihalomethanes in municipal drinking water and risk reduction strategy. *Science of the Total Environment* 463-464: 922–930

Chowdhury S., 2016. Effects of plumbing systems on human exposure to disinfection byproducts in water: A case study. *Journal of Water and Health* 14(3): 489-503

Chowdhury S., Chowdhury I. and Hasan Z., 2018. Trihalomethanes in desalinated water: Human exposure and risk analysis. Human and Ecological Risk Assessment (*HERA*): 1-23.

El Din, AMS, R.A. Arain, A.A. Hammoud, A contribution to the problem of trihalomethane formation from the Arabian Gulf water, Desalination. 85 (1991) 13–32.

Hildesheim, M.E., Cantor, K.P., Lynch, C.F., Dosemeci, M., Lubin, J., Alavanja, M., Craun, G., 1998. Drinking water source and chlorination byproducts II. Risk of colon and rectal cancers. Epidemiology 9, 29–35.

Ioannou P., Charisiadis P., Andra SS and Makris KC, 2016. Occurrence and variability of iodinated trihalomethanes concentrations within two drinking-water distribution networks. Science of the Total Environment 543: 505–513

Kim, D., G.L. Amy, T. Karanfil, Disinfection by-product formation during seawater desalination: A review, Water Res. 81 (2015) 343–355.

King, W.D., Marrett, L.D., 1996. Case-control study of bladder cancer and chlorination byproducts in treated water (Ontario, Canada). Cancer Causes Control 7, 596–604.

Kutty, PCM, A. Amin Nomani, T.S. Thankachan, 1989. Monitoring of organic compounds in feed and product water samples from MSF plants in the eastern coast of Saudi Arabia, Desalination. 74 (1989) 243–257.

Latif, NiAA, Trihalomethane Compounds in the Drinking Water of Kuwait : a Survey From Source To Consumer Daughter Jumana and, Brunei University, 1991.

Le Roux, J., N. Nada, M.T. Khan, J.P. Croué, Tracing disinfection byproducts in full-scale desalination plants, Desalination. 359 (2015) 141–148.

NL-DOE (Department of Environment and Climate Change), 2016. Newfoundland and Labrador M. Nihemaiti, J. Le Roux, J. Crou, Formation of emerging disinfection byproducts by chlorination / chloramination of seawater impacted by algal organic matter, (2016).

Plewa, M.J., Kargalioglu, Y., Vankerk, D., Minear, R.A., Wagner, E.D., 2002. Mammalian cell cytotoxicity and genotoxicity analysis of drinking water disinfection by-products. Environ. Mol. Mutagen. 40, 134–142.

Plewa, M.J.,Muellner,M.G., Richardson, S.D., Fasano, F., Buettner,K.M., Woo, Y.T., McKague, A.B., Wagner, E.D., 2008a. Occurrence, synthesis and mammalian cell cytotoxicity and genotoxicity of haloacetamides: an emerging class of nitrogenous drinking water disinfection by-products. Environ. Sci. Technol. 42, 955–961.

Richardson, S.D., Plewa, M.J., Wagner, E.D., Schoeny, R., DeMarini, D.M., 2007. Occurrence, genotoxicity, and carcinogenicity of regulated and emerging disinfection by-products in drinking water: A review and roadmap for research. Mutation Research 636, 178–242

Tawabini, B., M. Al-Mutair, A. Bukhari, Formation potential of trihalomethanes (THMs) in blended water treated with chlorine, J. Water rReuse Desalin. 1 (2011) 172–178.

USEPA (United States Environmental Protection Agency), Table of Regulated Drinking Water Contaminants, (2016). https://www.epa.gov/ground-water-and-drinking-water/table-regulated-drinking-water-contaminants.

USEPA (United States Environmental Protection Agency), Integrated risk information system: Online database, (2018). https://cfpub.epa.gov/ncea/iris2/atoz.cfm.

Wagner, ED, Michael J. Plewa, MJ. 2017. CHO cell cytotoxicity and genotoxicity analyses of disinfection by-products: An updated review. Journal of Environmental Sciences 58(2017): 64-76

Weinberg, H.S., Pereira, V.R.P.J., Singer, P.C., Savitz, D.A., 2006. Considerations for improving the accuracy of exposure to disinfection by-products by ingestion in epidemiologic studies. Sci. Total Environment 354 (1), 35–42.

Zhang L, Liang Xua,, Qiang Zenga,b, Shao-Hui Zhanga,b, Hong Xiea,b, Ai-Lin Liua,b,∗∗, Wen-Qing Lua,b,∗ Comparison of DNA damage in human-derived hepatoma line (HepG2) exposed to the fifteen drinking water disinfection byproducts using the single cell gel electrophoresis assay. Mutation Research 741 (2012) 89– 94.

Smart Sensing and Control of Thermoelectric Coolers for Atmospheric Water Generation

Natalie Ownby, Melina Mijares, Mark Summers, Bahram Asiabanpour
(Ingram School of Engineering, Texas State University, San Marcos, TX, USA)

ABSTRACT: The availability of drinking water is a major issue in some countries, especially during the summer season due to the lack of rain. Atmospheric Water Generation (AWG) is an alternate solution to collect fresh and clean water directly from condensation of air particles. The method examined in this paper uses a Thermoelectric Cooler (TEC) system. Several methods to optimize the system were tested, to compare which was most efficient. These methods include timed-cycle, and a smart sensing and control system that utilizes an Arduino. Environmental conditions were also examined using psychrometric analysis. With a ratio comparison method, feasible regions that would allow a reasonable amount of water to be collected with minimum energy were identified.

INTRODUCTION

Over 1 billion people outside the United States do not have access to clean, safe drinking water (Suryaningsih, 2015). Thousands of women and children spend hours trying to collect water from distant and contaminated places. This means billions of families around the world do not have the ability to drink, cook, or shower with clean water. Water is a basic natural resource, but in reality there is a large shortage all over the world in places such as India, Brazil, South Africa, and Cape Town among others ("The 11 Cities Most Likely to Run out of Drinking Water - like Cape Town.").

Atmospheric Water Generation (AWG) offers a possible solution to this problem. The condensation that forms on plant leaves and other surfaces can be replicated using an AWG system. Condensation forms when air at the saturation temperature comes in contact with a surface that is below this saturation temperature (Esfahani & Modirkhazeni, 2012). AWG systems cause a surface to be cooled and then pass air over it, allowing the moisture to be extracted from the air as it passes. Some researchers have applied biomimetic principles to utilize atmospheric water (Davidson, K., Asiabanpour, B., & Almusaied, Z., 2017).

Current AWG systems include refrigerant-based dehumidification, vapor compression and vapor absorption refrigeration, desiccant liquid, radiative cooling, thermoacoustic refrigeration, and induced or controlled convection (Almusaied, Asiabanpour, 2017). These systems cannot compete with desalinating other water sources on an energy cost of water basis, because it requires energy to change water from its gaseous state to a liquid state. Current AWG systems require a minimum of 6.8g/m^3 water vapor density and operate ideally at the World Standard Conditions of 26.7 °C and 60% RH to operate at an efficiency of 0.4 kWh/L (Wahlgren, 2016).

Thermoelectric coolers are a commonly used option for AWG systems. Their low volume, fast response time, and ease of use have made them a popular choice for cooling systems and have led to increased characterization of the devices (Gao, LV, Wang, Yan, 2017), (Wan, Deng, Su, Wang, 2017). For systems that use thermoelectric coolers, a relative humidity>= 60% and temperatures >= 20 °C, is required for the price value and convenient operating range of the system to converge (Milani, Abbas, Vassallo, Chiesa, Al Bakri, 2011). These ideal atmospheric conditions also help the system avoid frosting over. Outside of these operating conditions, the efficiency of these systems only decreases. In order for AWG systems to be feasible outside tropical environments, steps must be taken to ensure energy is not wasted and the device is kept in the proper temperature range.

This paper presents a study of the development of a smart sensing and control system used to decrease the power consumption of an AWG system by only running it during feasible operating conditions. The system uses a Peltier device, Arduino and simple sensors to determine the energy required to produce water at certain points on the psychrometric chart.

METHODS AND MATERIALS

Thermoelectric Cooler AWG System. The AWG system being examined is constructed using thermoelectric coolers called Peltier devices, shown in Figure 1, in a module that includes a heat sink, shown in Figure 2. The Peltier devices are thermoelectric coolers that operate by transferring heat between two electrical junctions creating a hot side and a cold side. These devices require a DC voltage and are low maintenance, and controllable ("How do Thermoelectric Coolers (TECs) Work").

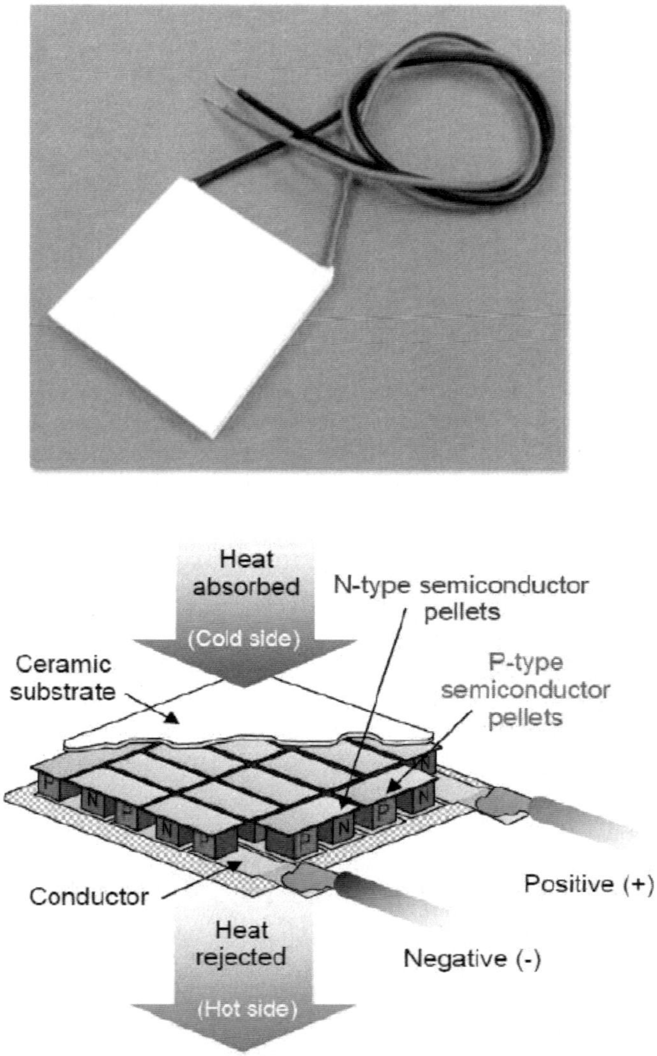

FIGURE 1. Peltier Device (Ya, G & Карапетян, Геворк & Dneprovski, Valeri, 2013)

The Peltier device is 40mm x 40mm and covered by a 40mm x 60mm aluminum plate. The system is rated at 12V 5A, but has been tested at a maximum voltage of 8.5V for this system.

In order for the system to continually produce water, the aluminum plate on the devices must be kept below dew point, but above freezing temperatures. This requires the device to turn off occasionally, so the water will not freeze, and turn on again as the device approaches dew point. This can also be done by providing a lower current through the device, but this method still leaves the device vulnerable to freezing in lower temperatures and may not be able to cool down enough in warmer atmospheric conditions.

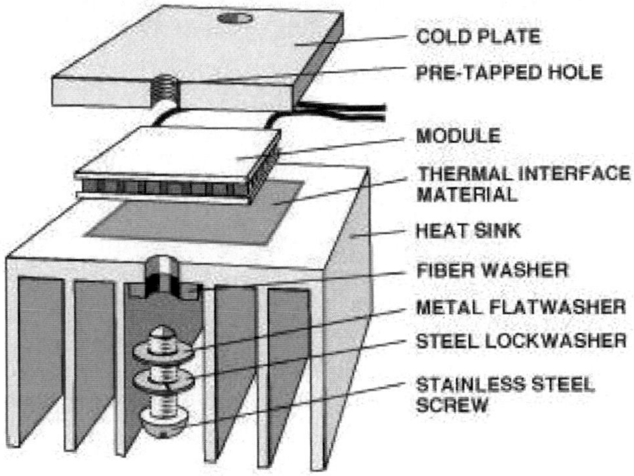

FIGURE 2. Peltier Device Module

These devices were originally put on a timed cycle so they would not freeze. This method, however, required testing to determine the length of the on and off cycle times. Different cycle times were also required depending on the power supplied to the system, and the atmospheric conditions. Given that the goal is to create a system suited for different conditions, the timed cycle would cause unnecessary testing and calibration and was also found to be rather ineffective at keeping the device in the ideal temperature range.

At this stage, the smart sensing and control system was implemented. The basis for this system is the Arduino Mega (Figure 3); an inexpensive microcontroller that easily interfaces with multiple sensors including those with an analog output. The relative ease of use makes it possible to test the system with different surface temperature sensors and atmospheric temperature and humidity sensors. The digital I/O pins are used to trigger the relay that turns the device on and off. The Arduino Mega could control up to 17 of the Peltier devices.

The smart sensing and control system uses the Peltier device system shown above, a thermistor, and an Arduino Mega. The thermistor is attached so it touches the center of the aluminum surface of the device. The thermistor is then connected to the Arduino, which takes the analog input of the thermistor and calculates the temperature using the Steinhart equation ("Steinhart & Hart Equations for 10k Thermistors"). When the surface temperature of the device is too cold, the device is turned off using a relay controlled by the Arduino until it warms up again.

This system was significantly more efficient than the timed cycle, yet is still cost effective and, once implemented, can run independently.

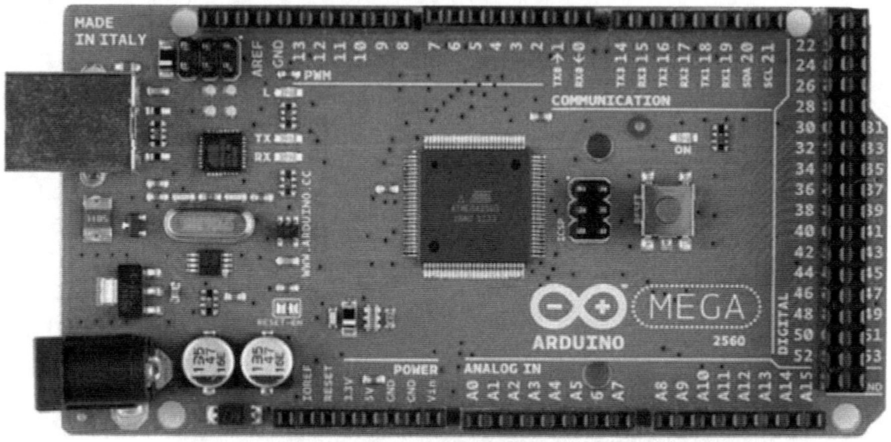

FIGURE 3. Arduino Mega Microcontroller

Psychrometric Analysis. Through a database obtained from a device that reads the environment conditions every five minutes of the day, an analysis was made in order to determine best operating conditions at which the system should operate. We refer to "best conditions" as the conditions that minimize the amount of energy consumed and maximize production of water. With a total of 7,000 data points per month per year, it was attempted to find a trend in the final graphs in order to check in which months the system would be most efficient.

The inaccuracy of this method due to high variability of environmental conditions over time that derive from external factors such as global warming, led to another method to make the analysis.

The psychromertic chart is a visual representation of the atmospheric dry-bulb temperature, relative humidity and dew point at these conditions, and amount of water in the air at these conditions. This chart allows the enthalpy (energy) to be calculated when moving from state A (initial temperature) to state C (final temperature) (American Society of Heating, Refrigerating, and Air-Conditioning Engineers, 2009). For this analysis, state A and state C were selected as conditions where moving from A to C resulted in a difference in humidity ratio of 0.005 lb of water/lb of dry air.

When starting at the World Standard Conditions defined earlier, the enthalpy from state A to state C was found to be 13 KJ/Kg. This calculation was performed for multiple points on the psychrometric chart. The calculated enthalpy required from state A to state C was then compared to the enthalpy required when starting at the World Standard Conditions.

RESULTS

In comparing the smart sensing and control system to the timed cycle, it is valuable to look at the time the device spent in the desired temperature range, above dew point and below the desired range. These times are compared in Figure 5.

In a 20 minute trial run, the timed cycle system spent 67.65% of the time in the desired range whereas the smart sensing and control system allowed the device to be in the desired range 97.33% of the time. The "Time Too Hot" represents time the device was above dew point and therefore not producing any water. The "Time Too Cold" represents time the device spent below freezing where energy was wasted because water could not be condensed but the system was still running.

Referring back to the method using the psychrometric chart to obtain the feasible areas, the shape coding was determined the following way: ratios of enthalpy from state A to state C to the enthalpy when starting at the World Standard Conditions between 1.2 - 1.4 are considered ideal conditions (triangle), 1.0-1.2 are low efficiency conditions (square), 0.8- 1.0 are non-ideal conditions (dashes), and 0.6-0.8 are unsuitable conditions (circle). These regions are shown in Figure 6 below.

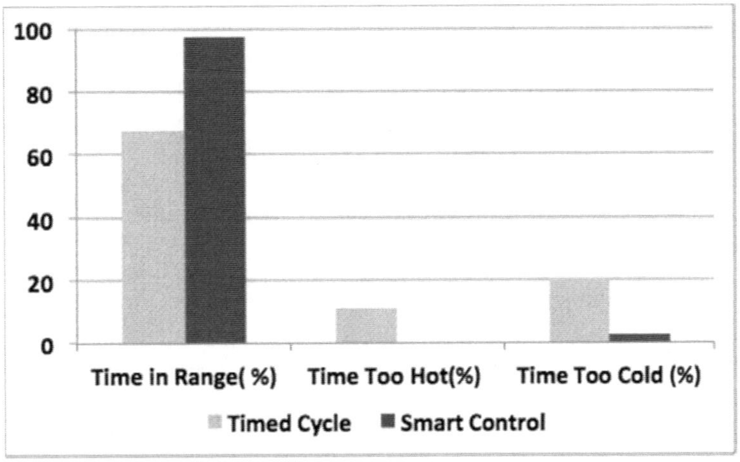

FIGURE 5. Timed Cycle vs. Smart Sensing

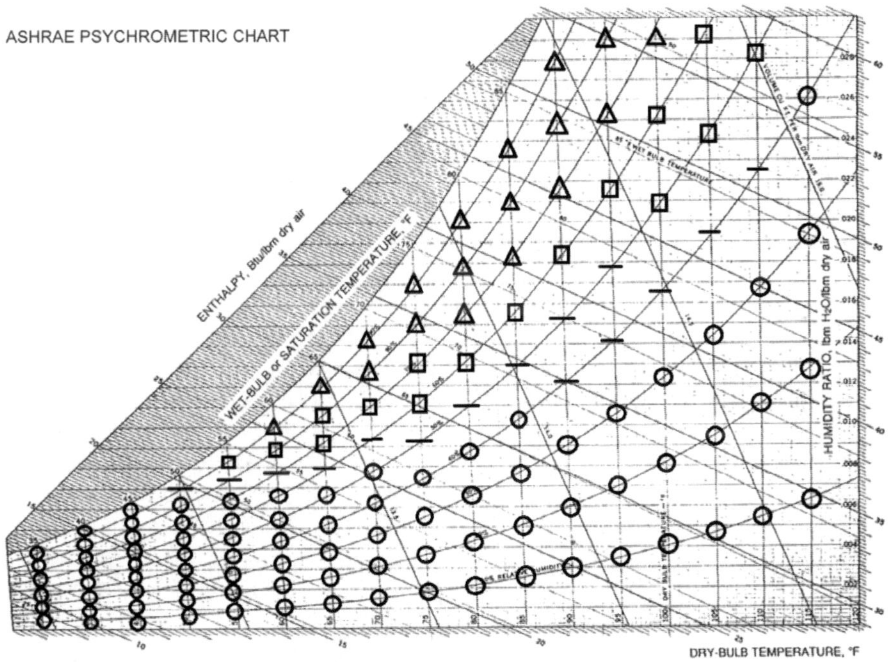

FIGURE 6. Operating Condition Ratios

DISCUSSION & CONCLUSION

The significant increase in the time the Peltier device spent in range using the smart sensing and control cycle can be attributed to the temperature sensing feedback loop. The small amount of time it spent out of the desired range could be a result of the time interval between taking readings from the thermistor. This could be improved by decreasing the time interval between readings or setting a temperature range slightly smaller than the actual desired range. Future improvements to this system involve selecting more accurate temperature sensors, and expanding the system to handle multiple Peltier devices at the same time. The sensors to be tested include thermal ribbons and infrared sensors. The infrared sensors would be particularly desirable because they would not need to be in contact with the surface and therefore would not interfere with water collection.

An atmospheric temperature and humidity sensor will also need to be integrated into the system. At this time, the results of the comparison of the World Standard Conditions to the atmospheric conditions can be used to determine if the system should be run. From here, the effect of the atmospheric conditions on the power consumption of the system can be examined with actual data from the system. The power consumed in kWh to run the device in different environments can be compared, and the kWh/L of water produced at different atmospheric conditions can be calculated.

By determining how and when to run the Peltier device system, the efficiency of atmospheric water generators that use thermoelectric coolers can be increased. The information about which atmospheric conditions to run the device at can also be integrated into existing AWG systems.

ACKNOWLEDGEMENTS

This work was completed with funding from the US Department of Agriculture (Grant # 2016-3842225540) and the Department of Education-MSEIP program (Grant #P120A140065). The authors would like to thank the USDA, Department of Education, and Texas State University for providing funding and access to both infrastructure and laboratories. The sponsors are not responsible for the content and accuracy of this article. The authors declare that there is no conflict of interest regarding the publication of this paper.

REFERENCES

Almusaiad, Z., Asiabanpour, B. 2017."Atmospheric Water Generation: Technologies and Influential Factors." Institute of Industrial and Systems Engineering (*IISE*) Annual *Conference* Proceeding, Pittsburgh, PA.

American Society of Heating, Refrigerating, and Air-Conditioning Engineers. 2009 ASHRAE Handbook: Fundamentals (I-P and SI Ed.). 2009.

Davidson, K., Asiabanpour, B., & Almusaied, Z. 2017. Applying Biomimetic Principles to Thermoelectric Cooling Devices for Water Collection, *7*(3), 27–35. https://doi.org/10.5539/enrr.v7n3px

Esfahani, J.A., Modirkhazeni, S.M. 2012. "Modeling of Laminar, Film-Wise Condensation." In Ahsan, Amimul, Water Condensation: Process, Modeling and Control, pp 1-70. Nova Science Publishers Inc., New York.

Gao, Wei., Lv, Hao., Wang, Xiao-Dong., Yan, Wei-Mon., "Enhanced Peltier Cooling of Two-staged Thermoelectirc Cooler Via Pulse Currents." *International Journal of Heat and Mass Transfer* 114: 656-663.

Milani, D. & Abbas, A., Vassallo, A., Chiesa, M., & Al Bakri, D. 2011. Evaluation of Using Thermoelectric Coolers in a Dehumidification System to Generate Freshwater from Ambient Air. Chemical Engineering Science - CHEM ENG SCI. 66. 2491-2501. 10.1016/j.ces.2011.02.018.

Suryaningsih, Sri., and N.Otong. 2015. "Optimal Design of an Atmospheric Water Generator (AWG) Based on Thermo-Electric Cooler (TEC) for Drought in Rural Area." Prepared for 2nd Padjadjaran International Physics Symposium 2015.

Wahlgren, Ronald V. 2016. "Water-from-Air Quick Guide: Second Edition". Atmoswater Research, North Vancouver, BC.

Wan, Qiushi., Deng, Yadong., Su, Chuqi., Wang, Yiping. 2016. "Optimization of a Localized Air Conditioning System Using Thermoelectric Coolers for Commercial Vehicles."Journal of Electronic Materials 46(5): 2990-2998.

Ya, G & Карапетян, Геворк & Dneprovski, Valeri. 2013. "Research Of Opportunity To Use Mism Structures For Cooling Of Light-Emitting Diodes". 10.13140/RG.2.1.4082.4162, Nova_Publisher: Nova Science Publishers, Inc.

"How Do Thermoelectric Coolers (TEC) Work?" Marlow, II-VI Incorporated, 2018, www.marlow.com/how-do-thermoelectric-coolers-tecs-work.

"The 11 Cities Most Likely to Run out of Drinking Water - like Cape Town." BBC News, BBC, 11 Feb. 2018, www.bbc.com/news/world-42982959.

"Steinhart & Hart Equations for 10k Thermistors" Water & Air, http://www.skyeinstruments.com/wp-content/uploads/Steinhart-Hart-Eqn-for-10k-Thermistors.pdf

Improvement of Vacuum Membrane Distillation Desalination Performance by Micro-Nano Bubbles Aeration

Yubing Ye, Shuili Yu, Li'an Hou, Baosen Liu and Qing Xia
(Tongji University, Shanghai, China)

Membrane distillation (MD) has been recognized as an alternative for seawater desalination. However, severe fouling caused by scaling ions restricts MD application. Several fouling control techniques were studied to alleviate or eliminate the negative effect of fouling in varying degree. In this study, micro-nano bubbles (MNBs) aeration was adopted to alleviate the membrane fouling during the vacuum membrane distillation (VDM) desalination process. A synthetic salty water was used. The results showed that the membrane fouling would be more serious at a higher concentration of feed water. However, with the addition of MNBs, the desalination performance was enhanced to some extent. The Zero time (the time when the permeate flux decreased to zero) was prolonged to at least 360min (with MNBs) from 150min (without MNBs). Moreover, compared to the performance without MNBs, the flux with MNBs was also higher at the same run time. Further studies demonstrated that higher steam and less precipitate produced with the addition of MNBs. It is demonstrated that the MNBs aeration could effectively improve the VMD desalination performance.

Reduction of Power Consumption in Atmospheric Water Generation

Natalie Ownby, Melina Mijares, Mark Summers, and Bahram Asiabanpour
(Texas State University, San Marcos Texas, USA)

Atmospheric water generation is the process of forming condensation from the moisture in the air. This can be manufactured using thermoelectric coolers to generate potable water. This research looks specifically at the use of a Peltier device to cool a steel surface that is mounted to it. In order to continually form water, the surface should remain in the dew point to ten degrees below dew point range. We will call this the "ideal range". If the steel surface is below this temperature, the device is unnecessarily cold and may begin to freeze the water. At temperatures above this range, there is no condensation forming.

This research looks at both the procedure for determining if it is feasible to run the system based on atmospheric conditions (temperature and humidity), and the control system to do this. The control system uses a thermistor and microcontroller to turn the Peltier device on and off to keep it in the desired range. This method also eliminates the need to determine the on/off time for the Peltier device at different atmospheric conditions. It makes it possible to compare the power used to keep the device in the ideal range to the potential amount of water the system can produce.

Currently, the control system for the Peltier device has been implemented and power consumption and temperature data can be collected. A heat map of a psychrometric chart has also been created by comparing the atmospheric conditions to the World Standard Conditions (80 F and 60% humidity) to show the ranges where running the system will generate water.

Assessment of Water Quality Degradation in the Drinking Water Sources Based on 18 Years Data from 441 Water Supply Systems

Shakhawat Chowdhury
(King Fahd University of Petroleum and Minerals, Dhahran, Saudi Arabia)

ABSTRACT: Source water quality degradation complicates water treatment processes, resulting in additional treatment cost and tap water quality deterioration. In this study, source water quality was investigated for 441 water supply systems (WSS) for 18 years (1999-2016) in Newfoundland and Labrador (NL), Canada. The investigation was performed on 21 water quality parameters (WQP) for groundwater (GWS) and surface water (SWS) sources. The averages of DOC, color and N were much higher in SWS than GWS while other 18 WQP (e.g., Alk. Cond., pH) were higher in GWS. In SWS, averages of DOC during 2000-2005, 2006-2010 and 2011-2015 were 6.08, 6.74 and 6.78 mg/L respectively. In these periods, pH were 6.39, 6.62 and 6.77 respectively. In GWS, averages of DOC in these periods were 1.43, 1.36 and 1.81 mg/L respectively while pH were 7.50, 7.69 and 7.89 respectively. The DOC in SWS and GWS were increasing at 0.0722 and 0.0491 mg/L/year respectively while pH were increasing at 0.0375 and 0.0441 units/year respectively. Trihalomethanes showed increasing trends in drinking water from SWS and GWS while haloacetic acids showed no trend. The higher DOC and pH, and their increasing rates could increase disinfection byproducts in drinking water, which can impart new challenges to WSS.

INTRODUCTION

The municipal water supply systems (WSS) obtain water from surface water and groundwater sources. Protecting water quality is essential to keep these sources useable, to satisfy regulatory limitations and to minimize water treatment cost (WHO, 2011). However, tempo-spatial variability and lack of understanding of water quality parameters (WQP) often make it difficult to control water quality degradation. In Newfoundland and Labrador (NL), Canada, approximately 441 WSS supply drinking water to about 436,000 populations. Among these, 304 and 137 WSS are surface water sourced (SWS) groundwater sourced (GWS) plants respectively. The SWS plants obtain water from brooks, canals, lakes, ponds, reservoirs, rivers and springs while the GWS plants obtain water from drilled and dug wells, springs and reservoirs (NL-DOE, 2016). The SWS and GWS supply water to approximately 400,000 (91.8%) and 36,000 (8.2%) populations respectively. These WSS apply different treatment and disinfection approaches depending on source water quality and regulatory guidelines on final water quality (Chowdhury, 2013, 2018; Health Canada, 2012; NL-DOE, 2016). The WSS monitored approximately thirty seven (37) WQP in these sources and published the data through the online web-portal (http://maps.gov.nl.ca/water/). These include alkalinity, color, conductivity, hardness, pH, total dissolved solids (TDS), turbidity, boron (B), bromide (Br), calcium (Ca), chloride (Cl), fluoride (F), potassium (K), sodium (Na), sulfate (SO_4), ammonia (NH_3), nitrate (NO_3), dissolved organic carbon (DOC), nitrogen (N), total phosphorus (TP) and magnesium (Mg). The WSS also monitored and reported other WQP including heavy metals, microbiological indicators and radiological parameters (NL-DOE, 2016).

The WQP in the sources and the treatment, and disinfection approaches affect the tap water quality. Use of chlorine with high DOC increase disinfection byproducts (DBPs) while many DBPs are possible or probable human carcinogens (USEPA, 2016). Higher pH, bromide and temperature can further increase the formation of some DBPs (e.g., trihalomethanes: THMs) in drinking water (Chowdhury et al., 2008). The regulatory agencies have the limits on some DBPs (USEPA, 2006; Health Canada, 2012). In NL, about 303 WSS supply drinking water to communities with 500 populations or less, which occasionally face challenges in satisfying the DBPs guideline. Due to public health relevance, there is a need to better understand the occurrences, concentrations and the changes of WQP with time.

In this study, source water quality was assessed for 441 WSS for 18 years (1999-2016) in NL, Canada. The water sources were divided into SWS and GWS sources and the periodic changes of 21 WQP (as noted above) were investigated. The WQP in SWS and GWS were compared and the time dependent changes were assessed. The seasonal variability of WQP in the SWS and GWS were investigated. Finally, implications of WQP changes on the performance of WSS were discussed. The possible strategy to minimize the effects were proposed.

METHODOLOGY

Quarterly data (four times a year) of the above-noted 21 WQP were obtained for the period of 1999-2016 (NL-DOE, 2016). These data were reported for 441 WSS, including those supplied with surface water and groundwater in NL, Canada. The locations of most of the WSS can be found in literature (NL-DOE, 2016; Chowdhury, 2018). A total of 130, 97 and 76 WSS supplied water to ≤100, 101-250 and 251-500 populations (NL-DOE, 2016; Chowdhury, 2018). The data were divided into two groups: SWS and GWS. Concentrations of WQP for SWS and GWS were analyzed and compared. The statistical analyses were performed separately to assess data patterns, percentile values and seasonal variability. The data were also divided based on the types of surface water sources and the WQP were analyzed. The variability of WQP since 1999 was investigated. The statistical analyses were performed using Minitab™ and JMP™ statistical software (Minitab Inc., 2016; SAS, 2016). The sample collection and analyses were performed following standard procedure and can be accessed through the Department of Environment and Climate Change, NL water resources portal [http://maps.gov.nl.ca/water/] (NL-DOE, 2016).

The DOC and pH are important WQP for DBPs. Trends were analyzed for these WQP, THMs and HAAs. In trend analysis, the Mann – Kendall (M-K) analysis is frequently used (Mann, 1945). The hypothesis: H_o is used for no trend, and the data (x_i) are randomly variable. The alternative hypothesis: H_1 for the decreasing or increasing monotonic trend. The P-value ≤ 0.05 supports for rejection of H_o and acceptance of H_1, meaning the decreasing or increasing trend. The method assumes the series of *n* records, and x_j and x_k as the subsets of data where $k = 1, 2, 3, \ldots, n-1$; and $j = k+1, k+2, k+3, \ldots, n$. The data analysis is performed as a time series in sequence. Each data point is compared with all of the subsequent data points. If the data for later period is greater than the prior one, the M-K statistics: S is decremented by 1. If it is smaller, S is decremented by -1. The value of *S* is computed as:

$$S = \sum_{k=1}^{n-1} \sum_{j=k+1}^{n} Sgn\left(x_j - x_k\right) \tag{1}$$

$$S = Sgn\left(x_j - x_k\right) = \begin{Bmatrix} +1 & \text{if } x_j - x_k > 0 \\ 0 & \text{if } x_j - x_k = 0 \\ -1 & \text{if } x_j - x_k < 0 \end{Bmatrix} \tag{2}$$

Where, x_j and x_k are monthly or annual data and $j > k$ respectively. The positive high S-value shows the increasing trend and the negative high value shows the decreasing trend. The statistics *S* is considered to be normally distributed when $n \geq 10$. For non-parametric test, Kendall's tau (τ) indicates the statistical relationships based on the ranks of the data. For two sets of data, Kendall's tau can be obtained as:

$$Tau(\tau) = \frac{\text{No. of concordant pairs - No. of discordant pairs}}{n(n-1)/2} \tag{3}$$

The magnitude of τ ranges between -1 to +1 showing the perfect negative to positive relationship among the data groups. The positive τ shows increasing trend and negative τ shows decreasing trend.

RESULTS

Overall. The summary of 21 WQP for SWS and GWS are shown in Table 1. The averages of DOC, color and nitrogen (N) were much higher in SWS than GWS while the averages of other 18 WQP were higher in GWS (Table 1). The averages of DOC in SWS and GWS were 5.91 and 1.45 mg/L respectively while the averages of color were 40.6 and 6.65 TCU respectively (Table 1). Averages of N in SWS and GWS were 0.24 and 0.08 mg/L respectively. In SWS, averages of DOC, color and N were 4.1, 7.1 and 2.95 times the averages in GWS respectively (Table 1). In contrast, averages of alkalinity, hardness and TDS in GWS were 6.7, 5.1 and 4.7 times the averages in SWS respectively. The averages of alkalinity, hardness and TDS in GWS were 104.2, 100.6 and 218.9 mg/L respectively and the ranges were 0.0-721, 0-750 and 0-2720 mg/L respectively. In SWS, these averages were 15.5, 19.6 and 46.7 mg/L respectively and the ranges were 0-289, 0-514 and 4.0-1100 mg/L respectively. Averages of pH in SWS and GWS were 6.49 and 7.65 respectively with ranges of 4.07-8.81 and 5.47-9.97 respectively. Averages of bromide ion (Br^-) were higher in GWS (0.05 mg/L) than SWS (0.01 mg/L). The coefficient of variation (CV) for WQP in SWS and GWS were 1.71 and 2.08 respectively and the ranges were 0.11-5.13 and 0.09 – 4.92 respectively (Table 1).

Percentile Distributions of WQP. The percentile distributions of WQP showed considerable variability. In SWS, medians of DOC, color, turbidity and N were higher than GWS while the medians of 17 WQP were higher in GWS than SWS. The medians of DOC in SWS and GWS were 5.8 and 0.8 mg/L respectively, and the ranges were 0.0 - 28.7 and 0.0 - 35.4 mg/L respectively. The 90 percentile DOC were 1.83 and 4.63 times the medians in SWS and GWS respectively while the maximum values were 4.95 and 44.3 times the medians respectively, indicating much higher skewness in GWS. The CV in GWS was also higher (1.62) than SWS (0.54) (Table 1). The color and turbidity also showed higher variability in GWS than SWS (e.g., higher CV). In GWS, medians of alkalinity, conductivity, hardness, pH and TDS were significantly higher than SWS. In addition, few other WQP including Cl, Ca, Na, K and Mg had higher medians in GWS than SWS. Higher levels of these WQP in GWS might be due to the dissolution of minerals from groundwater aquifers, which were also consistent to past studies (Shrestha and Kazama, 2007; Chowdhury and Al-Zahrani, 2014). Further, these WQP (e.g., Cl, Ca, Na, K and Mg) had lower CV in GWS than the SWS (Table 1), indicating lower variability and relatively consistent properties for GWS. The sources of WQP can be better explained through multivariate statistical investigation in future.

Table 1. Summary of Groundwater and Surface Water Data in Newfoundland and Labrador for 1999-2016

	GWS					SWS					Ratio
	Mean	Std.	Min.	Max	CV	Mean	Std.	Min.	Max	CV	GWS/SWS
Alkalinity (mg/L)	104.2	62.3	0.0	721.0	0.60	15.5	31.2	0.0	289.0	2.01	6.70
Color (TCU)	6.65	21.6	0.0	340.0	3.24	47.2	40.6	0.0	360.0	0.86	0.14
Conductivity (µs/cm)	342.8	254.8	32.0	4190.0	0.74	68.3	88.7	0.0	1420.0	1.30	5.02
Hardness (mg/L)	100.6	82.0	0.0	750.0	0.81	19.6	36.8	0.0	514.0	1.87	5.12
pH	7.65	0.70	5.47	9.97	0.09	6.49	0.74	4.07	8.81	0.11	1.18
TDS (mg/L)	218.9	166.9	0.0	2720.0	0.76	46.7	63.4	4.0	1100.0	1.36	4.69
Turbidity (NTU)	1.11	5.5	0.0	120.0	4.92	0.74	2.0	0.0	97.0	2.70	1.51
B (mg/L)	0.03	0.1	0.0	0.6	1.85	0.01	0.0	0.0	0.5	2.02	4.69
Bromide (mg/L)	0.05	0.2	0.0	3.4	3.79	0.01	0.0	0.0	1.3	3.07	3.79
Ca (mg/L)	30.32	26.5	0.0	269.0	0.87	5.09	9.4	0.0	102.0	1.85	5.96
Cl (mg/L)	34.84	52.4	0.0	1030.0	1.50	9.74	20.9	0.0	500.0	2.14	3.58
F (mg/L)	0.14	0.2	0.0	1.2	1.39	0.03	0.1	0.0	1.1	2.04	4.46
K (mg/L)	1.41	1.9	0.0	13.0	1.35	0.29	0.6	0.0	25.0	1.95	4.88
Na (mg/L)	30.28	44.2	0.0	969.0	1.46	5.99	12.9	0.0	410.0	2.15	5.06
Sulfate (mg/L)	17.69	46.9	0.0	800.0	2.65	2.90	4.5	0.0	100.0	1.55	6.11
Ammonia (mg/L)	0.03	0.1	0.0	1.1	1.92	0.02	0.1	0.0	7.0	5.13	1.18
DOC (mg/L)	1.45	2.4	0.0	35.4	1.62	5.91	3.2	0.0	28.7	0.54	0.25
Nitrate (mg/L)	0.26	0.5	0.0	7.9	1.97	0.03	0.1	0.0	4.8	3.93	9.74
K-Nitrogen (mg/L)	0.08	0.1	0.0	3.4	1.68	0.24	0.2	0.0	8.6	0.99	0.34
TP (mg/L)	0.03	0.0	0.0	1.3	1.86	0.01	0.0	0.0	0.9	4.00	2.48
Mg (mg/L)	6.04	5.2	0.0	56.0	0.86	1.57	3.2	0.0	63.0	2.05	3.86

Seasonal variability of WQP. The seasonal variability of WQP was assessed for four seasons: S1 (Jan-Mar); S2 (Apr-Jun); S3 (Jul-Sep); and S4 (Oct-Dec) representing winter, spring, summer and fall respectively. In GWS, most of the WQP showed the highest and lowest averages in S2 and S1 respectively. In addition, four WQP (DOC, color, turbidity and bromide) had the highest averages in S3. A total of 18 WQP had the highest averages in S2 and S3 (14 and 4 WQP respectively), demonstrating the highest and lowest averages during the warm and cold periods respectively (Table 2).

Table 2. Seasonal Variability of WQP in the GWS and SWS in NL, Canada

	GWS				SWS			
	S1	S2	S3	S4	S1	S2	S3	S4
Alkalinity (mg/L)	92.7 (54.8)	117.4 (63.2)	99.3 (65.0)	100.7(60.0)	9.04 (20.5)	15.13(30.2)	19.21(35.8)	17.2 (33.6)
Color (TCU)	5.43 (15.4)	6.27 (23.4)	10.11 (27.7)	5.18 (16.4)	48.5 (42.7)	43.5 (32.9)	47.3 (47.2)	54.2 (44.6)
Conductivity (μs/cm)	295.4 (158.6)	391.8(286.2)	333.8(253.4)	326.5(255)	58.6(65.6)	68.1 (72.3)	72.6(80.5)	71.1(77)
Hardness (mg/L)	82.3 (57.51)	120.8(94.2)	94.8(79.2)	93.8(77.2)	9.6 (20)	20.0 (36.9)	20.9(38.4)	20.2(37.5)
pH	7.67 (0.71)	7.73(0.67)	7.61(0.69)	7.60(0.72)	6.34(0.60)	6.52(0.77)	6.68(0.73)	6.46(0.72)
TDS (mg/L)	187.8 (103)	250.1(192.1)	215.7(163.7)	208.8(166.6)	37.5(40.8)	41.7(44.9)	45.9(51.4)	44.3(48.0)
Turbidity (NTU)	0.82 (2.69)	0.88(3.74)	2.03(9.97)	0.88(3.10)	0.82(2.67)	0.58(1.13)	0.79(1.26)	1.02(3.66)
B (mg/L)	0.02 (0.027)	0.04(0.07)	0.03(0.07)	0.03(0.05)	0.00(0.01)	0.01(0.01)	0.00(0.01)	0.01(0.02)
Bromide (mg/L)	0.05 (0.15)	0.03(0.12)	0.08(0.32)	0.05(0.10)	0.01(0.04)	0.01(0.04)	0.01(0.04)	0.01(0.03)
Ca (mg/L)	24.96 (17.9)	36.3(31.0)	27.9(25.5)	28.8(25.0)	3.48(5.81)	5.43(9.73)	5.88(10.7)	5.52(10.2)
Cl (mg/L)	28.8 (30.1)	38.2(56.4)	36.5(55.2)	33.6(55.4)	9.71(12.9)	9.17(10.0)	8.68(9.40)	9.04(9.48)
F (mg/L)	0.14 (0.22)	0.12(0.17)	0.14(0.21)	0.15(0.18)	0.02(0.05)	0.02(0.05)	0.01(0.03)	0.04(0.10)
K (mg/L)	0.91(1.23)	1.64(2.15)	1.52(2.01)	1.36(1.86)	0.24(0.39)	0.30(0.80)	0.12(0.27)	0.30(0.29)
Na (mg/L)	26.5 (25.3)	33.02(51.05)	29.7(38.2)	29.94(47.6)	5.57(7.36)	5.69(5.77)	5.37(5.83)	5.56(5.50)
Sulfate (mg/L)	12.22 (11.5)	22.95(65.3)	17.14(43.5)	16.22(40.3)	2.84(5.1)	2.60(4.6)	2.37(2.84)	3.23(5.1)
Ammonia (mg/L)	0.01 (0.035)	0.03(0.05)	0.03(0.05)	0.03(0.06)	0.02(0.04)	0.02(0.04)	0.02(0.04)	0.03(0.05)
DOC (mg/L)	1.31 (1.76)	1.33(1.98)	1.57(2.52)	1.53(2.67)	5.95(2.92)	5.38(2.40)	6.68(3.60)	7.09(3.46)
Nitrate (mg/L)	0.26 (0.57)	0.27(0.55)	0.23(0.44)	0.28(0.52)	0.03(0.07)	0.03(0.06)	0.01(0.05)	0.03(0.07)
K-Nitrogen (mg/L)	0.09 (0.11)	0.08(0.12)	0.07(0.11)	0.08(0.17)	0.22(0.19)	0.24(0.29)	0.25(0.19)	0.25(0.30)
TP (mg/L)	0.02 (0.026)	0.03(0.04)	0.02(0.05)	0.03(0.06)	0.01(0.01)	0.01(0.01)	0.01(0.02)	0.01(0.02)
Mg (mg/L)	4.84 (4.1)	7.27(5.45)	6.13(5.94)	5.38(4.56)	0.80(2.05)	1.46(3.26)	1.54(3.2)	1.61(3.1)

The ratios between the averages of WQP for S2/S1, S3/S1 and S4/S1 were 1.345, 1.354 and 1.24 respectively and the ranges were 0.71 – 2.42, 0.78 – 2.47 and 0.92 – 2.43 respectively, indicating the overall lowest values during S1. In S1-S4, averages of DOC and bromide were in the ranges of 1.31 – 1.57 mg/L and 0.03 – 0.08 mg/L respectively while averages of pH were in the ranges of 7.6 - 7.73. Despite the DOC in GWS were lower, higher bromide and higher pH can lead to higher DBPs formation and higher fractions of brominated DBPs in the tap water. The averages of alkalinity, hardness and TDS were in the ranges of 92.7-117.4, 82.3-120.8 and 187.8-250.1 mg/L respectively. The higher values of few WQP in summer might be explained by higher temperature, higher pollution and bio-geo properties of the aquifers.

In SWS, a total of 10 and 6 WQP showed the highest averages during S4 and S3 respectively. The ratios of averages of WQP for S2/S1, S3/S1 and S4/S1 were 1.24, 1.10 and 1.42 respectively, and the ranges were 0.71 – 2.71, 0.51 – 2.18 and 0.92 – 2.84 respectively, indicating the overall lowest averages during S1. Averages of DOC were in the ranges 5.38 - 7.09 mg/L with the highest value in S4 (Table 2). Averages of color were in the ranges of 43.5 – 54.2 TCU with the highest value in S4. Averages of alkalinity, hardness and TDS in S1-S4 were in the ranges of 9.04 – 19.21, 9.6 – 20.9, 37.5 – 45.9 mg/L respectively while these WQP had the highest averages during S3. The highest averages of DOC, color, alkalinity, hardness and TDS during S3 and S4 might be partially attributed to increased contamination of SWS due to snow melting, road salts, plants and leaves, and the bio-geo properties of the sources (Table 2).

Historical changes of WQP. The changes of WQP in GWS and SWS are summarized in Table 3. The duration was divided into three periods: D1, D2 and D3 representing 2000 – 2005, 2006 – 2010 and 2011 – 2015 respectively. In GWS, averages of DOC in D1, D2 and D3 were 1.43, 1.36 and 1.81 mg/L respectively. In these periods, averages of pH were 7.49, 7.69 and 7.89 respectively (Table 3). The DOC and pH were increasing at the rates of 0.0491 mg/L/year and 0.0441 units/year respectively. The increase of DOC and color could be explained by the increase of organic matter (OM) while increase of pH might be partially explained by the intrusion of seawater into the groundwater aquifers. Further study is warranted to determine the accumulation of OM in GWS and the effect of seawater intrusion. Few WQP were decreased from D1 to D3 while alkalinity, conductivity, TDS, turbidity and Na showed variable patterns.

In SWS, averages of DOC during D1, D2 and D3 were 6.08, 6.74 and 6.78 mg/L respectively. In these periods, averages of pH were 6.39, 6.62 and 6.77 respectively (Table 3). The DOC and pH were increasing at the rates of 0.0722 mg/L/year and 0.0375 units/year respectively. The DOC, color and pH showed higher averages during the later periods than the earlier periods (e.g., D3 > D2 > D1), indicating the time dependent increase of OM in these sources (Table 3). Several WQP including hardness, alkalinity, conductivity and Ca had higher averages during D3 than D1 and D2 (Table 3) while TDS, turbidity, bromide, Cl, F, Na, Ammonia and Mg showed variable patterns (Table 3). Few WQP including K, NO_3 and N showed decreasing patterns (Table 3).

Table 3. Historical changes of WQP in the GWS and SWS in NL, Canada

	GWS			SWS		
	D1	D2	D3	D1	D2	D3
Alkalinity (mg/L)	107.2	99.8	106.8	14.1	17.1	20.5
Color (TCU)	5.5	6.6	7.8	45.3	49.6	49.9
Conductivity (μs/cm)	358.1	326.5	352.6	65.6	68.7	77.4
Hardness (mg/L)	108.1	95.2	93.5	16.9	20.9	22.1
pH	7.49	7.69	7.89	6.39	6.62	6.77
TDS (mg/L)	232.6	203.3	228.1	43.3	39.1	50.2
Turbidity (NTU)	0.98	1.33	1.02	0.79	0.88	0.80
B (mg/L)	0.039	0.031	0.025	-	0.002	0.002
Bromide (mg/L)	0.059	0.043	0.034	0.020	0.001	0.005
Ca (mg/L)	32.83	28.38	28.67	4.81	5.67	6.38
Cl (mg/L)	38.13	32.46	35.29	9.50	7.96	9.30
F (mg/L)	0.17	0.12	0.12	0.05	0.00	0.00
K (mg/L)	1.56	1.37	1.10	0.31	0.21	0.05
Na (mg/L)	32.06	28.86	33.39	5.82	5.12	5.34
Sulfate (mg/L)	19.42	16.53	16.42	3.82	1.98	1.84
Ammonia (mg/L)	0.036	0.020	0.039	0.031	0.012	0.029
DOC (mg/L)	1.432	1.356	1.808	6.084	6.735	6.775
Nitrate (mg/L)	0.284	0.256	0.232	0.034	0.019	0.011
K-Nitrogen (mg/L)	0.104	0.072	0.067	0.271	0.248	0.215
TP (mg/L)	0.038	0.018	0.020	0.010	0.004	0.003
Mg (mg/L)	6.342	5.897	5.401	1.263	1.626	1.589

GWS: Groundwater sources; SWS: Surface water sources; D1: 2000-2005; D2: 2006-2010; D3: 2011-2015

The linear trends and M-K statistics for yearly averages of DOC and pH are shown in Figure 1a. The positive values of τ, high positive values of S and P-values < 0.05 indicated the presence of increasing trends for pH in GWS and SWS, and DOC in SWS respectively. For DOC in GWS, P-value were higher than 0.05. However, positive values of τ and S supported the presence of trends. Overall, DOC and pH in SWS and GWS showed increasing trends. Figure 1b shows the linear trends and M-K statistics for the yearly averages of THMs and HAAs in tap water from SWS and GWS. The SW-THMs and GW-THMs showed positive values of τ, high positive values of S and P-values of < 0.05, indicating the presence of increasing trends (Figure 1b). However, the SW-HAAs and GW-HAAs showed variable patterns (Figure 1b). The M-K statistics showed high P-values (>0.05), negative values of τ and low negative values of S, indicating no evidence of trend. Overall, increase of THMs were consistent to increase of DOC and pH in SWS and GWS However, HAAs could have been affected by higher pH and microbiological activities in water distribution systems. The increasing trends of DOC, pH and THMs can be a concern, which might need better treatment facilities.

SUMMARY AND CONCLUSIONS

In SWS, averages of DOC, color and N were much higher than GWS (4.1, 7.1 and 3.0 times respectively) indicating that SWS were more susceptible to the organic pollution (Table 1) while the averages of other WQP were higher in GWS (Table 1). In SWS and GWS, CV were in the ranges of 0.11-5.13 and 0.09 – 4.92 with averages of 2.08 and 1.71 respectively, indicating higher variability in SWS. In

GWS, 18 WQP including DOC, color, turbidity, bromide and pH had the highest averages during S2-S3. In contrast, SWS had the highest averages of 16 WQP during S3-S4. The higher average for SWS during S3-S4 might be partially linked to increased pollution in S3-S4 (e.g., road salts, snow melts, etc.) while the higher averages of GWS during S2-S3 might be linked to higher temperature, higher pollution and bio-geo properties of aquifers. Since 2000, DOC were increasing at the rates of 0.0722 and 0.0491 mg/L/year in SWS and GWS respectively while pH were increasing at the rates of 0.0375 and 0.0441 units/year respectively in these sources (Figure 1a). During these periods, THMs also showed the increasing trends (Figure 1b). In SWS, DOC and its rate of increase were higher, which could increase DBPs in tap water while several DBPs are possible or probable human carcinogens, and some DBPs are regulated. The increasing trends of DOC, pH and THMs could be a concern. In addition, other WQP and their variable patterns might impart new challenges, which can increase the treatment cost and/or deteriorate tap water quality. Alteration of sources to GWS could be a temporary solution to reduce DBPs in tap water. However, GWS had higher values for most of the WQP including pH and bromide ions, which would be complex and expensive for treatment. Source specific better understanding is needed for source alteration.

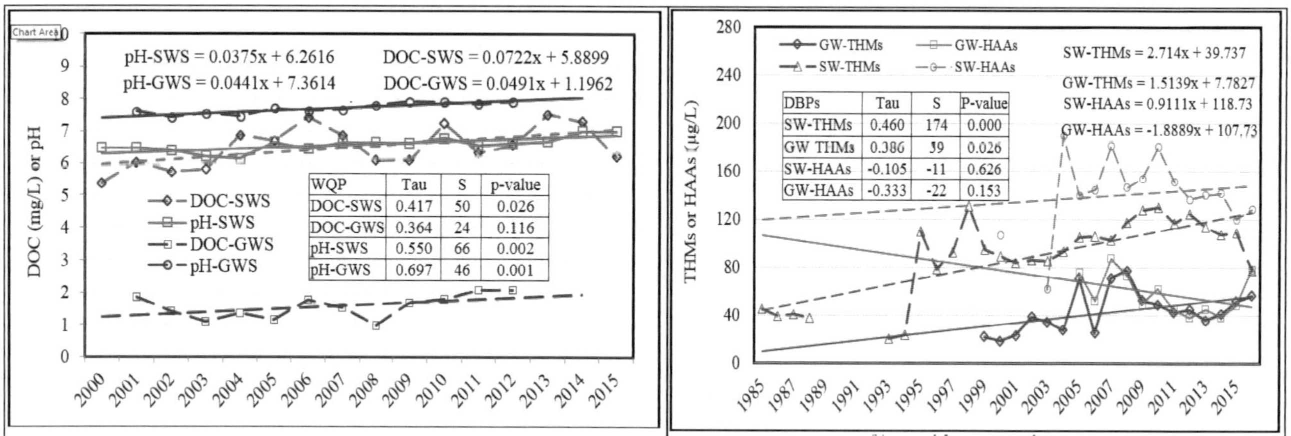

(a): Trends for pH and DOC

(b): Trend for THMs and HAAs

Figure 1. Mann-Kendal Trends of DOC, pH, THMs and HAAs in SWS and GWS; SW: Surface water tap; GW: Groundwater tap; Tau: Kendal's Tau; S: S-statistics: p-value: Significance probability; $\alpha = 0.05$

The data analyzed were quarterly data (4 times a year) reported by the municipal water systems and available through the Department of Environment and Climate Change (NL-DOE) web portal. As such, daily or monthly variability of WQP was not represented by these data. This study analyzed the passively collected data without performing the cause and effects analysis. Despite the fact that discernible trends in the occurrences of WQP in different sources can be identified from the historical data; understanding overall systems may require further study through cause and effect analysis.

REFERENCES

Chowdhury S., Champagne P. and McLellan P.J., 2008. Factors influencing formation of Trihalomethanes in drinking water: Results from multivariate statistical investigation of the Ontario Drinking Water Surveillance Program database. Water Qual. Res. J. of Canada 43(2/3): 189–199

Chowdhury S., 2013. Impact of source waters, disinfectants, seasons and treatment approaches on trihalomethanes in drinking water: a comparison based on the size of municipal systems. Water Environment Journal 27: 197-206

Chowdhury S. and Zahrani M., 2014. Water quality change in dam reservoir and shallow aquifer: Analysis on trend, seasonal variability and data reduction. *Environmental Monitoring and Assessment* 186: 6127-6143

Chowdhury S., 2018. Occurrences and changes of disinfection byproducts in small water supply systems. Environ Monit Assess (2018) 190:32; https://doi.org/10.1007/s10661-017-6410-8

Health Canada (2012). Guidelines for Canadian Drinking Water Quality. Prepared by the Federal-Provincial-Territorial Committee on Health and the Environment: March, Ottawa, Canada

Mann, H.B. 1945. Non-parametric test against trend. Econometrica. 13: 245–259.

Minitab Inc. (2016). The statistical software package. Available at: http://www.minitab.com/en-us/

NL-DOE (Department of Environment and Climate Change), 2016. Newfoundland and Labrador Water Resources Portal. http://maps.gov.nl.ca/water/

SAS Inc., 2016. Statistical discovery software. SAS Institute Inc. (http://www.jmp.com/), NC, USA

Shrestha S, Kazama F, 2007. Assessment of surface water quality using multivariate statistical techniques: a case study of the Fuji River basin, Japan. Environmental Modelling and Software 22: 464–475.

USEPA, 2006. National Primary Drinking Water Regulations: Stage 2 Disinfectants and Disinfection Byproducts Rule: Final Rule. Federal Register 71(2), January

USEPA (United States Environmental Protection Agency). 2016. The Integrated Risk Information System (IRIS) online database; USEPA. Available at: http://www.epa.gov/iris/subst/index.html, Washington D.C., USA

WHO (World Health Organization), 2011. Guidelines for drinking-water quality, 4th eds. Vol. 1, Recommendations, Geneva.

Quantifying Inland Surface Water Quality via Remote Sensing of Water Surface Reflectance

Rose S. Sobel, Amin Kiaghadi, and Hanadi S. Rifai
(University of Houston, Houston, TX, USA)

Conventional water quality monitoring requires resource intensive field campaigns, and produces sparse datasets. Surface reflectance, measured from satellite imagery, provides a wealth of data for land and ocean applications. Satellites with high spatial resolutions, like Landsat and Sentinel, can be used to assess water quality in narrower inland waters. Water quality constituents such as chlorophyll-a and suspended solids have distinct optical responses. In this study, remotely sensed surface reflectance is utilized to develop a regression equation that quantifies total suspended solids (TSS) in the San Jacinto River Basin. Historical TSS concentrations from 1999 – 2016 were correlated to surface reflectance in the visible and near infrared bands of Landsat 7. Median reflectance within a nine-pixel window surrounding each water quality monitoring (WQM) station was sampled for each band with a date tolerance of up to two days between scene and sample acquisition. Non-water pixels were removed using the USGS water mask product and only pixel windows containing at least three water pixels were used for regression. The developed approach produced a regression data set of 91 unique TSS values between 2 and 177 mg/L. Step-wise log-linear regression was used with most highly correlated spectral bands and band ratios. The resulting log-linear regression uses reflectance in the red and green bands along with four reflectance ratios to produce an R^2 of 0.75 and root mean square error (RMSE) of 19 mg/L. The resulting regression was applied to the satellite record to develop more rigorous historical TSS at a spatial resolution of 30 m x 30 m and temporal resolution of 16 days. The study methodology and results have yielded valuable insights into regional water quality over time for the studied basin.

Framework for Developing Groundwater Protection Management Model for Saudi Arabia

M.S. Al-Suwaiyan
(King Fahd University of Petroleum & Minerals, Dhahran, Saudi Arabia)

Saudia Arabia like many arid region countries depends on groundwater aquifers to meet the ever increasing domestic and agricultural demand. This vital source is threatened through quality deterioration as a result of human related activities including industrial and urban development. As such protecting groundwater aquifers from point source contamination and the assessment of such pollution is certainly very important if this natural source is to remain reliable in meeting required demand. Pollution from leaking fuel and chemicals storage tanks is a major groundwater contamination problem. Such leaks could happen as a result of pipe-fitting problems and or tank/pipe corrosion. Many commercial/industrial facilities are potential contributions to this problem. A water quality management protocol is proposed with the objective of the prediction and assessing groundwater pollution from such point sources. The model combines an ever increasing data base containing various hydrological parameters for local aquifers with mathematical models that simulate water flow and contaminant transport in both the groundwater as well as the vadose zone. Such models can predict the degree of groundwater pollution potential as well as develop an approximate clean up costs. Environmental protection agencies may consider making the use of such model to asses potential groundwater pollution be a required component of environmental impact assessment submitted by new businesses.

Chemical Leaching from Two Low Impact Development Techniques

Musa Akther, Anton Skorobogatov, ***Jianxun He***, Angus Chu
(University of Calgary, Calgary, Canada); Bert van Duin (City of Calgary, Calgary, Canada)

Low Impact Development techniques (LIDs) have been proposed to be alternatives to conventional stormwater management strategy, which manages urban stormwater in a centralized manner. LIDs such as green roofs and bioretention cells take advantages of both growing media and vegetation to enhance infiltration, water retention, and evapotranspiration. Green roofs and bioretention cells, both of which are often constructed using nutrient-rich growing media to support vegetation establishment and growth. Therefore, in contrast to the expectation the initial implementation of such LIDs could lead to elevated concentrations of some pollutant constituents in urban stormwater runoff due to chemical leaching. In this study, chemical leaching was observed from both green roofs and bioretention cells in laboratory and mesocosm facilities. Chemical leaching was characterized and its association with design and hydrological variables was also researched. As for green roofs, the degree of nutrient leaching decreased with the cumulative amount of water applied and it could be explained as a function of water volume applied. Chemical leaching behavior was also associated with growing media types, as leaching rate was found to be different among growing media types for both green roofs and bioretention cells. The results obtained can help in further understanding and thus developing modeling approach to quantify pollutant leaching and thus its negative impact on urban aquatic environment at the initial stage of their implementation.

Riverine Water Quality Response to Hydro-Meteorological Conditions

Sajjad Rostami, ***Jianxun He*** and Quazi K. Hassan
(University of Calgary, Calgary, Alberta, Canada)

The study of the variation in riverine water quality with hydro-meteorological variables has a long history but it is still continuing due to the site-specific nature of the problem as well as the increase of data. In this study, the long-term monitoring data (1988 – 2014) of 11 water quality parameters were collected from three selected rivers in Alberta, Canada, and used to study riverine water quality response to hydro-meteorological conditions. Similar to many previous studies, the relationships between water quality and two primary hydro-meteorological variables including river flow and precipitation were assessed. In addition, the linkage between precipitation and flow was investigated. The identified significant correlations among water quality, river flow, and precipitation confirmed that these hydro-meteorological variables drive water quality variations. On the other hand, the response of riverine water quality to hydro-meteorological conditions varies among water quality parameters. Different from previous studies, cluster analysis was employed to classify these water quality parameters into groups based upon their dependency on flow in different flow magnitudes in this study. The cluster analysis broadly categorized the water quality parameters into three groups. The water quality parameters in each group have similar responses to flow and might originate from similar pollution source(s). All these results from this study demonstrate the importance of managing surface stormwater runoff and highlight that more efficient management should be formulated considering the properties (sources and responses to influencing factors) of water quality parameters.

Nitrogen Removal by New-Type Rotating Biological Contactors with Anoxic Filter for Domestic Sewage Treatment in Rural Area

Xiao Zha, Jun Ma, and Xiwu Lu
(School of Energy and Environment, Southeast University, 2 Sipailou, Xuanwu District, Nanjing, Jiangsu Province, 210018, China)

ABSTRACT: Various parties, including governments of every province and the state as well as thousands of researchers, have made tremendous efforts in order to treat small-scale decentralized sewage in rural areas. A bio-ecological combined technology was proposed for rural areas of eastern China. Fore parts, anaerobic, anoxic, aerobic reactors and ecological system, was included. The aerobic part combined rotating biological contactors (RBC) with waterwheels which inspired by the traditional waterwheels. The waterwheels replaced motors of normal RBC and connected with discs by an axis to drive the discs rotating. As a result, pumps were the only energy-consuming part of the whole system which reduce energy consumption and operating costs. The connection of different levels was also improved. Every level was set at ranged heights and water fall from the former level to the later level by gravity both in discs section and waterwheel section. Waterwheel was driven by the gravity and acceleration of the dropping water. The water dropping process provided more oxygen than traditional RBC, by longer-period atmospheric re-oxygenation, collision- oxygenation between dropping water and water surface as well as oxygenation when discs rotating. Sufficient oxygen was beneficial for nitrogen oxidation. An anoxic filter was combined with the new-type RBCs (named waterwheel driving rotating biological contactors, wdRBCs) to insure nitrogen removal, especially the ammonia removal. The coupling device performed well on chemical oxygen demand (COD), total nitrogen (TN), and ammonia (NH_4^+) removal. Single-factor experiments were operated for determination of optimal operating parameters and a long-period monitoring was running under the optimal parameters. When running under the optimal parameters (150 % reflux ratio, 1 h HRT of each-stair wdRBC and 7.11 h HRT of ANF), the removal efficiency of COD, TN , NH_4^+ and total phosphorus (TP) were 88.40±2.16 %, 52.33±3.80 %, 88.14±2.30 % and 34.11±7.00 %, respectively. Besides, the NH_4^+/NO_3^- concentration ratio in effluent of the anoxic-aerobic system was approximately 3, which were beneficial to plants growth in ecological system. The entire system was proved to achieve excellent pollutant removal efficiencies, while costed low construction and operating cost and need simple management. The simple system was suitable for the application in Chinese rural areas.

INTRODUCTION

World widely, severe environmental problems has been caused by influx of human-induced nitrogen and phosphorus{Tong, 2017 #65}{Conley, 2009 #67}.It has been reported globally that discharged domestic sewage causes terribly environmental pollution in rural areas, especially in developing country (Massoud et al. 2009). Non-point sources are a significant source of nutrients to rivers in eastern China, Tong, etc. {Tong, 2016 #63} reported that non-point sources contributed 36% of the riverine TN discharge loading and 63% of the TP loading in Yangtze River. In order to ameliorate the rural environmental conditions and avoid eutrophication, great efforts has been made to explore appropriate rural domestic sewage treating methods. Bio-eco technology combined systems are considered as the most common options for rural domestic sewage treatment (Kavanagh and Keller 2007).

The common treating methods chose to deal with the mixed decentralized sewage, ignoring the contrasting characteristic of greywater (GW) and black water (BW), which resulted in resource and energy wasting. The problem would be magnified in treatment for low-strength sewage especially. In general, BW (originates from the toilets, including water, urine, feces and toilet paper) contains the main part of organic load (Paulo et al. 2013) and GW (originates from bath, shower, sinks, kitchen and laundry) retained low contaminant concentration (Li et al. 2009). Besides, greywater discharged much more than black water roughly (Li et al. 2009), despite that quantity ratio of black water and greywater ranged due to local economic conditions. Thus, treating sewage sorted from the source means an important step for domestic sewage treatment. Oppositely, there were minority of researches treating black water and greywater respectively in rural area nowadays. Based on the above points, a combined system, which separated domestic sewage from the source and pretreated black water by anaerobic baffled reactor (ABR), was proposed for Chinese rural area. An "anoxic filter/multi-stair waterwheel driving rotating biological contactor (ANF/ms-wdRBC)" coupling device was used to post-treat the mixed sewage of the pretreated black water and raw greywater. In addition, ecological treatment methods, various constructed wetlands as an example, were added to enhance removal of total nitrogen (TN) and total phosphorus (TP) to achieve tighter effluent standards. The plants in wetlands were selected carefully to bring about contribution to environmental protection as well as harmony of economic returns, some agricultural plants were picked expressly.

It has been proved that multi-stage aerobic filter with a continuous flow configuration provide outstanding dissolved oxygen condition for reducing nutrient (Wang et al. 2011). And many evidences suggest that RBCs performed well for decentralized sewage on land area, maintenance, energy etc.(Dutta et al. 2007, Hiras et al. 2004). As can be imagined, the combination of multi-stage aerobic filter and RBC would gain advantages in many aspects. Furthermore, waterwheels were connected with disks for driving the disks to rotate. The waterwheel-driving structure was helpful to energy conservation. The aggregate unit was named multi-stair waterwheel driving rotating biological contactor (ms-wdRBC). Oxygen was captured from kinds of modes, including atmospheric re-oxygenation, reoxygenation from water dropping and discs rotation. The multiple re-oxygenation modes produced ranged dissolved oxygen in different stages which was conductive to the growth of different microbial communities. The novel structure of ms-wdRBC also saved more land area and energy than regular RBCs or other kind of mechanical aerators. Biological reticulated polyurethane sponge fillers (BRPSF) were added under disks in each stair of wdRBC. BRPSF posses excellent space utilization and sort of denitrification ability inside.

There is a general view that anoxic-aerobic combined systems are an effective method for removing nitrogen by nitrification and denitrification (Ge et al. 2014, Ruiz et al. 2006). In above system, ammonia was oxidized in the DO-sufficient ms-wdRBC and an anoxic filter was added for providing hypoxic condition to denitrification. The objectives of this study were to evaluate the performance of "ANF-wdRBC" coupling device on low strength rural domestic sewage (mixed sewage of anaerobic treated BW and raw GW) treatment, especially focused on nitrogen removal and conversion.

MATERIALS AND METHODS

Set-up of Reactor Experiments. The experiments were conducted using a coupling device (figure 1), including an anoxic filter (material PVC, volume 160L) and an ms-wdRBC (material PVC). 8 outlet pipes were set on ranged height for adjusting HRT of ANF. ANF were packed with vertical combined packing to retain sludge and provide growth space for microorganisms. The ms-wdRBC contained 3-stair wdRBC (volume 9L each) and one effluent collection tank (volume 3L). Height difference among each wdRBC stair was maintained at 0.6m. Every stair wdRBC was divided to 2 parts, named discs area (15 discs with 150mm diameter) and waterwheel area (1 waterwheel with 150mm diameter), respectively. The bottom part of discs areas were packed with BRPFS.

Wastewaters and Inoculum. ABR was inoculated with sludge from secondary sedimentation tank in municipal sewage plant. Ms-wdRBC was started without seed-sludge, and got affluent biofilm after 1-month operation. BW influent was consisted of half toilet wastewater from teaching building in southeast university (Wuxi, Jiangsu, China) and half synthetic water (COD: glucose, sucrose, starch; TN and NH4+-: ammonium chloride, ammonium bicarbonate, urea; TP: potassium dihydrogen phosphate, dipotassium phosphate). GW influent was consisted of half wastewater from a canteen and dormitories in southeast university and half synthetic water. The influent of ms-wdRBC was consisted of ABR digestion effluent and raw GW influent, quantity ratio of them was 1:3.6~1:7.2.

Batch Assays. Batch assays were performed to estimate the coupling device's pollutant removal in ranged HRT and reflux ratio to search optimum operating parameters. Influence of reflux ratio was estimated in run 1, TN、TP、COD removal efficiencies under different reflux ratios (RR) (0, 50, 100, 150, 200, 250, 300 %) were evaluated while wdRBC's HRT (HRTwdRBC, HRT of each-stair wdRBC) and ANF's HRT (HRTANF) were kept at 40 min and 7.11 h, respectively. Influence of HRTANF was estimated in run 2, TN、TP、COD removal efficiencies under different HRTANF (6.67 h, 7.11 h, 8.89 h, 10.67 h) were evaluated while HRTwdRBC and RR were kept at 40 min and 150 %, respectively. Influence of HRTwdRBC was estimated in run 3, TN、TP、COD removal efficiencies under different HRTwdRBC (30 min, 40 min, 60 min, 80 min) were evaluated while HRTANF and RR were kept at 7.11 h and 150 %, respectively. A duration of steady-state operation under optimum conditions was lasted for 10 weeks in run 4, and contaminant concentrations in influents and effluents were monitored every 2 days. Temperatures during the whole experiments were between 15-32°C.

Analyses. Concentrations of COD, NH_4^+-N, NO_2^--N, NO_3^--N, TN and TP were measured quantitatively according to standard method (N° 2006). DO was measured with DO meter (YSI-DO200, YSI, Yellow Springs, OH, USA).

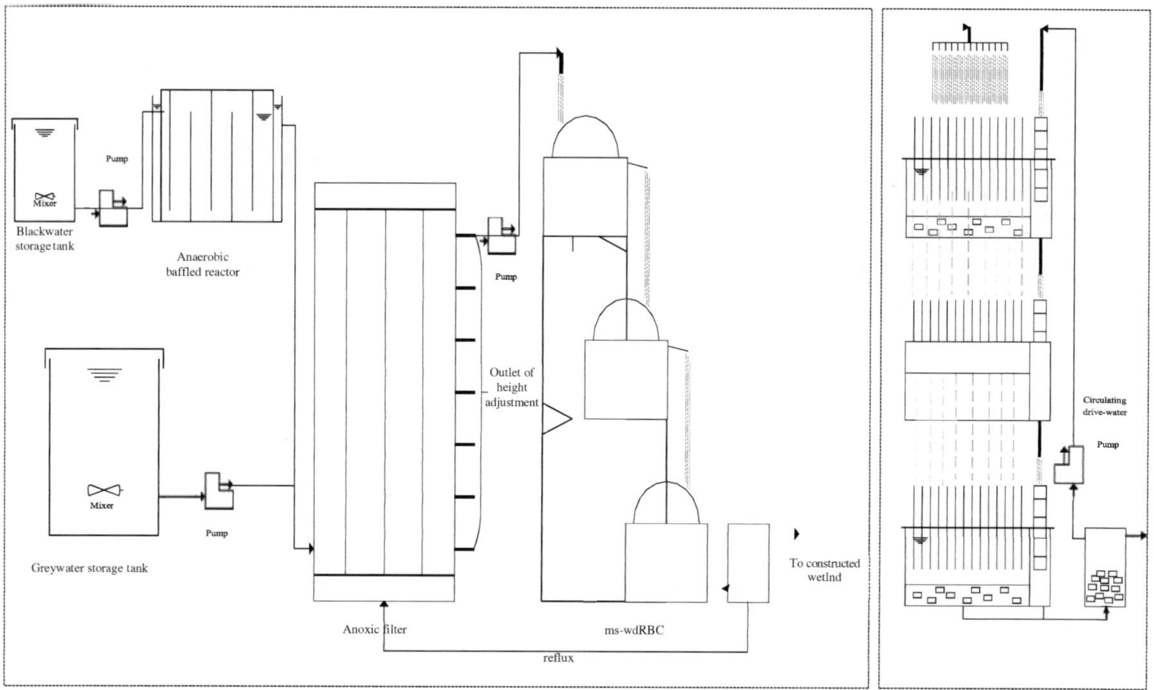

The flow chart of whole combined system ms-wdRBC

Figure 1. Schematic diagram of the whole system and ms-wdRBC

RESULTS

Reflux Ratio Dependence. Figure 2 shows the evolution of COD, TN and NH_4^+-N values in the influents and effluents from ANF and ms-wdRBC, as well as removal efficiencies under experimental reflux ratios.

The COD, TN, ammonia concentration ranges in the ANF influents were 99-116 mg/L, 30-33 mg/L and 24-27 mg/L, respectively.

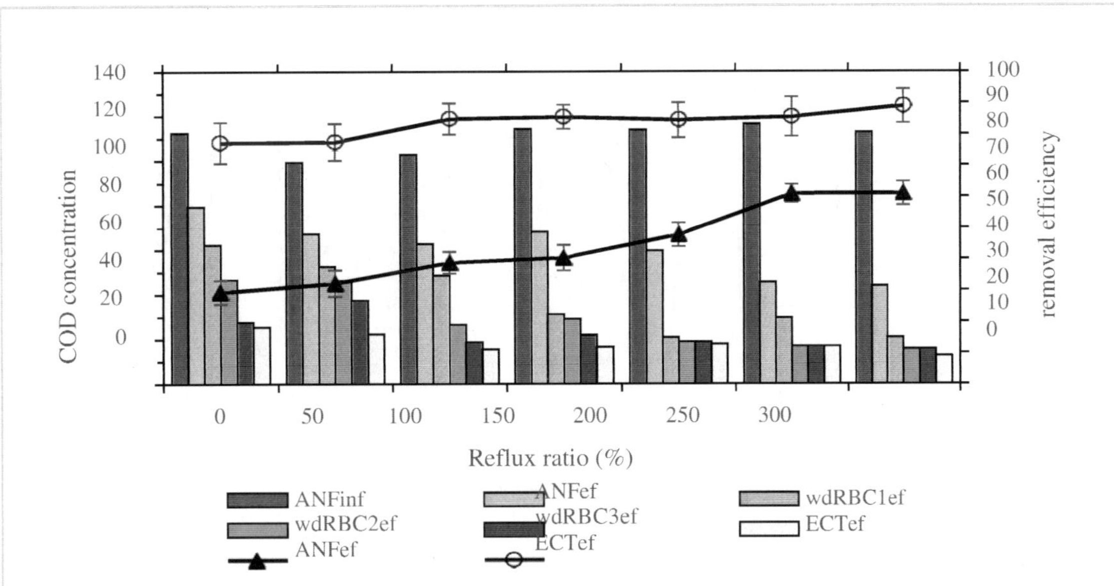

Figure 2(a). Effects of RR on removal efficiencies of COD

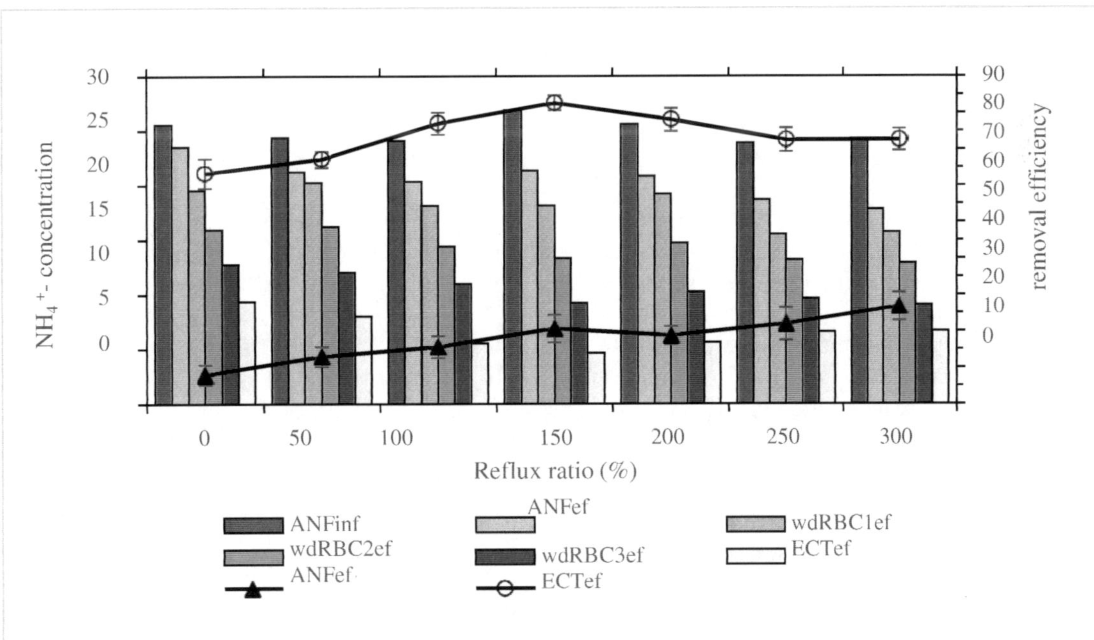

Figure 2(b). Effects of RR on removal efficiencies of ammonia

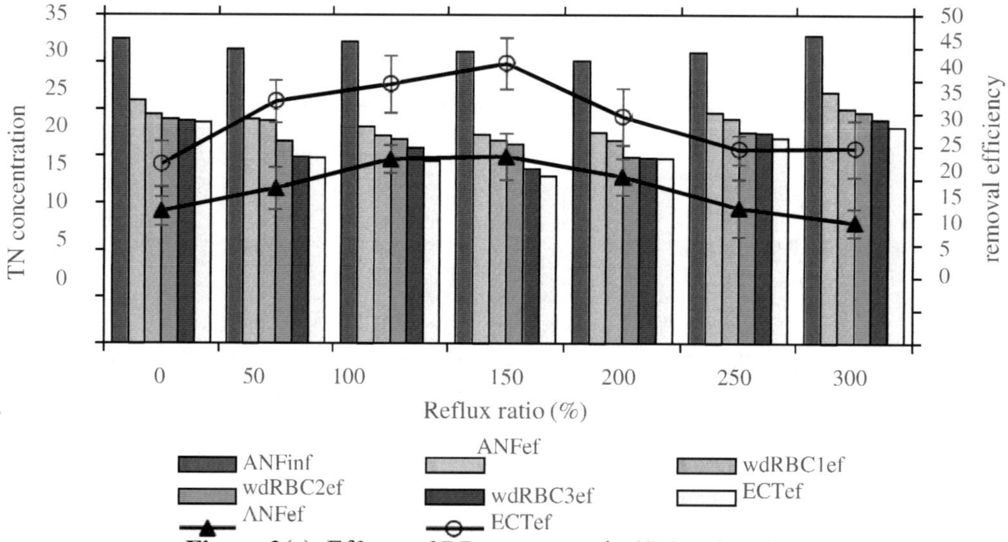

Figure 2(c). Effects of RR on removal efficiencies of TN

In general, the average COD removal efficiencies in the whole system increased following the rising reflux ratio, but the increasing was not manifest. The lowest average removal efficiency of COD was 77% at 0% RR and the highest was 89% at 300% RR, with COD concentration in effluent 26 and 13 mg/L, respectively. The average COD removal efficiencies in ANF also increased with reflux ratio climbing, and the trend was more obvious than that of coupling device. That means, COD removal ratio of ANF to ms-wdRBC increased with the growing of reflux ratio. The dilution effect of recycled flux and the biodegradation of organic matters by bacteria were the main reasons for COD removal. The recycled flux contained numerous nitrate, with the rising of reflux ratio, recycled nitrate-nitrogen to ANF increased and denitrifying bacteria needed more carbon source to execute denitrification. Generally speaking, higher COD removal efficiency could be achieved at higher reflux ratio if the carbon source wad provided sufficiently, but the dilution effect caused by reflux would decrease both the BOD_5/COD ratio and the COD/TN ratio which may have adverse effect on the denitrification (Shi et al. 2010). However, because COD concentrations in influents were low, most of the organic matter can be effectively removed at each reflux ratio.

Ammonia removal happened mainly in ms-wdRBC, where provided aerobic conditions for nitrifying bacteria to transform ammonia-nitrogen to nitrate-nitrogen. NH_4^+-N removal efficiency in the whole system reached the peak at 83% when reflux ratio is 150%, but NH_4^+-N removal efficiency in ANF increased with the reflux ratio rising continuously. NH_4^+-N removal efficiency in ANF was nearly none with non-recycling, because the low DO condition in ANF was not conductive to the growth of nitrifying bacteria and nitrification reaction. The increasing of NH_4^+-N removal efficiency in ANF mainly because the dilution of reflux and the DO rising caused by reflux recycling. It has been proved that ammonia removal can be achieved under low DO condition despite that the reaction rate was limited by DO {Wang, 2004 #69} {Hanaki, 1990 #71}. In ms-wdRBC, the highest NH_4^+-N removal efficiency point occurred at 150% reflux ratio. Although dissolved oxygen in effluent was taken back to the previous part with the backflow, but the relative DO concentration in per liter of water was diluted and the hydraulic retention time was decreased. Since that, the peak occurred in a middle reflux ratio. The changes of NH_4^+-N removal in ANF and ms-wdRBC influenced the removal rate of the entire system. Thus, there was also a peak in whole-system ammonia removal efficiency trend.

The TN removal efficiency of coupling device increased from 27% to 43% when the reflux ratio increased from 0% to 150%, but decreased when the reflux ratio continued to rise, and dropped to 30% when reflux ratio rise to 300%. The TN removal mainly happened in ANF due to the requirements of denitrification. All of the TN removal efficiencies in ANF, ms-wdRBC as well as entire system reached the highest point at 150% RR, and were 29%, 14% and 43%, respectively. Not only nitrate-nitrogen quantity but also carbon resource were affected by reflux ratios. And they were two important impact factors in denitrification process, nitrate accumulation in ANF led to COD consumption. Theoretically, 1 g of nitrate-nitrogen consumed 2.7 g of carbon source in the denitrification process. Thus, appropriate COD/NO $^{-}$-N was significant for achieving efficient denitrification. Denitrifying bacteria were generally considered as facultative bacteria, and hypoxia condition was conductive to denitrification. Compared to the disk surface, the bottom of wdRBCs contained lower DO concentration which was closer to hypoxia. In addition, the fillers gave denitrification bacteria more room for growth. So that, commendable TN removal efficiency was achieved in ms-wdRBC.

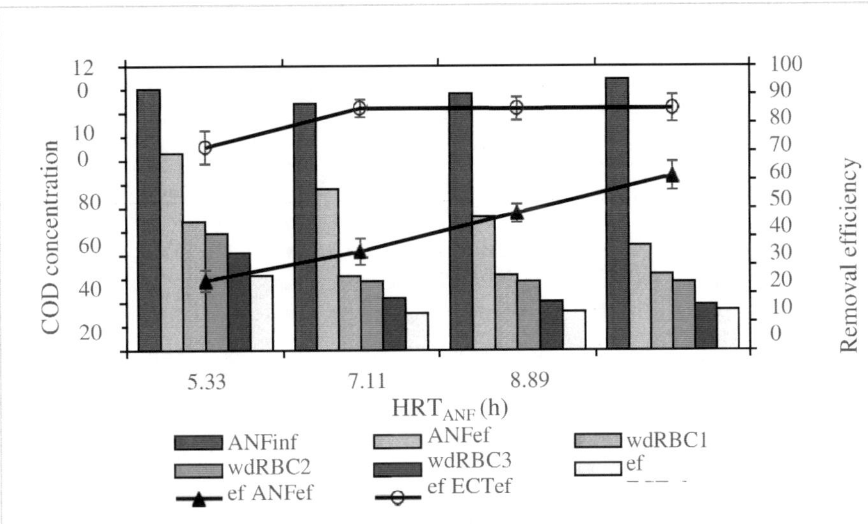

Figure 3(a). Effects of HRT$_{ANF}$ on removal efficiencies of COD

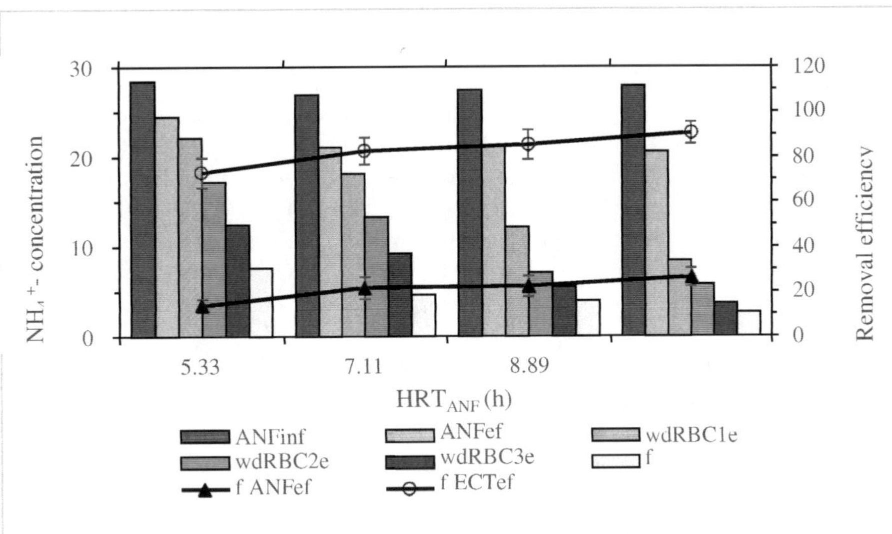

Figure 3(b). Effects of HRT$_{ANF}$ on removal efficiencies of ammonia

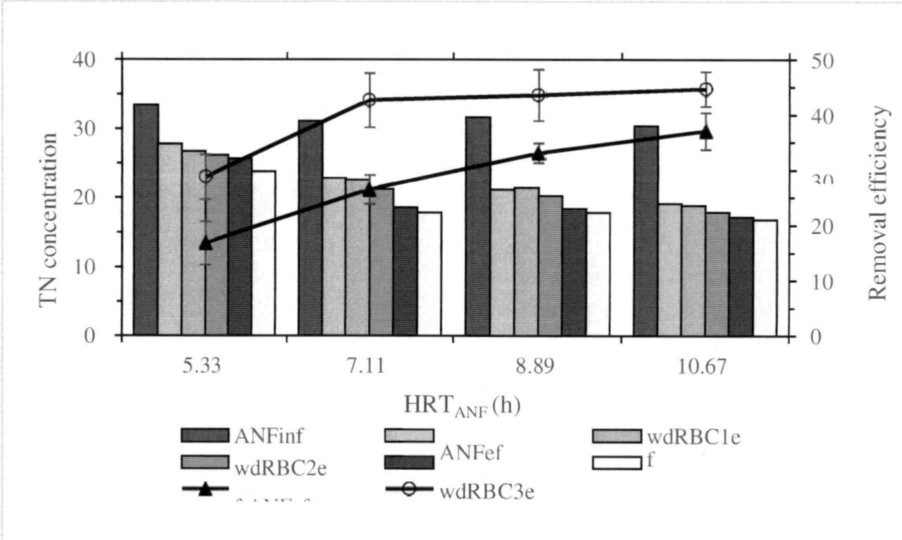

Figure 3(c). Effects of HRT_{ANF} on removal efficiencies of TN

Above all, the removal efficiency of COD increased with the increase of reflux ratio, but the removal efficiency of TN and ammonia reached the maximum at 150% RR. Because the low COD concentration of influent, COD concentrations of effluent at every RR achieved acceptable value. Therefore, 150% reflux ratio was desirable for the coupling device system according to the nitrogen removal.

HRT_{ANF} Dependence. Figure 3 shows the evolution of COD, TN and ammonia values in the influents and effluents from ANF and ms-wdRBC, as well as removal efficiencies under different HRT_{ANF}. The COD, TN, ammonia concentration ranges in the ANF influents were 104-114 mg/L, 30-34 mg/L and 27-29 mg/L, respectively.

The average overall-system removal efficiencies of COD were almost equal under all the HRT_{ANF} except a lower COD removal efficiency at 5.33 h HRT_{ANF}. Average COD removal efficiencies in the coupling device at 7.11 h, 8.89 h and 10.67 h were 85.00 %, 84.87 % and 85.01 %, respectively. However, the COD removal efficiency increased with the growth of HRT_{ANF} in ANF and decreased in ms-wdRBC contrarily. Obviously, longer HRT gave longer contact time for microorganisms with organic matters and reduce the DO concentration in reactors, which were both conductive to COD removal in anoxic condition. The average COD concentration of effluents at 7.11 h, 8.89 h and 10.67 h were 15.63, 16.37, 17.11 mg/L, respectively. In fact, most of the COD had been removed before dropping into the second stage wdRBC. It can be indicated that there was part of COD hard to be degraded in this combined biological system and there was small quantity of heterotrophic microorganisms in second and third stage wdRBC limited by carbon sources despite the low DO condition inside BRPSF. The COD concentration of effluents have reached the standard conditions in China, so the further analysis was not carried out.

NH_4^+-N removal efficiency increased with HRT_{ANF} in ANF, ms-wdRBC and entire system. Although ammonia-nitrogen was generally considered to remove by nitrosation and nitrification process, it can also be removed at low DO. The removal of ammonia-nitrogen in ANF could cause by dilution of reflux as well. Fluctuating DO caused by Prolonging of HRT_{ANF} did not affect nitrification in ms-wdRBC significantly. This indicated that novel structure of ms-wdRBC permitted it good reoxygenation ability.

The trend that TN removal efficiencies changed with HRT_{ANF} was similar with COD removal efficiencies trend, TN removal efficiency raised extremely slow after HRT_{ANF} reached 7.11 h. The removal efficiency increased with HRT_{ANF} in ANF, while increased first and then decreased in ms- wdRBC. As previously described, denitrification is the main reason for TN removal and nitrate- nitrogen as well as COD concentration were important for denitrification process. Sufficient contact time was conductive to TN removal, while the low concentration of nitrate-nitrogen and COD caused by long HRT were harmful to denitrification. Apparently, under the experimental HRT_{ANF}s, denitrification process in the bottom of ms-wdRBC was limited by the COD concentration but in ANF was not affected. Continued extension of HRT_{ANF} would not be beneficial to TN removal.

The impact of HRT in ANF on whole-system pollutant removal was not obvious after reaching 7.11 h. Thus, considered about the cost investment and energy consumption, 7.11 h of HRT_{ANF} was desirable.

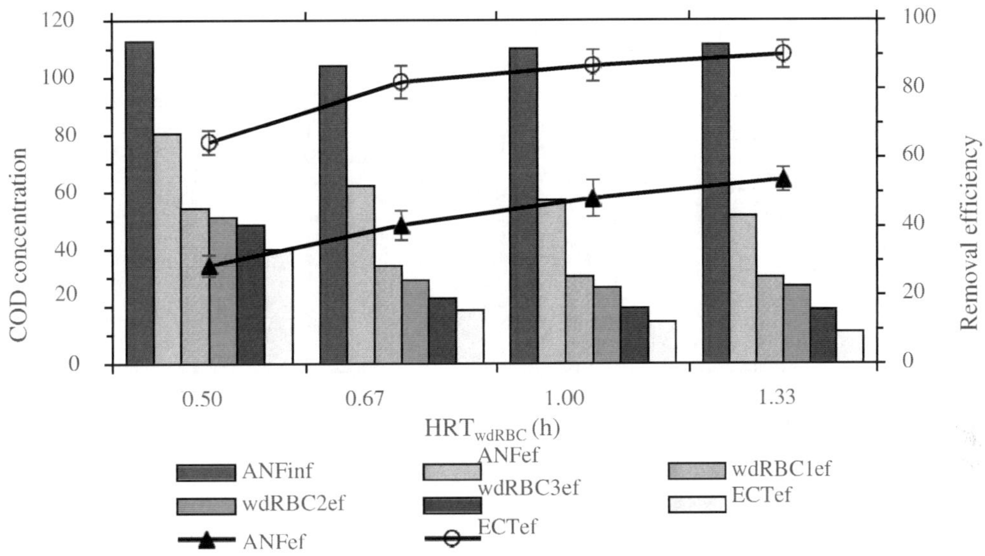

Figure 4(a). Effects of HRT_{wdRBC} on removal efficiencies of COD

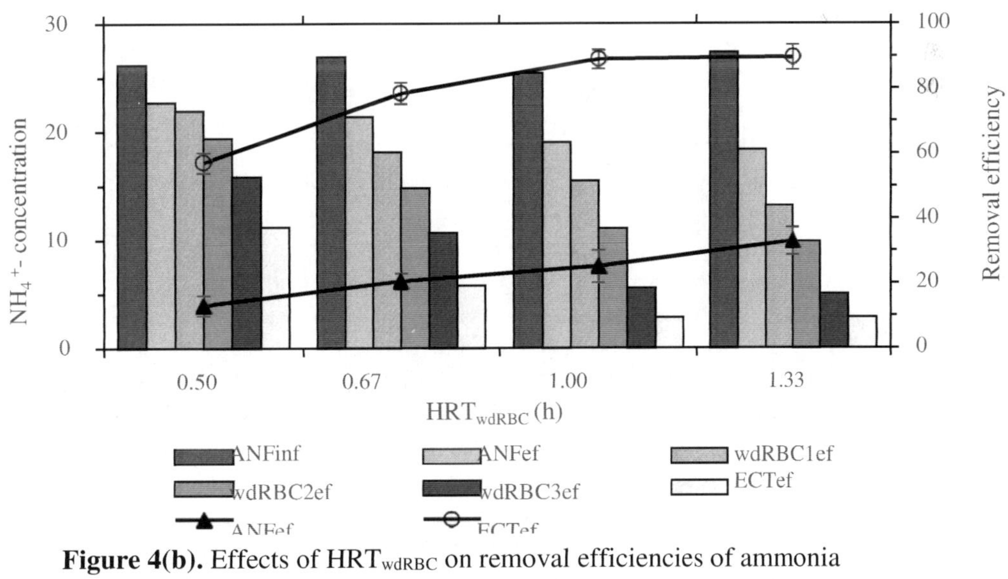

Figure 4(b). Effects of HRT_{wdRBC} on removal efficiencies of ammonia

HRT_{wdRBC} Dependence. As reported, HRT is a significant factor for water treatment reactions. Figure 4 illustrates the effect on the system performance under different HRT_{wdRBC}. The fruition of COD, TN and NH_4^+-N values in the influents and effluents from ANF and ms-wdRBC, as well as removal efficiencies under different HRT_{wdRBC} was presented in figure 3. The COD, TN and NH_4^+-N concentration ranges in the ANF influents were 104-113 mg/L, 30-31 mg/L and 25-27 mg/L, respectively.

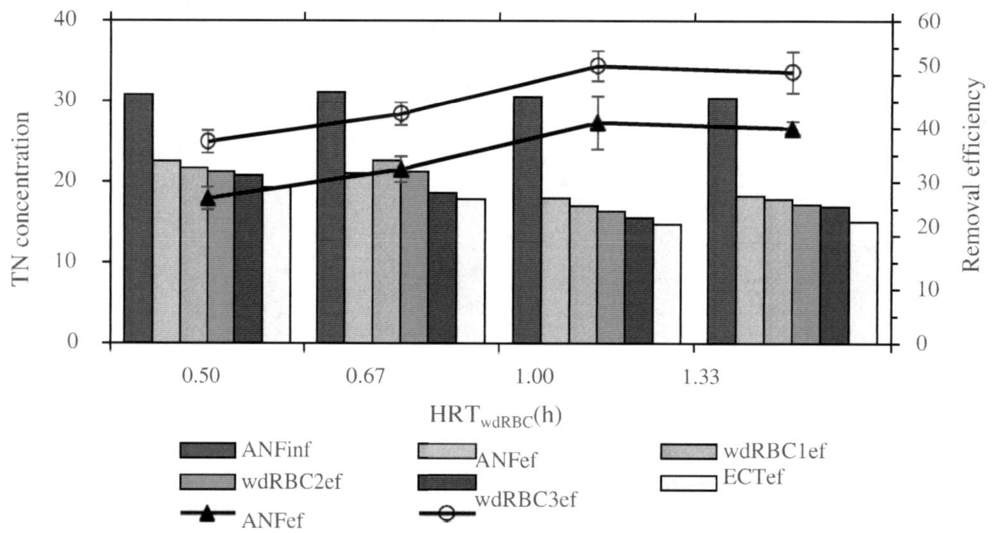

Figure 4(c). Effects of HRT_{wdRBC} on removal efficiencies of TN

Removal efficiency of COD in the entire system increased as the HRT_{wdRBC} increasing. Generally, to prolong HRT results in a rise of contact time for microorganisms with contaminants within the reactor, thereby increasing contaminant removal. However, COD removal didn't increase obviously with the increase of HRT_{wdRBC} in ms-wdRBC, but on the contrary, increased significantly in ANF. This may be due to the fact that the COD removal in hypoxia and aerobic conditions mainly relied on the denitrification by heterotrophic microorganisms, which chiefly occurred in ANF because the better environmental condition for heterotrophic microbial growth. When the HRT_{wdRBC} was up to 0.67 h, the COD removal efficiency in the coupling device was more than 80%, and the effluent concentration was below 20 mg/L.

The ms-wdRBC performed well on oxygenation due to the special configuration combined RBC with aerobic filter. Thus, it also performed well on nitrification. Atmospheric re-oxygenation, oxygenation through rotation of discs, and re-oxygenation by dropping water were three main oxygenation modes for ms-wdRBC. Oxygenation in ms-wdRBC would be affected by the single- width flow of dropping water with the changed HRT_{wdRBC}. As for ms-wdRBC, applicable HRT needed to be ensured in order to achieve a balanced state with suitable DO concentration and contact time for microorganisms. As shown in the figure 4, removal of ammonia achieved outstanding results both in ms-wdRBC and the coupling device under 1 h HRT_{wdRBC}. Nearly 90% ammonia-nitrogen was removed at this operational condition.

As presented in Figure 4, the TN removal efficiency in ms-wdRBC was non-significantly different under each HRT_{wdRBC}, but increased at first and then decreased as HRT_{wdRBC} raised in ANF, which means removal efficiency in the total system also increased first and then decreased. As denitrification occurred mainly in ANF, HRT_{wdRBC} dependence on ANF was more distinguished. Removal efficiency reached the peak at 1 h HRT_{wdRBC} both in ANF and the coupling device. According to the previous discussion, to prolong

HRT_{wdRBC} was beneficial to ammonia removal. Accordingly, nitrate-nitrogen concentration in effluent would also rise. The concentration of nitrate-nitrogen and redox potential in reflux would increase in the meantime. Higher concentration of nitrate-nitrogen provided more electron acceptors for denitrification. But for denitrification, the redox potential needs to be controlled at proper value. Based on above two aspects, HRT_{wdRBC} was not the longer the better.

The previous discussion about reactors performance on contaminant removal, especially nitrogen conversions, suggested that 1 h can be recommended for the HRT_{wdRBC}.

Steady-State Performance. A duration of continuous operation under optimum conditions (150 % RR, 7.11 h HRT_{ANF} and 1 h HRT_{wdRBC}) was taken for 10 weeks. Contaminant concentrations in influents and effluents were monitored every 2 days. The monitoring data were presented in figure 5 and table 1.

Table 1. Summary of the coupling device performances

	Concentration			Removal Efficiency	
	ANF_{inf}	ANF_{ef}	ECT_{ef}	ANF_{ef}	ECT_{ef}
COD	111.78±12.63	56.43±7.20	13.01±3.03	49.36±5.06	88.40±2.16
TN	31.86±2.45	18.88±1.63	15.17±1.48	40.06±4.82	52.33±3.80
NH_4^+	27.08±2.08	21.51±1.56	3.22±0.69	20.49±3.28	88.14±2.30
TP	4.93±0.69	3.59±0.56	3.24±0.52	26.99±6.78	34.11±7.00
NO_2^-	/	/	0.95±0.25	/	/
NO_3^-	/	/	9.98±1.47	/	/

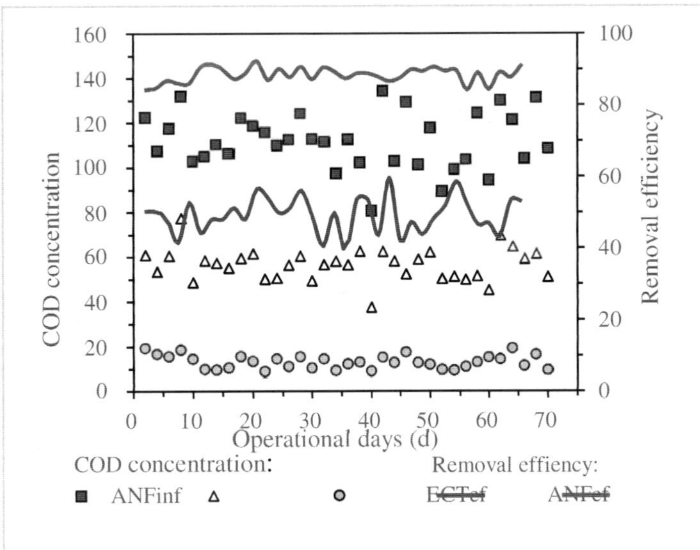

Figure 5(a). Influent, effluent, and removal efficiency of COD throughout the steady-state operation

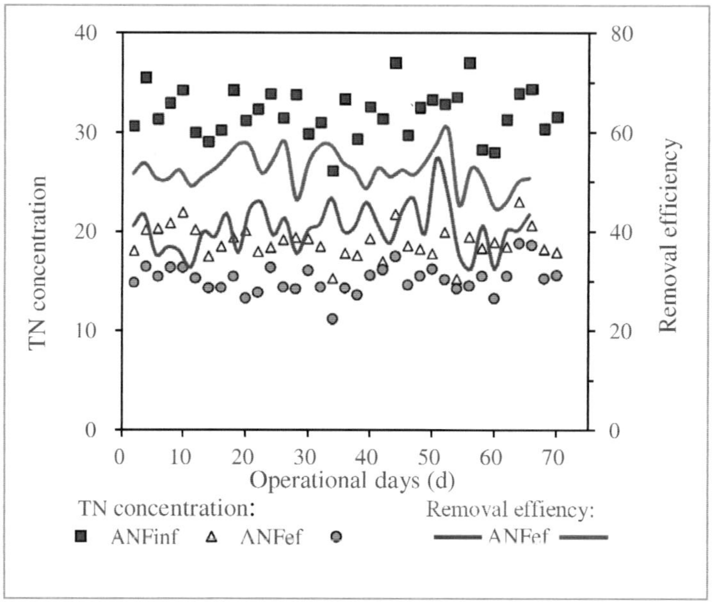

Figure 5(b). Influent, effluent, and removal efficiency of TN throughout the steady-state operation.

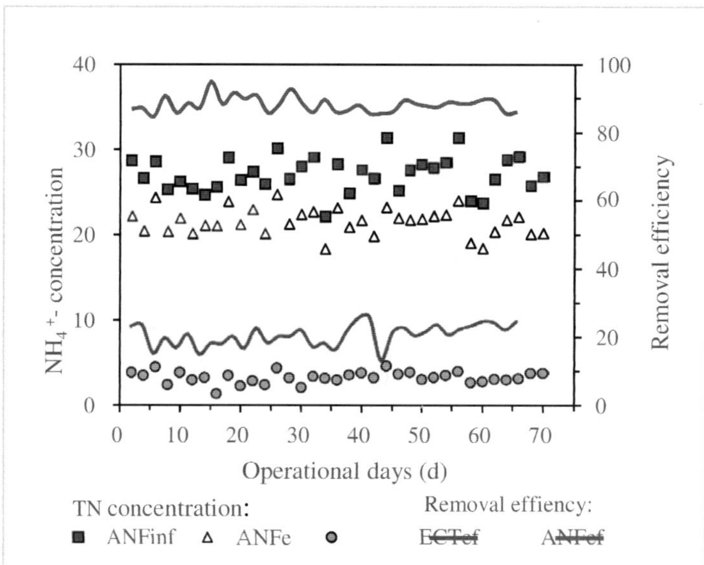

Figure 5(c). Influent, effluent, and removal efficiency of ammonia throughout the steady-state operation.

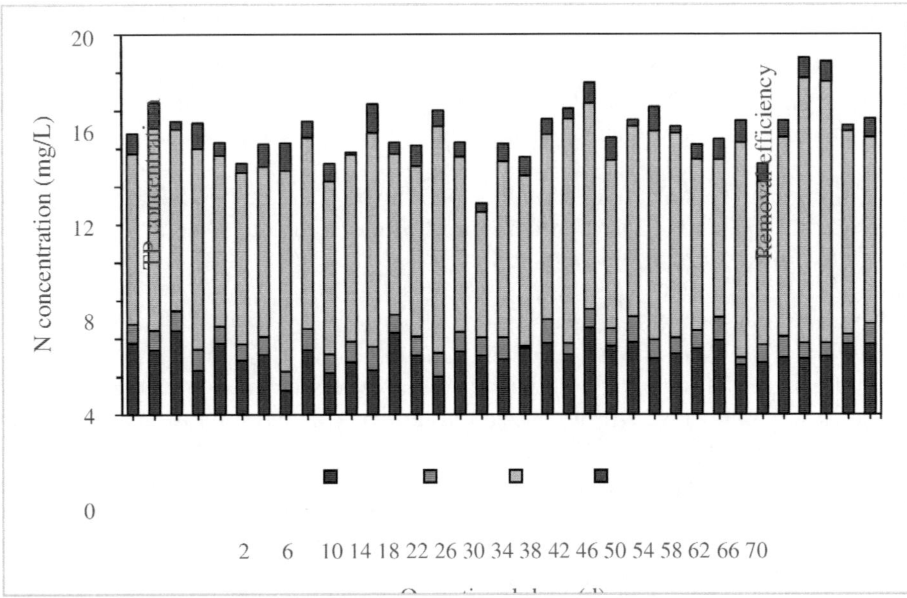

Figure 5(d). NH_4^+/ NO_3^- concentration ratio of effluent throughout the steady-state operation.

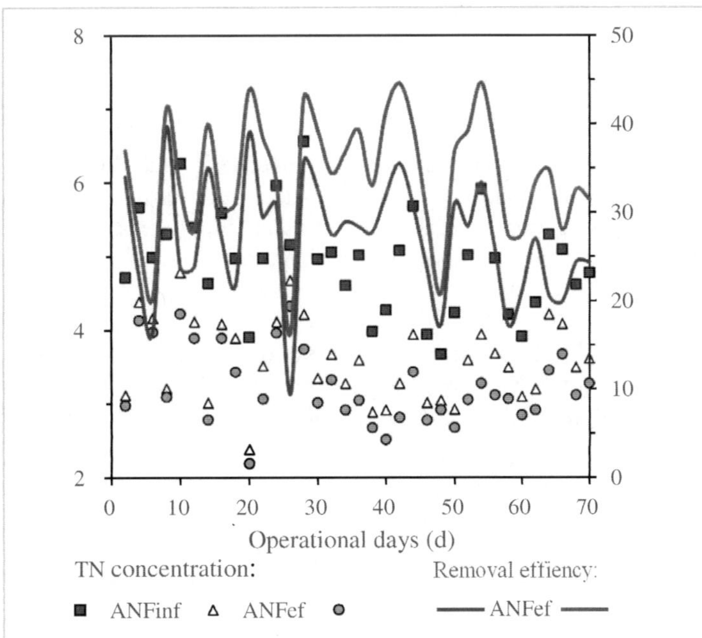

Figure 5(e). Influent, effluent, and removal efficiency of TN throughout the steady-state operation.

Approximately 50 % and 40 % of total COD was removed by ANF and ms-wdRBC, respectively. COD removal efficiency in aerobic part was higher than previous reports. Wu, etc. {Wu, 2013 #59} reported 35 % COD removal by WDASB and Li, etc. {Li, 2017 #61} reported 18.3 ± 3.3 % COD removal by four-stage WDSRBC. The increase in COD removal of aerobic part may be due to the filling of packing under the disks. Packings provided more space for microbial growth, especially the growth of heterotrophic microorganisms inside the packing at hypoxia condition. However, the possibility of COD removal by heterotrophic microorganisms under aerobic conditions can not be ruled out.

For nitrogen conversion, about 88 % ammonia and 52 % TN were removed in the whole system. Ammonia-nitrogen removal occurred mainly in ms-wdRBC but TN removal occurred mainly in ANF. Concentrations of TN, NH_4^+, NO_2^- and NO_3^- in effluent were 15.17±1.48, 3.22±0.69, 0.95±0.25 and 9.98±1.47 mg/L, respectively. NH_4^+/NO_2^- concentration ratio was approximately 3 which was reported as a proper ratio for constructed wetland (Colmer, 1998; Tylova-Munzarova, 2005; Munzarova, 2006; Yun, 2007).

34.11±7.00 % of TP was removed in the coupling device, and nearly 80 % of TP removal occurred in ANF. Considering the wetland portion comprised in the complete system, reservation of part nutrient was conductive to growth of plants in wetland.

DISCUSSION

Chinese rural sewage doesn't get sufficient treatment despite great efforts have done by the government and experts. This mixed treating process provides another potential method for large- scale application with low capital investment. The lab-scale experiment demonstrates the on-site feasibility to meet the relevant effluent standards.

However, there are still several unsolved points, including the limited denitrification rate and suitable C/N in every section. Besides, the operating parameters should be calculated. (Chang et al. 2004) reported that in wastewater treatment process the control strategy can be formed on the identification of the endpoint after biological reaction. (Wang et al. 2009) established a real-time control strategy for nutrient removal more efficiently in A2N-SBR. Thus, a real-time control strategy based on the endpoint after biological reaction may be a solution for higher pollutant removal rates.

CONCLUSION

The anoxic filter (ANF)/multi-stair waterwheel driving rotating biological contactors (ms- wdRBC) coupling device displayed high removal efficiency for nutrients, especially for COD and ammonia. 150 %, 1 h and 7.11 h were optimum parameters for reflux ratio, HRT_{wdRBC} and HRT_{ANF}, respectively. When operating at the optimum parameters, 88.40±2.16 % COD removal, 52.33±3.80 % TN removal, 88.14±2.30 % ammonia removal and 34.11±7.00 % TP removal were achieved. And the NH_4^+/NO_3^- concentration ratio in effluent was approximately 3 which was reported to benefit to various plants growing. Besides, the ms-wdRBC comsumed much less energy than traditional mechanical aerobic devices. And the whole system performed well no matter in low cost or simple management.

ACKNOWLEDGEMENTS

This research was funded by the Major Science and Technology Project of Water Pollution Control and Management in China (2012ZX07101005)(2017ZX07202004-002) and the National Science and Technology Support Program in China (2015BAL01B01).

REFERENCES

Massoud, M.A., Tarhini, A. and Nasr, J.A. (2009) Decentralized approaches to wastewater treatment and management: applicability in developing countries. Journal of Environmental Management 90(1), 652-659.

Kavanagh, L.J. and Keller, J. (2007) Engineered ecosystem for sustainable on-site wastewater treatment. Water Research 41(8), 1823-1831.

Paulo, P.L., Azevedo, C., Begosso, L., Galbiati, A.F. and Boncz, M.A. (2013) Natural systems treating greywater and blackwater on-site: integrating treatment, reuse and landscaping. Ecological Engineering 50(4), 95-100.

Li, F.Y., Wichmann, K. and Otterpohl, R. (2009) Review of the technological approaches for grey water treatment and reuses. Cheminform 407(11), 3439.

Wang, L., Guo, F., Zheng, Z., Luo, X. and Zhang, J. (2011) Enhancement of rural domestic sewage treatment performance, and assessment of microbial community diversity and structure using tower vermifiltration. Bioresource Technology 102(20), 9462.

Dutta, S., Hoffmann, E. and Hahn, H.H. (2007) Study of rotating biological contactor performance in wastewater treatment using multi-culture biofilm model. Water Science & Technology A Journal of the International Association on Water Pollution Research 55(9), 345-353.

Hiras, D.N., Manariotis, I.D. and Grigoropoulos, S.G. (2004) Organic and nitrogen removal in a two- stage rotating biological contactor treating municipal wastewater. Bioresource Technology 93(1), 91-98.

Ge, S., Peng, Y., Qiu, S., Zhu, A. and Ren, N. (2014) Complete nitrogen removal from municipal wastewater via partial nitrification by appropriately alternating anoxic/aerobic conditions in a continuous plug-flow step feed process. Water Research 55(2), 95.

Ruiz, G., Jeison, D., Rubilar, O., Ciudad, G. and Chamy, R. (2006) Nitrification-denitrification via nitrite accumulation for nitrogen removal from wastewaters. Bioresource Technology 97(2), 330- 335.

Nº (2006) Standard Methods for the Examination of Water and Wastewater, 21st Edition. Journal - American Water Works Association (1), 130.

Shi, X.L., Hu, X.B., Wang, Z., Ding, L.L. and Ren, H.Q. (2010) Effect of reflux ratio on COD and nitrogen removals from coke plant wastewaters. Water Science & Technology A Journal of the International Association on Water Pollution Research 61(12), 3017.

Chang, K.Y., Lee, D.S. and Vanrolleghem, P.A. (2004) Application of multiway ICA for on-line process monitoring of a sequencing batch reactor. Water Research 38(7), 1715-1732.

Wang, Y., Peng, Y. and Stephenson, T. (2009) Effect of influent nutrient ratios and hydraulic retention time (HRT) on simultaneous phosphorus and nitrogen removal in a two-sludge sequencing batch reactor process. Bioresour Technol 100(14), 3506-3512.

Strategies for Maintaining Nitritation in Membrane Bioreactors Operated under Long Solids Retention Time

Yishuai Jiang, Leong Soon Poh, and Wun Jern Ng
(Nanyang Technological University, Singapore)

The membrane bioreactor (MBR) combines biological treatment with membrane filtration, and so provides better effluent quality with a lower physical footprint. Additionally, nitritation based nitrogen removal can save up to 60% of the energy demand for aeration, and have lower requirements for the electron donor as well as reduced sludge production, compared with the conventional biological nitrogen removal process. A stable nitritation process relies on promoting ammonium-oxidizing bacteria (AOB) activities and suppressing nitrite-oxidizing bacteria (NOB) activities. This study investigated the strategies for long-term operation of a nitritation MBR with long solids retention time (SRT).

Based on our previous findings, the strategy of combining low dissolved oxygen (DO) with high free ammonia (FA) failed to maintain nitritation, or even partial nitritation. AOB was inhibited at FA concentration of 34.3 ± 6.85 mg NH_3-N/L while the threshold inhibiting concentration of FA was likely 23.6 ± 4.51 mg NH_3-N/L. Even though FA was reduced to 8.89 ± 0.95 mg NH_3-N/L in the following 37 days, after 70 days of operation under high FA, AOB cells became more sensitive to free nitrous acid (FNA). FNA concentration of 1.58×10^{-3} mg HNO_2-N/L caused a sharp decline in AOB cell numbers by 62.7%. The NOB had acclimated to the high FA conditions. This showed that long-term FA inhibition had adverse impact on process recoverability and subsequent failure threshold of the nitritation process.

The other strategy examined in this study was low DO and high FNA. High FNA concentration of up to 0.37-0.56 mg HNO_2-N/L was induced by gradually reducing pH to 6.1-6.3 through reduction of buffer dosing when the NOB population was higher than 10.0% as measured by qPCR. Nitritation was successfully achieved at the feed ammonium concentration of 240 mg N/L to 650 mg N/L, and this was maintained for 200 days. Maximum ammonium loading rate was reached at 1.81kgN/ (m^3. d), with ammonium removal ratio at above 95.0% and nitrite accumulation ratio at around 94.8%. qPCR results showed that after in-situ FNA treatment, NOB population decreased while AOB remained the dominant bacteria, varying from 45.8.0% to 82.0%. This is because that AOB had higher FNA tolerance than NOB and so recovered after pH was adjusted to 7.8 while NOB activity remained inhibited.

This study indicated that NOB would acclimate to both high FA and FNA, and proliferate under long-term operation. Despite this, compared to FA, in-situ FNA treatment was more efficient at maintaining nitritation in the MBR operated under long SRT.

The Deammonification for Nitrogen Removal from the THP (Thermal Hydrolysis Process)'s Sidestream with Immobilization Technology

Yifeng Yang and Siqing Xia (Tongji University, Shanghai, China)
Yuan Li and Slawomir Hermanowicz (University of California, Berkeley, Berkeley, USA)

The thermal hydrolysis process (THP) is an excellent pre-treatment process for anaerobic digestion. However, previous study showed the sidestream produced after anaerobic digestion of the THP treated sludge (THP-PS) could have inhibitory effect on deammonification. A novel deammonification up-flow reactor was established with AOB (ammonium oxidizing bacteria) and ANAMMOX (anoxic ammonium oxidizing) gel beads. The goal of this work was to facilitate the start-up of deammonification with low biomass concentration and optimize it for efficient nitrogen removal despite the inhibition effects caused by the change of microbial community. A comparison of nitrogen removal performance of sidestream derived from THP between non-immobilized deammonification and immobilized deammonification will be studied. This paper includes a discussion of the unique benefits and challenges of immobilization technology on the deammonification of sidestream.

Application of MgO Modified Diatomite for Simultaneous Recovery of Chlorella Vulgaris, Nitrogen and Phosphate: Interaction Mechanism

Jing Li and Xuejiang Wang
(Tongji University, Shanghai, China)

Microalgae have always been a nightmare with the uncontrollable growth in eutrophic waters with great amount of nitrogen and phosphate. Recently, attention has been drawn to the outstanding values of microalgae with its usage in food, feed, fuel and wastewater treatment. Consequently, recovery of microalgae from water has become a crucial topic. Furthermore, because of environmental concerns about eutrophication, it will be a good choice to get rid of both nitrogen and phosphate, which are usually in the form of ammonium-N and phosphate respectively, from water together with algae. Here, we explore the potential of autoflocculation induced by MgO modified diatomite for harvesting Chlorella vulgaris. Meanwhile, the removal of ammonia-N and phosphorus can be achieved synchronously. The focus of the presentation will be to elucidate the mechanism of total recovery. The influence with respect to adsorbent dosage, the concentration of algae, the concentration of ammonia-N and phosphorus, solution pH, contact time and coexisting ions on simultaneous recovery of microalgae and nutrients were evaluated under optimal condition. From the results, MgO-D demonstrates excellent removal capacities for microalgae, ammonia-N and phosphate in a wide pH range of 3-9, and the highest removal capacities of microalgae, ammonia-N and phosphate are 121.7 mg/g, 58.4 mg/g and 103 mg/g for 3 h with the dosage of 0.6 g/L.

Research on Interchange Ratio (IR) Parameter Optimization and Mechanism of A^2/O-MOSA Process with Source Sludge Reduction Characteristics

***Yuji Lin*[1]**, Zhixiong Fan[1], Weibin Guo[1], Chenyi Shi[1], Tingjin Ye[2], Wuzhen Guo[2], Lianpeng Sun[1,3,*]
(1. School of Environmental Science and Engineering, Sun Yat-Sen University, Guangzhou 510275, China; 2. Foshan Water Group, Guangdong, Foshan, 528000, China;3. Guangdong Provincial Key Laboratory of Environmental Pollution Control and Remediation Technology, Guangzhou 510275, China)

ABSTRACT: Two sets of A^2/O and A^2/O-MOSA process were built up in the laboratory to investigate their pollutant removal effect, sludge reduction effect in different Interchange Ratio (IR) situation, and to explore the sludge reduction mechanism in MOSA process. It showed that the A^2/O-MOSA process has high pollutant removal efficiency and the insert of anaerobic tank did not affect the sludge quality. When the IR=10%, the A^2/O-MOSA process reached the biggest sludge reduction of 24.11%. The result of High-throughput sequence (Illumina Miseq platform) showed that the inserting of MOSA anaerobic tank affected the microbial species abundance of activated sludge while the primary microorganisms were the same. The microorganisms which increased in the aeration tank were facultative anaerobic bacteria or anaerobic bacteria, mostly related to anaerobic ammonia oxidation and phosphorus removal, which had slow growth characteristics. It indicated that the MOSA anaerobic tank played a role of biological selector to retain the bacteria group with function of denitrifying and phosphorus removal and slow growth rate, which may be the key of source sludge reduction mechanism in MOSA process.

INTRODUCTION

Activated sludge process is widespread used in sewage treatment plants home and abroad, due to its high pollutant removal rate, steadily processing and low operating cost (Qian Y., 2000). However, with the expansion of water treatment scale, huge volumes of excess sludge is discharged, the ultimate disposal of which is hard and expensive. It has become an important constraint on the development of sewage treatment industry and solutions are badly needed.

Sludge reduction in the existing plants would be an ideal option to tackle the sludge problem. There are four possible solutions commonly to provide feasible engineering approach: (1) cell lysis or mineralization of excess sludge using physical or chemical methods such as ozonation(Pan Y P, 2005; Lee J W, et al, 2005; Lu ZL, et al, 2011), chlorination(Takdastan A, et al, 2009; Liu W Y, et al, 2016), UV-Fenton(Wang L, et al, 2011; Tokumura M, et al, 2009), ultrasound destruction(Li X, et al, 2016; Wang L, et al, 2014) and strong acid or base(Yuan GH, et al, 2012; Xie M, et al, 2014); (2)restriction of sludge growth by using chemicals like metabolic uncouplers (Li H C,et al, 2015; Chen GH, et al, 2002), high S_0/X_0 rate(Liu Y, et al, 1998; Liu YH, et al, 2016) according to energy uncoupling; (3)limitation of sludge growth by putting in predators such as tubificids (Zhang H, 2008) and oligochaetes (Zhu H, et al, 2008); (4)modification of activated sludge process by inserting an anaerobic tank in the sludge return circuit to form an anoxic-settling-anaerobic (OSA) process(Chudoba P, et al, 1992; Saby S, et al, 2003).

Oxic-settling-anaerobic process was firstly proposed in 1970s (Westgarth W C, et al, 1964). Compared to the regular activated sludge process, OSA process can reduce sludge production by 30%~80% steadily and efficiently by simply inserting an anaerobic tank, which costs low and has little effect on the original process (Wei Y, et al, 2003). The anaerobic zone is different from such zones in biological denitrifying and phosphorus removal process. Anaerobic or anoxic zones in regular process are often set in the front, where the microbe can decompose abundant organism and use the energy to finish denitrifying and phosphorus removal (Chen G H, et al, 2003).

As the core of OSA process, many researches on parameter optimization of the anaerobic tank like ORP (oxidation-reduction potential) and SRT (solid retaining time) have been carried out to promote the sludge reduction (Saby S, et al, 2003; Cui YH, 2013; Easwaran S P, 2006). Many approaches have been studied to modify the structure and operation mode of OSA process, including inserting pre-anoxic tank, changing the influent mode of sludge (Yang Y, 2010), and forming different operation mode of modified-oxic-settling-anaerobic (MOSA) process (Wang T J, 2012; Yu X Y, 2012; Tan JX, 2014; Chen L L, 2015). Interchange ratio (IR), which refers to the ratio of total dry sludge pumping into the anaerobic tank and that in the oxic zone, was announced as a parameter related to helping sludge reduction and decreasing the size of reaction tank (Yang Y, 2010). However, there are few reports about optimizing the IR of MOSA process to accelerate sludge reduction.

Several conjectures, mainly including theory of sludge recession (Chen G H, et al, 2003; Yu L, et al, 2001), theory of energy uncoupling (Chen G H, et al,2000; Feng Q, et al, 2012), theory of EPS (extracellular polymeric substance) dissociation (Novak J T, et al, 2007; Novak J T, et al, 2003; Park C, et al, 2006) and theory of slow-growth microbe (Chen G H, et al,2003), were proposed to explain the mechanism of OSA process but were still not consensus. The slow-growth microbe theory suggests that inserting an anaerobic tank changes the living environment of microbes and makes great difference on the microbial community and diversity, resulting in the emerging slow-growth microbes. Many microbes were related to phosphorus releasing, denitrification and sulfate reduction and had a relative slow growth rate, which led to the sludge yield decreased (Jin W B, et al, 2008). Some research suggested that some of the bacteria were screened in the inserted anaerobic tank by oxic and anaerobic alternate, while some specific species arised such as slow-growing Trichococcus (Zhou Z, et al, 2015; Jia L, et al, 2012). Goel and Noguera studied the Cannibal-EBPR process and pointed out that there were abundant slow-growth microbes of phosphorus accumulating organisms (PAOs) and zymophyte (Goel R K, et al, 2006). PAOs was also reported to be the dominant population in CAS-OSA process (He M H, et al, 2010). Zhou found that the relative abundance of β-proteobacteria in the oxic tank of OAS process was higher than that of the reference, especially the anaerobic bacteria like propionivibrio and sulfuritalea (Zhou Z, et al, 2015). However, Ning etc. found that the relative abundance of β-proteobacteria in the oxic tank of OAS process decreased (Ning X, et al, 2014). Due to the diversity of the combining processes and parameters, the dominant microoganisms were different in OSA process, which indicated the complexity of sludge reduction mechanism.

As an ideal solution for sludge treatment and disposal problem, MOSA process differs a lot from OSA process on the mode of operation, which also suggests that the mechanism of OSA is not entirely matched to that of MOSA. Thus, study on parameter optimization mechanism of MOSA process has great significance. In this research, two sets of A^2/O and A^2/O-MOSA process were built up in the laboratory to investigate their pollutant removal effect, sludge reduction effect in different Interchange Ratio (IR) situation. To explore the microbial mechanism of sludge reducing, high-throughput sequence was used to investigate the microbial abundance and diversity in the MOSA process. This study was based on the operation of Zhen`an sewage treatment plant, in where run three sets of full scale test of MOSA process.

MATERIALS AND METHODS

Source of Sludge and Wastewater. To start the experiment, the sludge in the two sets of A^2/O process and A^2/O-MOSA process was taken from the aerobic tank of the first period A^2/O process in Zhen`an sewage treatment plant to start the cultivation untill the systems run smoothly. Synthetic wastewater with the composition and characteristics as shown in Table 1 and Table 2 was continously fed to each setup.

Operation of Two A^2/O System. Based on the operation of A^2/O process in Zhen`an sewage treatment plant, two A^2/O process were set up in the lab and run under the same condition including HRT (hydraulic retention time), SRT (solid retention time), MLSS (mix liquor suspended solids) and DO (dissolved oxygen) during the caltivation period. After the 46-days caltivation, the sludge production and treated effluent quality in these two systems became stable and parallel. One of them was refitted as an MOSA system by inserting an anaerobic tank between the aeration tank and setteler while the other remained unmodified, serving as the reference system. The flow chart and controlled parameters are showed as Figure 1 and Table 3.

TABLE 1. Composition of the Synthetic Wastewater

Composition	Concentration (mg/L)
$KHSO_4$	20
KH_2PO_4	15
NH_4Cl	60
NaAC	145
Glocose	145
Urea	28
Becterial peptone	28
$CaCl_2$	105
$MgCl_2$	26
$FeCl_3$	13
$Al_2(SO_4)_3$	25

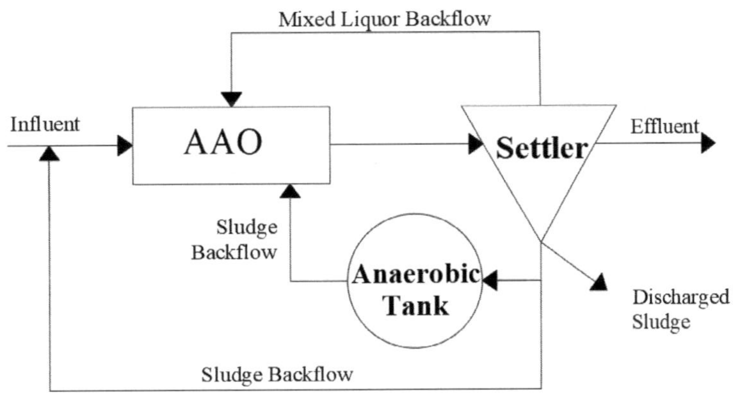

(a)

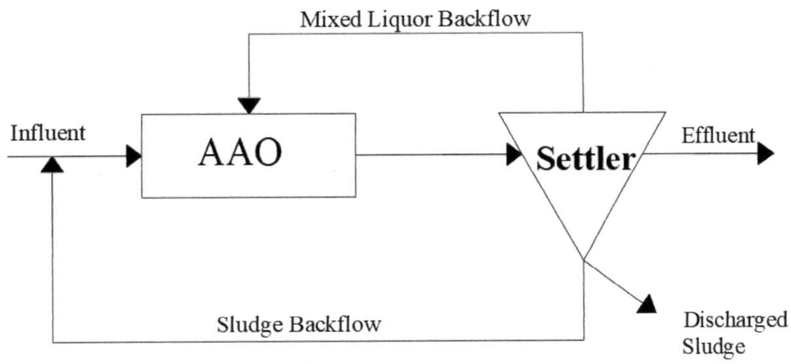

(b)

Figure 1. Flow Chart (a) A^2/O-MOSA; (b) A^2/O

The A^2/O-MOSA system operated 366 days totally, which could be divided into three periods. In the first period, stepwise chang in the interchange ratio of 5%, 10% and 15% was adopted and the system stably run for 50 days in each grad, which indicated that the maximun sludge reduction could be reached when IR is between 5% to 10%. During this period, sludge bulking occurred when the system run with IR=10% to the day 51 and lasted for 58 days, after working out which IR was changed into 15%. In the second period,

IR was changed to 8% and run stably for another 50 days, finally to find that IR=10% was the best fit. In the third period, the system was run with IR=10% for 62 days to study the sludge reduction efficiency. During the A^2/O-MOSA system operation, all the operating conditions and parameters maintained at the same levels as that in the reference system. The DO in the oxic tank in each setup and ORP in the anaerobic tank were monitored daily. The MLSS, mixed liquor suspended solid (MLVSS), NH_4^+-N, TN, TP, and SS of the influent wastewater and treated effluent were analyzed twice a week according to Stand Methods. Metal and soluble protein were detected by ICP-OES and Folin-phenol reagent method.

TABLE 2. Characteristic of the Synthetic Wastewater

Parameter	Value
COD (Chemical oxygen demand)	212.5 mg/L
TN (Total nitrogen)	31.3 mg/L
NH_4^+ (Ammonium nitrogen)	15.8 mg/L
TP (Total phosphorus)	3.5 mg/L
SS (Suspended solids)	50.5 mg/L
Fe^{3+}	3.75 mg/L
Al^{3+}	2.36 mg/L
Mg^{2+}	8.67mg/L
Ca^{2+}	56.46 mg/L
pH	7.0

Table 3. Operation Parameters of A^2/O and A^2/O-MOSA Process

Parameter	Value
HRT in the anaerobic tank	1.0h
HRT in the anoxic tank	2.4h
HRT in the oxic tank	5.7h
MLSS in the oxic tank	2000~3500mg/L
DO in the oxic tank	2~4mg/L
SRT in the anaerobic tank	5d
ORP in the anaerobic tank	-150mV
Interchande time of the anaerobic tank	4 times

Analytical Procedure: Chemical Analysis. The measurement of MLSS, MLVSS, SS, NH_4^+-N and TP followed the Standard Methods in "China's State Environmental Protection Administration. 2002. Water and wastewater monitoring and analysis methods". DO, ORP and pH were monitored by specialized instruments. All the samples were tested on the day of sampling.

Sluge Reduction Analysis: The microbial yield coefficient (Y_{obs}, kg MLSS/kg COD) was uesd to measure the sludge production from the microbial respective. The sludge reduction efficiency of MOSA process can be measured by comparing the value of Y_{obs} with that in A^2/O process. The computation formulas of Y_{obs} and sludge reduction rate are shown as follow.

$$Y_{obs} = \frac{(MLSS_2 - MLSS_1) \times V_{oxic} + U_{elimination}}{(COD_{in} - COD_{out}) \times Q \times T} \quad (1)$$

$$\text{Sludge reduction rate (\%)} = \frac{Y_{obs(1\#)} - Y_{obs(2\#)}}{Y_{obs(1\#)}} \times 100\% \quad (2)$$

$MLSS_1$ = the value of day1 of each period; $MLSS_2$ = the value of last day of each period; V_{oxic}= the effectinve volumn of oxic tank; $U_{elimination}$ = the value of discharged sludge; "1#" = MOSA; "2#" = A^2/O

Microbe Enumeration: High-throughput sequencing are widely used in science research espacialy in the field of microbiology. The sequencing technology on Illumina Miseq platform is based on sequencing by synthesis and features quick and accurate data analysis process, which has obvious superiority in microbial community characteristics and diversity analysis (Zhao W L, et al, 2015). The sludge samples were sent to the MAGIGEN Agency for testing. Considering the microbial abundance of the sludge, 16S rDNA was adopted and PCR amplification and sequencing were carried out in the V4 high-variable area of the sludge.

RESULTS AND DISCUSSION

Effect of MOSA System on Process Perfomance and Sludge Characteristic. In A^2/O-MOSA process, inserting an anaerobic tank had little influence on the original process but the living environment of bacteria. To investigate the effect of anaerobic tank on the MOSA process performance and sludge characteristics horizontally compared with the reference on the and longitudinally with different IR, quality indexes of water and sludge in anaeobic tank and oxic tank of the two setups were monitored regularly during the operation.

Figure 2. shows the removal of COD in A^2/O and A^2/O-MOSA system. During the operation period, the concerntration of COD in effluent fluctuated slightly but was stably meet the emission standards. The average removal rates of MOSA system and the reference were 89.07% and 86.22% respectively. Both systems had good organic pollutant removal ability and were hardly effected by different IR.

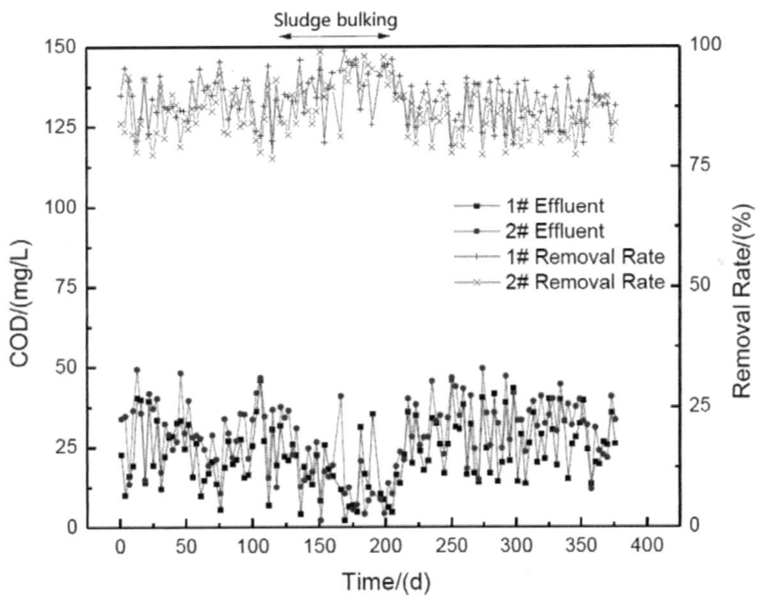

Figure 2. Changes of COD concerntration in A^2/O and A^2/O-MOSA system.

The removals of NH_4^+-N, TN, SS and TP in A^2/O and A^2/O-MOSA system are shown in Figure 3. The NH_4^+-N and TN concerntrations of effluent could meet the emission standard in most time of the operation, except the sludge bulking period and pipe plug occurrence. The average NH_4^+-N and TN concerntration and removal rate of A^2/O-MOSA were 1.48mg/L, 13.34mg/L and 92.6%, 56.31% respectively, slightly worst than that of the reference but still meet the standard. The removal of SS was stable at a high level of 86.14% and 86.98% in A^2/O -MOSA and A^2/O system severally but undulated a lot during the sludge bulking period when the sludge settling and spate separation property declined. The average concerntrations and removal rates of TP in A^2/O -MOSA and A^2/O process were 0.189mg/L,

0.199mg/L and 93.85%, 92.91% severlly, almost all met the standard. Fluctuation appeared during 125d~175d mainly related to the sluge bulking which increasd the concerntration of suspended solid, an important contributor to particulate phosphorus (Fernández D J, et al, 2003). The removal rate rised after bulking controlling measures were adopted. The concorntration of total phosphorus in sludge was 0.42, slightly higher than 0.39 mg/mg MLSS in the reference, which was just contrary to that in effluent. It indicated that A^2/O -MOSA process can guarantee good phosphorus removal property in the context of sludge reduction.

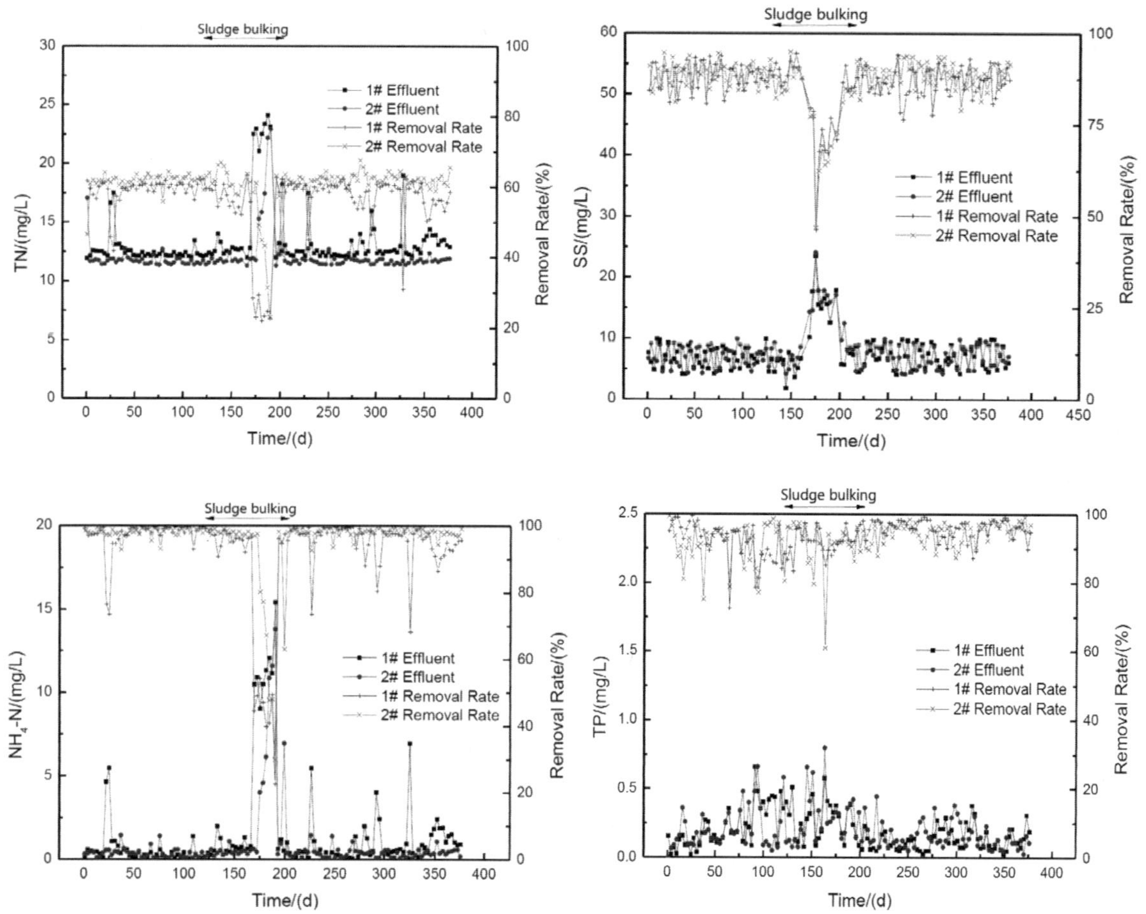

Fifure 3. Changes of TN, NH_4^+-N, SS and TP concerntration in A^2/O and A^2/O-MOSA system.

Figure 4. shows the changes of mixed liquor suspended solid concertration in the two systems. Fluctuation appeared especially at the beginning of the operation because of the uncertain volume and concerntration of discharged sluge. During the entire operation, the mean values in A^2/O -MOSA and A^2/O process were 2733.01mg/L and 3188.46mg/L, fluctuated around 2500mg/L. The ratio of MLVSS to MLSS were ranged in 0.6~0.8 while dropped to 0.4 during sludge bulking period, which indicated the two systems possessed similar biodegradability.

During the entire operation, the horizontal comparison shows that both systems performed close to each other, which demonstrates that inserting an anaerobic tank in the original A^2/O process had little effect to performance and sluge characteristics. Vertically, A^2/O -MOSA system stably maintained high pollutant removal efficiency and good sludge characteristics while different IR values were adopted, which signified that changing IR brought little effect to the process.

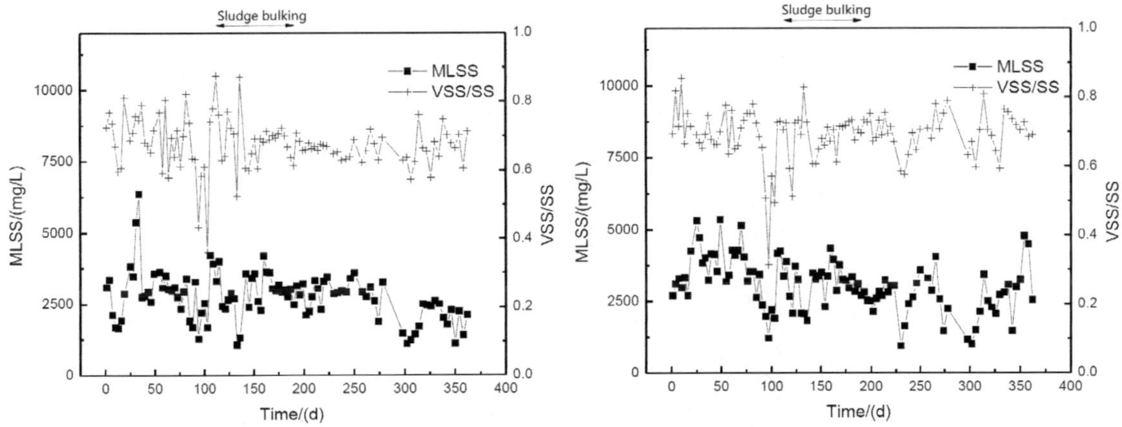

(a) (b)

Figure 4. Changes of MLSS over the entire operation period: (a) A^2/O-MOSA; (b) A^2/O.

Excess Sludge Prouction. Microbial yield coefficients was used to reflect the sluge production. The sluge yield of two parallel sets of A^2/O processes during the cultivation period were calculated as 0.3862 kg MLSSkg·COD^{-1} and 0.3852 kg MLSS·kg COD^{-1}, indicated their high comparability. Table 4 summurises the microbial yield coefficients of the two systems after one of the A^2/O system was modified as A^2/O -MOSA process running with different interchange ratio adopted.

Table 4. Microbial yield coefficients of the A^2/O-MOSA (1#) and A^2/O (2#) processes

Subject	1#	2#	1#	2#	1#	2#	1#	2#
IR(%)		5		8		10		15
COD_{in} (mg/L)	210.91	205.86	210.91	205.86	210.91	205.86	210.91	205.86
COD_{out} (mg/L)	20.14	36.02	26.64	33.10	22.52	38.36	25.18	35.78
Q (L/d)					100			
T (d)					50			
$MLSS_1$ (mg/L)	2178	3283	1872	2370	3195	3351	3885	3852
$MLSS_2$ (mg/L)	3518	3938	1766	3786	2979	2842	3160	3837
V_{oxic} (L)					24			
$U_{elimination}$ (kg)	202.25	226.89	257.67	250.88	259.19	308.93	268.64	266.94
Absolute yield of sluge (kg)	234.43	242.63	255.22	284.87	254.01	296.71	251.24	266.58
Y_{obs} (kg MLSS/kg COD)	0.2468	0.2857	0.2770	0.3298	0.2697	0.3553	0.2705	0.3134
Sluge reduction rate (%)		13.98		16.00		24.11		13.69

Compared to the A^2/O process, A^2/O -MOSA process obtained 13.98%, 16.00%, 24.11% and 13.69% sludge reduction at the IR of 5%, 8%,10% and 15% respecively. It was supposed that sludge reduction was lower at IR of 5% and 15%, mainly caused by sludge recession. When IR was 5%, the backflow sludge from anaerobic tank presented a small part of sludge in the oxic tank and had little effect on the living environment of microbe. When IR was 15%, the backflow sludge was too much for the anaerobic tank, causing excessive concerntration of sludge and microbial massive die-off. Sludge reduction rate increased at IR of 10% and 8% probably because the backflow volume was just appropriate for creating a suitable environment for slow-growth microbe, which was very likely to playing a big role of sludge reduction.

Microbe Community and Abundance. To investigate the mechanism of sludge reduction of MOSA process from the respective of slow-growth microbe, activated sludge in the oxic tank of the two systems and the anaerobic tank of A^2/O -MOSA system were sampled to examined by high-throughput sequencing on Illumina Miseq platform. At the level of classification fourteen categories were identified, accounted for 99.11% in A^2/O, 99.01% in A^2/O -MOSA and 94.36% in MOSA-A. Therefore, the sequencing results were representative. The relative abundance of species in the three tanks at the phylum classification level is shown as figure 5.

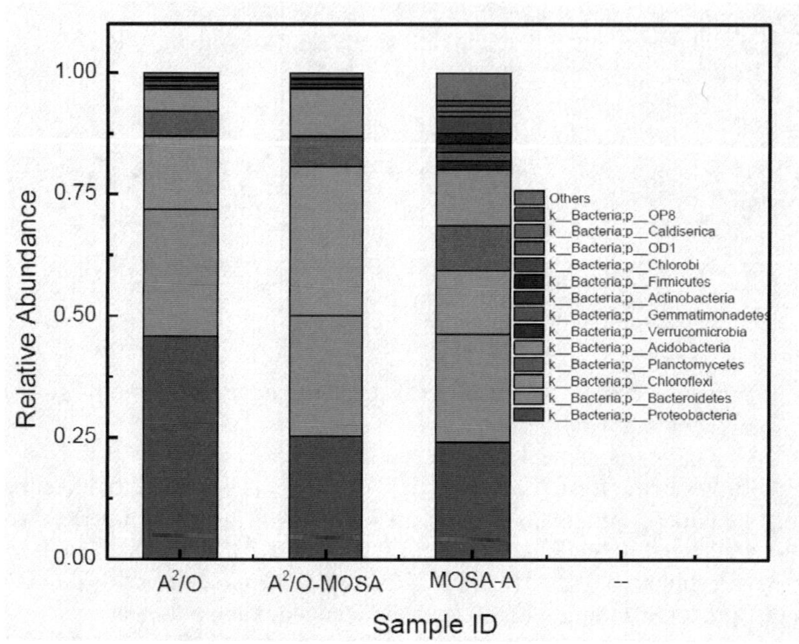

Figure 5. The Relative Abundance of Species in Phylum Classification Level. Samples from the oxic tank of A^2/O system, the oxic tank and the anaerobic tank of MOSA system were marked as A^2/O, A^2/O-MOSA and MOSA-A respectively.

Compared the microbe in the oxic tanks of A^2/O and A^2/O -MOSA process, it could tell that there was little difference in the species but the relative abundance. It signified that MOSA process screened the microorganisms to raise the abundance of that were adaptable to the oxic-anaerobic alternant. Proteobacteria, Bacteroidetes and Chloroflexi were dominant in all samples, accounting for more than 60%. Inserting an anaerobic tank caused the decrease of Proteobacteria and the significant increase of Chloroflexi compared to the reference (Ning X, et al, 2014). Chloroflexi growed by using the soluble products of microorganisms as substrate in anoxic and anaerobic zones, which indicated there were more soluble products in the anaerobic tank caused by sludge recession (Reed A J, et al, 2006; Xie Z, et al, 2014). Acidobacteria increased in MOSA process accounting for 11.5% in the anaerobic tank and 9.8% in the oxic tank, more than twice as much as that in the oxic tank in A^2/O system. It was generally considered to be phosphorus removal related and characteristic with slow growth (Jeon C O, et al, 2003).

There was also difference to some degrees between the microbial diversity and abundance in the anaerobic tank and the oxic tank of A^2/O -MOSA process. The percentage of Planctomycetes, Verrucomicrobia, Gemmatimonadetes, Actinobacteria, Frimicutesand Chlorobi popped up accounting for 20.6% totally. Planctomycetes and Chlorobi were reported to be related to anammox (Rui Z., 2014). Verrucomicrobia was a newly classified mesophilic hydrocarbon-degrading bacterium which performed nitrozation in anaerobic zone (Freitag T E, et al, 2003). Caldiserica, OD1 bacteria and OP8 bacteria were emerging in the MOSA process with low abundance. The newly-presented Caldiserica and OP8 bacteria, which were proved to exist with sufficient simple organics, confirmed that there was sludge recession happened in the anaerobic tank (Tao L, et al, 2003).

CONCLUSIONS

In this research, two sets of A^2/O and A^2/O-MOSA process were built up in the laboratory to investigate their pollutant removal effect, sludge reduction effect in different Interchange Ratio (IR) situation. To explore the microbial mechanism of sludge reducing, high-throughput sequence was used to investigate the microbial abundance and diversity in the MOSA process. There are three conclusions can be obtained:

(1) The A^2/O-MOSA process has high pollutant removal efficiency and the inserting of anaerobic tank did not affect the sludge characteristic. When the IR=10%, the A^2/O-MOSA process reached the maximum sludge reduction of 24.11%.

(2) The inserting of MOSA anaerobic tank affected the microbial species abundance of activated sludge while the primary microorganisms were the same. The microorganisms which increased in the aeration tank were facultative anaerobic bacteria or anaerobic bacteria, mostly related to anaerobic ammonia oxidation and phosphorus removal, which had slow growth characteristics. It indicated that the MOSA anaerobic tank played a role of biological selector to retain the bacteria group with function of denitrifying and phosphorus removal and slow growth rate, which may be the key of source sludge reduction mechanism in MOSA process.

(3) Further study on the the emerging species in the anaerobic tank may be a new direction to revealing the sludge reduction mechanism of MOSA process. Metagenomics and macro transcriptome can be used to investigate the microbial characteristics of the anaerobic tank at a deeper level.

ACKNOWLEDGEMENTS

This reaerch was supported by Strategic Emerging Industry Core Research Project in Guangdong province, China.

REFERENCES

Chen G H, Mo H K, Saby S, Yip W K, Liu Y. 2000. “Minimization of activated sludge production by chemical stimulated energy spilling.” *Water Science & Technology*. 42(12): 189-200.

Chen G H, Hau-Kwok M, Yu L. 2002. “Utilization of a metabolic uncoupler, 3,3′,4′,5-tetrachlorosalicylanilide (TCS) to reduce sludge growth in activated sludge culture.” *Water Research*. 36(8): 2077-2083.

Chen G H, Kyoung-Jin An, Saby S, Brois E, Djafer M. 2003. “Possible cause of excess sludge reduction in an oxic-settling-anaerobic activated sludge process (OSA process).” *Water Research*. 37(2003): 3855-3866.

Chen L L. 2015. *“Research of sludge reduction in a modified oxic-settling-anaerobic (MOSA) process based on A^2/O.”* M.S.Thesis, Sun Yat-sen University, Guangdong Province, China.

Chudoba P, Chudoba J, Capdeville B. 1992. “The aspect of energetic uncoupling of microbial growth in the activated sludge process: OSA system.” *Water Science & Technology*. 26:2477-2480.

Cui Y H. 2013. *“Research on the effect of sludge reduction in OSA process with different anaerobic hydraulic retention time (HRT).”* M.S.Thesis, Sun Yat-sen University, Guangdong Province, China.

Easwaran S P.2006. “Developing a Mechanistic Understanding and Optimization of the Cannibal Process: Phase II.”

Feng Q, Yu A, Chu L, Chen H, Xing X H. 2012. “Mechanistic study of on-site sludge reduction in a baffled bioreactor consisting of three series of alternating aerobic and anaerobic compartments.” *Biochemical Engineering Journal*. 67(1): 45-51.

Fernández D J, Ribas AJ, Freixó R A, Sánchez FA. 2003. “Characterisation of phosphorous forms in wastewater treatment plants”. *Journal of Hazardous Materials*. 97(1–3): 193-205.

Freitag T E, Prosser J I. 2003. Community structure of ammonia-oxidizing bacteria within anoxic marine sediments. *Applied & Environmental Microbiology*. 69(3): 1359-1371.

Goel R K, Noguera D R. 2006. “Evaluation of sludge yield and phosphorus removal in a cannibal solids reduction process.” *Journal of environmental engineering*. 132(10): 1331-1337.

He M H, Wei C H. 2010. “Performance of membrane bioreactor (MBR) system with sludge Fenton oxidation process for minimization of excess sludge production.” *Journal of Hazardous Materials*. 176(1–3): 597-601

Jeon C O, Lee D S, Park JM. 2003. “Microbial communities in activated sludge performing enhanced biological phosphorus removal in a sequencing batch reactor.” *Water Research*. 37(9): 2195-2205.

Jia L, Guo J S, Gao X, Fan Y, Zhai X M. 2012. “Microecology of an anoxic + oxic-settling-anaerobic sludge reduction process.” *Chinese Journal of Environmental Engineering*. 6(06): 2049-2054.

Jin W B, Wang J F, Zhao Q L, Lin JK. 2008. “Performance and mechanism of excess sludge reduction in an OSA (oxic-settling-anaerobic) process.” *Environmental Science*. 39(03): 726-732.

Lee J W, Cha H Y, Park K Y, Song K G, Ahn K H. 2005. "Operational strategies for an activated sludge process in conjunction with ozone oxidation for zero excess sludge production during winter season." *Water Research.* 39(7): 1199-1204.

Li H C, Yan H, Xiao B Y. 2015. "Research progress in sludge reduction technology by chemical uncoupler." *Industrial Water Treatment.* 35(01): 9-13.

Li X, Peng Y, He Y, Jia F, Wang S, Guo S. 2016. "Applying low frequency ultrasound on different biological nitrogen activated sludge types: An analysis of particle size reduction, soluble chemical oxygen demand (SCOD) and ammonia release." *International Biodeterioration & Biodegradation.* 112: 42-50.

Liu Y, Chen G H, Paul E. 1998. "Effect of the S_0/X_0 ratio on energy uncoupling in substrate-sufficient batch culture of activated sludge." *Water Research.* 32(10): 2883-2888.

Liu Y H, Xie Q L. 2016. "Research progress in metabolic uncoupler in activated sludge reduction technologies." *Water Purification Technology.* (S1): 70-74.

Liu W Y, Peng H, Chen Y H, Xiao H. 2016. "Effect of direct addition of chlorine dioxide into a SBR on its sludge reduction and pollutant removal performance." *Chinese Journal of Environmental Engineering.* 31(06): 506-510.

Lu Z L, Pang Z H, Wang H Y, Zhou Y X, Liu D S, Zhang N. 2011. "Research progresses in sludge reduction technologies." *Environmental Protection of Chemical Industry.* 31(06): 506-510.

Ning X, Qiao W, Zhang L, Gao X. 2014. "Microbial community in anoxic-oxic-settling-anaerobic sludge reduction process revealed by 454 pyrosequencing analysis." *Canadian Journal of Microbiology.* 60(12): 799-809.33.

Novak J T, Sadler M E, Murthy S N. 2003. "Mechanisms of floc destruction during anaerobic and aerobic digestion and the effect on conditioning and dewatering of biosolids." *Water Research.* 37(13): 3136-3144.

Novak J T, Chon D H, Curtis B A, Doyle M. 2007. "Biological Solids Reduction Using the Cannibal Process." *Water Environment Research.* 79(12): 2380-2386.

Pan Y P. 2005. *"Sludge reduction by ozonation and carbon source reuse."* M.S.Thesis, Harbin Institute of Technology, Heilongjiang Province, China.

Park C, Abuorf M M, Novak J T. 2006. "The digestibility of waste activated sludges." *Water Environment Research A Research Publication of the Water Environment Federation.* 78(1): 59-68.

Qian Y. 2000. "Appropriate process and technology for wastewater treatment and reclamation in China." *Water Science Technology.* 42(12):107-14.

Reed A J, Lutz R A, Vetriani C. 2006. "Vertical distribution and diversity of bacteria and archaea in sulfide and methane-rich cold seep sediments located at the base of the Florida Escarpment." *Extremophiles.* 10(3): 199-211.

Rui Z. 2014. "Microbial structure of the anaerobic denitrifying system." *Biotechnology.* 24(06): 91-94.

Saby S, Djafer M, Chen G H. 2003. "Effect of low ORP in anaerobic sludge zone on excess sludge production in oxic-settling--anaerobic activated sludge process." *Water Research.* 37(1):11–20.

Takdastan A, Mehrdadi N, Azimi A A, Torabian A, Bidhendi G N. 2009. "Investigation of intermittent chlorination system in biological excess sludge reduction by sequencing batch reactors." *Iranian journal of Environmental Health Science & Engineering.* 6(1): 53-60.

Tan J X. 2014. *"A pilot-scale study of sludge reduction in a modified oxic-settling-anaerobic (MOSA) process based on A/O."* M.S.Thesis, Sun Yat-sen University, Guangdong Province, China.

Tao L, Peng W, Pin X W. 2008. "Bacterial and archaeal diversity in surface sediment from the south slope of the South China Sea." *Acta Microbiologica Sinica.* 48(3): 323~329.

Tokumura M, Katoh H, Katoh T, Znad H T, Kawase Y. 2009. "Solubilization of excess sludge in activated sludge process using the solar photo-Fenton reaction." *Journal of Hazardous Materials.* 162(2-3): 1390-1396.

Wang L, Ma J, Liu T Z, Li C M, Zhang H Y. 2011. "Efficacy of ferrate oxidation and hydrolyze remnant activated sludge." *Environmental Science.* 32(07): 2019-2022.

Wang L, Qiang Z M, Dong H Y, Lv Y F, Nie Y F. 2014. "Sludge reduction by ozone in municipal wastewater treatment plants." *Acta Scientiae Circumstantiae.* 34(02): 363-369.

Wang T J. 2012. *"Research on the pilot scale test of MOSA process based on an UNITANK system for sludge reduction."* M.S.Thesis, Sun Yat-sen University, Guangdong Province, China.

Wei Y, Houten RTV, Borger AR, Eikelboom DH, Fan Y. 2003. "Minimization of excess sludge production for biological wastewater treatment." *Water Research.* 37(18): 4453-4467.

Westgarth W C, Sulzer F T, OKUN, D.A. 1964. "Anaerobiosis in the activated sludge process." *Proceedings of the Second IAWPRC Conference.* Tokyo, Japan.

Xie M, Huang C, Liu X B, Li M Y, Ou S H. 2014. "Effects of acid/alkali lysis on the characteristics and reduction efficacy of sludge in continuous running systems." *Industrial Water Treatment.* (05): 25-27+32.

Xie Z, Wang Z, Wang Q, Zhu C, Wu Z. 2014. "An anaerobic dynamic membrane bioreactor (AnDMBR) for landfill leachate treatment: Performance and microbial community identification." *Bioresource Technology.* 161(3): 29-39.

Yang Y. 2010. *"Research on the performance and mechanics of sludge reduction in SBR-OSA and SBR-OOSA process."* M.S.Thesis, Sun Yat-sen University, Guangdong Province, China.

Yu L, Tay J H. 2001. "Strategy for minimization of excess sludge production from the activated sludge process." *Biotechnology Advances.* 19(2): 97-107.

Yu X Y. 2012. *"Research on changes of bacterial community diversity in an oxic-settling-anaerobic (OSA) activated sludge process with sludge reduction effect".* M.S.Thesis, Sun Yat-sen University, Guangdong Province, China.

Yuan G H, Zhou X Q, Wu J D. 2012. "Enhancement of anaerobic digestion f excess sludge by acid-alkali pretreatment." *Environmental Science.* 33(06): 1918-1922.

Zhang B, Ji FY, Guan C, Gao XP, Ding CJ. 2006. "Research on municipal wastewater sludge reduction in OSA process and its influencing factors." *Journal of Chongqing Jianzhu University.* 28(01): 92-95.

Zhang H. 2008. *"Study on Performance of Sludge Reduction by Micro-Metazoa."* M.S.Thesis, Faculty of Urban Construction and Environmental Engineering, Chongqing University, China.

Zhao W L, Hao R X, Wang R Z, Du P. 2015. "Denitrification of composite carbon filler and character of microbial community." *China Environmental Science.* 35(10): 3003-3009

Zhou Z, Qiao W, Xing C, et al. 2015. "Microbial community structure of anoxic–oxic-settling-anaerobic sludge reduction process revealed by 454-pyrosequencing." *Chemical Engineering Journal.* 266(12): 249-257.

Zhu H, Wei Y S, Wang Y W, Liu J X. 2008. "Advances in Sludge Reduction Induced by Oligochaetes and Theirs GrowthKinetics." *The Chinese Journal of Process Engineering.* 8:1031-1040.

Impact of Reported Flocculative and Deflocculative Conditions on Laboratory Cultivated Sludge Flocs

Akshaykumar Suresh, Ewa Grygolowicz-Pawlak, Leong Soon Poh and Wun Jern Ng
(Nanyang Technological University, Singapore)

Proper sludge flocculation is vital in promoting effective liquids-solid separation in biological wastewater treatment processes. Many studies have been performed to establish the conditions promoting flocculation and deflocculation (ie. disruption and break-up of sludge flocs). Earlier studies performed in our laboratory had indicated extended low dissolved oxygen (DO) and anaerobic conditions did not promote deflocculation as was reported in the literature. Reflocculation was then observed in physically deflocculated sludge samples exposed to aerobic (high DO) conditions. In this study, we assessed if such reflocculation was specifically brought about by exposure to aerobic conditions and promotion of aerobic microbial activity. The results indicated reflocculation had occurred even if the deflocculated samples were exposed to extended anaerobic conditions. Supernatant turbidity and SMP levels decreased over the experimental period (20 hours) regardless of aerobic or anaerobic conditions. The anaerobic conditions also did not promote more deflocculation over the experimental period and this was consistent with our previous findings. Microbial activity was indicated in terms of ammonia removal. Differences in microbial activity, between anaerobically and aerobically conditioned samples, were observed in terms of the fluorescent SMP content using the 3D-excitation emission matrix. The differing microbial activities did not appear to have significant influence on the overall flocculation results (neither in terms of supernatant turbidity nor the SMP levels). Supernatant Mg levels were observed to be decreased after treatment (anaerobic/aerobic) of the deflocculated samples, which could be indicative of divalent cation bridging leading to the reflocculation event in those samples in terms of reduced supernatant turbidity and SMP content. Thus, sludge flocculation and reflocculation is influenced by more than just increased aerobic microbial activity but can also include factors like changes in multivalent cations and EPS interactions.

The Efficiency of the Multi-Plate Screw Press in the Sludge Dewatering Process, Is It the Best Solution?

Anne Wambui Mumbi and Li Fengting
(Tongji University Shanghai China)

Sludge dewatering is a crucial step in sludge treatment because it determines the amount of and quality of sludge produced. There are many mechanical methods of dewatering sludge. The paper aims at analyzing three major methods based on principle of operation and possible optimization techniques then compares it to multi-plate screw press which is being adapted as a model that aims at solving problems that arise from usage of the fore mentioned equipment while clearly offering energy efficient solutions. Ultimately the question of the best option can be answered after carefully analyzing available literature on other used equipment. The development of Multi-plate Screw Press which is adaptable to deal with sludge from a variety of industries such as petro chemistry, petrochemical chemical fiber, paper-making, pharmaceuticals, leather, food and beverage etc. It is easy to operate and more power efficient. It utilizes the principles of dewatering force water homo-direction and thin-layer dewatering. The machine allows automatic continuous operation of sludge flocculation thickening dewatering and filtrate discharging. It is equipped with moving rings that are self-cleaning it utilizes very little power as it relies on a low inner pressure of the filter chamber. Moreover the problem of noise pollution is eradicated as well as space as this technology can be connected directly to the aeration tank or the secondary sedimentation tank. While currently major technological advancements are occurring today, the treatment of water is a well-known process and is executed by state of the art techniques. The sludge resulting from this process represents the next challenge for the water treatment industry, in particular the minimizing of its volume. With the Multi plate screw press it is the best option for dewatering sludge and treating sludge as compared to the other equipment as explained and elaborated in the paper.

Dissolved Methane Recovery from Novel Anaerobic Attached Growth Reactor Treating Synthetic Domestic Wastewater

Brian Crone* and George Sorial (University of Cincinnati, Cincinnati, OH, USA)
Jay Garland (United States Environmental Protection Agency, Cincinnati, OH, USA)

Anaerobic Membrane Bioreactors (AnMBRs) are emerging as a sustainable option for wastewater treatment due to their production of methane through biological metabolism of Domestic Wastewater. The recovered methane can be combusted to generate electricity and offset Wastewater Plant Operational Costs. The largest technical barrier to the widespread application AnMBRs is dissolved methane losses in effluent streams which contribute to greenhouse gas emissions and result in significant reductions in energy recovery. This is exacerbated by low temperature operation, which increases methane solubility. Several authors have previously suggested in-situ degassing of the bulk reactor liquid as on option to improve COD removal, methane recovery, and increase alkalinity through the removal of carbonic acid. However, all of these improvements are reduced over time by membrane fouling. To counteract this, a novel approach was utilized which included culturing a biofilm stratified by carbon metabolism, i.e. hydrolosis, acidogenesis, acetogenesis, and methanogenesis from the outside of the biofilm to the surface of an in-situ degassing membrane in an AnMBR treating synthetic domestic wastewater. Effectiveness of this approach was evaluated by comparison of COD removal, methane yield, and stability of pH with a control reactor which contained membranes for structural support of the biomass but no active degassing. Stratification of biofilms was evaluated visually using Fluorescent In-Situ Hybridization on cross sections of membrane biofilms.

Isolation and Optimization of Native Microorganisms for the Aerobic Treatment of Oil and Grease Wastewaters

Zhiwei Liang, Guofang Liu, Ancheng Luo
(Zhejiang University, Hangzhou, China; zhiweiliangzg@zju.edu.cn)

The isolation of microorganisms from natural sources and optimization of inoculums to increase of the O&G removal efficiency was investigated by using synthetic O&G wastewater. A second-order polynomial model was to analysis the role of each variable and their second-order interactions. The results demonstrated that six strains were responsible for most of the O&G degradation, and the O&G removal efficiency of the selected strains (70.34%) was much higher than that of stains mixture before optimization (41.56%) after 72h degradation. The 16S rRNA gene sequence analysis of the six isolated strains belonging to *Burkholderia cepacia* strain 2Pe22, *Penicillium expansum* strain NRRL 2304, *Micrococcus*, *Bacillus licheniformis* strain DQgbc4, *Bacillus amyloliquefaciens* strain F6 and *Yarrowia lipolytica*, respectively. Biodegradation of the selected strains on synthetic and real O&G wastewater treatment suggests that optimization of inoculums is successful and FFD is an effective way to optimization of removal efficiency with mixed inoculums.

Keywords: biodegradation, fractional factorial design, consortium, oil and grease wastewater

Comparative Analysis of Persistent Organics Removal Based On the Aeration Source Supplied To Biological Treatment of Textile Industry Wastewater

Ninad Oke and S. Mohan
(Indian Institute of Technology Madras, Chennai, Tamil Nadu, India)

ABSTRACT: This paper presents a study on the removal of persistent organics from the textile industry effluents. The higher quantities of water and wider spectrum of dyes and auxiliary chemicals used impart a complex nature to combined effluent from textile manufacturing units. Standards for color, organics and dissolved solids are becoming stringent with time and regulators are in demand of zero liquid discharge units. The study suggests innovative solution to said issues by upgradation of the existing secondary treatment systems without need of additional area footprint. Present study investigates the use of pure oxygen to supply aeration instead of convention air diffusion system. The proposed aeration system is aimed at reducing required operation time to achieve the target organic removal percentage. A COD removal of 68 % and 73 % was observed for the treatment units supplied by conventional air and pure oxygen respectively. The present focus of the study on secondary treatment improvement can also contribute to reduced chemical use in primary treatment and reduced energy use in following tertiary treatment operations. The study attempts at reducing cost of recycled water for textile industry units and provide economically viable treatment options for these major polluting industry units.

INTRODUCTION

In the developing country like India, Textile Industry has a significant contribution towards economic development and growth of the country. Nearly 80% of the total dye stuff produced in India is consumed by the Textile Industries due to high and continuously growing demand (Holkar et al., 2016). Typically, 20 to 40% of initial dye load is present in the textile effluent because of hydrolysis of reactive dyes (Dasgupta et al., 2015). It is important to recycle the process water in textile industry and reduce the overall water consumption. The recycling of process water can further reduce wastewater generation leading to minimum impact of textile industry on the natural water bodies. Textile industries consume large volume of water and chemicals for wet processing of textiles. The water consumption in textile manufacturing is reported to be 200 L per kg for Indian scenario (Chavan 2001, Holkar et al., 2016). The integrated textile industry is engaged in production of yarn, fabric and finished goods from raw fibres. The wastewater obtained from the textile industry is usually rich in colour, chemical oxygen demand (COD), complex chromophores, inorganic salts, total dissolved solids (TDS) and heavy metals which underlines the importance of effluent treatment in textile industry.

The textile production stages can be briefly stated as yarn formation, fabric formation, wet processing and textile fabrication (Babu et al., 2007). Processing operations like desizing, scouring, bleaching, mercerizing, dyeing, printing is included in the wet fabric processing. The sizing operation is done to increase tensile strength and smoothness of the fibres for weaving. Different sizing materials used are starch, polyvinyl alcohol, carboxy methyl cellulose. In desizing step the sizing material is removed by sulphuric acid or enzyme action. Generally, 50 % of the water pollution is due to desizing effluent having high BOD load (Babu et al., 2007). The scouring and kiering operations are aimed at grease, fats, waxes removal. It involves boiling of cloth to remove waxes, natural fats and the effluent is highly alkaline leading to the alkaline nature of combined textile effluent. The natural color of fiber is removed, and brightness is increased by bleaching operation. In the dyeing stage, dye solution and fiber are mixed under highly saline conditions to enhance dye uptake by the fiber. Chromophore groups such as azo, carbonyl, nitro, quinoid groups and auxochrome groups like amine, carboxyl, sulphonate are responsible for the color. Salts like NaCl and Na_2SO_4 are widely used at the dyeing stage. Following Figure 1 gives various textile manufacturing stages and associated waste generation.

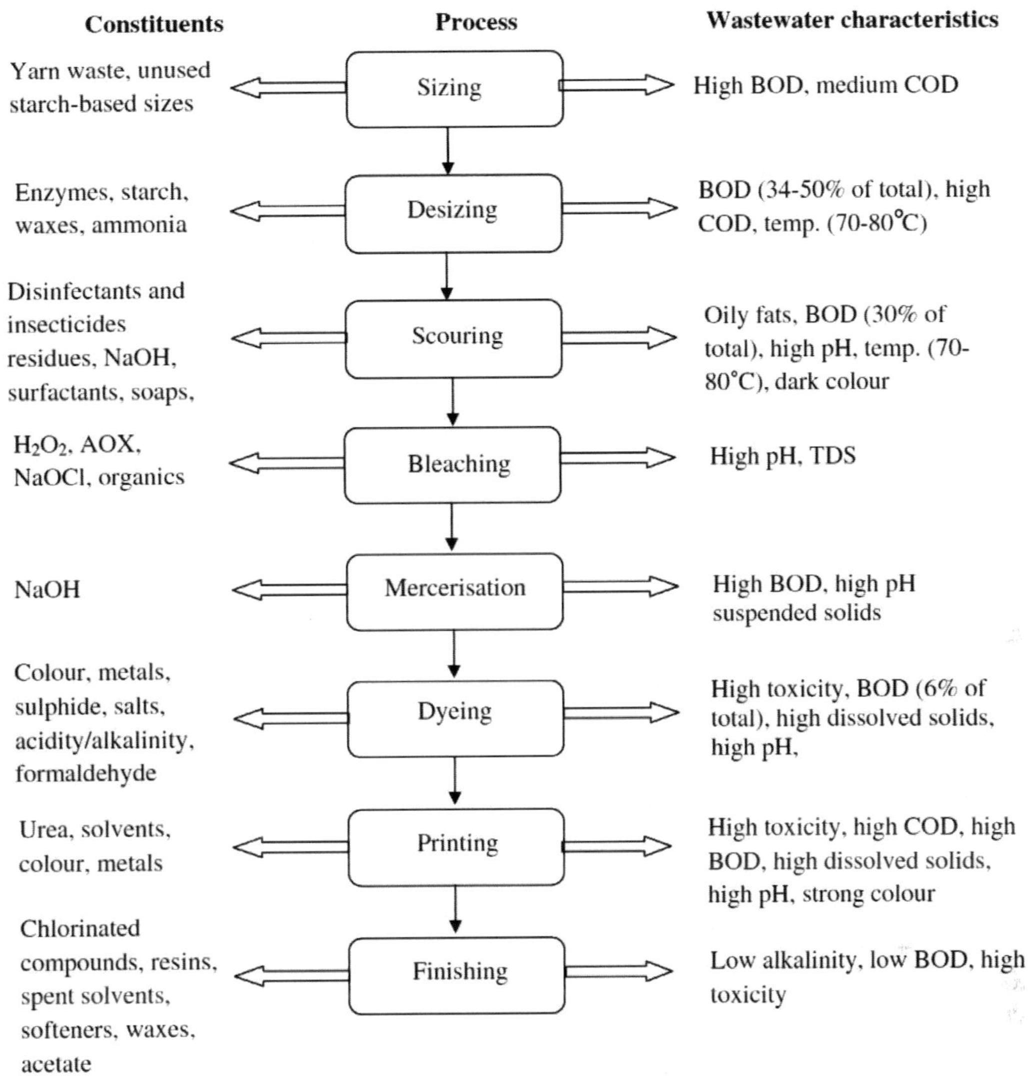

FIGURE 1. Flowchart of textile fabrication (Source: Ashwin et al., 2011)

The industrial effluent discharge standards in India are applicable for textile manufacturing units with effluent generation less than 25 KLD. The discharge standard for chemical oxygen demand (COD), total dissolved solids are 250 mg/L and 2100mg/L respectively. As per the recent amendments in 2016, all units having wastewater discharge more than 25 KLD must establish a zero liquid discharge effluent treatment plant. As per the recent amendments, all the common effluent treatment plants for textile industry cluster units should be upgraded to zero liquid discharge-common effluent treatment plants. It is clearly stated in the rules that member units should use recycled water supplied by the ZLD- CETP as a process water and limit ground water abstraction only for makeup water and drinking purpose. This underlines the need of improvement in efficiency of existing common effluent treatment plants to achieve zero liquid discharge status.

MATERIALS AND METHODS

Effluent. The combined textile industry effluent was collected from equalization tank of the effluent treatment plant within a zero liquid discharge facility situated at Erode district, Tamil Nadu, India. The collected effluent was well characterized using standard analytical methods. The collected samples were stored at 4 °C before characterization. Based on the characterization a stimulated synthetic textile effluent was prepared in lab using the components listed in Table 1. Red, yellow and blue dye powders were collected as used at industry level from the textile industry in Erode District, Tamil Nadu, India. They were described as reactive azo dyes with vinyl sulphonyl group by the plant operators.

TABLE 1. Simulated Textile Effluent Contents

Component	Concentration
Red dye powder	50 mg/L
Yellow dye powder	50 mg/L
Blue dye powder	50 mg/L
Starch	2000 mg/L
Glucose	1000 mg/L
NaCl	1200 mg/L
$NaHCO_3$	1500 mg/L

Biological Process Runs. A lab scale reactor model was assembled with a working volume of 10 L. The plan and front section view of the designed model are shown in Figure 2 (A) and Figure 2 (B) respectively. A submerged diffused aeration was provided from bottom of the reactor, above the level of sludge settling zone. Two such assembled models were used, one using the conventional air supply for aeration and other using pure oxygen as the aeration source. Each treatment run cycle consisted of 0.5 hr feeding, 6 hr reaction time, 2 hr settling time followed by 0.5 hr discharge. The gas flow rate for both air and pure oxygen was kept as 3.2 L/min during reaction time. The samples of 10 ml volume were collected periodically at 1 hr interval and analysed for organic content in terms of Chemical Oxygen Demand (COD). The sludge collected from municipal wastewater treatment plant inside university campus was used after an acclimatization period of 72 hours to initiate the biological treatment process run.

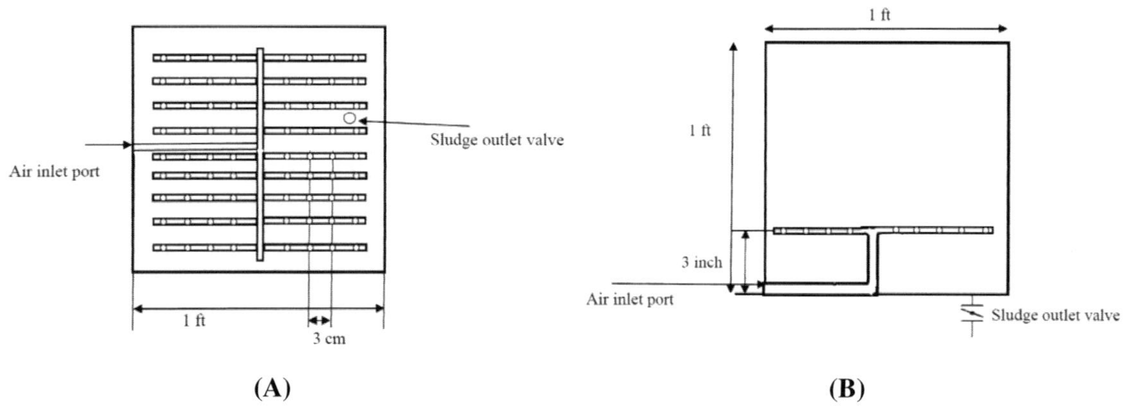

FIGURE 2. **(A)** Plan view of reactor, **(B)** Front section view of reactor

Analytical Methods. pH of the liquid samples was measured with a glass electrode attached to Microprocessor pH Stat/Analyser (Polman, Model: LP-139S). Chemical Oxygen Demand (COD) of the wastewater samples was measured according to the prescribed closed reflux dichromate titrimetric method (American Public Health Association (APHA) 2012) in which the wastewater samples were digested in a block digestor, preheated to 150 °C temperature before refluxing for 2 h. 5-day biochemical oxygen demand (BOD_5) of the wastewater was determined by modified Winkler's method as described in APHA (2012). Dissolved oxygen was measured using a Hach HQ30d meter. Total dissolved solids (TDS) and total suspended solids (TSS) were determined by the respective standard methods outlined in APHA (2012). Absorbance of wastewater samples was measured using a UV Visible spectrometer (UV-Vis Shimadzu 260, Japan). TOC of wastewater samples was measured on a TOC analyzer (TOC- VCSH, Shimadzu, Kyoto, Japan). Fourier transform infrared spectroscopy (FTIR) analysis of samples was performed on 3000 Hyperion microscope with vertex 80 FTIR system (Bruker, Germany).

RESULTS AND DISCUSSION

The results of characterization of combined textile industry effluent and the stimulated synthetic textile effluent generated in lab are shown in Table 2. The stimulated synthetic textile effluent was used as the wastewater for treatment runs.

Table 2. Characteristics of combined textile industry effluent

Parameter	Combined Effluent	Stimulated Effluent
pH	8.79	8.24
EC (µS/cm)	8550 ± 150	7990 ± 200
COD (mg/L)	2809 ± 180	2754 ± 40
BOD_5 (mg/L)	620 ± 82	670 ± 128
TOC# (mg/L)	1103 ± 95	1026 ± 87
TIC* (mg/L)	397 ± 40	328 ± 80
TSS (mg/L)	3672 ± 120	3124 ± 90
TDS (mg/L)	5728 ± 100	5244 ± 180
Absorbance at 436 nm	2.02	2.54
Absorbance at 525 nm	1.36	1.82
Absorbance at 620 nm	1.06	1.76

Total Organic Carbon
* Total Inorganic Carbon

The values reported in above table are based on the triplicate measurements. It can be observed that the stimulated textile effluent has similar characteristics that of the collected combined textile effluent.

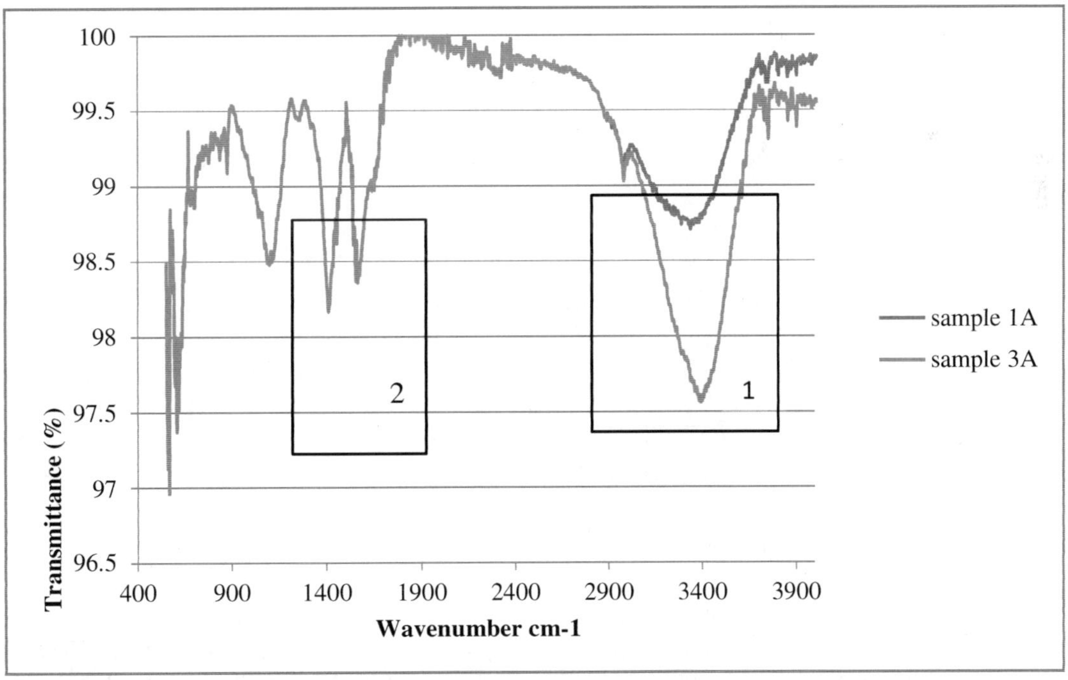

FIGURE 3. FTIR spectrum for collected effluent samples

The existing extent of treatment of persistent organics achieved by the secondary biological treatment at site was analyzed in present research work. Fourier Transform Infrared spectroscopy (FTIR) analysis was performed for combined effluent (sample 1A) and effluent after secondary treatment (sample 3A). The combined plot is shown in the following Figure 3. The samples were oven dried for 24 hrs to remove all the moisture and then analyzed in a powder form

Above spectrums were analyzed by identification of the various peaks using literature available on FTIR spectrums of textile dye solutions. Following Table 3 gives the wavenumber ranges and corresponding possible compounds present in the sample.

Table 3. Wavenumber range and corresponding compounds for FTIR spectrum (Ref. Yuen et al., 2005, Pinkert et al., 2011, Ekambaram et al., 2016)

Wavenumber Range (cm^{-1})	Corresponding compounds
3700 - 3100	Carboxylic acid, aldehyde, amides and amines
3000-2850	Alkane C-H stretching
1600-1500	Azo group (-N=N-)
1500-1400	Secondary amides
1200-1000	Aliphatic amines, Sulfones
800-400	Aromatic rings

From above spectrum and the Table 3 we can observe from region 1 that there is increase in simple organic compounds like aldehyde, carboxylic acid, amide and amines after secondary treatment. First peak in region 2 indicates presence of sulfone group along with some secondary amines. Second peak in region 2 indicates the present of azo group. It can be observed that secondary treatment leads to reduction in the peaks in region 2, but a complete disappearance could not be achieved. A comparable presence of aromatic rings in both the samples underlines the need of improvement in biological treatment stage.

The results of biological treatment runs are shown in following Figure 4. The plot of COD values vs time as presented in the figure shows higher organics removal for the system supplied with pure oxygen as the aeration source.

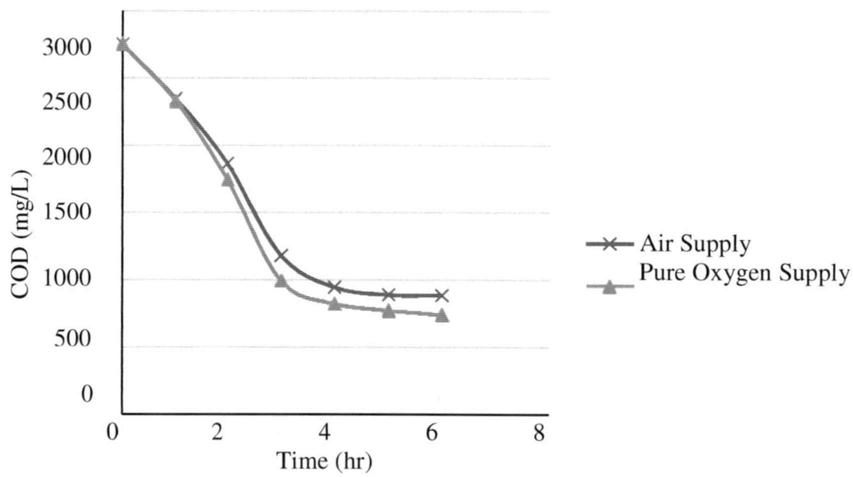

FIGURE 4. Variation of COD with time

It was observed that the biological treatment with conventional air supply for aeration could achieve COD removal of 67.9 % with final COD value of 885 mg/L. The biological treatment unit supplied by pure oxygen as the aeration source could achieve COD removal of 73.1 % with final COD value of 742 mg/L. It can be found that availability of more oxygen has a positive impact over the performance of microbial community towards degradation of complex aromatic compounds. As reported by Carmen and Daniela, 2012, in aerobic conditions, the mono- and dioxygenase enzymes catalyze the incorporation of dissolved oxygen into the aromatic ring of organic compounds prior to ring fission. Observed results suggest that required aeration time can be reduced by using pure oxygen as the aeration source. The reduction in required time for treatment directly leads to reduced energy consumption which can further lead to reduced operation cost for the treatment plant.

CONCLUSION

The study of the combined effluents from textile industry units underline the significance of improvements needed in the wastewater treatment sector for textile industries in developing countries. From the observed results there is significant improvement in the biological treatment stage by use of pure oxygen as the aeration source. The improvements in the existing treatment systems, with no additional area footprint are highly welcomed by the plant operators. Considering the legislative demands of setting up the 'zero liquid discharge' facilities, improvement in secondary treatment can reduce the energy requirement for tertiary treatment step and reject management system.

ACKNOWLEDGEMENTS

Authors would like to thank Pioneer Processing India (A Division of Janson Industries Ltd.), Erode for providing industry grade dyes and textile effluent for present research work. Authors would also like to thank the staff of Department Workshop facility, Department of Civil Engineering, IIT Madras for aiding in reactor model assembly.

REFERENCES

Ashwin, R., Athinthra, K. S., Sriramvignesh, M., Kathiravan, M. N., & Gopinath, K. P. (2011). Treatment of textile effluent using SBR with pre and post treatments: Kinetics, simulation and optimization of process time for shock loads. *Desalination*, 275(1), 203-211.

Babu, B. R., Parande, A. K., Raghu, S., & Kumar, T. P. (2007). Cotton textile processing: waste generation and effluent treatment. *Journal of cotton science.*

Carmen, Z., & Daniela, S. (2012). Textile organic dyes–characteristics, polluting effects and separation/elimination procedures from industrial effluents–a critical overview. In *Organic pollutants ten years after the Stockholm convention-environmental and analytical update*. InTech.

Chavan, R. B. (2001). Indian textile industry-environmental issues. *Indian Journal of Fibre and Textile Research*, *26*(1/2), 11-21.

Dasgupta, J., Sikder, J., Chakraborty, S., Curcio, S., & Drioli, E. (2015). Remediation of textile effluents by membrane based treatment techniques: a state of the art review. *Journal of environmental management*, *147*, 55-72.

Ekambaram, S. P., Perumal, S. S., & Annamalai, U. (2016). Decolorization and biodegradation of remazol reactive dyes by Clostridium species. 3 *Biotech*, 6(1), 20.

Holkar, C. R., Jadhav, A. J., Pinjari, D. V., Mahamuni, N. M., & Pandit, A. B. (2016). A critical review on textile wastewater treatments: possible approaches. *Journal of Environmental Management*, *182*, 351-366.

Pinkert, A., Goeke, D. F., Marsh, K. N., & Pang, S. (2011). Extracting wood lignin without dissolving or degrading cellulose: investigations on the use of food additive-derived ionic liquids. *Green Chemistry*, 13(11), 3124-3136.

Punzi, M., Nilsson, F., Anbalagan, A., Svensson, B. M., Jönsson, K., Mattiasson, B., & Jonstrup, M. (2015). Combined anaerobic–ozonation process for treatment of textile wastewater: Removal of acute toxicity and mutagenicity. *Journal of hazardous materials*, *292*, 52-60.

Yuen, C. W. M., Ku, S. K. A., Choi, P. S. R., Kan, C. W., & Tsang, S. Y. (2005). Determining functional groups of commercially available ink-jet printing reactive dyes using infrared spectroscopy. *Research Journal of Textile and Apparel*, 9(2), 26-38.

Environmental Science & Technology 2018

Experimental Study on Treating and Recycling of Steel Rolling Wastewater by Constructed Wetland Enhanced with Sulfur Autotrophic Denitrification

***Xiaohu Lin*[1]**, Jingcheng Xu*[1,2], Jie Ren[1], Wencong Lin[1], Haihua Cao[1], Wei Liu[1]
(1. College of Environmental Science and Engineering, Tongji University, Shanghai, China
2. State Key Laboratory of Pollution Control and Resource Reuse, Shanghai, China)

ABSTRACT: A series of ordinary constructed horizontal subsurface flow wetland devices (HF_{ord}) and constructed wetland enhanced with sulfur autotrophic denitrification (HF_{SAD}) were built and utilized to treat steel rolling wastewater with low organic carbon, exploring the feasibility of applying sulfur autotrophic denitrification technology to enhance denitrification capacity of constructed wetlands. Results showed that HF_{ord} had a good performance on removing turbidity, DO, COD and TP of steel rolling wastewater but unsatisfying effect on nitrogen pollutants removal. The average removal rate of TN in HF_{ord} was 25.2%, but on this basis, HF_{SAD} could remove 45.0% of TN contaminants subsequently, contributing to an average TN removal rate of 58.0%, with about 5 times the denitrification capacity equivalent to HF_{ord}. The analysis result of microbial community structure and observation results of fillers and monitored sulfate both provided proofs for the occurrence of sulfur autotrophic denitrification. Therefore, constructed wetlands combined with sulfur autotrophic denitrification can be applied in the treatment of steel rolling wastewater with low C/N ratio, with better reuse feasibility in processes of steel and iron production, reducing emitted pollutants and water consumption, which can provide a new technical route for ecological wastewater treatment and water recycling in iron and steel industry.

INTRODUCTION

With the improvement of wastewater effluent quality standards and the large demand of iron and steel enterprises for reclaimed water, it is of great significance to treat deeply and reuse steel rolling wastewater. In the field of advanced treatment of steel wastewater, the dual-membrane method, advanced oxidation and biological methods and alternative advanced treatment method have been studied and applied[1, 2],with satisfying performance but great difficulty, high investment and large energy consumption. As a type of ecological wastewater treatment technology, constructed wetlands have been applied in the field of advanced treatment of industrial wastewater, due to the advantages of stable effluent quality, low running cost, convenient maintenance and aesthetic appreciation[1-5]. Constructed wetlands have good removal effect on turbidity, COD, TP, NH_4^+-N and several other pollutants but not satisfying removal effect on NO_3^--N, difficult to live up to nitrogen standard especially when treating wastewater devoid of organic carbon to offer electron[6, 7].

Therefore, some researches have been undertaken to enhance denitrification in constructed wetlands to treat wastewater with low C/N ratio, which mainly include adding external carbon sources with economic feasibility and abundantly availability[8] and incorporating autotrophic denitrification into constructed wetlands[9-11]. For wastewater with insufficient carbon source, when heterotrophic denitrification is applied, acetic acid, formic acid, ethanol or other substances able to provide carbon source are required to be added into the water, which cost high and may cause secondary pollution[12]. However, sulfur autotrophic denitrification, as one kind of autotrophic denitrification technologies which apply inorganic matter rather than external carbon source to provide electron, utilize reductive sulfur such as S^{2-}, S^0, $S_2O_3^{2-}$ S^{2-}, S^0 and $S_2O_3^{2-}$ as energy source and electron donor and use nitrate as electron acceptor to accomplish denitrification[13-16]. Due to the advantages of needing no external organic carbon source, high removal performance in nitrogen pollutants, economic feasibility, sustainability and distinct reduction in nitrous oxide emissions[17-19], sulfur autotrophic denitrification has been paid much attention in recent years. At present, even though related researches on surface water and groundwater are available, there are still few researches on advanced treatment and resource recycling of steel wastewater especially steel rolling wastewater by incorporating sulfur autotrophic denitrification technology into constructed wetlands.

Therefore, steel rolling wastewater was taken as research object and sulfur autotrophic denitrification combined with constructed wetlands was applied in this study, to explore the feasibility in advanced treatment and recycling of steel rolling wastewater, providing technical support and experience in efficient denitrification of steel wastewater. In this study, a series of ordinary constructed wetland devices and constructed wetland enhanced by sulfur autotrophic denitrification were built, using steel rolling wastewater as influent and operating for a period of time, and the effluent quality of the two devices as well as water quality along the way of the second device is measured to explore the mechanism of sulfur autotrophic denitrification combining with microbial analysis.

MATERIALS AND METHODS

Experimental Device and Operation. Experimental device was a combination of ordinary constructed horizontal subsurface flow wetland device (HF_{ord}) and constructed wetland enhanced with sulfur autotrophic denitrification (HF_{SAD}). HF_{ord} simulated ordinary constructed horizontal subsurface flow wetland and HF_{SAD} imitated constructed horizontal subsurface flow wetland enhanced by sulfur autotrophic denitrification.

The experiment was conducted for three months during its normal operation period (not including the acclimation period of the device). During its operation period, the device was run under the following three kinds of conditions with different HRT in turn: (1) HRT_{ord} = 2 d, HRT_{SAD} = 0.5 d; (2) HRT_{ord} = 4 d, HRT_{SAD} = 1 d; (3) HRT_{ord} = 6 d, HRT_{SAD} = 1.5 d. Each condition was run for a month and water samples were collected every 3 days from the inlet of the device, the outlet of HF_{ord} and the outlet of HF_{SAD} effluent to monitor and analyze the indicators of water quality.

Realistic treated steel rolling wastewater was adopted as the influent of the device. Its characteristic and recycling standard was shown in TABLE 1.

TABLE 1. Ordinary Characteristic of Steel Rolling Wastewater

Index	pH	Turbidity	COD_{Cr}	TP	TN
Range	6.0~9.3	6.97~9.82	32.9~42.6	0.19~0.49	7.38~9.90
Average	7.15	8.38	37.9	0.28	9.01
Recycling Standard	7~8	≤5	≤30	≤0.8	≤5

Analysis Methods. Water samples were routinely collected and filtered through disposable Millipore filters (0.22 μm pore size) before either being analyzed immediately or stored in a refrigerator at 4 °C until the analyses were performed. Sample analysis was carried out less than 24 h after sample collection. Samples were analyzed for ammonium, nitrite and nitrate concentrations using standard methods. Dissolved oxygen and pH were monitored using a Hach HQ30D portable multi meter (HACH, USA).

RESULTS AND DISCUSSION

The Treatment Effect of Contaminants. The composite constructed wetland was operated continuously for over three months, in three kinds of operation conditions by turn as mentioned above.

High removal efficiency of turbidity, TP and COD was achieved, with the concentration meeting the reuse standard very well. pH, an important environmental factor which can influence the growth of microorganisms and plants, were stabilized to a normal range in HF_{ord}. The slight decrease of pH in HF_{SAD}

can attribute to the production of H^+ during the reaction of sulfur autotrophic denitrification.

The reaeration of plants in HF_{ord} supplemented oxygen into the substrate, which made DO in HF_{ord} maintain a relatively high level. However, the filter especially like sulfur hindered the downward growth of plant roots in HF_{SAD} so that the roots could just grow in shallow soil areas, which made supplementation of DO difficult in deep areas. DO was quickly consumed in the front part of HF_{SAD}, leading to the facultative anaerobic condition and low DO concentration in effluent. The facultative anaerobic condition was vital for the utilization of sulfur and sulfur autotrophic denitrification.

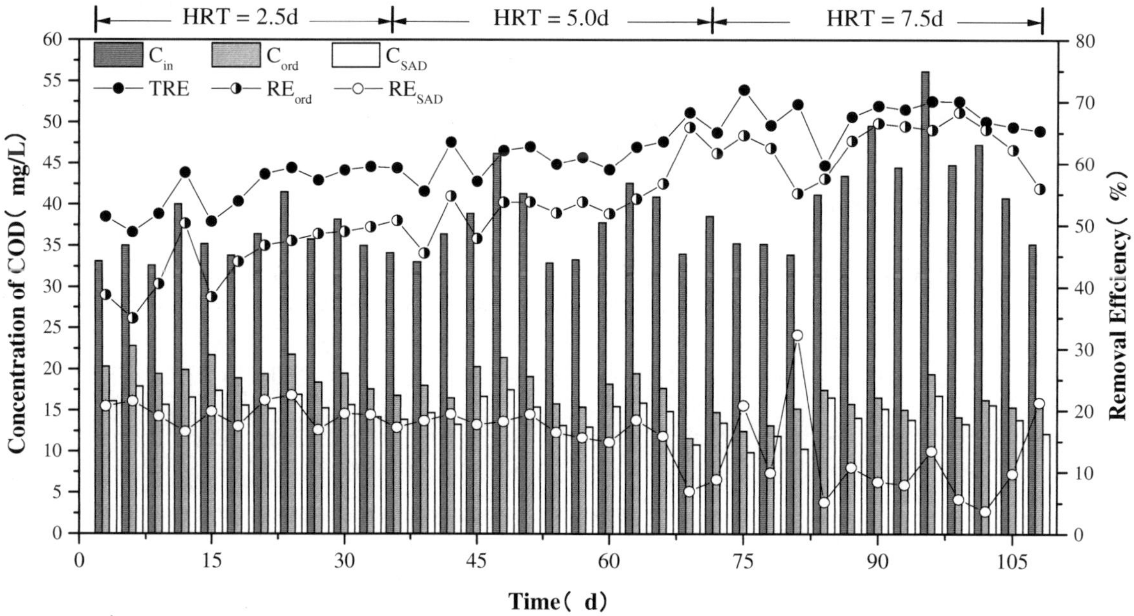

FIGURE 1. The Concentration Variation of COD

Because COD is an important index of wastewater, it is analyzed in detail as is shown in FIGURE 1. COD was removed greatly after treatment in HF_{ord} and the removal efficiency of COD was enhanced with the increase of HRT. When HRT increased from 4 d to 6 d, the removal efficiency of COD was not as good as that of HRT from 2 d to 4 d. What's more, the COD removal rate of HF_{SAD} was lower under three HRT conditions, which may be due to the decrease of COD to lower range (20mg/L or less). The residual COD biodegradability was poor and the water-soluble trace S may also be included in COD. Even though there existed difference in the composition of wastewater, the removal performance of HF_{ord} in this study was similar to many researches such as studies of Xu et al.[20] and Schulz et al[21]. The concentration of COD and some other regular pollutants could live up to reuse standard just after treatment in HF_{ord}, but nitrogen pollutants hindered the compliance to related criteria of wastewater, rendering the need of further efficient denitrification.

Removal of Total Nitrogen. The variation of concentrate of total nitrogen was shown in FIGURE 2. It could be concluded that the effect of TN concentration removal rate in HF_{ord} and HF_{SAD} increased with HRT increasing. The amount of TN removal rate HF_{SAD} was larger than in HF_{ord}. When HRT_{ord} was 6 d, TN concentration of HF_{ord} was able to meet the water quality standard for reuse. Therefore, it was hard to meet the standard by using HF_{ord} only on the condition of short HRT, while longer HRT would reduce the treatment flow of constructed wetlands. Combined with HF_{SAD}, constructed wetlands could perform better on denitrification and accomplish all treatment requirements perfectly.

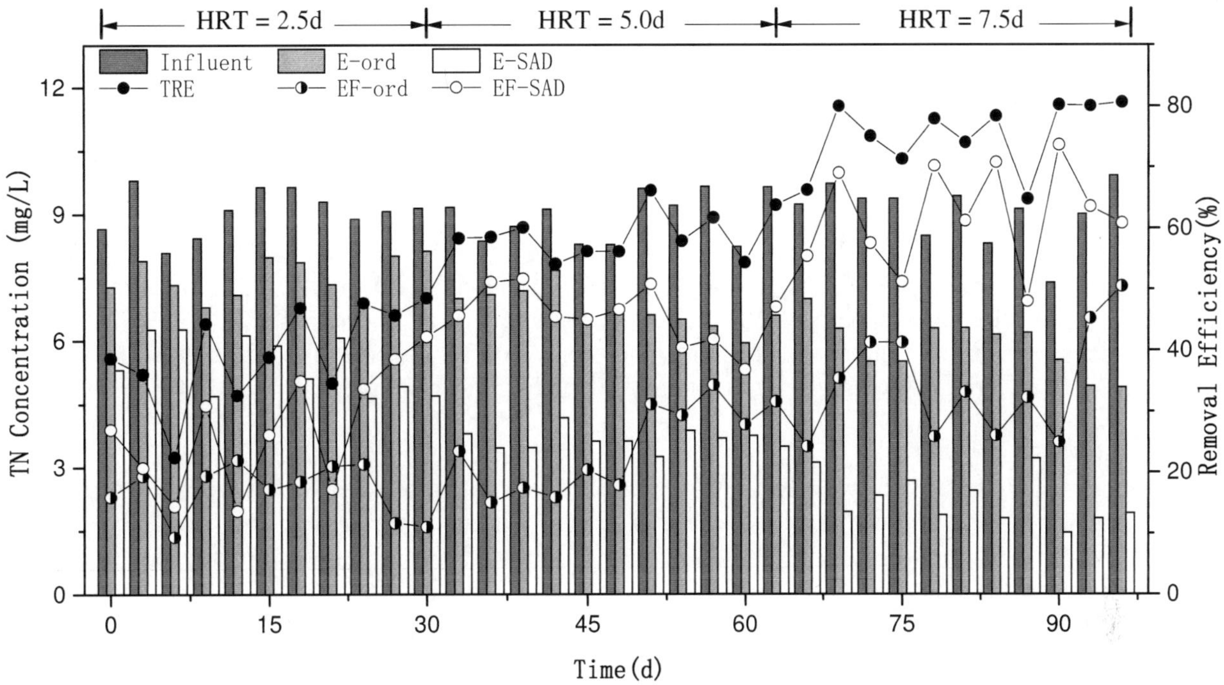

FIGURE 2. The Variation of Concentration of Total Nitrogen (TN)

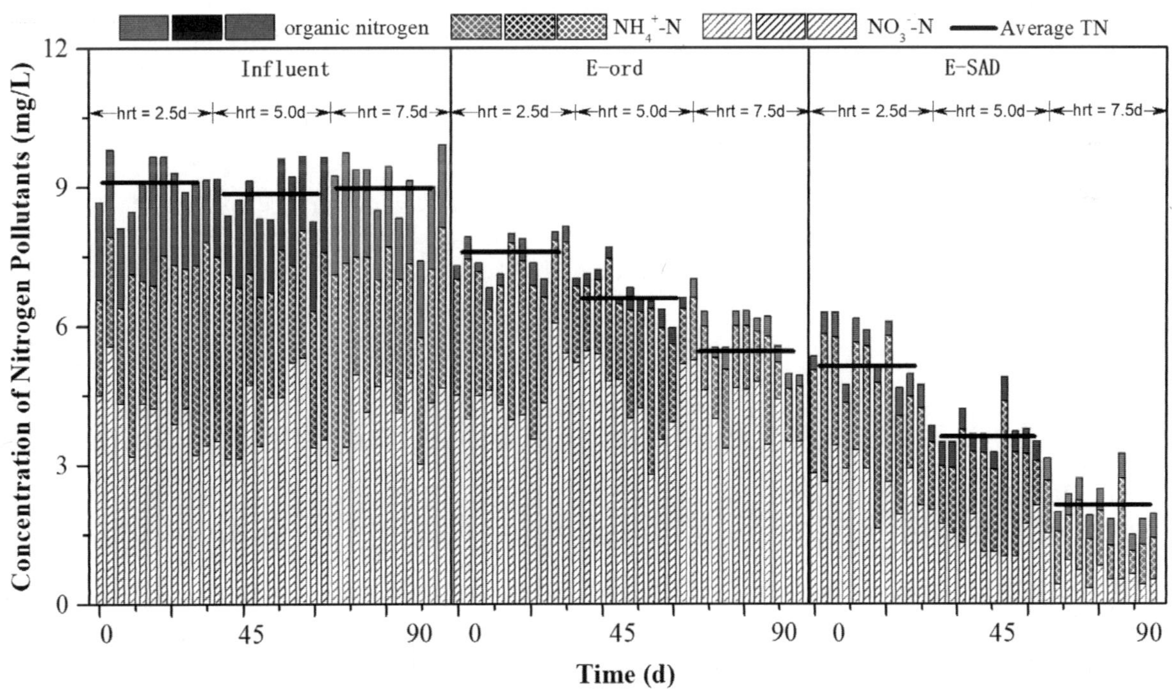

FIGURE 3. The Variation of Different Types of Nitrogen Contaminants

Transformation of Nitrogen Contaminants. Th□r□m□v□l □f nitr□g□n c□nt□min□nts w□s □n im□□rt□nt f□cus □f this study. Th□r□f□r□, NH_4^+-N, NO_3^--N, ON □nd TN c□nc□ntr□ti□n □f d□vic□influ□nt □nd □fflu□nt were monitored. The concentration variation of nitrogen pollutants in different forms is shown in FIGURE 3.

The operational effects of HF_{ord} and HF_{SAD} were analyzed to find out their ability to remove different kinds of nitrogen contaminants (the amount of contaminant removed per unit time and area) in this study. The removal capability of nitrogen contaminants (if it was negative, the absolute value was considered, indicating the degree of increase) decreased with increasing HRT. As HRT increased, the removal amount of nitrogen contaminants increased but its degree was not as high as that of HRT, which contributed to the decreasing trend of ratio. Relatively strong nitrification and weak denitrification caused that the concentration of nitrate in effluent was higher than that in influent in HF_{ord}, which was contrary to that in HF_{SAD}. The removal capacity of nitrate was positive in HF_{ord} and negative in HF_{SAD} but the decreasing degree of their absolute was different. Therefore, from the perspective of the whole device, the removal capacity of nitrate increased with the increasing of HRT.

CONCLUSIONS

Two combined wetlands worked well to treat steel rolling wastewater with low carbon source and high nitrogen removal requirements, with the results showing that the average removal efficiencies of DO, COD and TP were respectively up to 91.41%, 51.22% and 65.52%. HF_{SAD} had greatly better removal performance on nitrogen contaminants, whose nitrate removal capacity was nearly 5 times as great as that in HF_{ord}. The average removal rate of TN and nitrate was just respectively 25.2% and -46.1% in HF_{ord}, while respectively 45.0% and 63.6% in HF_{SAD} subsequently, with the highest removal rate of respectively 61.9% and 84.6% achieved in the longest HRT. The denitrification capacity of HF_{ord} was about 5 times as high as that of HF_{ord}. The analysis result of microbial community structure, observation results of fillers and monitored sulfate both provided proofs for the occurrence of sulfur autotrophic denitrification, which had an excellent performance on denitrification of wastewater with low carbon source. Therefore, combining sulfur autotrophic denitrification with constructed wetlands can improve the treatment effect of ordinary constructed wetland on steel rolling wastewater, which is of great significance for water conservation and efficient recycling in steel and iron enterprises.

ACKNOWLEDGEMENTS

This research was supported financially by the the National Major Science and Technology Project of Water Pollution Control and Management, China (Grant No. 2010ZX07319-001-02) and the key scientific and technological project of the Shanghai Science and Technology Committee, China (16DZ1202107).

REFERENCES

[1] XU J C, CHEN G, HUANG X F, et al. Iron and manganese removal by using manganese ore constructed wetlands in the reclamation of steel wastewater [J]. Journal of Hazardous Materials, 2009, 169(1–3): 309-17.

[2] HUANG X F, LING J, XU J C, et al. Advanced treatment of wastewater from an iron and steel enterprise by a constructed wetland/ultrafiltration/reverse osmosis process [J]. Desalination, 2011, 269(1-3): 41-9.

[3] ABOU-ELELA S I, GOLINIELLI G, ABOU-TALEB E M, et al. Municipal wastewater treatment in horizontal and vertical flows constructed wetlands [J]. Ecol Eng, 2013, 61(19): 460-8.

[4] VYMAZAL J. Constructed wetlands for treatment of industrial wastewaters: A review [J]. Ecol Eng, 2014, 73(724-51.

[5] VYMAZAL J, KRöPFELOVá L. Multistage hybrid constructed wetland for enhanced removal of nitrogen [J]. Ecol Eng, 2015, 84(202-8.

[6] ZHU H, YAN B, XU Y, et al. Removal of nitrogen and COD in horizontal subsurface flow constructed wetlands under different influent C/N ratios [J]. Ecol Eng, 2014, 63(58-63.

[7] WU S, KUSCHK P, BRIX H, et al. Development of constructed wetlands in performance intensifications for wastewater treatment: a nitrogen and organic matter targeted review [J]. Water Research, 2014, 57(5): 40-55.

[8] HANG Q, WANG H, CHU Z, et al. Application of plant carbon source for denitrification by constructed wetland and bioreactor: review of recent development [J]. Environmental Science & Pollution Research, 2016, 23(9): 8260-74.

[9] PARK J H, KIM S H, DELAUNE R D, et al. Enhancement of nitrate removal in constructed wetlands utilizing a combined autotrophic and heterotrophic denitrification technology for treating hydroponic wastewater containing high nitrate and low organic carbon concentrations [J]. Agr Water Manage, 2015, 162(1-14.

[10] SONG X, WANG S, WANG Y, et al. Addition of Fe 2+ increase nitrate removal in vertical subsurface flow constructed wetlands [J]. Ecol Eng, 2016, 91(487-94.

[11] XU D, XIAO E, XU P, et al. Bacterial community and nitrate removal by simultaneous heterotrophic and autotrophic denitrification in a bioelectrochemically-assisted constructed wetland [J]. Bioresour Technol, 2017, 245(Pt A): 993-9.

[12] WANG W, DING Y, WANG Y, et al. Intensified nitrogen removal in immobilized nitrifier enhanced constructed wetlands with external carbon addition [J]. Bioresource Technology, 2016, 218(1261-5.

[13] ZHANG T C, ZENG H. Development of a Response Surface for Prediction of Nitrate Removal in Sulfur–Limestone Autotrophic Denitrification Fixed-Bed Reactors [J]. J Environ Eng, 2006, 132(9): 1068-72.

[14] LV X, SHAO M, LI J, et al. Nitrate removal with lateral flow sulphur autotrophic denitrification reactor [J]. Environmental Technology, 2014, 35(21): 2692-7.

[15] WU D, EKAMA G A, CHUI H K, et al. Large-scale demonstration of the sulfate reduction autotrophic denitrification nitrification integrated (SANI((R))) process in saline sewage treatment [J]. Water Research, 2016, 100(496-507.

[16] MONTALVO S, HUILINIR C, GALVEZ D, et al. Autotrophic denitrification with sulfide as electron donor: Effect of zeolite, organic matter and temperature in batch and continuous UASB reactors [J]. Int Biodeter Biodegr, 2016, 108(158-65.

[17] WANG Y, BOTT C, NERENBERG R. Sulfur-based denitrification: Effect of biofilm development on denitrification fluxes [J]. Water Research, 2016, 100(184-93.

[18] YANG W, LU H, KHANAL S K, et al. Granulation of sulfur-oxidizing bacteria for autotrophic denitrification [J]. Water Research, 2016, 104(507-19.

[19] PARK J Y, YOO Y J. Biological nitrate removal in industrial wastewater treatment: which electron donor we can choose [J]. Applied Microbiology and Biotechnology, 2009, 82(3): 415-29.

[20] XU M, LIU W J, LI C, et al. Evaluation of the treatment performance and microbial communities of a combined constructed wetland used to treat industrial park wastewater [J]. Environ Sci Pollut R, 2016, 23(11): 10990-1001.

[21] SCHULZ C, GELBRECHT J, RENNERT B. Treatment of rainbow trout farm effluents in constructed wetland with emergent plants and subsurface horizonal water flow [J]. Aquaculture, 2003, 217(1–4): 207-21.

Inhibition of Nitrification and Degradation of the Heavy Metal Precipitation Agent Dimethyldithiocarbamate by Activated Sludge

Stephan Wasielewski, Ralf Minke, Peter Maurer, Bertram Kuch, Harald Schönberger
(University of Stuttgart, Stuttgart, Germany)

Dimethyldithiocarbamate (DMDTC) has been used as an industrial chemical for over 20 years e. g. as a herbicide, fungicide, nitrification inhibitor in fertilizers and biocide in pulp and paper production. In addition, it is also used in electroplating plants and by circuit board manufacturers as a means of precipitating heavy metals from waste water. It is used in the stoichiometric surplus. If excess DMDTC is not eliminated, e. g. by precipitation with iron ions, before the wastewater is discharged this can lead to considerable nitrification inhibition at the associated waste water treatment plant (WWTP) and consequently to fish death. In southern Germany and northern Switzerland, four fish deaths due to indirect discharges of DMDTC from electroplating plants have occurred in the past four years.

Although studies have shown that low concentrations of DMDTC in wastewater can be removed aerobically, anaerobically and anoxically, it is questionable to what extent the biocoenosis typical for WWTP reacts to an impact load with DMDTC-containing waste water.

For this purpose, investigations were carried out on nitrification inhibition and the retrieval of DMDTC after different contact times with activated sludge. Therefore activated sludge from a municipal WWTP was subjected to various loads of DMDTC and sampled at regular intervals according to ISO 9509:2006.

The effect concentration, which inhibits nitrification by 50%, is achieved with a sludge load (B_x) of 2 mg/g_{TSS}. It has been observed that with a sludge load of 3.2 mg/g_{TSS} the inhibition of nitrification decreases with increasing contact time. During the first 6 h contact time the nitrification inhibition was 60%, decreased after 8 h to 47% and after 24 h to 7%. In contrast, with a sludge load of 6.3 mg/g_{TSS}, which leads to an inhibition of nitrification of 80%, no decrease of nitrification inhibition can be observed.

Analyses of the samples using GC-MS show a low recovery of DMDTC (less than 10% in relation to the quantity used) in all samples. However, it can be observed that DMDTC is metabolized to dimethylthiocarbamate (DMTC). With increasing sludge loading, it can be observed that the proportion of DMTC in relation to the DMDTC used correlates with the resulting nitrification inhibition. Accordingly, the DMTC metabolite primarily causes the inhibitory effect of activated sludge. The pH value of the partially inhibited test preparations was not stabilized and decreased as a result of nitrification.

It could be demonstrated by tests according to ISO 9888:1999 that DMDTC is removed in spite of the nitrification inhibition of the activated sludge. Depending on the sludge load, between 7 d (B_x = 2.5 mg/g_{TSS}) and 23 d (B_x = 13.4 mg/g_{TSS}) were necessary to achieve a complete elimination of the DMDTC. In contrast to the nitrification inhibition tests, the pH value was stabilized to be within a range of 7 to 7.5.

It can be concluded that DMDTC is degraded by contact with activated sludge. The metabolite DMTC is significantly responsible for inhibiting nitrification. In the event of an impact load of a WWTP with waste water containing DMDTC, a pH value reduction to 6.5 is desirable, since DMDTC and the metabolite are hydrolysed at this pH value and are therefore no longer toxic to nitrifying bacteria.

Assessing the Effects of Operational Parameters on Membrane Fouling in Ultrafiltration of ASP Flooding Wastewater by Membrane Fouling Structure Model

Liumo Ren and Shuili Yu
(Tongji University, Shanghai, China)

A large amount of oily wastewater is generated by oil industry, especially for those using water injection technology. It is necessary to purify this water so that it can be reused to save water resources and to protect the environment. With the continuous development of flooding technology, the quality of oil wastewater is more and more complicated and the treatment is more and more difficult. Alkali - Surfactant – Polymer (ASP) ternary composite flooding is a chemical flooding technique used in Daqing Oilfield. Related studies showed that ASP flooding can improve oil recovery significantly, whereas the produced water from this method contains a large quantity of polyacrylamide, surfactant and alkali, which induce larger viscosity, smaller oil droplets and emulsified more severely. The wastewater treated by conventional treatments, which include gravity separation and skimming, dissolved air flotation, de-emulsification, coagulation and flocculation, does not meet required standard for discharged to environment or reuse. Therefore, membrane separation is applied to solve the problem.

According to meet the requirements of rejection water, ultrafiltration is implemented to treat the wastewater in this study. Ultrafiltration has high oil removal efficiency, low energy cost, no chemical additives, small space occupancy and compact design compared to conventional treatment methods. However, one of the major challenges or obstacles for the application of ultrafiltration in treat oily wastewater is the decline in permeate flux as a result of membrane fouling. Besides the physic and chemical cleaning, optimum operational condition also can remit membrane fouling partly to make system operate effectively and continuously.

In this study, a pilot-scale ultrafiltration system of PVDF membrane was set up. The influences of TMP, cross-flow velocity, volume reduction factor(VRF), temperature on membrane flux were investigated and optimal operational condition was determined by Orthogonal experiment . under the experiment conditions, the optimum parameters are as follows: the average membrane pressure is 2.12 bar, the cross-flow velocity is 3.00 m/s, the VRF is 5 and the temperature of feeding solution is 22°C. The quality of treated water meets the standards of low permeability layer water injection. The results provide references and guidances for actual production.

Then based on the hydrodynamic theory from perspective of membrane structure, an ultrafiltration fouling model that includes internal pollution factor and external pollution factor was established. The tests data of optimum operational parameters were used to simulate the factors. Judging from the size of the two factors, it can be seen which part of membrane fouling was influenced by the operational parameters. Combined with variation of permeate flux, the mechanism of the effect of operational parameters on membrane fouling can be confirmed.

Metagenomic Analysis of Relationships between Carbon and Denitrification Metabolisms in Tannery Wastewater Treatment Plant Bioaugmented with Novel Microbial Consortium BMS-1

Aalfin Emmanuel, Chaeyoung Rhee, and ***Sung-Cheol Koh***
(Korea Maritime and Ocean University, Busan, South Korea)
Woo-Jun Sul and Hoon-Je Seong
(Chung-Ang University, Anseong, South Korea)

The goal of this study was to seek dynamic relationships between carbon degradation and denitrification pathways in the full scale tannery wastewater treatment plant bioaugmented with the novel microbial consortium BMS-1. It was hypothesized that denitrification process could be closely related to the carbon pathways such as degradation of amino acids, fatty acids and carbohydrates which could generate electron donors for the nitrate reduction, and that an efficient removal of nitrogen could be linked to COD removal (and hence sludge reduction). Shotgun metagenomic reads were mapped to 'ChocoPhlAn' pan-genome database and MetaPhlAn2 database for organism specific functional profiling and formed as a gene abundance data using HUMAnN2. We then compared and selected the gene families and pathways using the extended databases UniProt Reference Clusters (UniRef90, http:/www.uniprot.org) and MetaCyc metabolic pathway database. During the functional analysis, 40,181 gene families and 196 pathways were revealed for the five treatment stages (I, B, PA, SA and SD). Of these, 32 metabolic pathways were involved in amino acid production whose pathways were dominantly found. Heat map analysis showed that nitrite reductase genes (*nirS* and *nirK*) were highly dominant together with genes encoding aspartate-semialdehyde dehydrogenase and transaminated amino acid decarboxylase, and genes encoding oleate hydratase and 1-deoxy-D-xylulose-5-phosphate synthase, and genes encoding N-acetyl-D-mannosamine dehydrogenase in I, B, PA and SA. It was also revealed through a pathway correlation analysis that *nirS*, *nirK* and *nrfA* genes were highly linked to genes encoding D-serine/D-alanine/glycine transporter, ornithine carbamoyltransferase and aspartate-semialdehyde dehydrogenase (amino acid metabolism), and gene encoding 1-deoxy-D-xylulose-5-phosphate synthase (fatty acid metabolism), and genes encoding phosphoenolpyruvate carboxylase, N-acetyl-D-mannosamine dehydrogenase and L-lactate dehydrogenase (carbohydrate metabolism). It was concluded that the denitrification process was strongly correlated with some specific functional genes involved in metabolisms for amino acids, fatty acids and carbohydrates in order. This indicates that denitrification (reduction of nitrates) may demand much of electron donors generated in SD and B, leading to a significant reduction of sludge. The metagenomic analysis could provide a comprehensive metabolic insight of the microbial community functions in the different treatment stages. These metagenomic insights will contribute to a successful monitoring and operation of the eco-friendly tannery wastewater treatment system which is a highly important issue in the tannery business.

This research was supported by a National Research Foundation of Korea (NRF) grant funded by the Korean government (MSIP) (No. 2017R1D1A3B03036376).

Synthesis & Characterization of PDMS (Polydimethylsiloxane) Membrane and Enhance its Dye Degradability Property by Incorporating the Silica Nanoparticle

Naresh K. Sethy, Zeenat Arif, P.K. Mishra, ***P.Kumar***
(Department of Chemical Engineering and Technology, IIT(BHU), Varanasi, India-221005)

In this study, 4 wt% of SiO_2/PDMS (polydimethylsiloxane) nanocomposite membrane was synthesized by solution casting method. The structure, morphology and physical property of the membrane were characterized by FTIR, SEM, Contact angle measurement and swelling test for organic-water mixture. The adsorption and the photocatalytic activity of the SiO_2 nano-additive in the membrane (PDMS) was investigated by measuring the degradation of a cationic dye, for 3 ppm of methylene blue (MB), under UV/Visible light illumination for 3h. The results were compared with the dye solution (controlled) and dye solution with membrane for the degradation of methylene blue. The stability of the PDMS membrane was checked after the degradation. The main advantage in this work, a used membrane which was incorporated with nanoparticle passed through a minor cleaning and washing is applied for dye degradation analysis. Results show that performance of the used membrane was nearly same as obtained for fresh membrane. Hence one can conclude that membrane can be reused efficiently for dye degradation.

Efficient Biodegradation of Azo Dye Acid Blue 113 by Bacterial and Yeast Consortium and Molecular Analysis of Bacterial Biodegradative Pathway

Chaeyoung Rhee, Aalfin Emmanuel, and ***Sung-Cheol Koh***
(Korea Maritime and Ocean University, Busan, South Korea)
Hong-gi Kim (Bayo, Inc., Jinju, South Korea)

The goal of this study was to select a best suited microbial consortium for an efficient bioremediation of the recalcitrant azo dye (Acid Blue 113) and to elucidate the biodegradative mechanisms in terms of a molecularmicrobial perspective. A selected consortium composed of two bacterial cultures (*Mesorhizobium* sp. and *Sphingomonas melonis*) and two yeast cultures (*Apiotrichum mycotoxinivarans* and *Meyerozyma guilliermondi*)turned out to achieve more than 80% of decolorization in 24 hours (50 and 100 mg/L dye). The concomitant chemical oxygen demand removal rate for the bacterial consortium reached 97% in 72 hours while the rates of the yeast consortium and the total consortium achieved 98.0 % and 97.5 %, respectively in 24 hours. The FT-IR spectra profiles for yeast and bacteria consortia showed rapid disappearance of the absorbance at the azo bond specific wavenumber1455cm^{-1} in 24 hours and 72 hours respectively. Metabolic products containing -NH_2 groups were also detected based on the absorbance at the 1300cm^{-1}wave number, reflecting an occurrence of the azo bond cleavage. RNA-seq technology has been applied to biodegradative pathway analysis of *Sphingomonas melonis*. It was concluded that all the data collectively provide an evidence for the biodegradation of the azo dye by the bacterial and yeast consortium. Bioaugmentation of the selected microbial consortium can be applied to an efficient bioremediation of the azo dye wastewater at an industrial scale.

This research was supported by a National Research Foundation of Korea (NRF) grant funded by the Korean government (MSIP) (No. 2017R1D1A3B03036376).

Decolorization Azo Dyes by Fungus Isolated from Industrial Sewage Contaminated Soil

Patel Bhavinee, Mehra Bhavna K, ***Tank Shantilal K***
(Department of Biosciences, Veer Narmad South Gujarat University, Surat-395007, Gujarat, India)

South Gujarat region of India is hub of textile industries. These textile unit generates substantial quantity of effluents. A variety of synthetic dyestuffs are released from the textile industries and made a threat to environmental safety. Dyes are the most difficult constituents to remove by conventional biological wastewater treatment. Colored wastewater is mainly eliminated by physical and chemical procedures, which are very expensive and have drawbacks. An alternative, microbiological treatment has several advantages. In this study, fungal strain was isolated from industrial sewage contaminated soil from the Surat region, Gujarat, India and studied for their efficiency for decolorization of three different azo dyes- Congo red, Disperse blue DBR and Disperse red G. Azo dyes were significantly removed by fungal strain that was tested with 96 h of experimental time. There were different concentrations from 100 ppm to 500 ppm of selected dyes used for decolorization study. Results showed high and stable degrees of decolorization of more than 75% in all systems. Shaking improved the efficacy and rate of the dye removal comparative to static samples. The results suggest that fungal isolate has potential for cleaning up azo dyes containing effluents and development of non-conventional and low-cost method of bioremediation. This treated water (effluent) can be used for agricultural purpose and gardening.

Biodegradation of Phenol by Bacterial Strain Isolated from Industrial Sewage

Desai Jigna ,Mehra Bhavna K, ***Tank Shantilal K***
(Department of Biosciences, Veer Narmad South Gujarat University, Surat-395007, Gujarat, India)

The region where study is taken up is full of textile industries. The environment has been contaminated by avariety of toxic compounds by many of ways, especially through industrial discharges. Phenol is one of the major aromatic pollutants among the variety of toxic compounds.There are different methods available for phenol removals such as solvent extraction, adsorption,chemical oxidation, etc. which have serious drawbacks such as high cost and formation of hazardous by-products. But biodegradation is one of the cheapest methods without any production of hazardous by-products. The aim of the study was acollection of asample from industrial sewage and isolation of phenol-degrading bacteria, which has the potential to degrade the hazardous phenolic compounds. Based on it`s morphological, physiological and biochemical characteristics, the bacterial strain was characterized as a gram-positive, strictly aerobic, non-motile and cocci-shaped bacterium that utilizes phenol as a sole carbon and energy source. The strain was efficient in removing 88% of the 500 mg/l phenol within 48 h.These result indicated that isolated bacterial strain possesses a promising potential in treating phenolic waste water and water can be usable for agricultural purpose purpose.

Evaluation of Absorption Process for Heavy Metals Removal Found in Pharmaceutical Wastewater

Jamal A. Radaideh
(Al Balqa Applied University/ Huson University College, Dept. of Civil & Environmental Engineering, Jordan. jamrad08@gmail.com)

ABSTRACT: This study focused on the adsorption/absorption process of natural volcanic tuff, its application, kinetics, efficiency, velocity and selectivity order in removing heavy metals found in pharmaceutical wastewater.

Series of experiments were conducted in batch-wise and fixed-bed columns to study the removal performance and selectivity sequence of mixed metal ions (Pb^{2+}, Cr^{6+}, Cu^{2+}, $Zn^{2+}$and Fe^{3+}) in aqueous solution using natural volcanic tuff material as adsorbent. The main purpose of this study is to highlight the economical application of the method in treatment of pharmaceutical wastewater.

Heavy metal salts were used to prepare synthetic pharmaceutical wastewater containing a mixture of different metal ions concentrations ranged from 1 to 20 mg/L were to be applied to columns packed with natural volcanic tuff rich in zeolite ranged between (0.50 – 3.50 mm) in grain size. Photometric procedure was used for sample analysis. The absorption experiments were carried out under changing conditions, such as, different pH-values of solutions (2, 4, 6 and 7), initial solute concentrations ranging from (1, 5, and 10) mg/L, under different room temperatures (10, 20, and 30 C°), and with use of varying tuff particle sizes (0.50 -3.50) mm.

The results showed that Freundlich model described satisfactorily sorption of Cu^{2+} and Fe^{2+}. Used volcanic tuff exhibit efficiency in removing heavy metals ranging from 45 – 99 % of the added Cr, Cu, Zn, Pb and Fe metal concentrations respectively. According to the percentage sorption and distribution coefficients values, the selectivity sequence of studied metals by volcanic tuff is strongly dependent of pH, however approximating all results a selectivity sequence can be given as Pb $\geq$ Fe > Cr > Zn > Cu at pH around 2 and a selectivity sequence as Fe > Cu > Cr > Zn > Pb at pH = 6-7 . The results confirms that natural volcanic tuff hold great potential to remove cationic heavy metal species from industrial wastewater.

Keywords: heavy metal ions, absorption, pharmaceutical wastewater, natural volcanic tuff, absorption isotherms.

INTRODUCTION

Effluents from various processing industries and other utilities such as electroplating industries, medical care centers and hospitals is reported to contain high amounts of heavy metal ions, such as nickel, iron, lead, zinc, chromium, cadmium and copper (Konstantinos et al., 2011). Heavy metals are naturally occurring elements and the multiple industrial, domestic, agricultural, medical and technological applications have led to their wide distribution in the environment; raising concerns over their potential effects on human health and the environment. Their toxicity depends on several factors including the dose, route of exposure, and chemical species, as well as the age, gender, genetics, and nutritional status of exposed individuals. For this reason, heavy metals and their toxicity to environment is being the subject for many studies (Inglezakis et al., 2003;Tchounwou et al., 2012; Dorne et al., 2011; Rezymski et al., 2015;

Núñez et al., 2016; Stern, 2010). Due to mutagenic and carcinogenic properties of heavy metals, much attention has been paid for occurrence, concentrations, movement, fate and the anticipated health and environmental risks that may arise from these heavy metals since they have direct exposure to humans and other organisms (Anyakora, et al., 2011; Khan et al., 2008). The demand for pharmaceuticals has resulted in a consequent increase in pharmaceutical manufacturing companies in the world and hence increased pharmaceutical waste which most times contain substantial amount of heavy metals. Pharmaceutical residues in the environment, and their potential toxic effects, have been recognized as one of the emerging research area in the environmental chemistry (Hernandoa et al., 2006, Arnold et al., 2014). Pharmaceutical effluents are usually discharged into the environment and when improperly handled and disposed, they affect both human health and the environment (Osaigbovo, et al., 2006; Meena et al., 2008).

The presence of pharmaceutical compounds in waters comes from two different sources: production processes of the pharmaceutical industry and common use of pharmaceutical compounds resulting in their presence in urban wastewaters (Roberts et al., 2006; Verlicchia et al., 2012; Idris et al., 2013; Bulinski et al., 2009). The pharmaceutical wastewaters generated in different processes in the manufacture of pharmaceuticals and drugs contain a wide variety of dangerous compounds including heavy metals (Ashfaqa, 2017). Further, reuse of water after removal of contaminants, whether pharmaceuticals or otherwise, is required by industry and agriculture. In view of the scarcity of water resources, it is necessary to understand and develop methodologies for treatment of pharmaceutical wastewater as part of water management.

In order to remove toxic heavy metals from water systems, conventional methods have been used such as chemical precipitation, coagulation, ion exchange, solvent extraction and filtration, evaporation and membrane methods (Panayotova, 2003). These conventional treatment technologies for the removal of these toxic heavy metals are not efficient and generate huge quantity of toxic sludge. Adsorption of heavy metals on conventional adsorbents such as activated carbon have been used widely in many applications as an effective adsorbent, and the activated carbon produced by carbonizing organic materials is the most widely used adsorbent. However, the high cost of the activation process limits its use in wastewater treatment applications (Amarasinghe et al., 2007).

Bio-sorption using natural materials locally available in certain regions is emerging as a potential alternative to the existing conventional technologies for the removal and/or recovery of metal ions from aqueous solutions. Low cost and availability, high metal binding capacity, minimum production of chemical or biological sludge, possible regeneration of bio-sorbents count for the major advantages of using bio-sorption methodology over conventional treatment methods.

Zeolite-based technology can provide a cost effective solution for pharmaceutical wastewater treatment for the removal of toxic heavy metals under increasing demand of safe water from alternative sources (Babel, 2003 ; Inglezakis et al., 2006; Ziyath et al., 2011). Numerous previous studies have investigated the removal efficiency of zeolite tuff when placed in fixed bed columns confirmed their excellent performance on the removal of metal cations from wastewaters (Al-Haj Ali et al., 1997; Erdem et al., 2004; Gadipelly et al., 2014).

The efficiency of water treatment by using natural and modified zeolites depends on the type and quantity of the used zeolite, the size distribution of zeolite particles, the initial concentration of contaminants (cation/anion), pH value of solution, ionic strength of solution, temperature, pressure, contact time of system zeolite/solution and the presence of other organic compounds and anions. For water treatment with natural zeolites, standard procedures are used, usually a procedure in column or batch process. Ion exchange and adsorption properties of natural zeolites in comparison with other chemical and biological processes have the advantage of removing impurities also at relatively low concentrations and allows conservation of water chemistry, if the treatment is carried out in the column process (Caputo, 2007; Fu et al., 2011; Wang et al., 2010; Margeta, 2011).

There are several regions lacking water resources and this necessitates that much effort is put into water conservation and environmental protection. The expansion of industrial activities, including metal-based industries, and the excessive use of chemicals enhance the pollution of waters with heavy metals. All these require the availability of low-cost technology and materials for wastewater treatment. However, volcanic tuff deposits, rich in phillipsite and zeolite are natural occurring deposits, which they can be found in huge reserves in many regions (Barakat, 2011).

In this study, the sorption behavior of natural occurring volcanic tuff rich in zeolite with respect to chromium (Cr), copper (Cu), lead (Pb), zinc (Zn) and iron (Fe) ions has been thoroughly examined to confirm its performance for application to purify industrial wastewaters. The batch method has been employed, using competitive sorption system with metal concentrations in solution ranging from 1- 20 mg/l. Parameters such as pH, contact time, and initial metal concentration, were considered. The equilibrium adsorption capacity of volcanic tuff for studied heavy metals was measured and extrapolated using linear Freundlich and Langmuir isotherms. The results indicate the potentially practical value of this method in industries and also provide strong evidence to support the adsorption mechanism proposed.

The present research work aims to study the absorption/adsorption behavior of different heavy metal ions Cr^{6+}, Pb^{2+}, Zn^{2+} , Cu^{2+} and Fe^{2+} at natural untreated volcanic tuff material. These heavy metal ions are among most common ions found in pharmaceutical wastewater effluents. The results showed that Freundlich model described satisfactorily sorption of all metals. Used volcanic tuff exhibit efficiency in removing heavy metals ranging from 45 – 99 % of the added Cr, Cu, Zn, Pb and Fe metal concentrations respectively. According to the percentage sorption and distribution coefficients values, the selectivity sequence of studied metals by volcanic tuff is strongly dependent of pH, however approximating all results a selectivity sequence can be given as Pb ≥ Fe > Cr > Zn > Cu at pH around 2 and a selectivity sequence as Fe > Cu > Cr > Zn > Pb at pH = 6-7 . The results show that natural volcanic tuff hold great potential to remove cationic heavy metal species from industrial wastewater.

MATERIALS AND METHODS

Glass Columns. Glass columns with different heights (40 -70 cm, and Φ 5 cm) were used in the batch and column experiments. As pack and ion exchanger material, Jordanian natural tuff grains were used. Continuous flow fixed-bed column was applied to study the efficiency of the tuff bed rich in zeolite in removing heavy metal ions from pharmaceutical wastewater. The operation was controlled through the variables (flow rate, bed depth, and column internal diameter).

Table 1. The grain size distribution of used aggregates

Weight (g)	**Grain size (mm)**	**Percentage (%)**
1000	≤ 4	100
414.3	≥ 3	41.4
309.1	≥ 2	30.9
211.5	≥ 1	21.1
65.1	≥ 0.5	6.5

Absorbent. Natural occurring and untreated volcanic tuff rich in zeolite is used as absorbent. The conventional mineral processing techniques of tuff are starting with crushing the materials followed by autogenously tumbling mill and then low intensity magnetic and gravity separation. The crushed original tuff was ground and passed through 4- 0.5 mm sieves and was dried in an oven at 104.5 ◦C for 24 h.

The absorbent grains undergo a chemical treatment to eliminate any biological and oxidizable contents, batches of 50 gr of reagents were boiled separately with 50 mL of HCl acid or H_2SO4 acid with concentration of 0.1 M or 1.0 M for 30 minutes to destroy organic matter (Stylianou et al., 2007), then filtered under vacuum and washed thoroughly with distilled water until a pH of 7 is maintained. The acidified reagents were dried overnight at 80 °C. Volcanic tuff was grounded and then passed through 4-0.5 mm sieves and dried in an oven at 104.5 °C for 24 h. The grain size distribution of the used volcanic tuff is presented in table 1.

The tuff was not exposed to any pre - treatment or modification. Samples were just washed and dried at 103.5 °C and kept in desiccators for 24 h to ensure complete drying out.

Absorbate. The initial aqueous solution concentrations of absorbate (1, 5, 10, and 20) mg/L were prepared using standard solutions for each metal. Standard solutions of metal chlorides of 500 mg/l and 1000 mL were used as adsorbate. Solutions of various concentrations were obtained by diluting the standard solution with distilled water.

Chromium – stock solution (100 ppm) was prepared by dissolve 0.1923 g of chromium (VI) oxide (CrO_3) in water. When solution is complete, acidification with 10 mL concentrated nitric acid and dilute to 1000 mL with water (1.00 mL = 100 μg Cr). The stock solutions of copper, lead and zinc were prepared by dissolving 2.1368 g of copper chloride, 1.3557 g of lead chloride and 2.1237 g of zinc chloride in 1000 mL volumetric flasks, respectively. A stock solution for iron was prepared by adding 7.3073 g of Fe (NO3)3 in a 1000 ml volumetric flask.

Heavy metal concentrations were determined by spectrophotometer. All the chemicals used were of analytical grade reagent and all experiments were carried out in 500 ml glass bottles at the laboratory ambient temperature of 25 ± 2 °C.

Characterization and Identification of Absorbent (Natural Volcanic Tuff Material). Representative samples of used volcanic tuff were investigated by using optical microscopy, scanning electron microscopy (SEM), X-Rays diffraction (XRD).

Figure 1. SEM images of used zeolite tuff shows tabular crystals associated with small fibrous crystals.

The instrument used is a Philips XL30 FEGSEM with Oxford CT1500HF Cryo stage and EDAX DX4 integrated Energy Dispersive X-ray Analyzer. Micro-structural and elemental constituents are imaged and analyzed with magnifications up to 50,000; i.e. image clarity at <200 nm (nanometers) scale. The X-ray diffraction analysis indicated that the examined raw tuff material is rich in zeolite beside other non-zeolite materials.

The bulk chemical composition of volcanic tuff was measured using XRD; the data for chemical composition of the used fractions are given in Table 2. The main components were SiO_2, Al_2O_3, and Fe_2O_3 with others found in low concentrations.

Figure 2 shows an energy dispersive X-ray spectroscopy (EDS, EDX or EDXRF) for the used tuff crystals. It is an analytical technique used for the elemental analysis of a sample.

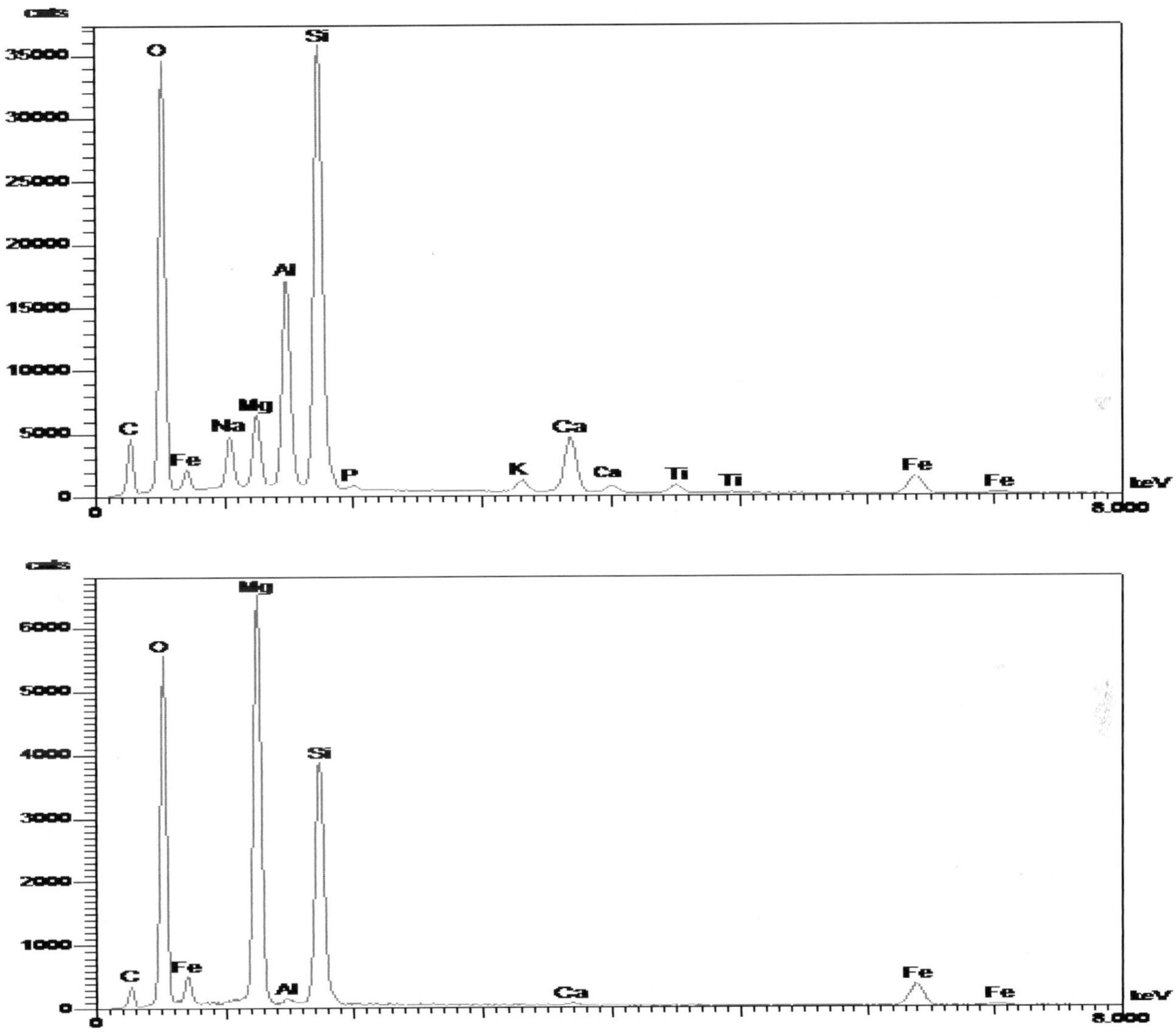

Figure 2. X-Rays diffraction spectrum of the used zeolite tuff

Scanning electron microscope (SEM) as shown in Figure 4 illustrates that the tuff material exhibit significant surface area with a large number of holes and pores, layers and sheets which allow water to pass through and due to the high surface area, the adsorption process will be promoted to occur quickly and easily. Its characterization capabilities are due in large part to the fundamental principle that each element has a unique atomic structure allowing x-rays that are characteristic of an element's atomic structure to be identified uniquely from each other. Identification of the principal elements; C, O, Fe, Na, Mg, Al, Si, Ca, P, K, Ti. The presence of these elements give the tuff the property of being an excellent ion exchanger.

The structures of volcanic tuff consist of three-dimensional frameworks of SiO_4^+ and AlO_4^+ tetrahedral. They were characterized by X-ray diffraction (XRD) and chemical analysis (Chutia et al., 2009). Al_2O_3, Fe_2O_3, CaO, and MgO were analyzed using titrimetric methods and SiO_2 was analyzed with a gravimetric method. Na_2O and K_2O were found by flame photometry. The results of chemical analysis are presented in Table 2.

Table 2. Chemical composition and physical properties of natural tuff sample (wt %)

Compound	Weight %	Compound	Weight %
SiO_2	65.18	MnO	0.44
Al_2O_3	12.24	Na_2O	1.55
CaO	3.18	K_2O	3.78
MgO	1.96	TiO_2	0.26
Fe_2O_3	1.20	SO_3	0.23

Si/Al = 5.32 water content ~ 5 %

Adsorption Experiments.

Design of Experiments. In the present work, a batch-scale system using coarse and fine zeolite tuff packed columns was operated to investigate the continuous removal of heavy metals. In these experiments, the effects of flow rate, contact time, particle size, column height, initial solute concentration, pH and temperature on removal efficiency were investigated.

In one-dimensional experiments with volcanic tuff aggregates filled columns of different sizes (fine, middle and coarse size), were fed with aqueous solution having different initial concentrations from top. The fixed bed columns, which will allow the most experimental variations, are to be filled with a layer of fine grains < 1mm at the bottom, over which a layer of tuff (different grain sizes) comes to slow down water passage. The hydraulic conductivity can be controlled by recording the time required to collect the injected sample outflow (contact time). Samples from inflow and outflows should be taken for chemical analyses, whereas zeolite tuff samples are to be taken for analyzing to record changes in chemical composition.

Synthetic wastewater samples were prepared to give heavy metal solutions with concentrations of 1 mg/L, 5 mg/L, 10 mg/L, and 20 mg/L by adding an appropriate standard solutions to deionized water. The dry mass of volcanic tuff used in one column is about 79.52 g which forms a height of 10 cm. The volume of wastewater sample applied in the columns each run is 40 ml.

Sampling Procedure. The uptake of metal ions by tuff material was studied to assess and understand the maximum removal efficiency under considering changing experimental variables. For these investigations, several 50 mL glass test tubes were used. Each test tube received 40 mL of a metal ion solution (each run has different initial concentration) and adjusted to the desired pH from 2.0 to 7.0 and controlled using a pH meter. The pH of the solution was adjusted using dilute solution of 0.1 N hydrochloric acid HCl or sodium hydroxide NaOH as necessary. Each test tube is filled with fixed amount of volcanic tuff (10 mg). The solution was introduced at a constant volumetric flow rate (Q = 40 mL) and initial concentration (Ci).

A continuous shake of tubes were performed to maintain an equilibrium and complete mixing. The effluent suspension was diluted (if necessary) to an appropriate concentration range and filtered through Whatman filter paper for the elemental analysis by atomic absorption spectroscopy (A.A.S). The contact time denoted as CT in minutes which reflects the real detention time of the aqueous solution within the fixed bed. Each batch consists of 4 effluent samples with detention times varying from zero to 10 minutes. The first sample (sample # 1) is added to fixed bed and collected immediately at the bottom, practically with a detention time zero (CF0). The second sample (sample # 2) is collected after lasting in the bed for 1 minute (CF1), the third sample (sample # 3) lasts 5 minutes (CF5), while the fourth sample (sample# 4) is allowed to last 10 minutes (CF10) within the bed prior to discharge. The experiment was conducted under a controlled room temperature (Air Conditioning unit controlled the temperature) in order to make the working temperature constant at 25 °C for a predetermined time (24 h) with continuous stirring. A peristaltic pump every 10 minutes feeds the column with 40 mL of aqueous solution. The small aliquots of samples were withdrawn from the reaction.

Experimental data were used to determine the equilibrium time, equilibrium concentrations, amounts adsorbed at equilibrium, optimum initial pH, and temperature influence over the bio-sorption process. Preferable removal sequence series were obtained. Also the experimental data were used to establish isotherm (linear and nonlinear regression), kinetics models and to calculate thermodynamic parameters. All the experiments were repeated three times, the values presented were calculated using averaged concentration readings.

RESULTS AND DISCUSSIONS

In experimental performance by the batch method, due to the small ratio of volumes of volcanic tuff/solution, the kinetics of the removal is determined by diffusion through the natural volcanic tuff particles. The experiments are carried out at different pH values (2, 4, 6 and 7) and with different initial concentrations (1 mg/L, 5 mg/L and 10 mg/L). The results show that the ionic exchange reactions took place for all samples in the experiments. The most obvious result obtained for all metals is the observed decrease in the initial concentration indicating that volcanic tuff is an active material in the absorption/ionic exchange process and can be used with high efficiency for removal of heavy metals from aquatic solutions.

In addition, a fundamental finding is that the heavy metal ions uptake is mainly correlated to solution pH, also that the removal of heavy metals is selective and selectivity order for removal could be established.

Effect of pH on Metal Uptake, Efficiency and Selectivity Sequence. Ambient pH was likely to be a major factor in the quantity of metal ion bio-adsorption owing to cations competition effects with hydrogen ions. Solution pH has a significant impact on heavy metal removal by zeolitic tuff since it can influence metal speciation, integrity of zeolite (mineral's surface properties) and also H^+ ions are considered competitive in ion exchange (Dimirkou, 2007; Hui et al., 2005; Trgo et al., 2003; Inglezakis et al., 2003; Cho et al., 2005). pH is one of the most important parameters influencing not only site dissociation, but also the solution chemistry of the heavy metals since hydrolysis, complexation by organic and/or inorganic ligands, precipitation and availability of heavy metals are all influenced by it (Esposito et al., 2002).

In this study, the adsorption capability of the volcanic tuff was observed to be strongly influenced by pH change. The experiments and obtained results exhibit clearly the decrease in initial concentrations of metal ions as a direct function of pH.

Table (3) represents the obtained results regarding the preferable uptake sequence of heavy metals in association with pH. It illustrates a summary of the optimal pH scale by which the highest uptake rate of each metal is occurring. In addition, it demonstrates the removal efficiency for each metal in correlation with the contact time between metal ions and tuff particles. For instance, at pH 4, adsorption percentage was 73.3% for Cr, 95.7% for Fe, 97% for Cu, 89% for Zi and 95% for Pb. At pH 6, the corresponding values rose to 83%, 98%, 98%, and 97% for Cr, Fe, Cu and Zi respectively, but decreased to 92 % for lead. These results are similar to results reported by (Omar et al., 2011; Orodu et al., 2014) for heavy metal ions sorption onto agricultural waste sorbents. Obtained results of this study revealed that adsorption efficiency of volcanic tuff is optimum at pH 4 to pH 7. Low adsorption rates of metal ions at higher pH >7 could be attributed to the formation of their hydroxides which build precipitate and prevent further adsorption as suggested by (Lisa et al., 2004 and Xiano and Ju-Chang, 2009; Olayinka, 2009).

In similar studies the use of natural and modified zeolites has been further examined for the simultaneous removal of Fe^{+2} and Mn^{+2} from underground water samples and results of Fe^{+2} and Mn^{+2} removal levels were suggested to be between 22-90% and 61-100% for natural zeolite – clinoptilolite (Inglezakis, 2010). In addition several selectivity sequences have been reported in several similar adsorption studies (mainly conducted in a single solution) for natural zeolites (Erdem et al., 2004;): $Co^{2+} > Cu^{2+} > Zn^{2+} > Mn^{2+}$. The selectivity series of clinoptilolite in the sodium form was determined by (Zamzow et al., 1990) as follows: $Pb^{+2} > Cd^{2+} > Cs^{+} > Cu^{2+} >$ Co2+> $Cr^{3+} > Zn^{2+} > Ni^{2+} > Hg^{2+}$.

Table 3. Preferable removal sequence (selectivity order series) of heavy metals and the efficiency of removal for each element in relation to pH and initial concentrations.

pH	Ci (mg/l)	Preferable removal sequence	Removal efficiency* (%) Cr	Fe	Cu	Zn	Pb
2	1	Pb > Fe> Cr > Cu > Zn	69.6	98.2	57	10	98
	5	Pb ≥ Fe > Cr > Zn > Cu	91	92.4	48.8	51.0	95.6
	10	Fe > Pb > Cr > Cu > Zn	85.5	95.4	51.2	40.2	96.7
4	1	Cu > Pb > Cr > Fe > Zn	91.9	97.2	98.5	58	98
	5	Pb > Cu > Zn > Cr > Fe	88.1	88.0	98	96	99
	10	Fe > Cu > Pb > Zn > Cr	73.3	95.7	97	89	95
6	1	Cu > Fe > Pb > Cr > Zn	90	98	99	71.6	97
	5	Cu > Zn > Fe > Cr > Pb	83	98	98	97	92
	10	Fe > Cu > Cr > Zn > Pb	73	99	99	95	88
7	1	Fe > Cu > Cr > Zn > Pb	86	99	91	67	93
	5	Cu > Fe > Pb > Cr > Zn	90	97	97	71	94
	10	Fe > Cu > Pb > Zn > Cr	83	97	97	95	96

*removal efficiency = $(C_i - C_f)/C_i \times 100$ %

Table 4. Optimal pH for absorption of heavy metals on volcanic tuff

Heavy Metal ions	Initial Conc. (C_i) (mg/L)	Optimal pH for adsorption/absorption
Cr	1	4 > 6 > 7 >2
	5	4 > 6 > 7 > 2
	10	4 = 6 = 7 > 2
Zn	1	7 > 6 > 4 > 2
	5	4 = 6 > 2 > 7
	10	6 = 7 > 4 > 2
Cu	1	4 = 6 > 7 > 2
	5	4 = 6 = 7 >>2
	10	4 + 6 = 7 > 2
Fe	1	6 >7 > 2 > 4
	5	6 = 7 = 4 > 2
	7	2 = 4 = 6 = 7
Pb	1	4 > 2 > 6 = 7
	5	4 > 2 > 6 > 7
	10	7 > 2 = 4 > 6

Removal Efficiency. The removal efficiency was calculated using the equation:

Removal efficiency = $(C_i - C_f)/C_i \times 100$ % which exactly reflects the percent of adsorption/absorption

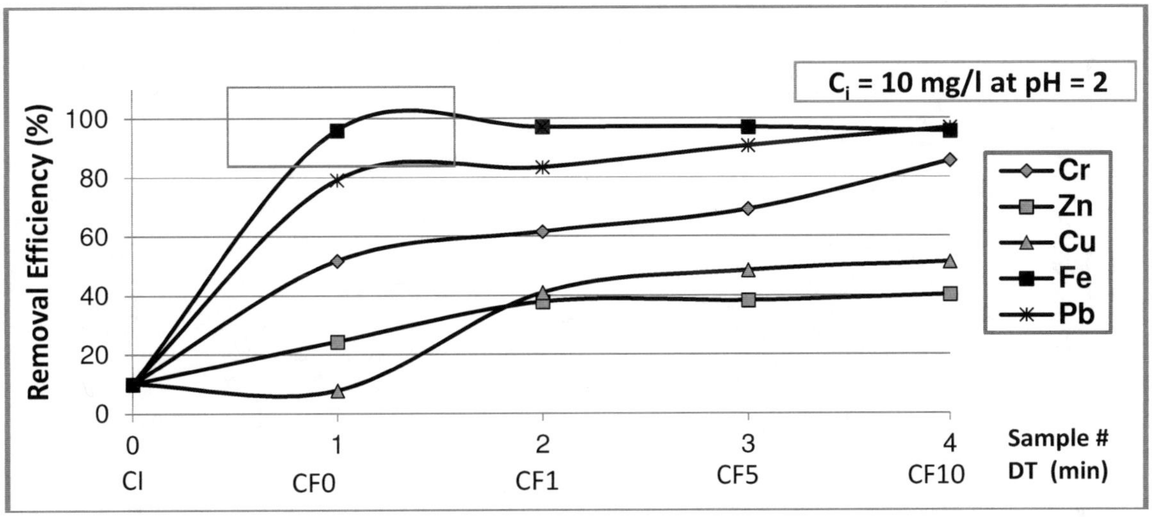

Figure 3. Removal efficiency of heavy metals in relation to initial solute concentration and contact time. Data based on calculations presented in table 3. Removal efficiency of Fe ions in average = 95.4 % with no value > 100 % as shown in fig.3.

The removal efficiency for each heavy metal (Cr, Zn, Cu, Fe and Pb) was calculated in relation to different acidic conditions (pH = 2, 4, 6, and 7) with respect to initial solute concentration C_i (1, 5 and 10 mg/L). As result of absorption analysis, a huge data sets were obtained (table 3 and table 4) which describe the absorption/ adsorption process of heavy metals on absorbent in relation to pH of solution, contact time and the initial concentration of solute used. Based on data obtained, it is to confirm that the absorbent behaves differently in regard to absorbing heavy metal ions in relation to pH, contact time and the initial concentrations of ions applied. Figure 3 shows the removal efficiency of investigated heavy metals in relation to contact time time (CF0 to CF10) when applied to absorbent with an initial concentration of 10 mg/L at pH = 2. It is clearly demonstrated that at these pre-defined conditions, volcanic tuff has great affinity to absorb iron ions (98% removal efficiency) followed by lead and chromium. As presented in table 3 and 4, when one or more of these prior defined experimental conditions change, the affinity and abundance of tuff material to absorb metals varies, sometimes significantly.

Effect of Contact Time. The effect of contact time on the removal efficiency of different heavy metals copper, zinc, chromium, iron and lead ions was studied under consideration of the ambient acidic condition (prevailing pH).

Figure 4 and Figure 5 describe the uptake of applied metal ions on the tuff material at varying contact time between absorbate and absorbent, while the pH is hold constant. The rate of uptake of metal ions was quite rapid for all ions as soon as a contact between tuff grains and ions solution occurs; the metal removal in the first run (zero contact time, pH = 4) ranges from 65 – 90 % when ions pass through tuff bed with an initial concentration of 5 mg/L. For iron, copper and zinc 90% of the ions are removed immediately as ions came in contact with tuff bed, while for chromium and lead 10 minutes contact time was required for an 82 % and 87 % removal, respectively (Figure 4). At pH = 6, the uptake of cupper on tuff grains occurred rapidly followed by iron when the solution passed through with an initial concentration of 1 mg/L

(Figure 5). It has been observed that a change in pH results an immediate change of the order of removal of metals.

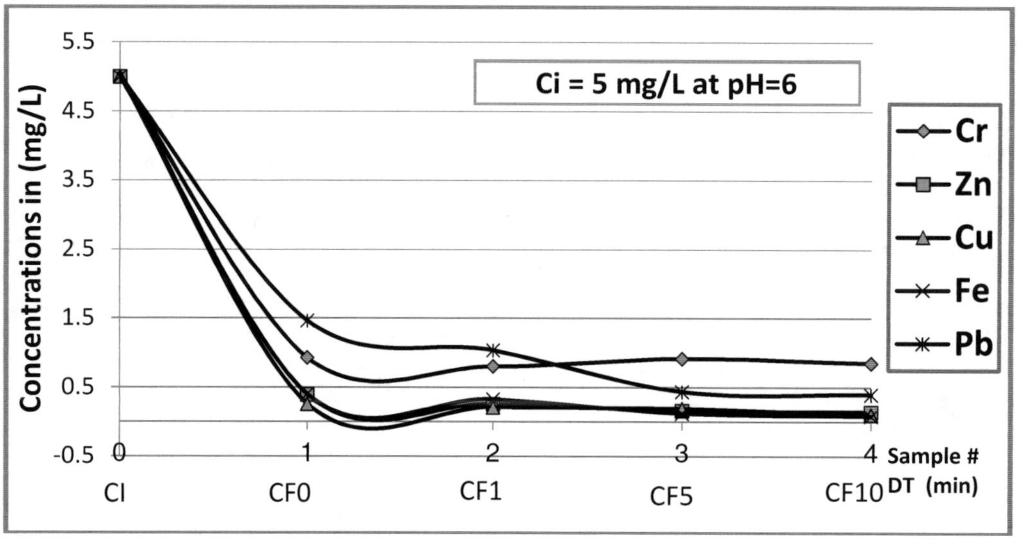

Figure 4. Effect of contact time on the removal of heavy metals on volcanic tuff grains

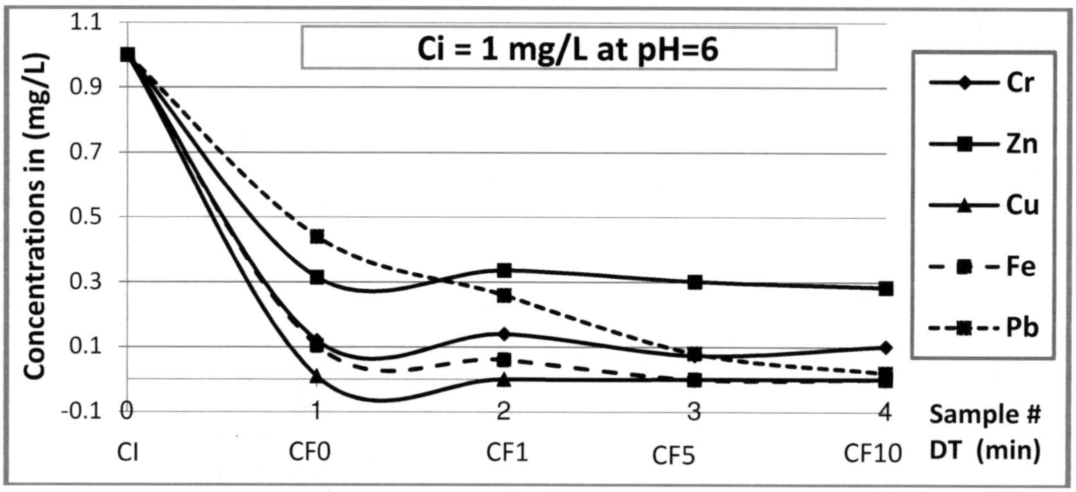

Figure 5. Effect of contact time on the removal of heavy metals on volcanic tuff grains

This is in agreement with the results obtained by (Bernard et al., 2013) for removal of heavy metals using orange peel activated carbon, (Sharma et al. 2008) for remediation of chromium rich waters and wastewaters by fly ash. These results suggest that the adsorption capacity increases with the increase of concentration up to a limit. This fact indicates that if the metal ion concentration in solution increases, the difference in concentration between bulk solution and surface also increases, intensifying the mass transfer processes (Kannan, et al., 2005).

Effect of Initial Concentrations. Results obtained from both column and batch technique confirmed that natural occurring volcanic tuff can remove heavy metal ions with high performance when applied at different concentrations ranged between 1 to 20 mg/L at varying acidic conditions as illustrated in table (3).

In general it can be observed that for some metal ions, the adsorption increases with increasing initial metal concentration, but under certain acidic conditions, and that leads to an increase in the amount of heavy metal ion adsorbed on to tuff grains. Whilst for other metals the adsorption decreases with increasing initial concentrations which emphasizes that the acidic condition is controlling the process. These results may be simply explained on the basis that the increase in the number of ions competing for the available binding spaces and also because of the lack of active sites on the absorbent at higher concentrations. Therefore, more metal ions were left in solution at higher levels of concentrations.

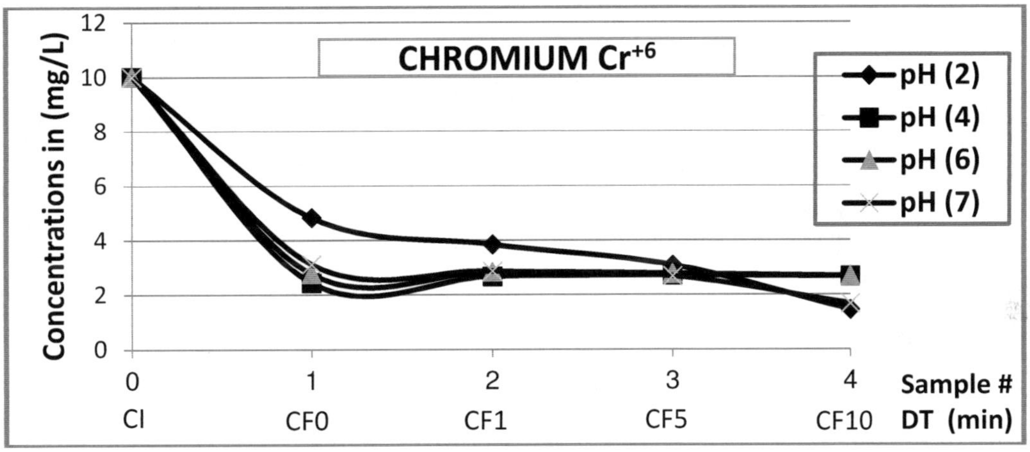

Figure 6. Removal of chromium Cr^{+6} from aqueous solution in association with contact time and pH value.

Figure 6 shows clearly that pH influence the process of absorption of Cr^{+6} on tuff material. At pH = 2 the ability of tuff material to absorb Chromium ions is low.

Uptake Capacity of Volcanic Tuff on Metal Ions. In equilibrium, a certain relationship prevails between solute concentration in solution and adsorbed state (i.e., the amount of solute adsorbed per unit mass of adsorbent). Their equilibrium concentrations are a function of temperature. Therefore, the adsorption equilibrium relationship at a given temperature is referred as adsorption isotherm.

The amount adsorbed (qe) was calculated from the formula:

$$qe = V\,(Ci - Ce)/m$$

where, C_i and C_e are the initial and equilibrium liquid-phase concentrations (mg/L) of adsorbates; V is the volume of the solution (mL); and m is the mass of adsorbent (g). This equation assumes that the change in volume of the bulk liquid phase is negligible as the solute concentration is small and the volume occupied by the adsorbent is also small. The amount of heavy metals adsorbed on the sample was calculated using a previously determined calibration curve.

The experimental results of the binding of the heavy metals (Cu^{+2}, Pb^{2+}, Cr^{2+}, Fe^{2+} and Zn^{2+}) ions on natural volcanic tuff indicate a higher degree of ion removal at (lower/higher) initial concentration area of metal ions.

The plot in Fig. 7 shows the uptake degree of each heavy metal ion as a function of the initial concentration and the prevailing acidic condition. The uptake degree is defined as:

$$\alpha = (Ci - Ce)/wt \quad [mg/l.\ g]$$

C_i being the initial concentration, and C_e final concentration of the particular ion, wt is the weight of tuff (g). α: uptake degree relates the decrease in concentration to the *wt* of absorbent. The analysis of obtained data shows the uptake of heavy metals depends mainly on: pH, initial concentration, contact time and which element is being absorbed.

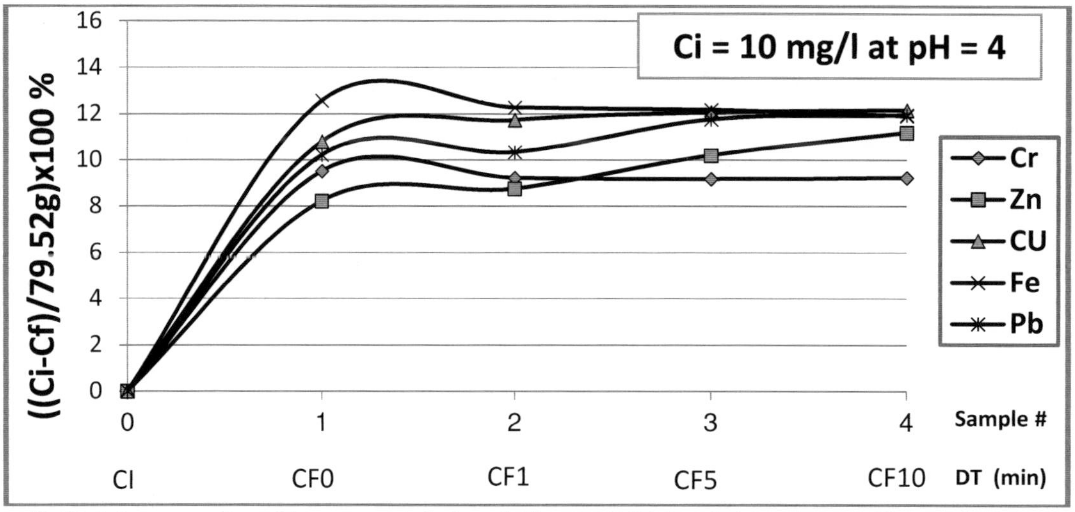

Figure 7. Uptake capacity of heavy metals in relation to initial solute concentration and contact time.

Figure 7 indicates that the volcanic bed behaves differently in respect to individual heavy metals is quite different. Each gram of the volcanic bed absorbed different amounts of heavy metals on selective basis and the velocity of uptake depends mainly on which heavy metal is being absorbed. At pH = 4, the absorption of chromium and iron is almost completed during the first minute of contact time in solid-liquid phase, while for zinc, lead and copper, the uptake process takes almost 10 minutes to be completed.

For comparison purposes, Figure 8 for instance, shows the uptake of applied heavy metal ions on material tuff at pH = 6. Under these circumstances, absorbent exhibits great affinity to absorb copper followed by iron and chromium. Another prove that prevailing acidic conditions have great role on controlling the absorption/adsorption process. In similar studies the use of natural and modified zeolites has been further investigated for the simultaneous removal of Fe+2 and Mn+2 from underground water samples. In particular, Fe+2 and Mn+2 removal levels were suggested to be between 22-90% and 61-100% for natural zeolite – clinoptilolite (Inglezakis, 2010).

Breakthrough Curves. Based on the fact that in equilibrium, a certain relationship exists between solute concentration in solution and the amount of solute adsorbed per unit mass of adsorbent. This equilibrium state depends on many factors such as, temperature, pH amount of absorbent and the initial concentration of solute. To examine the uptake capacity of absorbent (saturation state), two metals iron (Fe) and copper (Cu) are applied to a fixed bed column filled with 10 g of volcanic tuff in order to examine the time for reaching saturation of material with these two metal ions and the total amount absorbed. These two metals were selected based on absorption results presented in table 3 and in table 4. Analysis results suggested that both metal ions experience 100 % uptake by volcanic bed at pH = 6 and after 10 minutes contact time.

The batch adsorption experiment were conducted (started first with iron solution) with an initial concentration of 10 mg/l. 40 ml of this solution is applied to the column containing 10 g of bed material every 10 minutes. The small aliquots of samples were withdrawn from the reaction and analyzed.

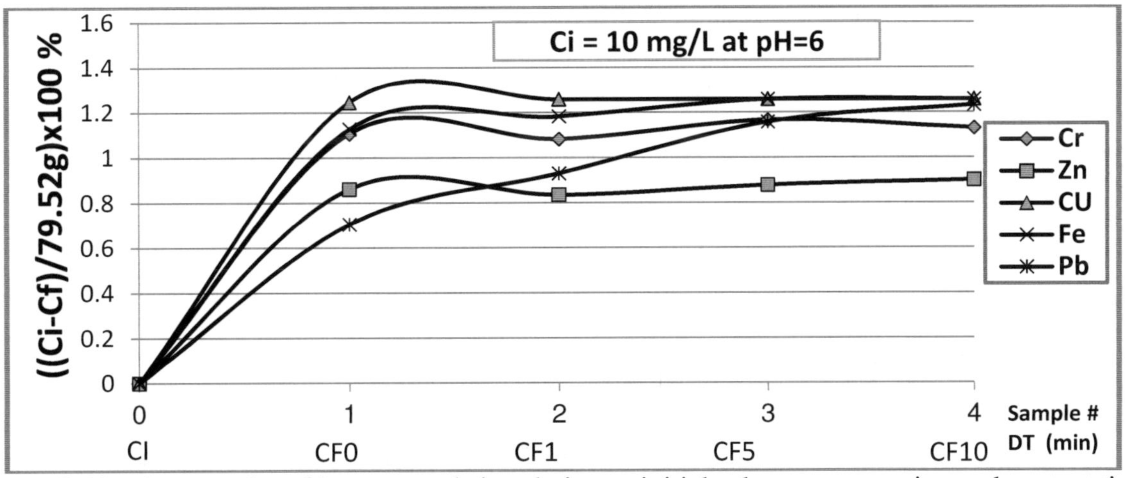

Figure 8. Uptake capacity of heavy metals in relation to initial solute concentration and contact time.

After exactly 335 minutes the analysis of the samples shows that the final concentration of iron in samples is still low, which means that the adsorption capacity of material is maintained high. In order to accelerate the saturation of the bed material, the experiment is then continued with an initial concentration of iron = 20 mg/l, also here with 10 minutes tact. After exactly 570 minutes the final concentration stabilizes and starts to increase to a final concentration very close to feed concentration, this gives indication that the material is saturated with iron ions and reaches its ultimate absorbing capacity.

It is to emphasize that iron is selected with the believe that because iron is a major component of the natural material and the absorbing capacity of the material for iron can be limited when compared to the absorbing capacity of tuff on other metal ions that are less present.

The experiment is repeated for copper solution with an initial concentration of 10 mg/l using the same bed that is previously brought to saturation with iron ions. It is to observe that after exactly 192 minutes the material is completely saturated with copper ions.

Uptake Capacity Factor for Iron and Copper. The Uptake capacity, UC, to absorb contaminants in this case metal ions, can be determined by using following mass balance equation:

$$UC = (V/m) \, x \, (C_i - C_f)$$

Where:

UC = contaminant loading at a given time period until saturation point [mg/g]

V = solution volume (mL). m = mass of the adsorbent material (g).

C_i = solute initial concentration in solution (mg/L), C_f = solute final concentration in solution [mg/L].

In this study, the iron uptake capacity and the breakthrough curve experiment was conducted and involves the application of 35 batch runs, 40 mL for each run with an initial concentration of 10 mg/L. In addition, there are 24 applications each 40 mL of solution with an initial concentration of 20 mg/L following the first batch. This iron loading is completely absorbed by the 79.52 g bed material until reaching the saturation point. An uptake capacity factor can be calculated using these data as follows:

Iron loading: 35 x 0.040 L x $\sum$(Ci –Cf) mg/L + 24 x 0.040 L x $\sum$(Ci –Cf) mg/L = 17.24 mg

Uptake capacity factor, UC_{Fe} = 17.24/79.52 g = 0.217 mg Fe/g tuff.

For copper and in order to reach the saturation point with copper, there are 30 applications with initial concentration = 10 mg/L needed.

Copper loading: 30 x 0.04 L x $\sum$(Ci –Cf) mg/L = 8 mg.

Uptake capacity factor, UC_{Cu} = 8.0 mg/79.52g = 0.10 mg Cu/g tuff.

Accordingly, the maximum absorption capacities for volcanic tuff were found as 0.217mg Fe/g tuff and 0.100 mg Cu/g tuff. An uptake ratio of iron uptake capacity and copper uptake capacity can be determined as UC_{Fe}/UC_{Cu} = 2.17. This result indicates that volcanic tuff is capable to absorb more iron rather than copper. Similar studies also confirms these results (Boldizsar et al., 2014). For copper uptake capacity, a study (Adaikalam et al., 2015) reported comparative results.

This experiment confirms that the uptake of iron ions Fe^{+3} (mg Fe^{+3} /g tuff) by tuff material declines gradually with time as more solution is added until a saturation (equilibrium) state is achieved. A correlation curve in Figure (9) demonstrates the association of declining uptake capacity with advancing time until saturation. Similar trend shows the uptake behavior of tuff material in regard to copper (Cu^{+2}).

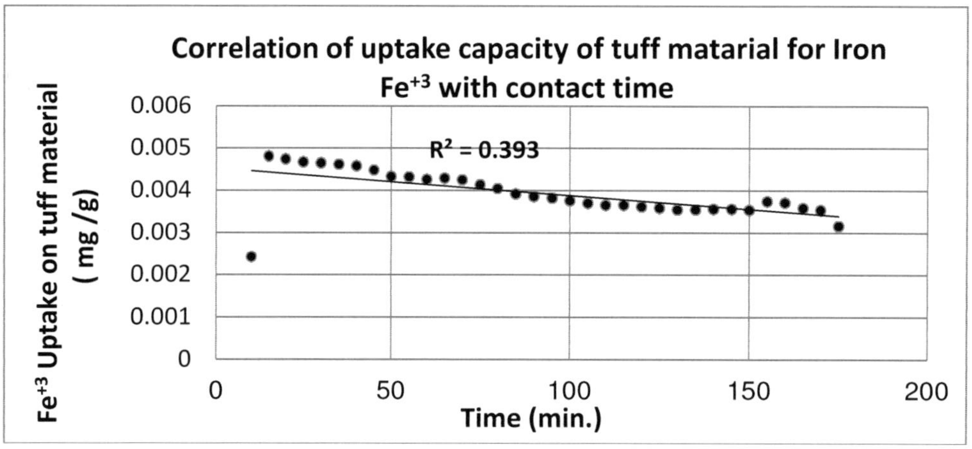

Figure 9. Correlation of uptake of iron ions on tuff material and contact time.

Sorption Isotherm Models. The term sorption is generally involves both expressions, absorption and adsorption. Adsorption is mostly estimated using the equation:

% Adsorption = $(C_i - C_f)/C_i) \times 100$ %

Where C_i *and* Cf are the concentrations of the metal ion in initial and final solution respectively (Hua et al., 2012; Fu et al., 2011).

Equilibrium studies that give the capacity of the adsorbent and adsorbate are described by adsorption isotherms, which is usually the ratio between the quantity adsorbed and the remaining in solution at fixed temperature at equilibrium (Hashem, 2007).

Several isotherm models are available to describe this equilibrium sorption distribution. Two of these isotherm equations were adopted in this study, as follows.

The Langmuir equation is used to estimate the maximum adsorption capacity corresponding to complete monolayer coverage on the adsorbent surface and is expressed by:

$$qe = (q_{max}\, K_L\, Ce) / (1 + K_L\, Ce)$$

where KL (dm^3g^{-1}) is a constant related to the adsorption/desorption energy, and q_{max} is the maximum sorption upon complete saturation of the biomass surface (Horshfall et al., 2004). The linearized form of the above equation after rearrangement is given by:

$$Ce / qe = 1/q_{max}\, K_L + Ce / q_{max}$$

The experimental data is then fitted into the above equation for linearization by plotting Ce/qe against Ce.

Freundlich Isotherm Equation. The Freundlich sorption isotherm, one of the most widely used mathematical descriptions, gives an expression encompassing the surface heterogeneity and the exponential distribution of active sites and their energies. The Freundlich isotherm is defined as:

$qe = KC_e^{1/n}$ and in linearized form is: $log\ q_e = log\ k + 1/n\ log\ C_e$

Where C_e is the equilibrium concentration in mg/l, qe = amount of adsorbate adsorbed per unit weight of adsorbent (mg/g). k is a parameter related to the temperature and "n" is a characteristic constant for the adsorption system under study.

Adsorption Experiments. After each adsorption, the residual metal ions of (Cr, Zi, Fe, Cu and Pb) were determined using following equations:

$$\%\ Adsorption = (C_i - C_f)/C_i) \times 100\ \%$$

$$X/M = V\,(C_i - C_f)/m$$

Where: V = volume of solute (mL), M = mass of adsorbent (mg), Ci = initial concentration (mg/L), Ce = final metal concentration at equilibrium (mg/L), X = the adsorbate concentration/adsorption capacity of tuff material (mg/g).

Five mixed solutions with concentrations 25 mg/L, 50 mg/L, 75 mg/L, 100 mg/L and 125 mg/L of Fe and Cu were made by proper dilution of stock solution for both metals. The pH was adjusted and maintained at 6 throughout the experiment. 25 mL of the prepared samples was poured into five flasks. 10 g of the adsorbent was introduced to each flask and was agitated for 1.25 hours. Thereafter the concentrations of Fe and Cu was determined.

The equilibrium data obtained from this experiment were observed to fit with Langmuir and Freundlich isotherm models. Plotting Ce/(x/m) against Ce for both metals as shown in Figures (10, 11) indicates that the equilibrium data follows the Langmuir model.

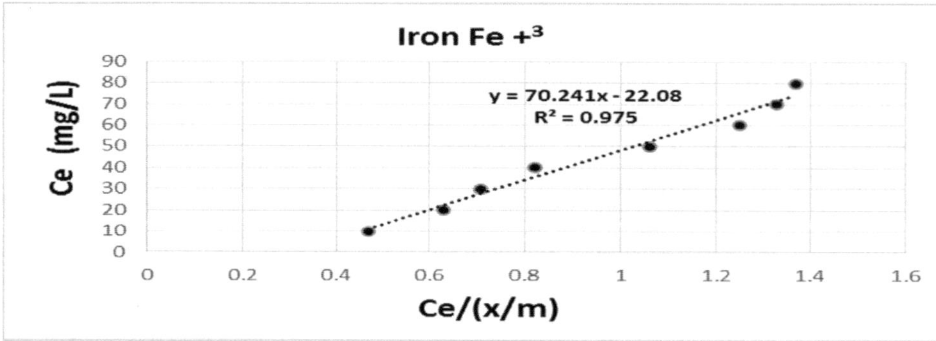

Figure 10. Removal of iron ions Fe^{+3} by volcanic tuff at pH = 6 fitted to Langmuir isotherm.

Langmuir parameters and correlation coefficient square R^2. The essential characteristics of the Langmuir isotherm can be expressed in terms of a dimensionless constant separation factor (R_L) which is defined by $R_L = 1/(1+ K_L C_e)$. Where R_L is Langmuir constant, C_e is the highest final concentration (mg/L). The value of R_L suggests the type of the isotherm to be either unfavorable (as $R_{L>1}$), linear $R_L = 1$, favorable $0< R_L< 1$ 0r may be irreversible if $R_L = 0$ (Malik, 2004). In this case, Langmuir isotherm is favorable as R_L = 0.966 and 0.944 for Fe^{+3} and Cu^{+2} respectively as shown in Figures 12 and 13.

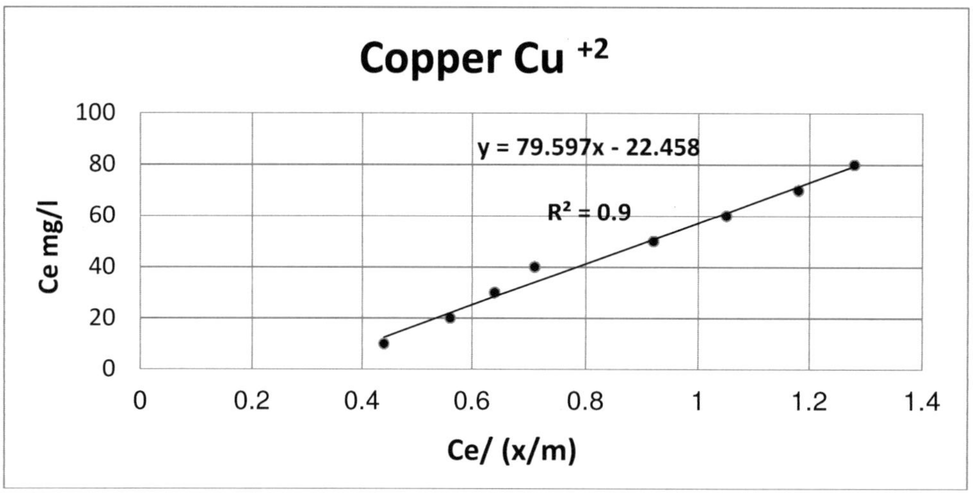

Figure 11. Removal of copper ions Cu^{+2} by volcanic tuff at pH = 6 fitted to Langmuir isotherm.

The correlation between *log x/m* and log *Ce* are presented in fig. and the adsorption of iron ions onto the adsorbent gave a straight line; values of "*n*" between 2 and 10 show good adsorption.

The kinetic studies of adsorption of heavy metals Fe^{+3} and Cu^{+2} ions onto volcanic tuff material was carried out using the second order models, the Freundlich isotherm constants and their correlation coefficients R^2 are listed in table 5.

The calculations presented in table 5 which are derived from Figures 10, 11, 12 and 13 indicate that the absorption process in following Langmuir and Frundlich isotherms with high certainty. Comparative results are reported by (Karnib, et al., 2014).

Relating absorbed concentrations of Cr^{+6} and Zi^{+2} to their applied initial concentrations as presented in Figure 14 and Figure 15 emphasizes the hypothesis that the absorption process of chromium and zinc ions also best fit to Langmuir and Freundlich isotherms.

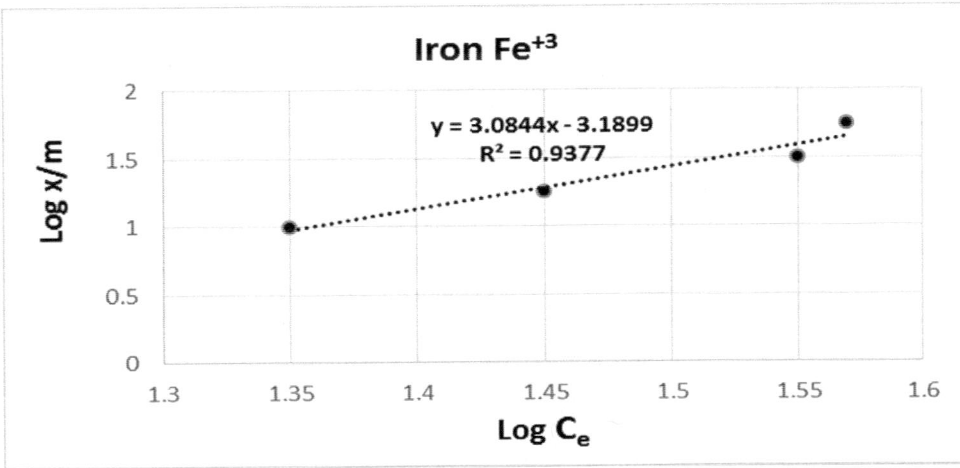

Figure 12. Removal of iron ions Fe^{+3} by volcanic tuff at pH = 6 fitted to Freundlich isotherm.

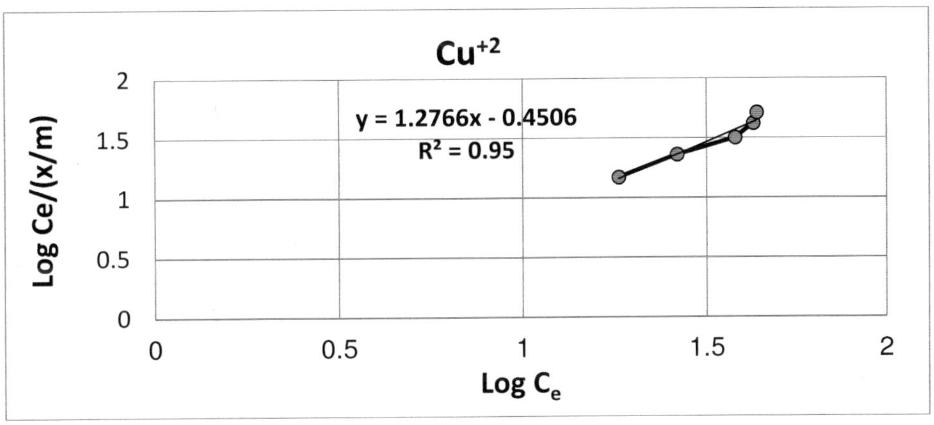

Figure 13. Removal of iron ions Cu^{+2} by volcanic tuff at pH = 6 fitted to Freundlich isotherm.

Table 5. Values for Langmuir and Freundlich adsorption models

Best-fit isotherm models	Fe^{+3}	Cu^{+2}
Langmuir model Plot (C_e/x/m) vs. C_e Correlation coefficient Regression equation	$R^2 = 0.975$ $y = 70.241\ x - 22.08$ $1/n = 0.51,\ \ k = 6.39$	$R^2 = 0.9874$ $y = 79.597\ x - 22.458$ $1/n = 0.51,\ \ k = 5.89$
Langmuir model Plot Log Ce/x/m vs. log C_e Correlation coefficient Regression equation	$R^2 = 0.9399$ $y = 3.0844\ x - 3.1899$ $q_{max} = 63.76\ (mg/g)$ $b = 0.42\ \ (1/mg)$	$R^2 = 0.95$ $y = 1.2766\ x - 0.4506$ $q_{max} = 68.46\ (mg/g)$ $b = 0.42\ \ (1/mg)$

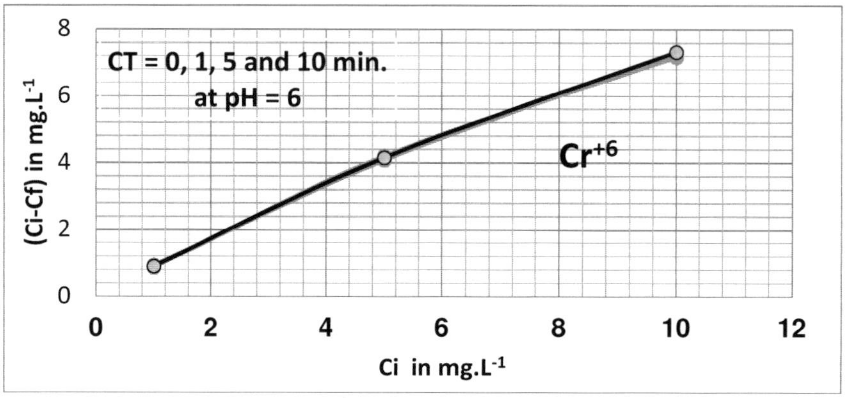

Figure 14. Removal of chromium ions Cr^{+6} by volcanic tuff at pH = 6 fitted to Freundlich isotherm.

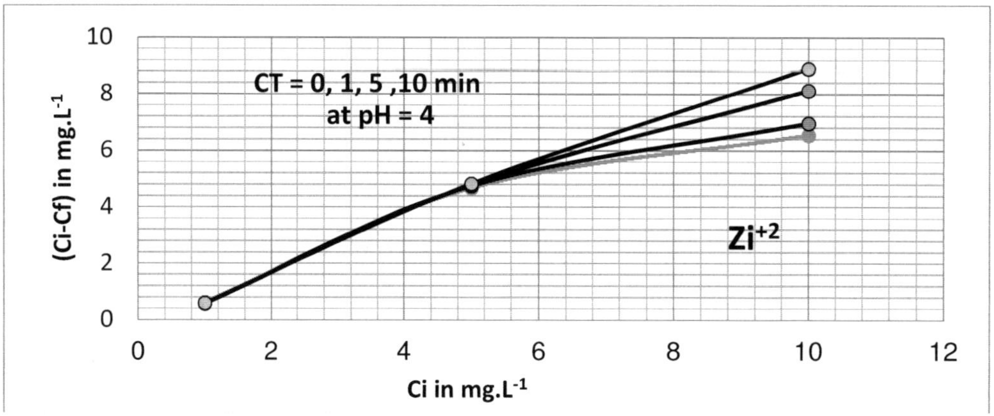

Figure 15. Removal of zinc ions Zi^{+2} by volcanic tuff at pH = 6 fitted to Freundlich isotherm.

CONCLUSIONS

On controlling and optimizing all the conditions studied in this work, it was confirmed that untreated natural volcanic tuff is an effective and inexpensive absorbent for the removal of Cr^{+6}, Fe^{+2}, Cu^{+2}, Zn^{+2} and Pb^{+2} from contaminated pharmaceutical wastewater and although unknown, it can compete well with commercial adsorbents such as treated zeolite and activated carbon.

The column experiments provide useful information about the transport behavior of the heavy metal contaminants and the relevant parameters for effective removal are contact time, ambient pH, initial concentration and temperature.

The sorption behavior of the used samples towards Cr^{+6}, Fe^{+2}, Cu^{+2}, Zn^{+2} and Pb^{+2} indicated high initial rate of metal ions uptake. The percentage removal of ions found to be between 10 – 99 % influenced mainly by ambient pH. Most probably, acidic metal solutions reduce the performance of natural volcanic tuff in retarding the movement of heavy metals and affect the mineralogical composition of the materials.

The natural volcanic tuff possess good retention capacity for cationic metals showing selectivity for metal uptake affinity For instance, when a solution containing two different metals is flowing through a fixed bed, the tuff material usually favors one ion over another. It showed that the selectivity sequence of metal ions by the adsorbents was dependent on the system employed, and was mainly influenced by the initial concentrations of the metal ions, but controlled also by the initial pH of the solution.

Obtained results showed that volcanic tuff has the greatest ability to remove Pb^{+2} at lower pH whereas at higher pH it favors the uptake of Fe and Cu ions. Obtained correlation coefficients $R^2 > 0.94$ suggested high fitting of the experimental data to Langmuir isotherm and Freundlich isotherm for both Cu and Fe ions. Results also confirm that the uptake of chromium and zinc ions best fit to Langmuir and Freundlich isotherm regimes.

Breakthrough curves for iron and copper shows that the uptake capacity of iron per gram of tuff material equals 0.217 mg Fe^{+2}/g tuff, while the uptake capacity for copper equals 0.1 mg Cu^{+2}/g tuff, suggesting the high ability of tuff material to eliminate iron rather than copper from aqueous solutions.

REFERENCES

Adaikalam S, Malairajan S, 2015. Removal of Cu(II) ions from synthetic waste water by using a novel biocarbon. Elixir Appl. Chem. 78, 29654-29656.

Ahmad Al-Haj Ali, A., Ribhi El-Bishtawi, R., 1997. Removal of Lead and Nickel Ions Using Zeolite Tuff. J. Chem. T ech. Biotechnol., 69, 27-34.

Amarasinghe B.M.W.P.K., R.A. Williams.2007. Tea waste as a low cost adsorbent for the removal of Cu and Pb from wastewater. Chem Eng J, 132 (1–3) , 299–309.

Anyakora C., Kenneth Nwaeze K., Awodele O., Nwadike C., Arbabi M., Coker H., 2001. Concentrations of heavy metals in some pharmaceutical effluents in Lagos, Nigeria. Journal of Environmental Chemistry and Ecotoxicology Vol. 3(2), 25-31, 2011. http://www.academicjournals.org/jece. ISSN-2141-226X.

Arnold K.E., Ross Brown A., Ankley G.T., Sumpter J.P., 2014. Medicating the environment: assessing risks of pharmaceuticals to wildlife and ecosystems. Philos. Trans. R. Soc. Lond., B, Biol. Sci. 369 (1656), 20130569. doi: 10.1098/rstb.2013.0569. PMCID: PMC4213582

Ashfaqa M., Khujasta N. K., Ur Rehman M.S., Mustafaa G., Faizan M.N., Sun Q., Iqbal J., Mulla S.I., Yu C-P. 2017. Ecological risk assessment of pharmaceuticals in the receiving environment of pharmaceutical wastewater in Pakistan. Ecotoxicology and Environmental Safety. Volume 136, 31–39.

Babel S., Kurniawan T.A., 2003. Low-Cost Adsorbents for Heavy Metals Uptake from Contaminated Water: A Review. Journal of Hazardous Materials, Vol. 97, No. 1-3, 219-243. doi:10.1016/S0304-3894(02)00263-7

Barakat M.A., 2011. New trends in removing heavy metals from industrial wastewater Arabian Journal of Chemistry, Volume 4, Issue 4, 361–377. http://dx.doi.org/10.1016/j.arabjc.2010.07.019

Bernard. E., Jimoh A., 2013. Adsorption of Pb, Fe, Cu, and Zn from industrial electroplating wastewater by orange peel activated carbon. International Journal of Engineering and Applied Sciences. Vol. 4, No. 2.

Boldizsar N., Mânzatu C. , Măicăneanu A., Tudoran C.I.L.B , Majdik C., Linear and nonlinear regression analysis for heavy metals removal using Agaricus bisporus macrofungus. Arabian Journal of Chemistry. http://dx.doi.org/10.1016/j.arabjc.2014.03.004

Bulinski R., Bloniarz J., Libelt J., 2009. Presence of some Trace Elements in Polish Food Products. XV. Contents of Lead, Copper, Cadmium, Nickel, Chromium, Zinc, Cobalt, Manganese, Copper and Iron in some Milk. Products. Bromatologia i. Chemia Toks, 26, 23-27.

Caputo D, Pepe F., 2007.Experiments and data processing of ion exchange equilibria involving Italian natural zeolites: a review. Micropor.Mesopor.Mater. 105, 222-231

Cho H., D. Oh, K. Kim, K., 2005. A study on removal characteristics of heavy metals from aqueous solution by fly ash. J Hazard Mater, 127 (1-3), 187–195

Chutia P., Kato, S., Kojima, T., Satokawa, S., 2009. Arsenic adsorption from aqueous solution on synthetic zeolites. J. Hazard Mater, 162 (1), 440–447

Dimirkou A., 2007. Uptake of Zn^{2+} ions by a fully ironexchanged clinoptilolite. Case study of heavily contaminated drinking water samples, Water Res., 41 , 2763–2773.

Dorne J.L., Kass GE, Bordajandi LR, Amzal B, Bertelsen U, Castoldi AF, Heppner C, Eskola M, Fabiansson S, Ferrari P, Scaravelli E, Dogliotti E, Fuerst P, Boobis AR, Verger P., 2011. Human risk assessment of heavy metals: principles and applications. Met Ions Life Sci. 8, 27-60.

Erdem E., Karapinar N., Donat R., 2004. The removal of heavy metal cations by natural zeolites. Journal of Colloid and Interface Science 280 , 309–314. www.elsevier.com/locate/jcis

Esposito A., Pagnanelli, F. Vegliò., 2002. pH-related equilibria models for biosorption in single metal systems, Chem. Eng. S

Fu, F, Qi, W., 2011. Removal of heavy metal ions from wastewaters: A review Journal of Environmental Management. Volume 92, Issue 3, 407–418

Gadipelly, C., Pérez-González, A., Yadav, G.D., Ortiz, I., Ibáñez, R., Rathod, V.K., Marathe, K.V., 2014. Pharmaceutical Industry Wastewater: Review of the Technologies for Water Treatment and Reuse. Ind. Eng. Chem. Res., 53 (29), 11571–11592. DOI: 10.1021/ie501210j

Hashem M.A., 2007. Adsorption of lead ions from aqueous solution by okra wastes Int J Phys Sci, 2 (7) , 178–184

Hernandoa M.D., Mezcuaa, M., Fernández-Albaa, A.R., Barceló b, D., 2006. Environmental risk assessment of pharmaceutical residues in wastewater effluents, surface waters and sediments. Talanta, Volume 69, Issue 2, 334–342

Horsfall, M. J., Abia, A. A., and Spiff, A. I. 2006. Kinetic Studies on the Adsorption of Cd2+, Cu2+ and Zn2+ Ions from Aqueous Solutions by Cassava (Manihotesculenta) Tuber Bark Waste. Journal Bioresource Technology, 97(35), 283-291.

Hua Ming, Shujuan Zhang, Bingcai Pan, Weiming Zhang, Lu Lv, Quanxing Zhang. 2012. Heavy metal removal from water/wastewater by nanosized metal oxides: A review. Journal of Hazardous Materials. Volumes 211–212, 317–331

Hui K.S., Chao,C.Y. Kot, S.C., 2005. Removal of mixed heavy metal ions in wastewater by zeolite 4A and residual products from recycled coal fly ash. Journal of Hazardous Materials, B127 , 89–101.

Idris M.A., Kolo, B. G., Garba, S. T. and Waziri, I., 2013. Pharmaceutical Industrial Effluent: Heavy Metal Contamination of Surface water in Minna, Niger State, Nigeria. Bull. Env. Pharmacol. Life Sci. Volume 2 (3) , 40-44.

Inglezakis V.J., Loizidou, M.D., Grigoropoulou, H.P., 2003. Ion exchange of Pb2+, Cu^{2+}, Fe^{2+}, and Cr^{2+} on natural clinoptilolite: selectivity determination and influence of acidity on metal uptake, J. Colloid. Interface Sci. 261, 49–54.

Inglezakis V.J., Doula MK, Aggelatou V, Zorpas AA., 2010. Removal of iron and manganese from underground water by use of natural minerals in batch mode treatment. Desalin. Water Treat. 18 , 341-346.

Inglezakis V.J., Poulopoulos, S., 2006. Adsorptio n, Ion Exchange and Catalysis: Design of Operations and Environmental Applications, Elsevier Science Ltd, 2006.

Inglezakis V. J., Stylianou, M.A., Loizidou, M., Zorpas, A.A., 2016. Experimental studies and modeling of clinoptilolite and vermiculite fixed beds for Mn^{2+}, Zn^{2+}, and Cr^{3+} removal. Desalination and Water Treatment. Vol. 57, Issue 25.

Kannan, N., and Rengasamy, G. 2005. Comparison of cadmium ion adsorption on various activated carbons, Water Air Soil Pollut. 163, 185–201.

Karnib M., Kabbani, A., Hanafy H., Olama, Z., 2014. Heavy Metals Removal Using Activated Carbon, Silica and Silica Activated Carbon Composite Energy Procedia 50, 113 – 120. doi: 10.1016/j.egypro.2014.06.014

Khan S., Cao, Q., Zheng, Y.M., Huang Y.Z., Zhu, Y.G., 2008. Health risks of heavy metals in contaminated soil and food crop irrigated with wastewater in Beijing, China. Environmental Pollution 152 , 686-692.

Konstantinos, D., Achilleas, C., and Valsamidou, V., 2011. Removal of Nickel, Copper, Zinc and Chromium from Synthetic and Industrial Wastewater by Electrocoagulation. International Journal of Environmental Sciences, 1(5), 698-703.

Lisa, N., Kanagaratnam, B., Trever, M., 2004. Biosorption of Zinc from aqueous solutions using biosolids. Adv. Env. Res., 8, 629-635.

Malik, P., 2004. Dye Removal from Wastewater using activated Carbon Developed from Sawdust. Adsorption equilibrium and kinetics. Journal of Hazardous Materials, Vol. 113, 81-88.

Margeta, K., Vojnović, B., Zabukovec Logar, N., 2011. Development of natural zeolites for their use in water-treatment systems. Recent patents of nanotech. 5:2, 89-99.

Meena A.K., Kadirvelu, K., Mishra, G.K., Rajagopal, C., Nagar, P.N., 2008.

Adsorptive removal of heavy metals from aqueous solution by treated sawdust (Acacia arabica). J Hazard Mater, 150 (3) , 604–611

Nalan, E.A ., Gulhayat, N. S., 2009. Column experiments to remove copper from wastewaters using natural zeolite. International Journal of Environment and Waste Management, Volume 3, Number 3-4 , 319 – 326.

Núñez, O., Fernández-Navarro, P., Martín-Méndez, I., Bel-Lan, A., Locutura, J.F., López-Abente, G., 2016. Environmental Science and Pollution Research International. 23(17), 17664-17675

Abdel Salam,O.E., Reiad, N. A., ElShafei, M.M., 2011. A study of the removal characteristics of heavy metals from wastewater by low-cost adsorbents. Journal of Advanced Research. Volume 2, Issue 4, 297–303

Olayinka, KO., Oyedeji, OA., Oyeyiola, OA., 2009 . Removal of chromium and nickel ions from aqueous solution by adsorption on modified coconut husk. Afr. J. Env. ScI. Tech., 3(10), 286-293.

Orodu, V.E., Olis, de me., Okpu R.C., 2014. Removal of heavy metals from aqueous solutions using snail shell powder as a vailable adsorbent. IJST, Vol. 3, No. 7.

Osaigbovo, A.U., Orhue, E. R., 2006. Influence of pharmaceutical effluent on some soil chemical properties and early growth of maize (Zea mays L). African Journal for Biotechnology. Vol.5 (18), 1612-1617. DOI: 10.5897/AJB06.032. ISSN: 1684-5315

Panayotova M., B. Velikov, B., 2003. Influence of zeolite transformation in a homoionic form on the removal of some heavy metal ions from wastewater. J. Environ. Sci. Health A Toxic Hazard Subst Environ Eng, 38 (3) , 545–554

Roberts P.H., Thomas, K.V., 2006. The occurrence of selected pharmaceuticals in wastewater effluent and surface waters of the lower Tyne catchment. Science of The Total Environment. Volume 356, Issues 1–3, 143–153

Rzymski, P., Tomczyk, K., Rzymski, P., Poniedziałek, B., Opala, T., Wilczak, M., Agric, A., 2015. Impact of heavy metals on the female reproductive system. Environ Med., 22(2), 259-64.

Sharma Y.C., Uma, U.S.N., Weng, C.H., 2008. Studies on an economically viable remediation of chromium rich waters and wastewaters by PTPS fly ash Colloids Surf A Physicochem Eng Aspects, 317 (1-3), 222–228

Stern BR., 2010. Essentiality and toxicity in copper health risk assessment: overview, update and regulatory considerations. Toxicol Environ Health A. 73(2), 114–127.

Stylianou M.A., Kollia D., Haralambous K.J., Inglezakis V.J., Moustakas, K.G. Loizidou M.D., 2006. Effect of acid treatment on the removal of heavy metals from sewage sludge. Desalination, 215, 73-81. Doi: 10.1016/j.desal.2006, 11.015.

Tchounwou P.B., Yedjou, C.G., Patlolla, A.K., Sutton, D. J., 2012. Heavy Metals Toxicity and the Environment. EXS. 101, 133–164. doi: 10.1007/978-3-7643-8340-4_6. PMCID: PMC4144270

Trgo M., Peri´c, J., 2003. Interaction of the zeolitic tuff with Zncontaining simulated pollutant solutions, J. Colloid Interface Sci., 260, 166–175.

Verlicchia P., M. Al Aukidya, E. Zambelloa, Occurrence of pharmaceutical compounds in urban wastewater: Removal, mass load and environmental risk after a secondary treatment—A review Science of The Total Environment. Volume 429, 1 July 2012, Pages 123–155

Wang, S., Peng, Y., 2010. Natural zeolites as effective adsorbents in water and wastewater treatment. Chem. Engin.J. 156, 11-24.

Xiao, F., Ju-Chang, HH., 2009. Comparison of biosorbents with organic sorbents for removing Copper (II) from aqueous solutions. J. Environ. Man., 90, 3105-3109.

Zamzow, M. J., Eichbaum, B. R., Sandgren, K.R., Shanks, D.E., 1990. Removal of heavy metals and other cations from waste water using Zeolites; Sep. Sci. and Technol.; 25 (13-15), 1555-1569

Ziyath, M., Mahbub, P., Goonetilleke, A., Adebajo, M., Kokot, S., Oloyede, A., 2011. Influence of Physical and Chemical Parameters on the Treatment of Heavy Metals in Polluted Stormwater Using Zeolite—A Review. JWARP. Vol.3 No.10. DOI: 10.4236/jwarp.2011.310086.

In situ Co-Precipitation Preparation of a Superparamagnetic Graphene Oxide/Fe_3O_4 Nanocomposite as an Adsorbent for Wastewater Purification: Synthesis, Characterization, Kinetics and Isotherm Studies

Shengyan PU[1,2*#], Shengyang XUE[1#], Zeng YANG[1], Yaqi HOU[1], Rongxin ZHU[1], Wei CHU[2]

1. State Key Laboratory of Geohazard Prevention and Geoenvironment Protection, Chengdu University of Technology, 1 Dongsanlu, Erxianqiao, Chengdu 610059, Sichuan, China.
2. Department of Civil and Environment Engineering, The Hong Kong Polytechnic University, Hong Kong, China.

*: Corresponding author. E-mail addresses: pushengyan@gmail.com; □ush□ngy□n13@cdut.cn.
#: The authors declare that there is no conflict of interests regarding the publication of this paper. S. Xue and S. Pu contributed equally to this work.

ABSTRACT: A superparamagnetic GO/Fe_3O_4 nanocomposite (MGO) was synthesized to remove different type of dye-contaminated wastewater. The resultant adsorbent was prepared by a facile in-situ co-precipitation strategy. The adsorption behaviors of MGO's performance towards common cationic dye methylene blue (MB) and anionic dye congo red (CR) were investigated, including pH effect, initial dye concentration, adsorption kinetics, adsorption equilibrium and thermodynamic study. The maximum adsorption capacity were ca. 546.45 mg g^{-1} for MB and ca. 628.93 mg g^{-1} for CR, respectively. The adsorption process fits the pseudo-second-order kinetic model well (R^2>0.9998) and the adsorption equilibrium of MB and CR on MGO followed the Langmuir model (R^2>0.9995). Meanwhile, the obtained magnetic adsorbent had a wide pH adaptation and showed good recycle stability after five cycles its removal rate could still maintain at 87.23%. The resultant MGO shows great potential as a promising adsorbent for the organic contaminant removal in wastewater treatment.

Keywords: In situ co-precipitation; graphene oxide; magnetic absorbent; dye wastewater

INTRODUCTION

Pollution from dye wastewater has drawn worldwide public concern due to its high biological toxicity, carcinogenicity, teratogenicity and mutagenicity in living organisms, especially human beings (Belpaire et al. 2015, Forgacs et al. 2004). Various methods such as chemical oxidation (Lucas &Peres 2006, Pu et al. 2017a), photochemical catalysis(Pu et al. 2017b), membrane separation (Ma et al. 2017, Netpradit et al. 2003), electrochemical degradation (Riera-Torres et al. 2011, Szpyrkowicz et al. 2001) and biodegradation (Malachova et al. 2013) have been developed for the purification of dye wastewater. However, the application of these techniques has been limited due to their inevitable drawbacks, such as high energy consumption, secondary pollution and high operating costs. Beyond the techniques listed above, absorption is considered to be one of the strategies with the most potential due to its advantages of easy operation, low energy consumption, few harmful byproducts, the possibility of adsorbent recycling, etc.(Ma et al. 2018). Adsorption approaches may be based on various adsorbents such as chitosan (Pu et al. 2017d), activated carbon (Greenwald et al. 2015), clay (Abidi et al. 2015), and bio-waste (Kyzas et al. 2013). Unfortunately, the practical application of adsorbents has been limited by their poor utilization rate, difficult recovery and low adsorption capacity (Akbari et al. 2002).

Graphene oxide (GO) and graphene-based composite materials (Pu et al. 2017c) are attracting tremendous research interest as a novel adsorbent for environmental purification due to their enormous specific surface area (1000-1217 $m^2 g^{-1}$) (Tang et al. 2010), large number of active sites (Dreyer et al. 2010) and excellent dispersibility (Cheng et al. 2017). Its hydrophilic oxygen-containing functional groups (e.g., carboxyl, hydroxyl and epoxy functional groups, etc.) (Dreyer et al. 2010, Huang et al. 2016, Lai et al. 2013) allow GO to combine easily with hydrophilic and hydrophobic compounds (e.g. heavy metals, dyes and sediment), which gives GO great potential in wastewater treatment (Liu et al. 2012). GO has been proved to possess great potential as superior adsorbent for removing many kinds of organic and inorganic contaminants including heavy metal ions (Peng et al. 2017b), uranium(VI) (Zhang et al. 2016b), fluoride (Barathi et al. 2014), methylene blue (Zhang et al. 2016a), dibenzothiophene (Dizaji et al. 2016) polyphenols (Peng et al. 2017a), etc. Recently, magnetic grapheme oxide has been widely used as a new magnetic solid-phase extraction adsorbent in environmental purification applications (Aliyari et al. 2016). In general, magnetic adsorbents are often composed of magnetic nanoparticles dispersed in a crosslinked polymer matrix, including natural and synthetic polymers (Yao et al. 2012). However, these methods suffer many drawbacks such as high cost, complicated operation and easy clogging.

In this study, we primarily focused on developing a recyclable, multi-functional and high efficiency adsorbent for dye wastewater purification. We prepared a superparamagnetic graphene oxide/Fe_3O_4 hybrid nanocomposite (MGO) via an in situ co-precipitation strategy, which was successfully obtained through a facile and green in situ synthetic method. The common dyes congo red (CR, anionic dye) and methylene blue (MB, cationic dye) were used to evaluate its adsorption properties. Batch experiments under varied influential parameters, such as initial pH, dye concentration, temperature and coexisting ions, were conducted to assess the properties of MGO. The synthesized MGO demonstrated good performance for dye removal and a wide range of abilities to adsorb many types of dyes, including both anionic and cationic dyes. The results demonstrate potential for the application of graphene oxide in wastewater treatment.

MATERIALS AND METHODS

Materials. The Graphite powder was purchased from Aladdin Reagent Factory (Shanghai, China). Sodium nitrate ($NaNO_3$, ≥ 99.9%), sulfuric acid (H_2SO_4, 98%), potassium permanganate ($KMnO_4$, 65-68%), hydrogen peroxide (H_2O_2, 30 wt%), ammonia (NH_4OH, 25-28%), ethanol (C_2H_5OH, ≥ 99.9%), methylene blue (MB) and congo red (CR) were provided by Chengdu Kelong Chemical Reagent Company (Sichuan, China). $FeCl_2 \cdot 4H_2O$ and $FeCl_3 \cdot 6H_2O$ were purchased from Tianjin Zhiyuan Chemical Reagent Company (Tianjin, China). All chemical reagents were analytical grade and used without further purification. Deionized water used for all experiments was generated from Milli-Q water purification system (Ulupure Corporation, CA).

Experimental Section

Preparation of Graphene Oxide. Graphene oxide was synthesized from graphite powder via a modified Hummers method (Stankovich et al. 2006). Typically, 2.0 g graphite powder and 1.2 g sodium nitrate were added into 60 mL concentrated sulfuric acid (98%) solution in a three-neck flask immersed in an ice bath while maintaining mechanical stirring. A total of 8.8 g potassium permanganate was slowly added into the flask and the temperature was cooled to 278.15 K for 12 h. Subsequently, 72 mL distilled water was added dropwise into the mixture at 293.15 K, and the mixture was stirred at 323.15 K overnight. After the color of the mixture turned bright yellow, 22 mL H_2O_2 solution was added gradually into the mixture for 2 h. The mixture was washed successively with 5% HCl solution and deionized water until the pH of the rinse water became neutral, after which it was freeze-dried (Scientz-50F, China) for more than

32 h. Finally, 0.3 g GO was dispersed in 100 mL water by an ultrasonic cell disruptor to obtain a GO dispersion (3 mg mL^{-1}).

Preparation of Magnetic Graphene Oxide Nanocomposite. The magnetic graphene oxide nanocomposite (MGO) was synthesized by a chemical in situ co-precipitation method. Briefly, 1.3 mmol $FeCl_2 \cdot 4H_2O$ and 2.6 mmol $FeCl_3 \cdot 6H_2O$ were dissolved in 400 mL deionized water under an N_2 atmosphere. An appropriate molar ratio of $FeCl_2$ and $FeCl_3$ was adopted (1:2) to meet the stoichiometric requirement. To prevent the aggregation of the GO sheets in a low pH (pH = 1.89) environment, 0.7 mL $NH_3 \cdot H_2O$ was used to adjust the pH before adding 100 mL GO. Then 40 mL $NH_3 \cdot H_2O$ was continuously added with vigorous stirring for 3 h at 353.15 K. Finally, the black precipitate was washed several times with deionized water and ethanol to remove the residual alkali solution, which was heated at 333.15 K in vacuum conditions for 8 h.

Batch Adsorption Experiments. Two typical cationic and anionic dyes, MB and CR, were used as representative contaminants to evaluate the adsorption properties of MGO. All batch adsorption experiments were performed in a constant temperature shaker. Briefly, adsorption experiments were carried out by adding 25 mg MGO powder into 50 mL dye solution in a shaker (298.15 ± 0.5 K) for 24 h to reach adsorption equilibrium. The adsorbent was removed by magnetism and the remaining concentration of dye was measured by a visible spectrophotometer (V-1100D, Mita, Shanghai) at wavelengths of 664 nm for MB and 497 nm for CR. The adsorption capacity (q_e) of MGO was calculated using equation (1) (Ren et al. 2013) as the following:

$$q_e = \frac{(C_0 - C_t)V}{m} \quad (1)$$

The removal rates (η) of the dyes were determined by using equation (2):

$$\eta = \frac{C_0 - C_t}{C_0} \quad (2)$$

where, C_0 (mg L^{-1}) is the initial concentration of the test solution, C_t is the concentration of the test solution (mg L^{-1}) at time t (min), V (L) is the volume of the test solution, and m (g) is the weight of the adsorbent. All experiments were performed in three parallel groups for accuracy.

To assess the effects of initial pH, different temperatures and initial concentration of dye on adsorption, consistent conditions were maintained except when changing the experimental variable. To investigate the effect of initial pH on adsorption, pH was adjusted by adding a small amount of HCl (1 mol L^{-1}) and NaOH (1 mol L^{-1}) while the solution volume and other conditions were constant. To assess the effects of different temperature, the reaction temperatures were adjusted by the water bath from 298.15 to 328.15 K ± 0.5 K. The effect of the initial dye concentration was measured by adsorbing different concentrations of dye (from 50 mg L^{-1} to 1500 mg L^{-1}). The effect of coexisting ions on adsorption was studied by adding known amounts of Na^+, Ca^{2+}, K^+, NO^{3+}, Cl^- and HPO^{4-}. The concentration of the residual dye was analyzed at established time intervals during the adsorption process.

Desorption Studies. To evaluate the stability and reusability of the adsorbent, the adsorbent was collected from solution after reaching equilibrium and washed several times with ethanol, NaOH (1 mol L^{-1}) and deionized water before being reused for the next adsorption. The above process was repeated five times.

Characterization. The crystal structure of MGO was characterized by XRD (Ultima IV, Japan) employing Cu K_α radiation (λ = 0.154 nm, 40 KV) in the range of 3° to 80°. FTIR (Nicolet-1170 SX, USA) was used to analyze the functional groups of MGO with the KBr wafer technique. The scanning wavelength range was from 400 to 4000 cm^{-1}. The surface morphology was characterized by SEM (SIGMA300, Germany) at a voltage of 20 kV. The micromorphology was observed by TEM (JEM2100, Japan) operated at 200 kV. The magnetic properties were recorded at room temperature with a vibrating sample magnetometer (VSM) (730T Lakeshoper, America). Samples were pretreated by heating under vacuum at 403.15 K for 10 h.

RESULTS AND DISCUSSION

Preparation and Characterization of MGO. The MGO composite was synthesized through a chemical in-situ co-precipitation approach (Fig. 1a). The color of the mixture rapidly changed from dark brown to black after $NH_3{\cdot}H_2O$ addition (Fig. 1b), as the in situ co-precipitation reaction between the Fe^{3+}, Fe^{2+} mixture and alkali solution formed magnetic Fe_3O_4 nanoparticles with the following reaction (Yang et al. 2012):

$$FeCl_2 + 2FeCl_3 + 8NH_4OH \rightarrow Fe_3O_4 \text{ (on GO)} + 8NH_4Cl + 4H_2O \qquad (3)$$

The color of mixture turned from brown to black, indicating the formation of Fe_3O_4. The resulting MGO powder was a black sand-like powder (**Fig. 1b**).

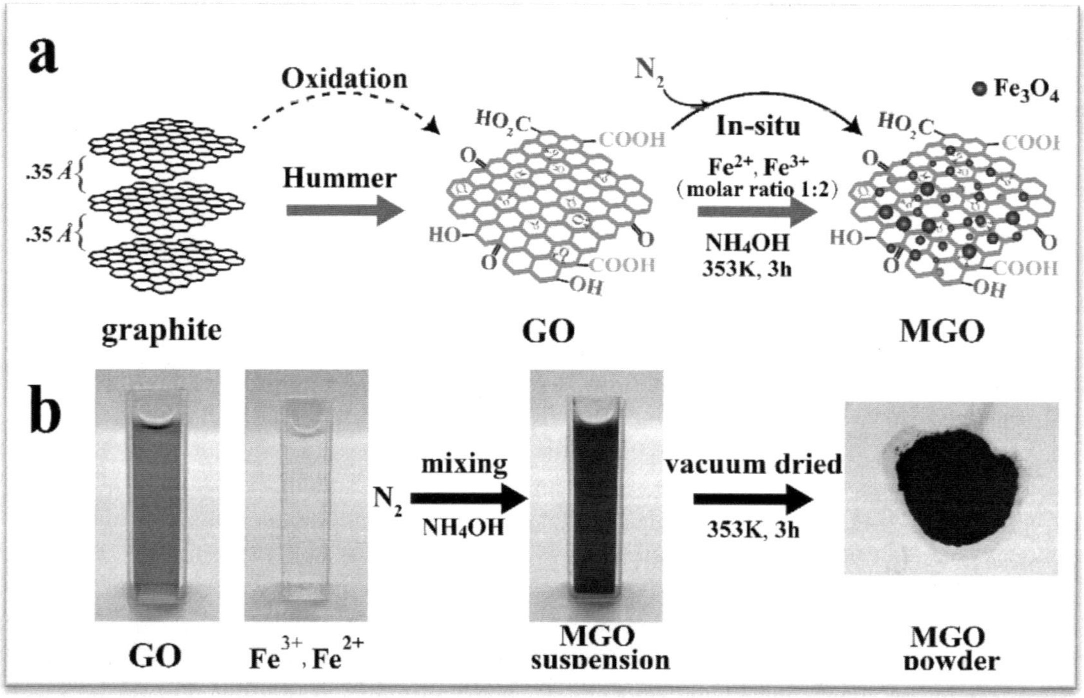

Figure 1. a) Schematic illustration of MGO preparation via in-situ co-precipitation. The graphene oxide was prepared by a modified Hummer's method; **b**) Photographs of GO solution, combined Fe^{3+} and Fe^{2+} solution, MGO suspension and the obtained MGO powder.

The crystalline phases of the pure Fe_3O_4, GO and synthesized MGO were identified with XRD (**Fig. 2a**). The characteristic diffraction peaks at $2\theta = 30.10°$, 35.4°, 43.11°, 56.28° and 62.74°, corresponded with the planes at (220), (311), (400), (511), and (440), respectively, which is in strong accordance with Fe_3O_4 nanoparticles (JCPDS No. 19-0629) (Wang et al. 2015). The GO displayed a broad diffraction peak ranging from 15° to 30°. The XRD pattern of MGO exhibited similar XRD patterns to Fe_3O_4 and the characteristic diffraction peaks matched well with those of Fe_3O_4. There was a weak diffraction peak of GO detected in the MGO, possibly due to the strong diffraction peaks of Fe_3O_4, which made the GO peak (25°) appear weak, indicating that the crystalline structure of GO was not changed substantially during synthesis. The EDX element distribution of MGO is shown in **Fig. S1**. The peaks of O, C and Fe indicated that Fe compounds were successfully introduced into the GO surface. Based on XRD analysis, we can describe the composite formed by Fe_3O_4 and GO.

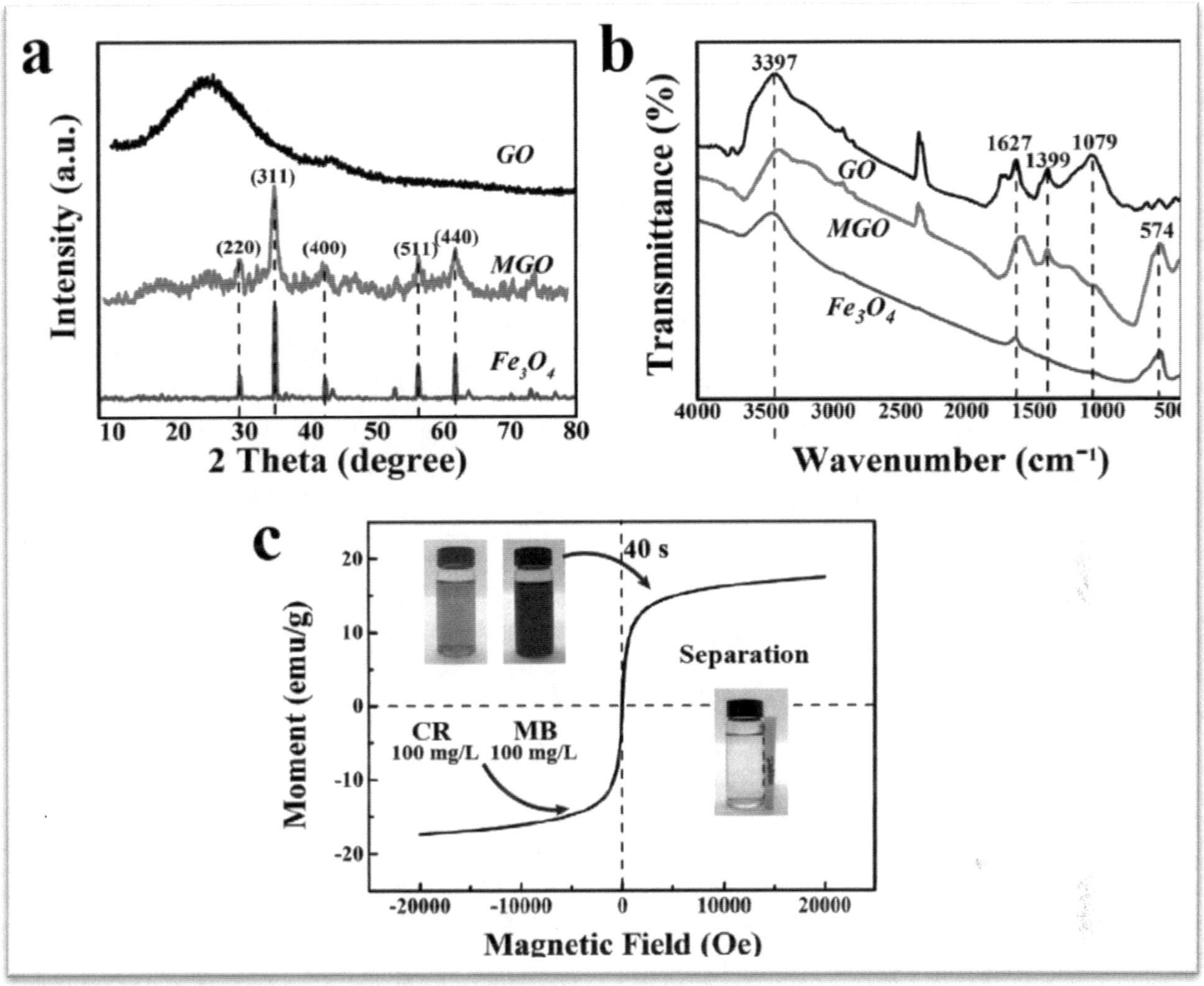

Figure 2. a) XRD patterns of GO, MGO and Fe_3O_4; **b**) FTIR spectra of GO, MGO and pure Fe_3O_4 NPs; **c**) Magnetic hysteresis loops of MGO. Conditions: initial MB/CR concentration 100 mg L^{-1}; adsorbent dosage 0.5 g L^{-1}; temperature 298 ± 0.5 K; stirring rate 180 rpm

The FTIR spectra of GO, Fe_3O_4 and MGO are shown in **Fig. 2b**. The pure GO curve showed obvious characteristic absorption peaks at 3397 cm^{-1}, 1737 cm^{-1}, 1627 cm^{-1}, 1399 cm^{-1} and 1079 cm^{-1}. The adsorption bands at 3397 cm^{-1} and 1627 cm^{-1} were attributed to -OH stretching and C=C stretching vibrations. The bands at 1737 cm^{-1} and 1399 cm^{-1} were mainly ascribed to C=O and O-C=O stretching vibrations. The band at 1079 cm^{-1} was attributed to the stretching vibration of sp^2 hybridized of C-O-C. A comparison of GO to MGO showed that the characteristic peaks at 1737 cm^{-1} and 1079 cm^{-1} disappeared, and a peak at 574 cm^{-1} emerged which was assigned to the Fe-O stretching vibration. Data analyses indicated that the Fe_3O_4 nanoparticles were successfully embedded in the graphene oxide. Therefore, it was evident that some functional groups of GO were reduced via the in situ co-precipitation process (Wan &Wang 2017). Moreover, the dye adsorption by MGO could be attributed to hydrogen bonding interactions between O-C=O, -OH and the dye's characteristic groups. Carboxyl O-C=O groups were responsible for interactions between the D-SO_3^- (D = dye molecule) and the MGO.

The magnetic performance of the prepared MGO nanocomposite was determined by VSM. Typical magnetization hysteresis loops were recorded and are illustrated in **Fig. 2c**. The saturation magnetization was found to be 17.38 emu g^{-1}. The magnetization curve of MGO was close to an "S" curve and the coercivity was negligible (nearly zero), indicating that no magnetization remained after removing the external magnetic field (Ren et al. 2013). Therefore, the superparamagnetic behavior of nanocomposite MGO was established. Moreover, the inset in **Fig. 2c** illustrates the color change in the dye solution before and after adsorption. The adsorbent dosage was 0.5 g L^{-1} and the initial dye concentration was 100 mg L^{-1}. After adsorption reaction, the MGO was easily separated from aqueous solution within 40 s under a magnetic field, and the supernatant was colorless, as shown in **Fig. 2c**. Thus, the obtained MGO had highly efficient adsorption (removal rate can reach 99%) properties and could be readily magnetically separated.

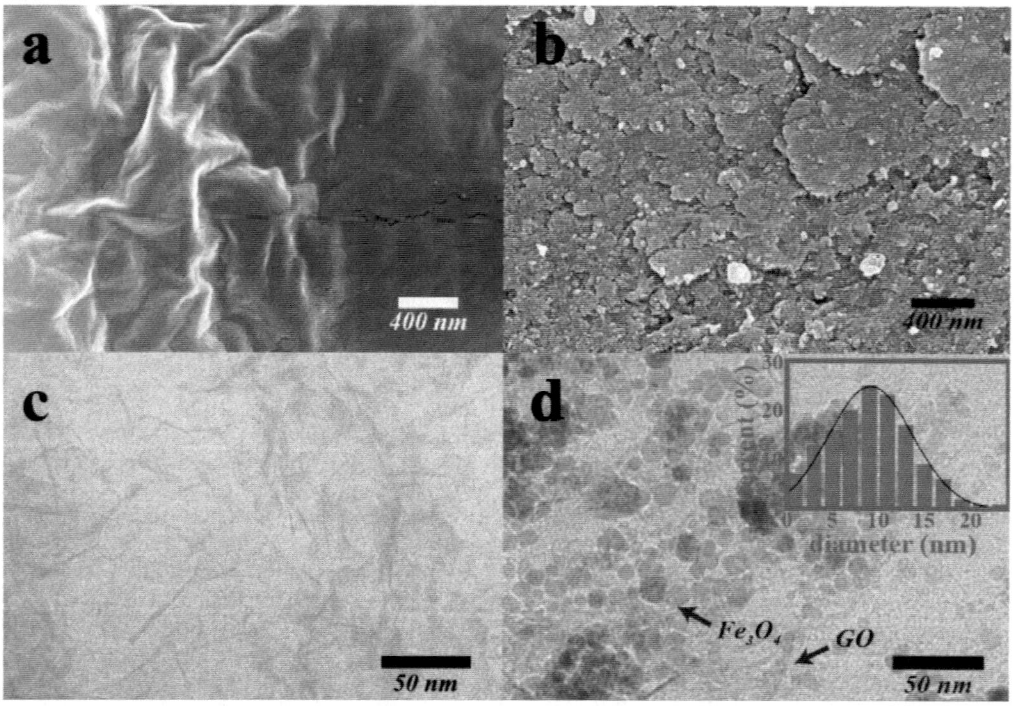

Figure 3. SEM images of (a) GO, (b) MGO; TEM images of (c) GO, (d) MGO

Subsequently, the morphology of GO and MGO was studied by SEM and TEM. SEM images of GO (**Fig. 3a**) revealed that the GO had a smooth surface with many wrinkles, which led to its large specific surface area and abundant active sites. The prepared GO presented a typical multi-layered structure and the surface was relatively rough (**Fig. 3b**). Transmission electron microscopy (TEM) was performed to elucidate the morphology and structure of the MGO compound. It was clear that the elliptical and circular Fe_3O_4 nanoparticles were uniformly dispersed on the translucent graphene oxide substrate with an average diameter of *ca.* ~9.8 nm (**Fig. 3d, inset**), which indicated that the Fe_3O_4 nanoparticles were successfully loaded on the GO sheets.

Adsorption Experiments. The initial pH was one of the most important factors in the adsorption process, influencing not only the surface charge of the adsorbent but also the functional groups on the GO and the chemical properties of the dye solution. The anionic dye CR and cationic dye MB were used to investigate the pH sensitivity of MGO, and the experimental pH values varied from 3 to 11. The results indicated that the adsorption of MGO on MB was insensitive to pH value changes. The equilibrium adsorption capacity of MB was higher than 180.2 mg g^{-1}, and increased smoothly upon increasing the solution pH (Fig. 4a).

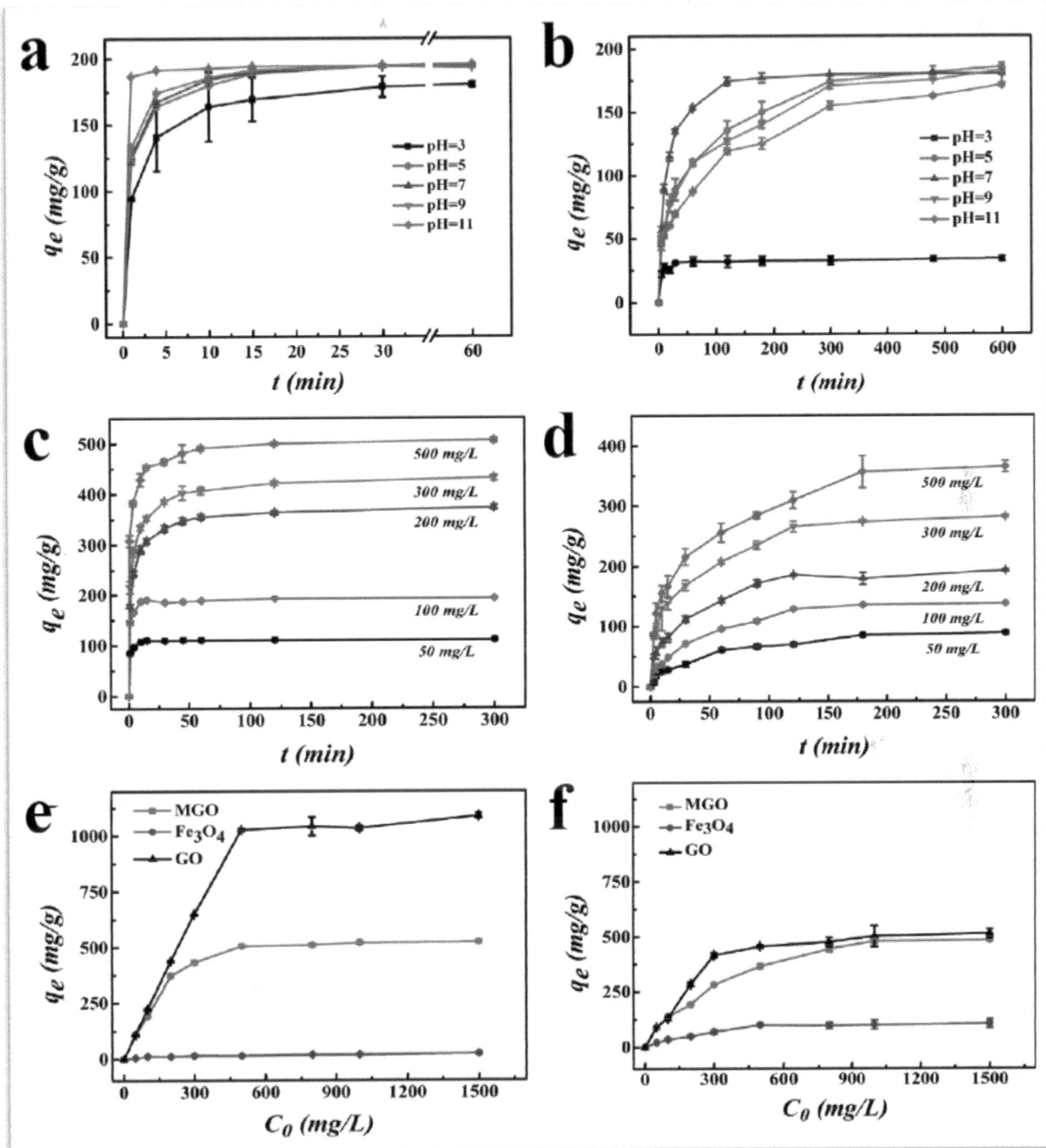

Figure 4 Effect of initial pH on MB (a) and CR (b) adsorption. Conditions: initial MB/CR concentration 100 mg L^{-1}; adsorbent dosage 0.5 g L^{-1}; temperature 298 ± 0.5 K; stirring rate 180 rpm. Effects of adsorption times and initial dye concentrations of MB (c) and CR (d) on MGO adsorption capacity; Effect of initial MB (e) and CR (f) dye concentration on adsorption of MGO, Fe_3O_4 and GO. Conditions: adsorbent dosage 0.5 g L^{-1}; temperature 298 ± 0.5 K; stirring rate 180 rpm; initial concentrations: 50, 100, 200, 300, 500, 800, 1000, 1500 mg L^{-1}

There are two acid–base equilibriums of MB and GO, represented by GO-H→GO^- + H^+ and MBH^{2+} → MB^++ H^+ (Zhang et al. 2011). The amount of hydrogen ions increased with decreasing pH values, thus competing more with the MB cations. Due to the abundance of H^+, GO-H was the main form of GO^- which decreased the amount of MB adsorbed. In contrast, there was more GO^- at higher pH values, which may have enhanced the electrostatic attraction between MGO and MB, thus increasing the adsorption capacity for MB. However, it can be clearly seen from Fig. 4b that the adsorption curve of CR showed different pattern from that of MB. The equilibrium adsorption capacity decreased sharply at low pH values (pH = 3), dropping from 189.4 to 34.1 mg g^{-1}. The reason for this phenomenon might be because CR is a reactive indicator dye, and its structure changed with the change in pH value. At a low initial pH range (pH < 3.5), the $-NH_2$ of CR would protonate to $-NH_3^+$ (Fig. S2), which decreased the electrostatic repulsive force between CR and MGO. The decrease of electrostatic repulsion in acidic condition was favorable to physical adsorption. However, according to the adsorption kinetics, it was found that the adsorption were mainly chemical adsorption, and these protonated functional groups would not be conducive to chemical adsorption. When the pH was lower than the acid dissociation constant of oxygen-containing groups, these groups became positively charged ($C-OH_2^+$, $C-COOH_2^+$), as a result of which the main interactions between CR and MGO were chemical bonds(Ahmad &Kumar 2010). The active sites at the surface of the adsorbent may determine the adsorption capacity of MGO.

With increasing pH, the zeta potential of MGO becomes negative. The zeta potential value decreased steadily from 12 mV to -50 mV (**Fig. S3**) as the pH increased from 2 to 11. It was observed that the adsorption capacity of MB increased with decreasing zeta potential value, due to the increase of the coulombic force between MB molecules and MGO.

The effect of initial concentration was systematically investigated. **Fig. 4c** and **d** compare the adsorption performance at various initial concentrations (i.e., 50, 100, 200, 300, 500 and 800 mg L^{-1} of MB and CR). The results showed that the dye adsorption capacity gradually increased as the initial concentration increased. The adsorption capacity of MGO increased rapidly during the first 5 min, and the initial concentration of the solution had little effect on the adsorption rate (**Fig. 4c, d**). As the concentration of the dye solution increased, there was a significant increase in adsorption capacity. After a long period of time, the adsorption capacity increased slowly and eventually reached equilibrium. The surface sites may have been occupied by dye molecules, causing a decrease in adsorption rate.

Furthermore, **Fig. 4e, f** compare the different adsorption properties of MGO, Fe_3O_4 and GO for dye removal. The theoretical maximum equilibrium adsorption capacity of MGO, GO and Fe_3O_4 were 1.4615, 2.8974, and 0.0591 mmol g^{-1} for MB and 0.9028, 0.7476, and 0.1721 mmol g^{-1} for CR (**Table S3**), respectively. The adsorption capacity of CR (36.2674 mg g^{-1}) on Fe_3O_4 was better than that of MB (12.8027 mg g^{-1}) because the Fe_3O_4 was positively charged in aqueous solution and therefore favored CR adsorption. The theoretical maximum adsorption capacities of MB (193.0502 mg g^{-1}) and CR (150.3759 mg g^{-1}) on MGO have similar values. However, upon comparing the dyes' K_L values and rates of adsorption (r), the MB adsorption rate (K_L = 0.0885, r = 0.75) was much faster than that of CR (K_L = 0.0178, r = 0.02) (**Table S3**). The adsorption of MB (225.225 mg g^{-1}) on GO was significantly higher than that of CR (138.1215 mg g^{-1}), which may be due to difference in the dyes' structures and characteristics.

The BET surface areas of GO and MGO were measured as 1115.3468 m^2 g^{-1} and 407.1267 m^2 g^{-1} respectively, suggesting that the introduction of Fe_3O_4 to GO decreased the surface area of the GO. However, MGO could be conveniently separated from the solution by magnetic field which increases its application prospects.

Adsorption Kinetics. The adsorption kinetics reflect the state of the adsorption process throughout the whole adsorption reaction (Ho &McKay 1998). The mechanism of adsorption can be explored according to adsorption kinetic models. To reveal the most suitable mechanism for the adsorption process, pseudo-first-order model and pseudo-second-order model were used to analyze the adsorption data. The pseudo-first-order kinetic model is expressed as follows (4) (Vadivelan &Kumar 2005):

$$ln(q_e - q_t) = lnq_e - k_1t \quad (4)$$

The pseudo-second-order kinetic model is represented by the following (Ho 2006, Ho &McKay 1999, Vadivelan &Kumar 2005) (5):

$$\frac{t}{q_t} = \frac{1}{k_2 q_e^2} + \frac{t}{q_t} \quad (5)$$

$$r = k_2 C \quad (6)$$

where q_e and q_t are the amounts of MB and CR adsorbed (mg g^{-1}) at equilibrium and at time t (min), respectively; k_1 is the pseudo-first-order rate constant (min^{-1}); k_2 is the pseudo-second-order rate constant (g mg^{-1} min^{-1}); C is the concentration of the solution (mg L^{-1}); and r is the rate of adsorption (g L^{-1} min^{-1}).

The correlation coefficients and rate constants of the pseudo-first-order and pseudo-second-order model are shown in **Table S2,** and **Fig. S4**. It can be seen that the R^2 values of the pseudo-first-order model resulted in low correlation coefficients (0.5897 to 0.9835 for MB, 0.8988 to 0.9802 for CR). The R^2 values of the pseudo-second order kinetic correlation coefficient were close to 0.9999, which is higher than the pseudo-first order kinetic correlation coefficient (**Table S2, Fig. S4**). There were great discrepancies between the theoretical adsorption capacities and the experimental data. The maximum theoretical adsorption capacities (505.05 mg g^{-1} for MB, 492.61 mg g^{-1} for CR) of the pseudo-second order kinetics model were consistent with the experimentally measured data. Therefore, it can be concluded that the adsorption of MB and CR by MGO fitted the pseudo-second-order kinetic model well, and the adsorption process may be chemical adsorption.

Furthermore, as the initial concentration increased from 50 to 500 mg L^{-1}, the second-order rate constants (k_2) of adsorption for MB and CR were found to decrease from 3.95×10^{-2} to 1.0×10^{-3} (g mg^{-1} min^{-1}), and from 3.0×10^{-4} to 1.0×10^{-4} (g mg^{-1} min^{-1}), respectively. The reaction rate was related to the solution concentration, which was consistent with previously reported studies (Demirbas et al. 2009, Hanafiah et al. 2012).

Adsorption Isotherms. The two adsorption models were used to interpret the adsorption isotherm at different concentrations and temperature ranges. The Langmuir and Freundlich sorption models were utilized to match the adsorption isotherms. The Langmuir isotherm model assumes that all adsorption sites are subjected to the same monolayer surface adsorption and that the adsorbed molecules are completely independent. The Freundlich model applies to multi-layer adsorption, and the adsorption capacity of varies from site to site.

The equation of the Langmuir isotherm is given as follows (7):

$$\frac{C_e}{q_e} = \frac{1}{k_L q_m} + \frac{C_e}{q_m} \quad (7)$$

where q_e is the amount of dye adsorbed (mg g^{-1}) at equilibrium, C_e is the equilibrium concentration of the solution (mg L^{-1}), q_m is the maximum adsorption capacity (mg g^{-1}), and k_L is the adsorption equilibrium constant (L mg^{-1}), which is related to the affinity of the binding site.

The Freundlich isotherm equation is expressed as follows (8):

$$log\ q_e = log\ k_F + \frac{1}{n} logC_e \quad (8)$$

where k_f is the Freundlich constant, and n is the heterogeneity factor.

The theoretical correlation coefficient (R^2) parameters of the adsorption isotherms are presented in **Table S3,** and **Fig. S5**. The R^2 values of the Langmuir and Freundlich models are close to unity: 0.997 and

0.90 for MB, 0.995 and 0.94 for CR. These values imply that the Langmuir model may better describe the adsorption isotherm for the adsorption of these two different dyes. The correlation coefficient R^2 for the Langmuir equation is higher than the linear correlation coefficients for the Freundlich model. Furthermore, the theoretical maximum equilibrium adsorption capacity of the Langmuir equation for MB and CR was 613.50 and 909.09 mg g^{-1} at 328.15 K, respectively, which was closer to the experimental data. It was concluded that the adsorption of MB and CR by MGO was better described with the Langmuir model. The adsorption process was mainly monolayer uniform adsorption and mainly occurred on the functional groups of adsorption sites on the surface of MGO, such as hydroxyl, carboxyl and epoxy functional groups. The monolayer coverage of dye molecules on the surface of the adsorbent may be explained at the molecular level. The structure of graphene oxide is layered, and there was no evidence of substitution taking place between different layers.

The maximum adsorption capacity of MGO increased from 546.45 mg g^{-1} (at 298.15 K) to 613.50 mg g^{-1} (at 328.15 K) for MB and from 628.93 mg g^{-1} (at 298.15 K) to 909.09 mg g^{-1} (at 328.15 K) for CR, indicating that the adsorption capacity increased with increasing temperature (**Table S4**). This trend could be due to the enhanced mobility of the MB and CR molecules with rising temperature (Dawood 2012). The dye molecules may this obtain adequate energy to interact with active sites at the surface of the adsorbent. The adsorption equilibrium constant k_L increased with increasing temperature and with greater adsorption capacity. This can be explained by the Brownian motion, which causes liquid molecules to move more intensely at higher temperature. Thus, the impact forces on dye particles by liquid molecules from different directions increase with rising temperature, resulting in increased adsorption capacity.

The suitability of the predicted adsorption model was informed by the value of the separation factor constant (R_L) which indicates the probability of the adsorption process. The separation factor (R_L) can be inferred from the Langmuir equation and indicates whether the adsorption process is favorable. The values of this parameter mean that the adsorption isotherm is unfavorable ($R_L > 1$), linear ($R_L = 1$), favorable ($0 < R_L < 1$) or irreversible ($R_L = 0$) (Chang et al. 2007, Vimonses et al. 2009). The separation factor (R_L) is defined as follows:

$$R_L=\frac{1}{1+k_L C_0} \qquad (9)$$

where k_L (L mg^{-1}) is the Langmuir constant, and C_0 (mg L^{-1}) is the initial dye concentration. In this study, the values of R_L were within the range of 0-1, indicating that the adsorption of MB, and CR onto MGO was a favorable process.

As an interesting material with enormous specific surface area, GO has abundant active sites that can effectively interact with dyes. The possible mechanism can be explained by electrostatic interact, hydrogen bonds, Van der Waals force and hydrophobic–hydrophobic interactions; a more in-depth explanation is as follows: (i) hydrogen bonds occur between the dye molecule and oxygen-containing hydroxyl (-OH) and carboxyl (-COOH) groups. (ii) Ion-exchange occurs between the oxygen-containing groups (-OH, -COOH) in MGO and D-SO_3^- (D- SO_3^-, D = dye molecule)), forming MGO–COO–SO_2–D and MGO–C–O–SO_2–D(Guo et al. 2016). Guo (Guo et al. 2016) have reported that the -SO_3^- groups of CR are involved in the adsorption process. (iii) The protonation of –COOH and –OH occurred on the surface of MGO in acidic solution (i.e., MGO–OH+H^+↔MGO–OH_2^+) (Ahmad &Kumar 2010).

Thermodynamic Study. To evaluate the effect of temperature on adsorption process, the thermodynamic parameters Gibbs free energy (ΔG^0: kJ mol^{-1}), enthalpy (ΔH^0: kJ mol^{-1}) and entropy (ΔS^0: kJ mol^{-1} K^{-1}) can be calculated by the following equations:

$$ln(k_L)=\frac{\Delta S^0}{R}-\frac{\Delta H^0}{RT} \qquad (10)$$

$$k_L=\frac{q_e}{c_e} \qquad (11)$$

$$\Delta G^0=\Delta H^0-T\Delta S^0 \qquad (12)$$

where k_L is the Langmuir constant (L g^{-1}), R is the universal gas constant (8.314 J mol^{-1} K^{-1}) and T is the absolute temperature (K).

Table 1. Thermodynamic parameters for MB and CR adsorption onto MGO

Dye	Temperature (K)	ΔG^0 (kJ mol^{-1})	ΔH^0 (kJ mol^{-1})	ΔS^0 (J K^{-1} mol^{-1})
MB	298.15	-28.23	113.14	474.16
	308.15	-32.97		
	318.15	-37.71		
	328.15	-42.46		
CR	298.15	-21.87	41.83	213.65
	308.15	-24.00		
	318.15	-26.14		
	328.15	-28.28		

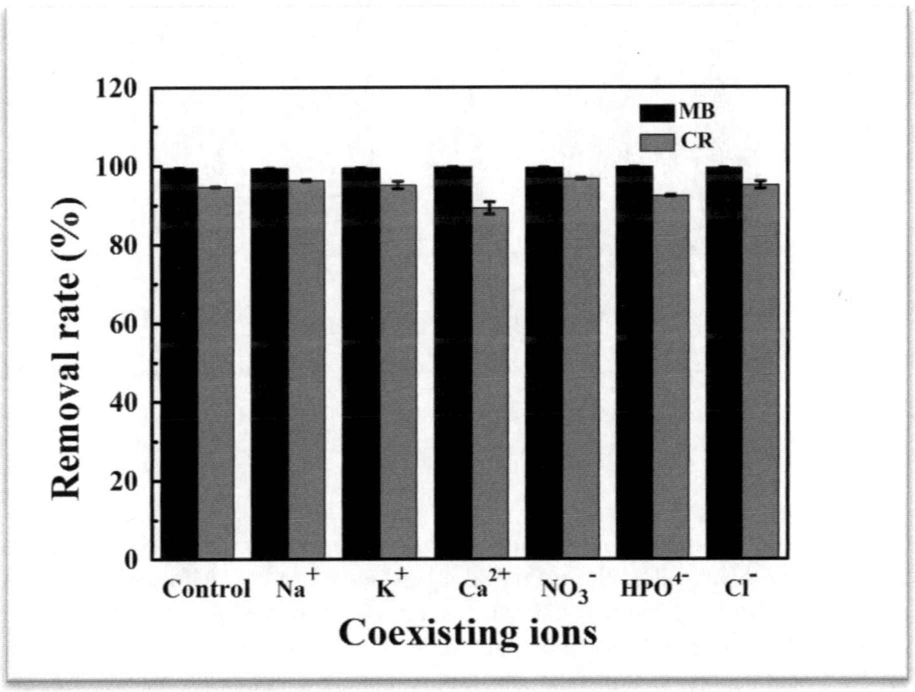

Figure 5. Effect of coexisting ions. Conditions: adsorbent dosage 0.5 g L^{-1}; temperature 298±0.5 K; stirring rate 180 rpm; initial dye concentrations 100 mg L^{-1}; Na^{+}, Ca^{2+}, K^{+} and NO^{3+}, Cl^{-} and HPO^{4-} concentrations 1 mol L^{-1}.

Table 1 shows the thermodynamic parameters of the adsorption. The negative values of ΔG^0 indicated that the adsorption reaction can be carried out at room temperature and the decrease of ΔG^0 with increasing temperature indicated that the adsorption reaction was spontaneous at higher temperatures. The values of ΔH^0 and ΔS^0 were found to be 113.14 kJ mol^{-1} and 474.16 J K^{-1} mol^{-1} for MB and 41.83 kJ mol^{-1}

and 213.65 J K^{-1} mol^{-1} for CR, respectively. The positive values of ΔH^0 and ΔS^0 showed that the adsorption of MB and CR by MGO is endothermic. Therefore, the prepared MGO could be effectively used in dye adsorption without harsh reaction conditions, indicating that the adsorbent has the potential possibility to be used in practical wastewater treatment.

Effect of Coexisting Ions. In practical applications, the large amount of dissolved metal cations and inorganic anions, such as Na^+, Ca^{2+}, K^+; NO^{3+}, Cl^-, HPO^{4-} and so on, may greatly affect the adsorption process. Therefore, the effect of individual coexisting ions (1 mol L^{-1}) on the adsorption of MB and CR was investigated in series. Fig. 5 summarizes the effect of the common coexisting metal cations Na^+, Ca^{2+}, and K^+ and inorganic anions NO^{3+}, Cl^-, and HPO^{4-}, all of which showed no significant effects on MB/CR adsorption. Hence, MGO composites could serve as an effective adsorbent for dye wastewater remediation.

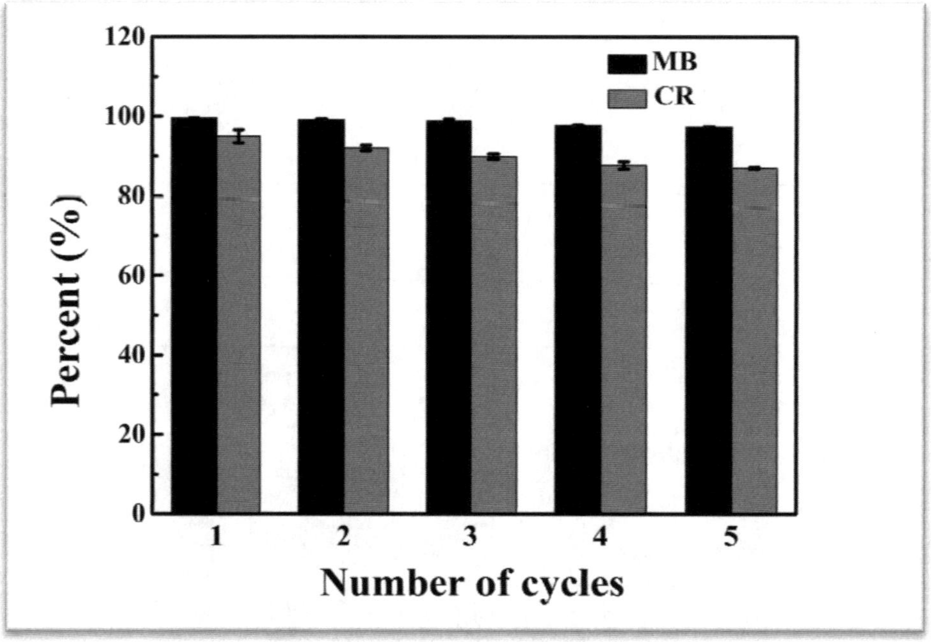

Figure 6. Regeneration and reuse of MGO. Conditions: adsorbent dosage 0.5 g L^{-1}; temperature 298 ± 0.5 K; stirring rate 180 rpm; initial concentrations 50 mg L^{-1}

Desorption and Regeneration. The continuous recycle experiments were conducted to evaluate the stability and the application potential of the prepared adsorbent. The obtained MGO particles were added to the MB/CR solution (50 mg L^{-1}), and the residual concentration was measured after reaching adsorption equilibrium. Distilled water and NaOH were used to desorb dye several times until no more leaching dye could be detected. The adsorption rates of MB and CR still were maintained at 97.43% and 87.23% after five successive adsorption-desorption cycles (Fig. 6). These results indicated that the performance of MGO was not appreciably impaired after five cycles, which illustrated its good stability and practical application potential.

Table 2 compares the maximum adsorption capacity of graphene-based adsorbents for dye removal under different experimental conditions. The maximum adsorption capacities of dyes on MGO in this study were better than those of most previously studied graphene-based materials. MGO has a slight advantage in adsorption capacity. Since it could also be efficiently recycled and reused, MGO has broad practical application prospects in the treatment of organic pollutants.

Table 2. Comparison of MGO with other graphene-based adsorbents

Adsorbents	Adsorbate	Qm (mg g^{-1})	References
Cellulose/graphene oxide	MB	480.77	(Chen et al. 2016)
Hydroxo Aluminum/Graphene	CR	497.51	(Wu et al. 2014)
Graphene oxide/chitosan fibers	CR	294.12	(Du et al. 2014)
Beta-cyclodextrin and GO gel (SCGG)	MB	133.24	(Wu et al. 2017)
Fe3O4@SiO2@CS-TETA-GO	MB	529.1	(Wang et al. 2017)
Magnetic GO/poly(vinyl alcohol) (PVA) composite gels	MB	231.12	(Cheng et al. 2015)
Magnetic rGO-loaded hydrogels	MB	111.9	(Halouane et al. 2017)
Magnetic graphene oxide	MB/CR	546.45/ 628.93	This study

CONCLUSIONS

In summary, a feasible, economical and green route for the removal of two typical dyes by a magnetic graphene oxide/Fe_3O_4 (MGO) nanocomposite was demonstrated. The results showed that the MGO exhibited a great adsorption ability for dye removal and could adapt to a wide pH range from 5 to 11. The adsorption behavior was best suited for the Langmuir adsorption isotherm equation, showing calculated maximum adsorption capacity of 546.45 mg g^{-1} and 628.93 mg g^{-1} for MB and CR, respectively. The adsorption process fit the pseudo-second-order kinetic model well, which revealed that it might be the chemical interaction between adsorbate and adsorbent. Furthermore, the prepared MGO nanoparticles maintained good stability after five successive adsorption-desorption cycles. The results show that this "green" magnetic graphene oxide adsorbent has a good stability and high adsorption capacity and may serve as a promising absorbent for wastewater purification.

ACKNOWLEDGMENTS

The authors thank Prof. Anatoly ZINCHENKO for a helpful discussion. This work was supported by the National Natural Science Foundation of China (Nos. 41772264, 51408074) and the Research Fund of State Key Laboratory of Geohazard Prevention and Geoenvironment Protection (SKLGP2017Z009). Dr. S.Y. Pu is grateful for support from the Hong Kong Scholars Program (No. XJ2015005 and G-YZ80).

REFERENCES

Abidi N, Errais E, Duplay J, Berez A, Jrad A, Schaefer G, Ghazi M, Semhi K, Trabelsi-Ayadi M (2015): Treatment of dye-containing effluent by natural clay. Journal of Cleaner Production 86, 432-440

Ahmad R, Kumar R (2010): Adsorptive removal of congo red dye from aqueous solution using bael shell carbon. Appl Surf Sci 257, 1628-1633

Akbari A, Remigy JC, Aptel P (2002): Treatment of textile dye effluent using a polyamide-based nanofiltration membrane. Chem. Eng. Process. 41, 601-609

Aliyari E, Alvand M, Shemirani F (2016): Modified surface-active ionic liquid-coated magnetic graphene oxide as a new magnetic solid phase extraction sorbent for preconcentration of trace nickel. Rsc Advances 6, 64193-64202

Barathi M, Kumar ASK, Kumar CU, Rajesh N (2014): Graphene oxide-aluminium oxyhydroxide interaction and its application for the effective adsorption of fluoride. Rsc Advances 4, 53711-53721

Belpaire C, Reyns T, Geeraerts C, Van Loco J (2015): Toxic textile dyes accumulate in wild European eel Anguilla anguilla. Chemosphere 138, 784-791

Chang Y-H, Huang C-F, Hsu W-J, Chang F-C (2007): Removal of Hg2+ from aqueous solution using alginate gel containing chitosan. J Appl Polym Sci 104, 2896-2905

Chen L, Li YH, Hu S, Sun JK, Du QJ, Yang XX, Ji Q, Wang ZH, Wang DC, Xia YZ (2016): Removal of methylene blue from water by cellulose/graphene oxide fibres. J. Exp. Nanosci. 11, 1156-1170

Cheng Z, Liao J, He B, Zhang F, Zhang F, Huang X, Zhou L (2015): One-Step Fabrication of Graphene Oxide Enhanced Magnetic Composite Gel for Highly Efficient Dye Adsorption and Catalysis. Acs Sustain Chem Eng 3, 1677-1685

Cheng ZL, Li W, Wu PR, Liu Z (2017): A Strategy for Preparing Modified Graphene Oxide with Good Dispersibility and Transparency in Oil. Ind Eng Chem Res 56, 5527-5534

Dawood SS, T. K. (2012): Removal of anionic dye Congo red from aqueous solution by raw pine and acid-treated pine cone powder as adsorbent: equilibrium, thermodynamic, kinetics, mechanism and process design. Water Res 46, 1933-46

Demirbas E, Dizge N, Sulak MT, Kobya M (2009): Adsorption kinetics and equilibrium of copper from aqueous solutions using hazelnut shell activated carbon. Chem Eng J 148, 480-487

Dizaji AK, Mortaheb HR, Mokhtarani B (2016): Noncovalently functionalized graphene oxide/graphene with imidazolium-based ionic liquids for adsorptive removal of dibenzothiophene from model fuel. Journal of Materials Science 51, 10092-10103

Dreyer DR, Park S, Bielawski CW, Ruoff RS (2010): The chemistry of graphene oxide. Chem. Soc. Rev. 39, 228-240

Du Q, Sun J, Li Y, Yang X, Wang X, Wang Z, Xia L (2014): Highly enhanced adsorption of congo red onto graphene oxide/chitosan fibers by wet-chemical etching off silica nanoparticles. Chemical Engineering Journal 245, 99-106

Forgacs E, Cserhati T, Oros G (2004): Removal of synthetic dyes from wastewaters: a review. Environment International 30, 953-971

Greenwald MJ, Redding AM, Cannon FS (2015): A rapid kinetic dye test to predict the adsorption of 2-methylisoborneol onto granular activated carbons and to identify the influence of pore volume distributions. Water Res. 68, 784-792

Guo YF, Deng J, Zhu JY, Zhou C, Zhou CY, Zhou XJ, Bai RB (2016): Removal of anionic azo dye from water with activated graphene oxide: kinetic, equilibrium and thermodynamic modeling. RSC Adv. 6, 39762-39773

Halouane F, Oz Y, Meziane D, Barras A, Juraszek J, Singh SK, Kurungot S, Shaw PK, Sanyal R, Boukherroub R, Sanyal A, Szunerits S (2017): Magnetic reduced graphene oxide loaded hydrogels: Highly versatile and efficient adsorbents for dyes and selective Cr(VI) ions removal. J Colloid Interf Sci 507, 360-369

Hanafiah M, Ngah WSW, Zolkafly SH, Teong LC, Majid ZAA (2012): Acid Blue 25 adsorption on base treated Shorea dasyphylla sawdust: Kinetic, isotherm, thermodynamic and spectroscopic analysis. J. Environ. Sci. 24, 261-268

Ho YS, McKay G (1998): Kinetic models for the sorption of dye from aqueous solution by wood. Process Saf Environ 76, 183-191

Ho YS, McKay G (1999): Pseudo-second order model for sorption processes. Process Biochemistry 34, 451-465

Ho YS (2006): Second-order kinetic model for the sorption of cadmium onto tree fern: A comparison of linear and non-linear methods. Water Res. 40, 119-125

Huang GJ, Chen ZG, Li MD, Yang B, Xin ML, Li SP, Yin ZJ (2016): Surface Functional Modification of Graphene and Graphene Oxide. Acta Chim. Sin. 74, 789-799

Kyzas GZ, Fu J, Matis KA (2013): The Change from Past to Future for Adsorbent Materials in Treatment of Dyeing Wastewaters. Materials 6, 5131-5158

Liu T, Li Y, Du Q, Sun J, Jiao Y, Yang G, Wang Z, Xia Y, Zhang W, Wang K, Zhu H, Wu D (2012): Adsorption of methylene blue from aqueous solution by graphene. Colloids Surf B Biointerfaces 90, 197-203

Lucas MS, Peres JA (2006): Decolorization of the azo dye Reactive Black 5 by Fenton and photo-Fenton oxidation. Dyes Pigments 71, 236-244

Ma H, Pu S, Ma J, Yan C, Zinchenko A, Pei X, Chu W (2018): Formation of multi-layered chitosan honeycomb spheres via breath-figure-like approach in combination with co-precipitation processing. Materials Letters 211, 91-95

Ma X, Chen PL, Zhou M, Zhong ZX, Zhang F, Xing WH (2017): Tight Ultrafiltration Ceramic Membrane for Separation of Dyes and Mixed Salts (both NaCl/Na2SO4) in Textile Wastewater Treatment. Ind Eng Chem Res 56, 7070-7079

Malachova K, Rybkova Z, Sezimova H, Cerven J, Novotny C (2013): Biodegradation and detoxification potential of rotating biological contactor (RBC) with Irpex lacteus for remediation of dye-containing wastewater. Water Res. 47, 7143-7148

Netpradit S, Thiravetyan P, Towprayoon S (2003): Application of 'waste' metal hydroxide sludge for adsorption of azo reactive dyes. Water Res. 37, 763-772

Peng R, Chen X, Ghosh R (2017a): Preparation of graphene oxide-cotton fiber composite adsorbent and its application for the purification of polyphenols from pomegranate peel extract. Separation and Purification Technology 174, 561-569

Peng W, Li H, Liu Y, Song S (2017b): A review on heavy metal ions adsorption from water by graphene oxide and its composites. Journal of Molecular Liquids 230, 496-504

Pu S, Xiang C, Zhu R, Ma H, Zinchenko A, Chu W (2017a): An efficient heterogeneous Fenton catalyst based on modified diatomite for degradation of cationic dye simulated wastewater. Desalination and Water Treatment 79, 378-385

Pu S, Zhu R, Ma H, Deng D, Pei X, Qi F, Chu W (2017b): Facile in-situ design strategy to disperse TiO2 nanoparticles on graphene for the enhanced photocatalytic degradation of rhodamine 6G. Applied Catalysis B-Environmental 218, 208-219

Pu S, Zhu R, Ma H, Deng D, Pei X, Qi F, Chu W (2017c): Facile in-situ design strategy to disperse TiO 2 nanoparticles on graphene for the enhanced photocatalytic degradation of rhodamine 6G. Applied Catalysis B: Environmental 218, 208-219

Pu SY, Ma H, Zinchenko A, Chu W (2017d): Novel highly porous magnetic hydrogel beads composed of chitosan and sodium citrate: an effective adsorbent for the removal of heavy metals from aqueous solutions. Environ Sci Pollut R 24, 16520-16530

Ren Y, Abbood HA, He F, Peng H, Huang K (2013): Magnetic EDTA-modified chitosan/SiO2/Fe3O4 adsorbent: Preparation, characterization, and application in heavy metal adsorption. Chem Eng J 226, 300-311

Riera-Torres M, Gutierrez-Bouzan MC, Morell JV, Lis MJ, Crespi M (2011): Influence of electrochemical pre-treatment in dyeing wastewater reuse for five reactive dyes. Text. Res. J. 81, 1926-1939

Stankovich S, Dikin DA, Dommett GHB, Kohlhaas KM, Zimney EJ, Stach EA, Piner RD, Nguyen ST, Ruoff RS (2006): Graphene-based composite materials. Nature 442, 282-286

Szpyrkowicz L, Juzzolino C, Kaul SN (2001): A comparative study on oxidation of disperse dyes by electrochemical process, ozone, hypochlorite and Fenton reagent. Water Res. 35, 2129-2136

Tang ZH, Shen SL, Zhuang J, Wang X (2010): Noble-Metal-Promoted Three-Dimensional Macroassembly of Single-Layered Graphene Oxide. Angew Chem Int Edit 49, 4603-4607

Vadivelan V, Kumar KV (2005): Equilibrium, kinetics, mechanism, and process design for the sorption of methylene blue onto rice husk. J Colloid Interf Sci 286, 90-100

Vimonses V, Lei S, Jin B, Chow CWK, Saint C (2009): Kinetic study and equilibrium isotherm analysis of Congo Red adsorption by clay materials. Chem Eng J 148, 354-364

Wan Z, Wang JL (2017): Degradation of sulfamethazine using Fe3O4-Mn3O4/reduced graphene oxide hybrid as Fenton-like catalyst. J Hazard Mater 324, 653-664

Wang F, Zhang LJ, Wang YY, Liu XJ, Rohani S, Lu J (2017): Fe3O4@SiO2@CS-TETA functionalized graphene oxide for the adsorption of methylene blue (MB) and Cu(II). Appl Surf Sci 420, 970-981

Wang H, Yuan XZ, Wu Y, Chen XH, Leng LJ, Wang H, Li H, Zeng GM (2015): Facile synthesis of polypyrrole decorated reduced graphene oxide-Fe3O4 magnetic composites and its application for the Cr(VI) removal. Chem Eng J 262, 597-606

Wu Y, Luo HJ, Wang H (2014): Efficient Removal of Congo Red from Aqueous Solutions by Surfactant-Modified Hydroxo Aluminum/Graphene Composites. Sep. Sci. Technol. 49, 2700-2710

Wu YC, Qi HJ, Shi C, Ma RX, Liu SX, Huang ZH (2017): Preparation and adsorption behaviors of sodium alginate-based adsorbent-immobilized beta-cyclodextrin and graphene oxide. RSC Adv. 7, 31549-31557

Yang CH, Wang CY, Huang KS, Yeh CS, Wang AHJ, Wang WT, Lin MY (2012): Facile Synthesis of Radial-Like Macroporous Superparamagnetic Chitosan Spheres with In-Situ Co-Precipitation and Gelation of Ferro-Gels. Plos One 7, 7

Yao YJ, Miao SD, Liu SZ, Ma LP, Sun HQ, Wang SB (2012): Synthesis, characterization, and adsorption properties of magnetic Fe3O4@graphene nanocomposite. Chem Eng J 184, 326-332

Zhang LY, Zhang W, Zhou Z, Li CM (2016a): gamma-Fe2O3 nanocrystals-anchored macro/meso-porous graphene as a highly efficient adsorbent toward removal of methylene blue. Journal of Colloid and Interface Science 476, 200-205

Zhang WJ, Zhou CJ, Zhou WC, Lei AH, Zhang QL, Wan Q, Zou BS (2011): Fast and Considerable Adsorption of Methylene Blue Dye onto Graphene Oxide. Bull. Environ. Contam. Toxicol. 87, 86-90

Zhang Z-b, Qiu Y-f, Dai Y, Wang P-f, Gao B, Dong Z-m, Cao X-h, Liu Y-h, Le Z-g (2016b): Synthesis and application of sulfonated graphene oxide for the adsorption of uranium(VI) from aqueous solutions. Journal of Radioanalytical and Nuclear Chemistry 310, 547-557

A Comparative Study of Cod Removal from Wastewater Using Biochar and Activated Carbon from Hardwood and Pine Tree

Hafiz Ahmad, Korhan Adalier, Douglas Brown and Andres Rodrigues
(Florida State University, Panama City, FL, USA)
Brandon Madden (Mott MacDonald, Panama City, FL, USA)

ABSTRACT: Production of large scale commercial carbon-products essentially requires examination of various feed stocks, manufacturing process/steps and the performance of the product to remove the target pollutant. This study evaluates the removal performance of Chemical Oxygen Demand (COD) from municipal wastewater using biochar and activated carbons (having different activation levels) produced from hardwood and pine tree. First, biochars were produced from hardwood (chips and pellets) and pine (pellets) by pyrolysis process at 850 °F. The biochars were then physically (steam) activated at 1500 °F with different activation (residence) times. The iodine numbers (IN) of biochar from hardwood chips, hardwood pellets and pine pellets were determined to be 76 mg/g, 128 mg/g and 121 mg/g respectively. On the other hand, the iodine numbers (IN) of activated carbon from hardwood chips, hardwood pellets and pine pellets were determined to be 1001 mg/g, 878 mg/g and 826 mg/g respectively, indicating significant activation level. Batch kinetic experiments and isotherm studies were performed to evaluate the effect of activation of biochars (i.e. activated carbons) on COD removal. Batch kinetic studies showed that the biochars were not effective at all in removing COD and even released COD. However, activated carbon produced from these biochars showed excellent COD removal in all cases. The results showed that up to 47% COD removal can be achieved with a target dosage 300 mg/L (with 30-minute reaction time) using the activated carbon (with IN of 1001 mg/g) from hardwood chips. The equilibrium data represented by Freundlich isotherm model also supported the above result showing the enhanced performance of this activated carbon (from hardwood chips) compared to the other carbon products for COD removal.

INTRODUCTION

Organic matter is the major target pollutant especially in municipal wastewater treatment plants. The organic content in greywater is frequently expressed as chemical oxygen demand (COD). There are various treatment technologies available for COD removal from grey water. Besides biological treatment, they include filtration, ion exchange, coagulation-flocculation, sedimentation, reverse osmosis (El-Naas et al., 2010) etc. One treatment method that is becoming very popular in recent years is the use of activated carbon especially for final polishing of wastewater in municipal wastewater treatment plants. Besides activated carbon, scientists (e.g. Yao et al., 2012, Kang et al., 2017) in recent years were also interested in evaluating the use of biochar (an intermediate product during activated carbon production) for removal of various pollutants from aqueous phase of wastewater. However, there are very few studies available where feasibility of commercial production of biochar and activated carbon is evaluated for their application in municipal wastewater treatment plants.

The adsorption of organics (i.e. COD) depends on various factors such as surface areas, micro-pore volumes, adsorption capacities of the carbon. These characteristics are controlled by the operating conditions (e.g. temperature, heating rate, residence time etc.) during their production/activation. This operating condition is controlled to achieve the desired activation level (i.e. quality) of the product for its end-use. As for example, for the targeted application of carbon to remove COD from aqueous solution, various qualities of carbon products can be produced by controlling two major operating conditions (temperature and residence time) during pyrolysis and activation process. However, the prediction of optimum activation level for most efficient removal of COD from aqueous solution is quite difficult without laboratory tests and/or model study. Therefore, it is essential to test the effectiveness especially for the products which can be derived easily from low-cost, commercially available feed-stocks like pine and hardwood.

In this research, initially biochars were produced by pyrolysis method and these biochars were further physically activated to produce activated carbon of various activation levels by controlling the activation residence times. Activation level and porosity of these carbon products were assessed by iodine numbers, a popular parameter used to estimate relative surface area, adsorptive capacity (Wang et al., 2013). To evaluate the performance of COD removal from water, biochars produced from hardwood chips, hardwood pellets and pine pellets (iodine number 76 mg/g, 128 mg/g and 121 mg/g respectively) were used as a base-line. Then, three different activated carbons (iodine numbers 1001 mg/g, 878 mg/g and 826 mg/g) produced from above mentioned biochars were used for COD removal to compare their effectiveness with the base-line (i.e. biochars).

MATERIALS AND METHOD

Adsorbent Preparation. Biochar (BC) and activated carbon (AC) from three feedstocks (hardwood chips, hardwood pellet and pine pellet) were prepared by the procedure described in detail elsewhere (Mayers, 2014). Figure 1 illustrates the summary of the manufacturing process of BC and AC.

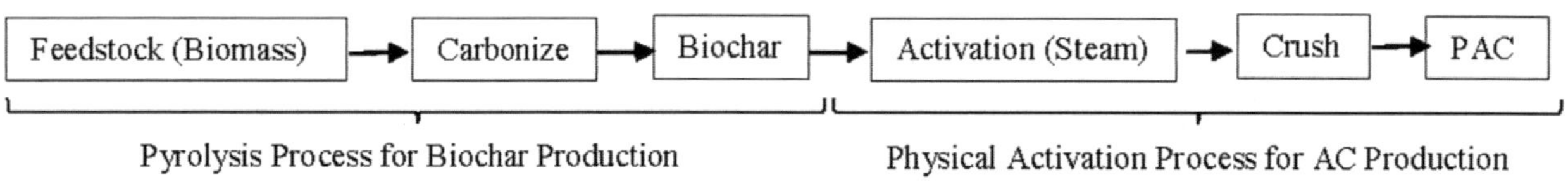

FIGURE 1. Flow diagram for the manufacturing of BC and AC materials

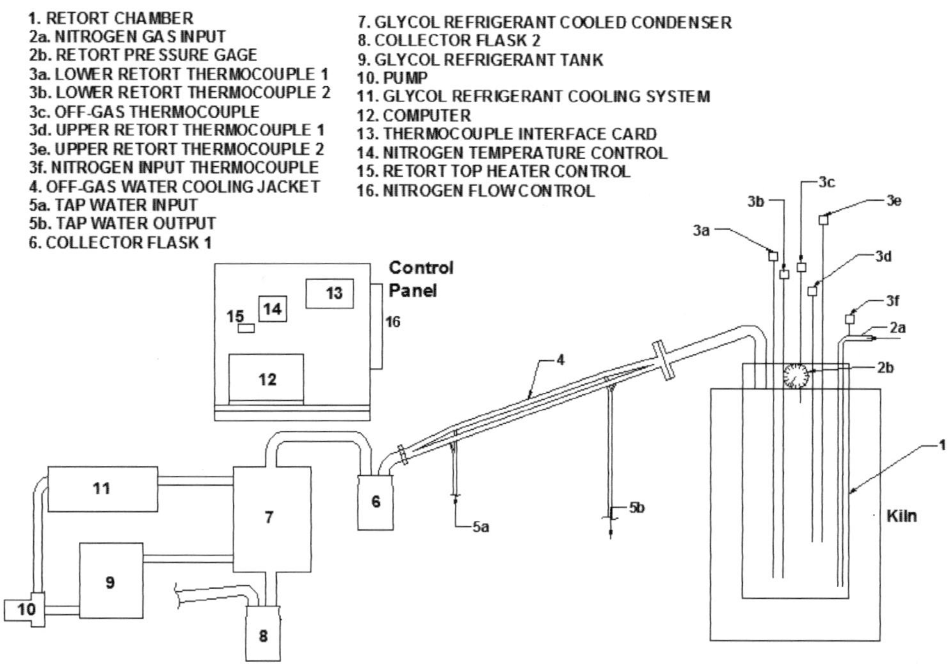

FIGURE 2. Schematic diagram of pyrolysis apparatus

Initially, the feedstock materials were cleaned with brushes, all foreign materials were removed and then oven dried. The dried material was then pyrolysed at 850 °F using a pyrolysis apparatus schematic diagram of which is shown in Figure 2. The BC produced after pyrolysis (in retort chamber #1) is then physically activated (steam activation) at 1500 °F for different residence times (35 minutes and 6 minutes) and three different grades of activated carbons were produced. The process diagram and the details of the activation process is available in Mayers (2014). The iodine numbers (IN) of each of these three BC and AC were determined. The surface areas of the same products were estimated using the model proposed by Mianowski et al. 2007. The values of iodine number (IN), surface area of each product and their production operating conditions are shown in Table 1.

TABLE 1. Iodine Number, Surface Area and Major Preparation Conditions of Adsorbents

Feed stock	Adsorbent (Name referred afterwards)	Preparation Conditions			Adsorbent Quality	
		Pyrolysis Temp. (°F)	Activation Temp. (°F)	Activation Time (minute)	Iodine Number (mg/g)	Surface Area (m^2/g)
Hard wood Chip	Biochar (BC#76)	850	-	-	76	75
	Activated Carbon (AC#1001)	850	1500	35	1001	916
Hard wood Pellet	Biochar (BC#128)	850	-	-	128	126
	Activated Carbon (AC#878)	850	1500	6	878	804
Pine Pellet	Biochar (referred BC#121)	850	-	-	121	119
	Activated Carbon (AC#826)	850	1500	6	826	814
-	Commercial Activated Carbon (AC#611)	-	-	-	611	602

Adsorption Procedure. All carbon products (BC and AC) were first ground using a standard mortar and pestle and then passed through a #200 sieve (75µm diameter). The adsorbents were then dried at 212 °F (100 °C) for a period of 24 hours to remove moisture. Different amount of adsorbents (dose ranging from 75 mg/L to 475 mg/L) were then added to the real wastewater samples collected from the clarifier (pH 6.5-7.4, temperature 25 °C) at the St. Andrews Wastewater Treatment Facility, Panama City, Florida. The carbons were mixed with wastewater in compact mixer at 350 rpm (30 minutes for the AC and 120 minutes for the BC). The samples were then filtered with filter paper (Fisher Scientific P8 filter paper) and analyzed for the residual COD. The COD was determined using Orion AQ4000 Advanced Colorimeter, Orion COD thermoreactor, and Orion AQUAfast COD Chemistries test kit. The COD removal percentage (R%) was calculated as follows (Eq. (1)):

$$R\% = [(C_0 - C_e)/C_i] \times 100 \qquad (1)$$

The amount of adsorption at equilibrium, q_e (mg/g), was calculated by (Eq. (2)):

$$q_e = [(C_0 - C_e)V]/M \qquad (2)$$

Where, C_0 and C_e (mg/L) are the liquid-phase concentrations of COD at initial and equilibrium, respectively. V (L) is the volume of the solution, and M (g) is the mass of dry adsorbent used.

RESULTS AND ANALYSIS

Adsorbent Treatment Time. Figure 3(a) shows the effect of contact time on COD removal from the wastewater samples using 136 mg/L dose of AC and BC. Our main goal for these tests was to determine the equilibrium treatment time for batch tests (described in next section). The commercial AC#616 (Norit CN1, Acros Organics) was used for comparing our results with a standard carbon product available in the market. It was observed that AC can remove COD from wastewater whereas BC gives negative COD removal. The maximum COD reduction was 33% and 30% using AC#1001 and AC#616 respectively. On the other hand, biochar (BC#76) showed no COD removal ability, and even released COD into the solution (up to 14.5% release). This result is in agreement with other studies (e.g. Yao et al., 2012, Kang et al., 2017) who found limited and negative removal (i.e. release) of pollutants such as nitrate, ammonium, phosphate using biochar. The presence of significant volatile organic carbon (VOC) occupying the pore spaces in biochar is the possible major barrier to achieve COD removal using biochar. In fact, complete removal of VOC from BC is difficult unless pyrolysis temperature is more than 1400 °F. The results (Figure 3(a)) suggest that activation (oxidation at high temperature) can improve the sorption ability of BC to adsorb COD. Conversely, BC cannot be used for COD removal from wastewater.

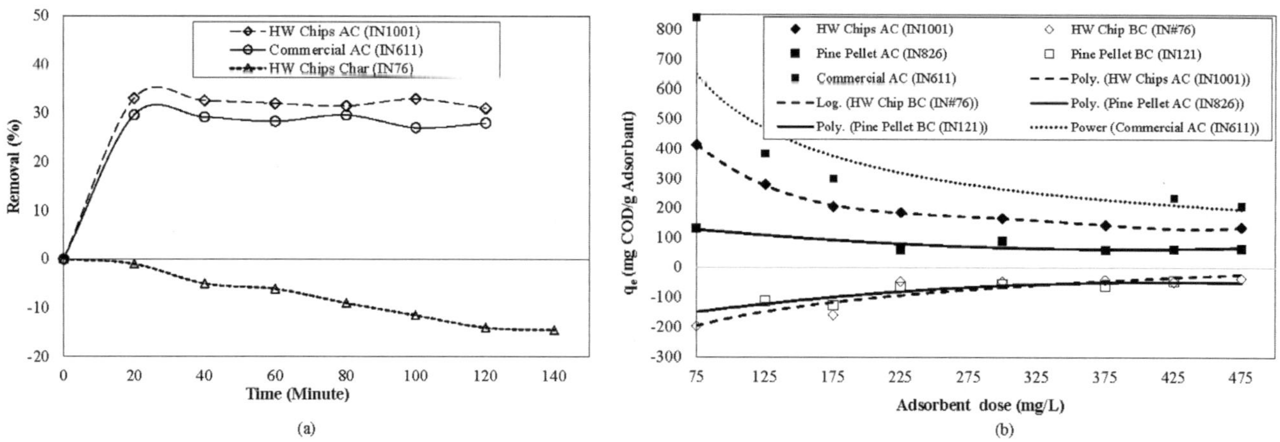

FIGURE 3. (a) Effect of treatment time on % COD reduction using AC and BC, (b) Effect of adsorption dose on COD uptake

For AC, as the treatment time progressed (Figure 3(a)), the curves become horizontal indicating the adsorption sites had the tendency towards saturation and reaching equilibrium (Devi et al., 2008). The COD removal reaches to the equilibrium condition approximately after 30 minutes for AC while it was approximately 120 minutes for BC. This leads us to consider the treatment time of 30 minutes and 120 minutes for batch tests with AC and BC respectively described in the next section.

Effect of Adsorption Dose. To confirm the above results, a batch test program was taken where various doses (from 75-475 mg/L) of AC and BC were applied to the wastewater samples. COD adsorption onto five different adsorbents were calculated and trend-lines were drawn as shown in Figure 3(b). Comparing the capacity of COD adsorption, it is clear that AC produced from hardwood and pine (AC#1001, AC#826) are effective adsorption performer while BC from the same hardwood and pine (BC#76, BC#121) do not work at all. This again reveals that activation is needed to enhance quality of BC to capture COD from wastewater. Besides the AC#1001, AC#826, the results of commercial AC#616 also showed significant adoption of COD, supporting our theory and the requirement of activation (that open-up the pore spaces of BC increasing its surface area by removing tars and volatiles from void-spaces).

The uptake of COD decreases with the increasing doses of AC (Figure 3(b)). This may be attributed to many factors such as availability of solutes, interference between binding sites, electrostatic interaction and reduced mixing due to high adsorption concentration in the solution (El-Nass et al., 2010, Fourest and Roux, 2009). The figure also confirms that increasing the dose beyond 300 mg/L has little impact on COD reduction (for the optimum 30-minute treatment time) by AC. Hence, a dose of 300 mg/L with a treatment time of 30 minute could be the optimum condition for COD removal by AC.

Comparison of COD removal. Figure 4 compares the COD removal (percentage) using various doses of AC (AC#1001, AC 878, AC#826) and biochar (BC#76, BC#128, BC#121) from the wastewater. The batch test results clearly show that the ACs are effective for the quantitative removal of COD (initial COD concentration of wastewater ranged 215-227 mg/L). Conversely, tested biochars were not effective in removing COD and even released COD with 2-hour reaction time (initial COD was 120-132 mg/L). Therefore, BC must be activated for the COD removal.

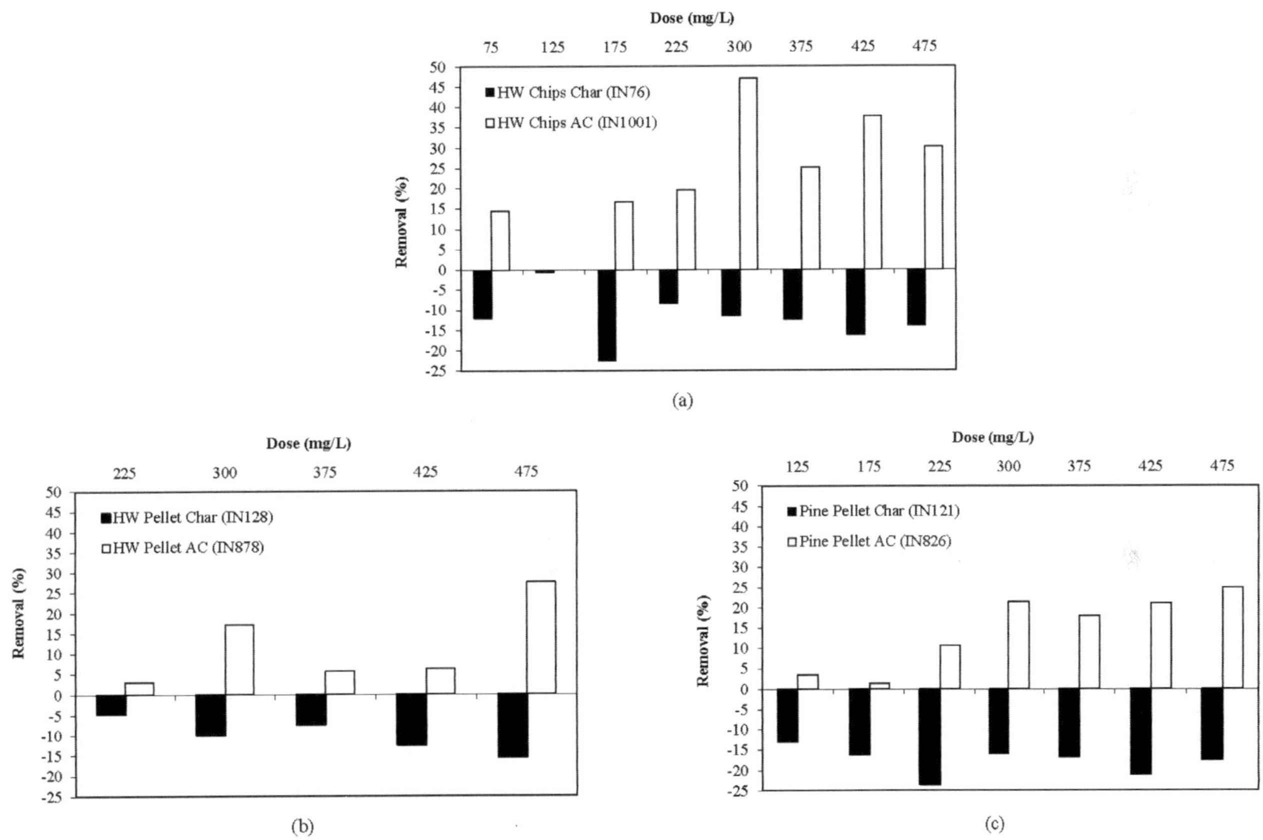

FIGURE 4. COD removal by AC and BC at different doses (a) Hardwood Chips, (b) Hardwood pellets, (c) Pine pellets

The removal of COD using AC#1001 and AC#826 significantly improves at/after the doses of 300 mg/L (Figure 4). The maximum removal is achieved beyond this dose (300 mg/L) for each type activated carbon. Using a dose of 475 mg/L, both AC#878 and AC#826 achieves maximum COD removal (28% and 25% respectively. It is evident from the results that among these three products, the best COD removal can be achieved by the activated carbon with IN of 1001 mg/g (i.e. AC#1001). In other words, the activated carbon produced from hardwood chips using 1500 °F with a 35-minute activation (residence) time is the best performing adsorbent for COD removal (Table 1) possibly due to its high quality (high IN and surface area). Hence, using this AC#1001, with a dose of 300 mg/L and 30-minute treatment time will be most effective for COD removal in the treatment plant.

Adsorption Isotherm. Adsorption isotherm plays an important role for the analysis and design of adsorption systems as well as model prediction (Mukherjee et al., 2007). It describes how solutes interact with adsorbents, and is critical in optimizing the use of adsorbents (Hameed and Rahman, 2008). The Freundlich isotherm models can be used to describe the relationship between the amount of COD adsorbed and its equilibrium concentration in solutions. The Freundlich isotherm is given by (Eq. (3)):

$$q_e = K_F C_e^{1/n} \qquad (3)$$

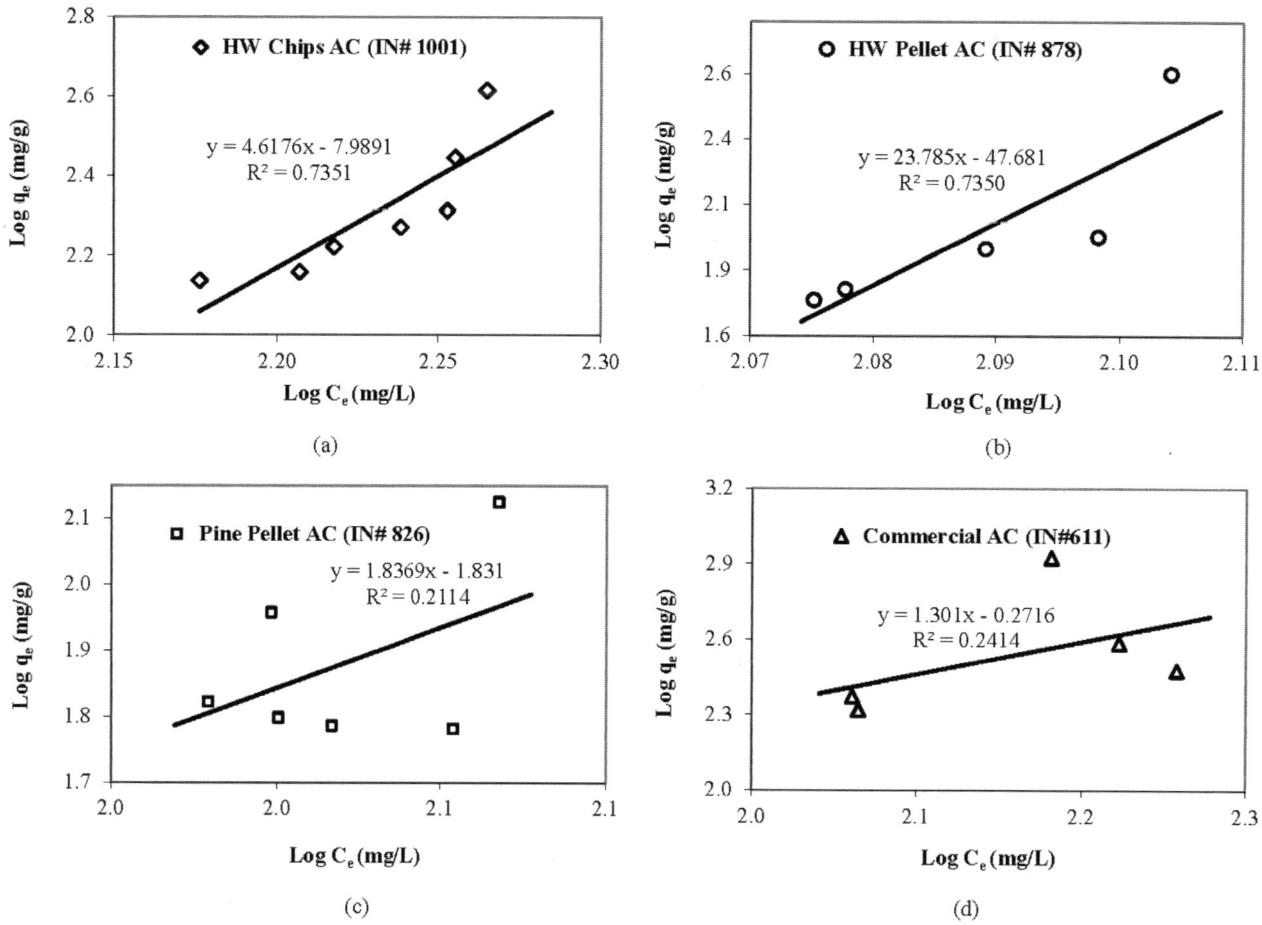

FIGURE 5. Freundlich adsorption isotherm of COD onto (a) AC#1001, (b) AC#878, (c) AC#826 and (d) AC#611.

where q_e (mg/g) amount of adsorption at equilibrium, C_e (mg/L) is the liquid-phase concentrations of COD at equilibrium. K_F and n are Freundlich constants. The linear form of the above Equation is (Eq. (4)):

$$\log q_e = \log K_F + (1/n) \log C_e \qquad (4)$$

which will have a straight line with a slope of 1/n and an intercept of log(K_F) when log(q_e) is plotted against log(C_e). K_F ((mg/g)(L/mg)$^{1/n}$) is an indicator of the adsorption capacity. The larger K_F value indicates that the adsorbent has a larger adsorption capacity and a favorable adsorbent (Mukherjee et. al., 2007). On the other hand, the low value of 1/n (0 and 1 range is preferred) represents better adsorbent (McKay, et al., 1982).

The data obtained in this study were used to plot log q_e vs. log C_e and presented in Figure 5. Based on the collected data, only AC#1001 and AC#878 seems fitting Freunlich isotherm model (R^2>0.5) while the other two adsorbents (AC#826, AC#611) shouldnot be interpreted using Freunlich isotherm model as R^2<0.5. Besides this, the plotting of the data for BCs was not possible due to negative q_e. The values of constants K_F and 1/n for AC#1001 and AC#878 are also calculated and shown in Table 2. A comparison of the values of K_F reveals that AC#1001 has a higher K_F value (1.03×10^{-8}) indicating it has a better adsorption capacity and is more favorable adsorbent among the two products. Furthermore, the lower 1/n value of 4.6176 was obtained for AC#1001 suggests that it has the high adsorption capacity compared to AC#878. This model study again suggests that activation of hardwood chips biochar (BC#76) to an optimum level (making AC#1001) would be the best for COD adsorption.

TABLE 2: Freundlich Isotherm Coefficients

Absorbent	K_F ((mg/g)(L/mg)$^{1/n}$)	1/n	R^2
Activated Carbon (AC#1001)	1.03×10^{-8}	4.6176	0.7351
Activated Carbon (AC#878)	2.08×10^{-48}	23.785	0.7350

CONCLUSIONS

The COD removal performance by various grades carbon products from hardwood and pine tree was investigated using batch tests. The study demonstrates that it is not possible to remove COD from wastewater using biochars. The biochars must be activated and converted to activated carbon for COD removal. However, preparation method of these carbon products has significant effect on COD adsorption capacity. Activated carbon produced with a 35-minute activation (residence) time (i.e. AC#1001) at temperature 1500 °F proved to be most effective for COD removal. The removal efficiency of this carbon product is more than that of commercial activated carbon (AC#611) with a target dosage of 300 mg/L and a treatment time of 30 minutes.

REFERENCES

Devi R., V. Singh, A. Kumar. 2008. "COD and BOD reduction from coffee processing wastewater using Avacado peel carbon." *Bioresource Technol.* 99: 1853–1860

El-Naas, M.H., S. Al-Zuhair, and M.A. Alhaija. 2010. "Reduction of COD in refinery wastewater through adsorption on date-pit activated carbon." *J. Hazard. Mater.* 173: 750–757.

Fourest, E. and J.C. Roux, 1992. "Heavy metal biosorption by fungal mycelial biproducts: mechanisms and influence of pH." *Appl. Microbiol. Biotechnol.* 37:399-403.

Hameed, B.H., A.A. Rahman, 2008 "Removal of phenol from aqueous solutions by adsorption onto activated carbon prepared from biomass material." *J. Hazard. Mater.* 160: 576–581.

Kang J., M. Davila, S. Mireles and J. Ho. 2017. "Nitrate Leaching from Sand and Pumice Geomedia Amended with Pyrogenic Carbon Materials." *Environments.* 4(70): 1-7

Mayers, J. 2014. *Pyrolysis and Activation of an Invasive Species.* M.S. Thesis, Florida State University, Tallahassee, Florida.

McKay, G., H.S. Blair, J.R. Gardner. 1982. "Adsorption of dyes on chitin-1: equilibrium studies." *J. Appl. Polym. Sci.* (27): 3043–3057.

Mianowski, A., M. Owczarek, and A. Marecka. 2007. "Surface Area of Activated Carbon Determined by the Iodine Adsorption Number." Energy Sources, Part A, 29.9, pp 839-850. Taylor & Francis, LLC, Abingdon, UK.

Mukherjee, S., S. Kumar, A. K. Misra, M. Fan. 2007. "Removal of phenol from water environment by activated carbon, bagasse ash and wood charcoal." *Chem. Eng. J.* 129: 133-142.

Wang X., D. Li, W. Li, J. Peng, H. Xia, L., Zhang, S. Guo and G. Chen. 2013. "Optimization of mesoporous activated carbon from coconut shells by chemical activation with phosphoric acid." *BioResources*, 8(4): 6184-6195.

Yao, Y., B. Gao, M. Zhang, M. Inyang and A. R. Zimmerman. 2012. "Effect of biochar amendment on sorption and leaching of nitrate, ammonium, and phosphate in a sandy soil." *Chemosphere*. 89: 1467–1471.

Kinetics and Equilibrium Study of Resorcinol Adsorption onto Surface Modified Ordered Mesoporous Carbon

Zaki Uddin Ahmad*, Mark E. Zappi, Daniel Dianchen Gang
(University of Louisiana at Lafayette, Lafayette, LA70504, USA)

ABSTRACT: Ordered Mesoporous Carbon (OMC), a relatively new member of carbonaceous family, has been able to attract substantial attention due to its unique attributes such as high BET surface area, narrow pore size distribution, large pore volume and thermal stability. A novel ordered mesoporous carbon was synthesized using COK-19 silica scaffold for the first time and doped with cerium (III) chloride for the removal of resorcinol from aqueous solution. Traditional multistage technique was employed for synthesizing OMC. Nitrogen Adsorption-Desorption Isotherms, Scanning Electron Microscope (SEM), Transmission Electron Microscope (TEM), Fourier Transformed Infrared Spectroscopy (FTIR), Energy Dispersive X-ray Spectroscopy (EDS) were employed to characterize the modified OMCs. Highest removal efficiency was achieved for 5% wt. cerium loading followed by a reduction in the removal efficiency for 7.5% and 10% wt. cerium loading. Results from adsorption kinetics and isotherms suggest that Pseudo-Second-Order Model and Langmuir Isotherm well described the experimental data. Adsorption capacity increased by 76.8% with increased cerium content from 0 to 5%.

INTRODUCTION

Rapidly increasing global population and their ever-growing demands are exerting pressure on the natural state of environment and influencing harmony among environmental variables. One of the important elements of natural environment sensitive to contamination is water which has been irreversibly degraded over the decades by various organic and inorganic pollutants. Phenols and their derivatives are widely used and identified in the effluents of a number of industries such as oil refineries, paper, textile, synthetic rubber, petrochemical units, and pharmaceutical and steel industries (Bilgili, 2006; Ahmad et al., 2018a). Even trace amount of phenolic compounds can potentially have severe impacts on living system since they are known as carcinogenic (Ku & Lee, 2000; Kumar et al., 2003; Shou et al., 2016).

Diversified technologies, such as coagulation and precipitation, photocatalytic degradation, chemical oxidation, ion exchange and adsorption (Ahmad et al., 2018b), have been developed and tested over the last few decades for the treatment of wastewater contaminated with organic and inorganic pollutants. Ordered Mesoporous Carbon (OMC), a new member of the carbonaceous family, has been able to demonstrate greater promise for their adsorption ability because of its distinct attributes, such as high BET surface area, narrow pore size distribution, well defined ordered structure, and large pore volume (Chao et al., 2017). Two factors affecting the adsorption efficiency of OMCs are the interaction between the adsorbate and the adsorbent, including electrostatic, hydrogen bonding, acid-base interaction, and textual parameters of the adsorbent such as BET surface area, pore size, and pore volume. It has been reported that the interaction between the f-orbitals of Lanthanides and different functional groups can assist the formation of organo-metal complexes (Goscianska et al., 2014). The main objective of this research work was to synthesize OMCs modified with cerium employing COK-19 silica template and study their effectiveness for resorcinol adsorption from aqueous solution. Surface modified OMCs were characterized using different techniques and the adsorption behavior of resorcinol was studied following batch adsorption method.

MATERIALS AND METHODS

Synthesis of COK-19. COK-19 template was synthesized using Pluronic F-127 triblock copolymer surfactant and sodium silicate solution as the silica source followed by reported literature (Kerkhofs et al., 2015). In a typical synthesis, 2.6 g of Pluronic F-127 was dissolved in 107.5 g of deionized water. To this solution, 3.7 g of citric acid monohydrate and 2.55 g of trisodium citrate dihydrate were added. This buffered surfactant solution was stirred overnight to dissolve all the components. After that, 10.4 g of sodium silicate solution (26.5% SiO_2) diluted in 30 g of water was added to the buffered F-127 solution at room temperature and the solution was stirred vigorously for 5 min. Precipitation of white solid occurred instantaneously and the milky white suspension was kept under quiescent conditions for aging for 24 hr at 70ºC in a constant temperature water bath. The white precipitation was then filtered off, washed with deionized water and dried in the oven at 105ºC overnight. The resultant solid was calcined at 350ºC for 24 hr using a heating rate of 0.5ºC per minute. The white COK-19 silica template was stored in the desiccator for the next step.

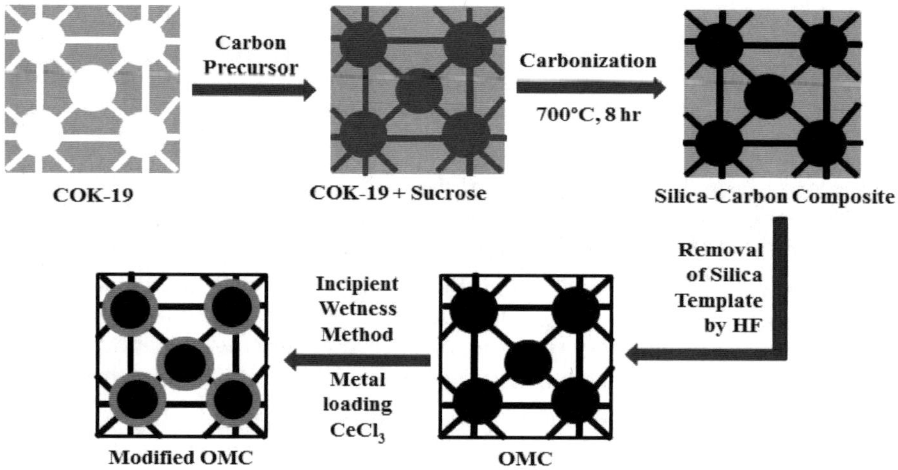

FIGURE 1. Step by step synthesis procedure of modified OMC

Fabrication of OMC. OMC was prepared via hard template method using sucrose as the carbon precursor and COK-19 as the silica scaffold. Typically, 1.5 g of sucrose ($C_{12}H_{22}O_{11}$) and 15 drops of sulfuric acid (H_2SO_4) were dissolved in 15 mL of DI water. 2.0 g of COK-19 was then added to the solution and stirred for 2 hrs. The mixture was heated at 100°C for 6 hr and subsequently at 160°C for an additional 6 hr. The resulting composite was impregnated again with an aqueous solution of 0.5 g sucrose mixed with 10 drops of a sulfuric acid solution dissolved in 10 mL of water. After the second heat treatment at 100°C and 160°C, respectively, the composite was carbonized at 700°C for 8 hr under a nitrogen flow of 100 mL min^{-1}. Afterwards, the silica template was removed using approximately 50 mL of 48% HF solution at room temperature. The product obtained was filtered and washed with 4,000 mL of DI water to obtain final OMC.

Sample Modification of OMC. Incipient wetness method was employed to impregnate ordered mesoporous carbon with an aqueous solution of cerium(III) chloride ($CeCl_3$) in accurately calculated amount to obtain 0.5, 1, 3, 5 wt% metal loading. The samples were successively dried at 100°C for 5 hr and heated in nitrogen for 3 hr at 400°C. The samples obtained were denoted as OMC-0.5Ce, OMC-1Ce, OMC-3Ce, OMC-5Ce, respectively. The step by step synthesis procedure of OMC is shown in **Fig. 1**.

RESULT AND DISCUSSION

Characterization of Adsorbents. The corresponding textual parameters of the modified OMCs are summarized in **Table 1**. The parent ordered mesoporous carbon shows the largest BET surface area of 1014.31 $m^2 g^{-1}$, with pore size of 6.9 nm and pore volume of 1.71 $cm^3 g^{-1}$. Modification with cerium(III) chloride causes to decrease BET surface area and pore volume. The highest decrease in BET surface area was observed for OMC-5Ce which was 932.46 $cm^3 g^{-1}$. The change in BET surface area for OMC-5Ce was 8% in comparison to that of the parent OMC. This phenomenon can be attributed to the pore blocking, from the interaction between $CeCl_3$ and carbon structure, and subsequent pore collapse. The increased amount of $CeCl_3$ nanoparticle introduced on the support surface caused a small increase in pore size, from 7.05 nm for OMC-0.5Ce, to 7.21 nm for OMC-5Ce. Narrow pore size distributions are centered at 6.9-7.2 nm for the all OMCs as depicted in **Fig. 2**.

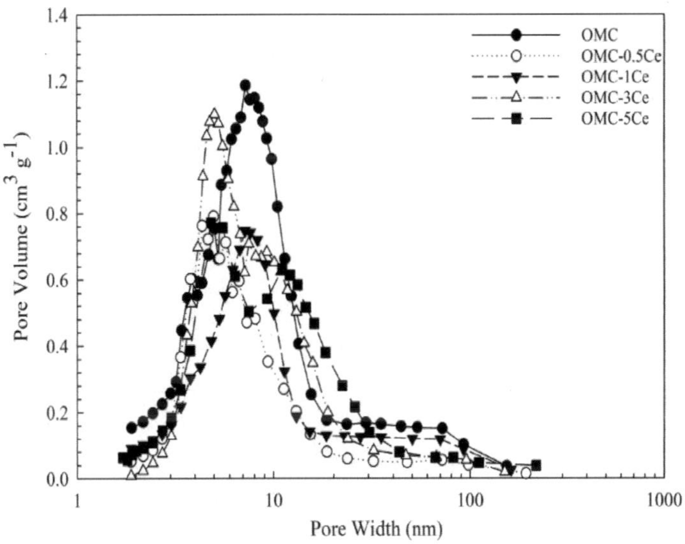

FIGURE 2. Pore size distributions of OMCs

TABLE 1. Textual parameters of modified OMCs

Sample	BET Surface Area ($m^2 g^{-1}$)	Total pore volume ($cm^3 g^{-1}$)	Pore size (nm)
OMC	1014.31	1.71	6.91
OMC-0.5Ce	1002.62	1.65	7.05
OMC-1Ce	993.40	1.63	7.12
OMC-3Ce	957.76	1.48	7.18
OMC-5Ce	932.46	1.44	7.21

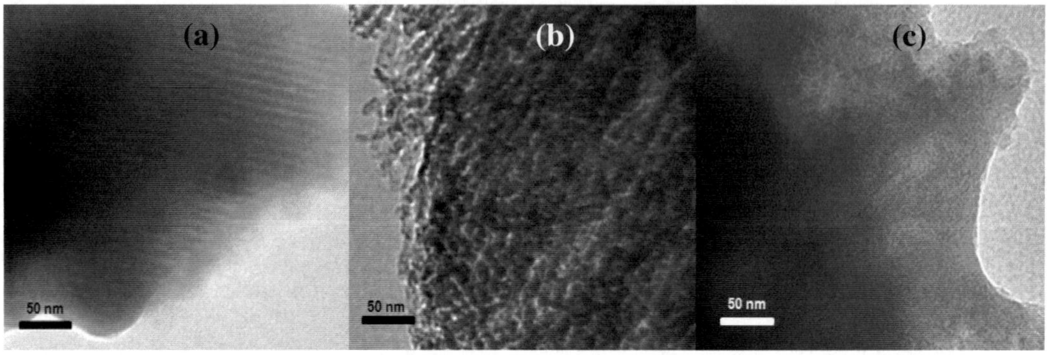

FIGURE 3. TEM micrographs of (a) OMC, (b) OMC-1Ce, (c) OMC-5Ce.

TEM micrographs of modified OMCs are shown in **Fig. 3(a)-(c)**. OMC synthesized using COK-19 silica template by the hard template was found to have ordered structure, the same as the modified OMCs. As the cerium content increased form 0.5% to 5%, more cerium was deposited on the OMC surface and cavities were further clogged, resulting in smooth surfaces (**Fig.3(c)**). The decrease in BET surface area can also be attributed to this phenomenon.

TABLE 2. Coefficients of Pseudo-First-Order, Pseudo-Second-Order and Elovich Kinetic Models for Resorcinol Adsorption onto OMCs

Models	Parameters	OMC	OMC-0.5Ce	OMC-1Ce	OMC-3Ce	OMC-5Ce
Pseudo-First-Order	$q_{e(exp)}$ (mg g^{-1})	28.85	37.25	43.14	48.04	55.88
	$q_{e(cal)}$ (mg g^{-1})	22.06	18.25	24.38	27.98	39.21
	k_1 (hr^{-1})	0.136	0.140	0.150	0.160	0.181
	R^2	0.9893	0.9975	0.9848	0.9850	0.9888
Pseudo-Second-Order	$q_{e(cal)}$ (mg g^{-1})	31.34	38.50	44.85	50.00	58.48
	k_2 (g mg^{-1} hr^{-1})	0.0090	0.0170	0.0130	0.0119	0.0110
	R^2	0.9986	0.9998	0.9997	0.9998	0.9996
Elovich	α (mg g^{-1} hr^{-1})	0.5258	15.8680	8.1280	6.0565	2.0940
	β (g mg^{-1})	6.2265	4.7750	5.8658	6.7179	8.8392
	R^2	0.9374	0.9470	0.9569	0.9404	0.9279

Adsorption Behavior. The time dependence of resorcinol adsorption onto the ordered mesoporous carbons was investigated to determine the adsorption equilibrium. As it can be seen from Fig. 4(a), the adsorption of resorcinol was accomplished in two distinct steps: a relatively faster step (steep slope) continuing up to 10 hr, followed by slower one to the equilibrium state. A contact period of 24 hr was required to reach the equilibrium state. The high initial uptake rate is attributed to the availability of higher number of active sites for resorcinol adsorption at the onset of the process. In addition, textual properties of OMCs allow resorcinol to penetrate though the pores, also known as intra-particle diffusion (Borah et al., 2009), and prompt the adsorption process at the active sites on the pore surfaces. Data generated from the batch experiment study were fitted to the kinetics Models. All the kinetic parameters are listed in Table 2. All these data suggest that resorcinol adsorption could be better represented by the Pseudo-Second-Order kinetic model (Fig. 4(b)) implying that the resorcinol adsorption on cerium modified OMCs is likely to be chemisorption. The efficiency of cerium modified OMCs on the equilibrium adsorption capacity of resorcinol were examined as a function of equilibrium concentration as depicted in Fig. 5. The Langmuir Model fitted well with the experimental data with a higher correlation coefficient (R^2) (Kaveeshwar et al., 2018).

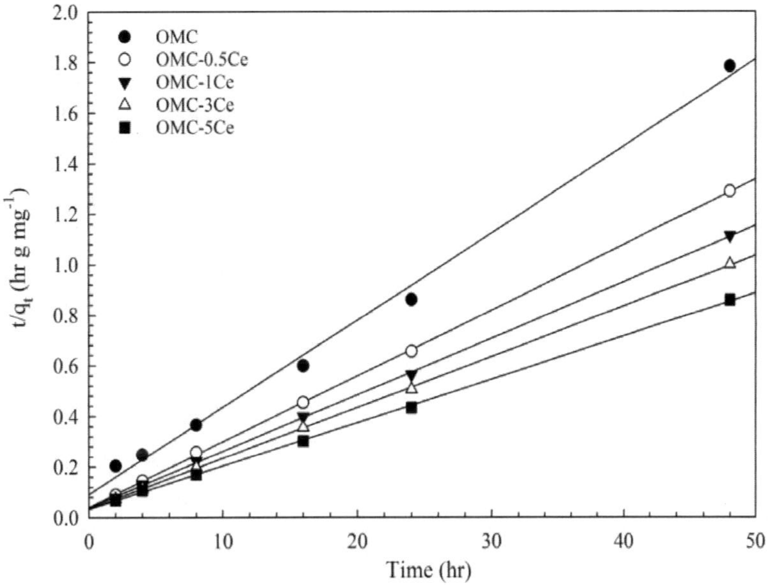

(a)

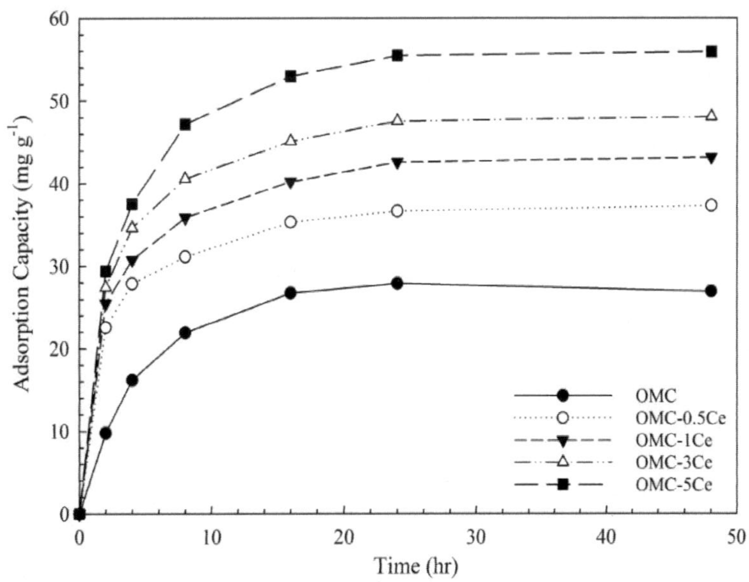

(b)

FIGURE 4. (a) Effect of contact time on resorcinol uptake by OMCs (C_0: 20 mg L^{-1}, T: 25°C, pH: 6.5, Adsorbent dosage: 0.1 g L^{-1}) (b) Data fitted to Pseudo-Second-Order model.

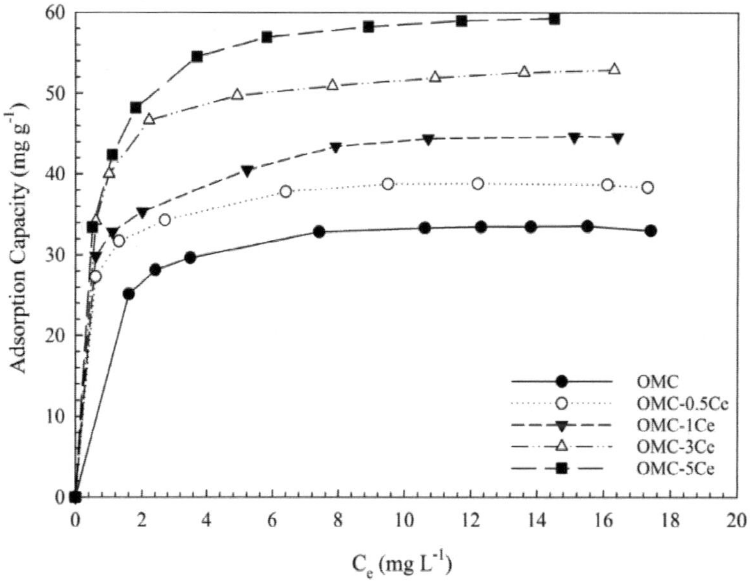

FIGURE 5. Adsorption isotherms of resorcinol onto ordered mesoporous carbon (C_0: 20 mg L^{-1}, T: 25°C, pH: 6.5, Contact time: 24 hr).

CONCLUSION

OMC was synthesized by the hard template method employing COK-19 silica template for the first time and then doped with cerium (III) chloride. The synthesized materials were used for the adsorption of resorcinol from synthetic wastewater as a model compound. The ordered mesoporous carbons modified with cerium showed better adsorption capacities than the parent OMC towards resorcinol which can be attributed to the physiochemical properties. The maximum monolayer adsorption capacities of the OMCs decrease in the following order: OMC-5Ce (60.98 mg g^{-1}) > OMC-3Ce (54.05 mg g^{-1}) > OMC-1Ce (46.08 mg g^{-1}) > OMC-0.5Ce (39.4 mg g^{-1}) > OMC (34.5 mg g^{-1}). Adsorption kinetics was found to follow closely the Pseudo-Second-Order kinetic model and Langmuir model.

ACKNOWLEDGEMENTS

This was supported by Louisiana Board of Regents (LEQSF(2015-16)-RD-C-15) and University of Louisiana at Lafayette.

REFERENCE

Ahmad, Z. U., B. Chao, M. I. Konggidinata, Q. Lian, M. E. Zappi, and D. D. Gang. 2018a. "Molecular Simulation and Experimental Validation of Resorcinol Adsorption on Ordered Mesoporous Carbon (OMC)." *Journal of Hazardous Materials*. 354: 258-265.

Ahmad, Z. U., Q. Lian, M. E. Zappi, P. R. Buchireddy, and D. D. Gang. 2018b. "Adsorptive Removal of Resorcinol on a Novel Ordered Mesoporous Carbon (OMC) Employing COK-19 Silica Scaffold: Kinetics and Equilibrium Study." *Journal of Environmental Sciences*.

Bilgili, M. S. 2006. "Adsorption of 4-chlorophenol from aqueous solutions by XAD-4 resin: isotherm, kinetic, and thermodynamic analysis." *Journal of Hazardous Materials*. 137(1): 157-164.

Borah, D., S. Satokawa, S. Kato, and T. Kojima. 2009. "Sorption of As(V) from aqueous solution using acid modified carbon black." *Journal of Hazardous Materials*. 162(2-3): 1269-1277.

Chao, B., M. I. Konggidinata, L. Lin, M. Zappi, and D. D. Gang. 2017. "Effect of carbon precursors and pore expanding reagent on ordered mesoporous carbon for resorcinol removal." *Journal of Water Process Engineering*. 17: 256-263.

Goscianska, J., M. Marciniak, and R. Pietrzak. 2014. "Mesoporous carbons modified with lanthanum(III) chloride for methyl orange adsorption." *Chemical Engineering Journal*. 247: 258-264.

Kaveeshwar, A. R., P. S. Kumar, E. D. Revellame, D. D. Gang, M. E. Zappi, and R. Subramaniam. 2018. "Adsorption properties and mechanism of barium(II) and strontium(II) removal from fracking wastewater using pecan shell based activated carbon." *Journal of Cleaner Production*. 193: 1-13.

Kerkhofs, S., T. Willhammar, H. V. D. Noortgate, C. E. A. Kirschhock, E. Breynaert, G. V. Tendeloo, S. Bals, and J. A. Martens. 2015. "Self-Assembly of Pluronic F127-Silica Spherical Core–Shell Nanoparticles in Cubic Close-Packed Structures." *Chemistry of Materials*. 27(15): 5161-5169.

Ku, Y., and K. C. Lee. 2000. "Removal of phenols from aqueous solution by XAD-4 resin." *Journal of Hazardous Materials*. 80(1-3): 59-68.

Kumar, A., S. Kumar, and S. Kumar. 2003. "Adsorption of resorcinol and catechol on granular activated carbon: equilibrium and kinetics." *Carbon*. 41(15): 3015-3025.

Shou, W., B. Chao, Z. U. Ahmad, and D. D. Gang. 2016. "Ordered mesoporous carbon preparation by the in situ radical polymerization of acrylamide and its application for resorcinol removal." *Journal of Applied Polymer Science*. 133(19): 43426-43436.

Adsorption of Lead (II) onto Phosphate Modified Ordered Mesoporous Carbon: Kinetics, Equilibrium and Thermodynamic Study

Qiyu Lian, ***Zaki Uddin Ahmad***, Mark Zappi and Daniel Dianchen Gang*
(University of Louisiana at Lafayette, Lafayette, LA 70504, USA)

ABSTRACT: Adsorption of Pb^{2+} onto nitric and phosphoric acid Modified Ordered Mesoporous Carbon (MOMC-NP) was investigated with variations of the parameters such as pH, contact time, temperature, and initial Pb^{2+} concentration. Optimum Pb^{2+} adsorption was observed at pH value of 5. Modified OMC were characterized by Nitrogen Adsorption-Desorption Isotherm, SEM, TEM, FT-IR, EDS, and Boehm Titration. The time required to reach equilibrium was between 20 min and 30 min. The equilibrium adsorption capacity of modified OMC (MOMC-NP) increased with the increase of temperature indicating that the process of Pb^{2+} adsorption is an endothermic process. The equilibrium adsorption capacity of MOMC-NP was found to be 56.06, 68.82, and 77.80 mg g^{-1} at an initial concentration of 60, 80, and 100 mg L^{-1} of Pb^{2+} respectively. Kinetics studies suggested better applicability of Pseudo-Second-Order kinetics model. Linear and non-linear isotherm studies suggested that the experimental data better fitted to Langmuir isotherm model.

INTRODUCTION

Since rapid urbanization and industrialization have been coupled with increased population growth, the demand for potable water is ever increasing. Heavy metal contamination attributable to industrial and domestic practices contaminates ground and surface water and has been identified as a major environmental threat (Ma et al., 2018) Lead is a heavy metal often discharged into waterbodies from various industries such as metal plating, painting, mining operations, and battery manufacturing. The solders of distribution system are the primary source of lead being dissolved into the drinking water. In the buildings built between the 1930's and early 1980's, copper pipes were widely used and, however, the solders were used to connect the pipes which contained elevated levels of lead (Actions You Take To Reduce Lead In Drinking Water (Report), EPA, 1993). Owing to the non-biodegradable nature of Pb(II), accumulation of this metal in living organisms can cause detrimental effect to health (Bai et al., 2018). Long term exposure to Pb(II) beyond the acceptable limits has been identified to cause health problems such as abortion, stillbirths, sterility, mental disturbance, and liver and kidney damage (Kadirvelu et al., 2010). For example, the Flint water crisis began in 2014 when the drinking water source for the city of Flint in Michigan was changed to the Flint River. Due to non-effective corrosion inhibitors, over 100,000 residents were potentially exposed to high levels of lead in the drinking water. The permissible amount of Pb(II) should be less than 0.01 mg L^{-1} in safe drinking water as recommended by World Health Organization (WHO) with the permissible level of Pb(II) in drinking water set by United States Environmental Protection Agency (USEPA) being 0.015 mg L^{-1} (Guyo et al., 2014).

Ordered Mesoporous Carbon (OMC), a new adsorbent, has gained considerable attention due to its unique features, such as high BET surface area, large pore volume, and high mechanical strenth. All these special features make OMC a suitable adsorbent to remove pollutants (Konggidinata et al., 2017). However, OMC has had limited success for metal ion removal and some organic pollutants (Chen at al., 2015). Therefore, chemical modifications of OMC are gaining considerable interests for improving sorption performance. In general, the OMC modified by chemical compounds have been shown to yield higher adsorption capacities and removal efficiencies for various aquatic pollutants (Mahmoud et al., 1999). For example, the researchers prepared modified OMC with different chemicals for mercury (Liu at al., 2016), orange(II) (Peng et al., 2014), BTEX and resorcinol removal (Guo at al., 2013). In this research, phosphate function groups were incorporated onto the OMC surface for improving Pb(II) ion removal.

MATERIALS AND METHODS

Synthesis of OMC. Initially, 1.5 g of sucrose ($C_{12}H_{22}O_{11}$) and 0.2 mL of concentrated sulfuric acid (H_2SO_4) were added to 20 mL of DI water with continuous stirring to completely dissolve the sucrose. Then, 2 g of SBA-15 template was added to the mixture and stirred for 90 mins. After stirring, the mixture was heated sequentially at 100 °C for 6 h and 160 °C for another 6 h. Finally, the obtained composite was then carbonized at 700 °C for 6 h under a nitrogen (N_2) flow of 100 mL min^{-1}. Afterwards, the black product was washed with approximately 50 mL of hydrofluoric (HF) solution at room temperature and stirred for at least 24 h to remove the silica template completely. After silica removal, the obtained product was filtered and washed with at least 4000 mL of DI water until the pH value of filtered water was close to 7.0.

Synthesis of Modified OMC (MOMC-NP). First of all, 2 g of OMC were added to 1M nitric acid with continuous stirring at room temperature for 2 h. The reflux condenser was applied to contain the mixture at 140 °C for 4 h. And then, the mixture was washed with DI water and placed in the oven to be dried at 80 °C for overnight. The product was labeled as MOMC-N. Next, 65 mL of 85% phosphoric acid was added and the mixture was stirred at room temperature for 2 h. The reflux condenser was also used to keep the mixture at 140 °C for 8 h. Afterwards, the obtained product was washed with 2,000 mL of DI water and dried in the oven at 80 °C for overnight. The final product was labeled as MOMC-NP.

Result and discussion

Transmission Electron Microscopy (TEM). The TEM images of SBA-15 template, OMC and MOMC-NP are shown in **Fig. 1**. The TEM image of SBA-15 shows the uniformly arranged pores that corresponded to the p6mm structure typically observed. The TEM images of OMC show the ordered structures and the TEM image of MOMC-NP indicates that the ordered structure disappeared after chemical treatment. The reason is probably that the acid treatment shrinks the ordered structure of OMC, which also explains the lower surface area and pore size after modified modification.

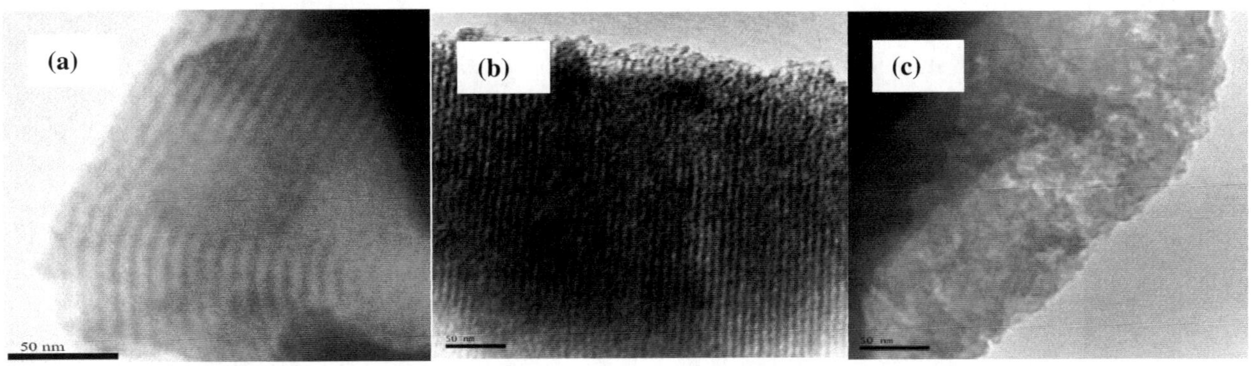

FIGURE. 1 TEM images of (a) SBA-15 template, (b) OMC and (c) MOMC-NP.

Fourier Transform Infrared Spectroscopy (FT-IR). The FT-IR spectra of OMC, MOMC-NP, and MOMC-NP after lead ion adsorption are shown in **Fig. 2**. As observed, several peaks ascribed to -OH groups, C-O groups, and alkenes bend were aroused at 3435 cm^{-1}, 2370 cm^{-1}, 1384 cm^{-1} and 1640 cm^{-1} in all of these three samples. The peaks observed at 1710 cm^{-1} was ascribed to C=O groups in MOMC-NP and the peak at 1230 cm^{-1} is attributed to the stretching vibrations of hydrogen-bonded P=O, stretching vibrations of O–C in P–O–C linkage, and P=OOH (Momcilovic et al., 2011). The disappearance of the peak at 1710 cm^{-1} and the reduction of the peak at 1230 cm^{-1} can be observed significantly in MOMC-NP after Pb(II) adsorption when compared to the MOMC-NP. The shifts occurred after Pb(II) adsorption possibly suggesting that the metal binding process was taking place on the active sites of MOMC-NP during the adsorption process. Accordingly, it can be further indicated that the chemical adsorption as opposite to physical adsorption is the major process in the adsorption process since the significant difference in the FTIR spectrum of MOMC-NP after Pb(II) adsorption compared with the FTIR spectrum of MOMC-NP.

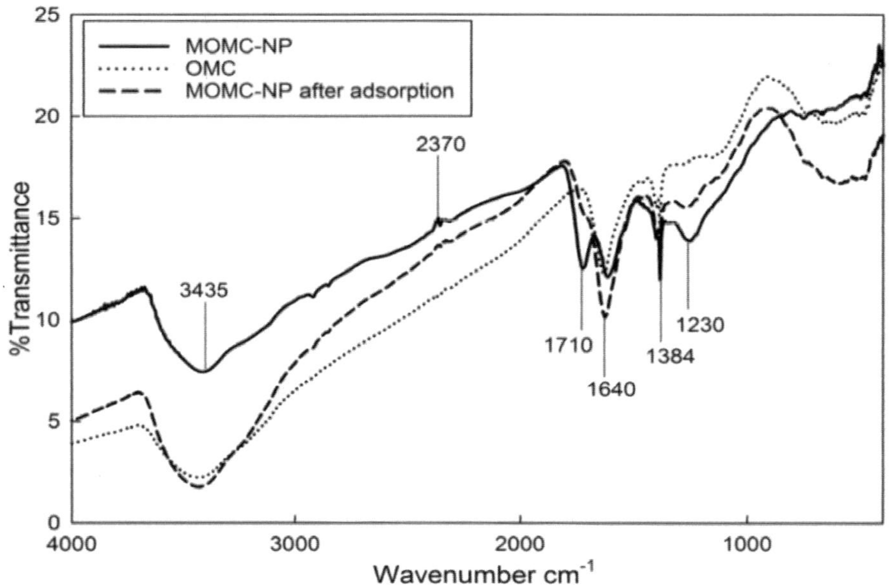

FIGURE. 2 FTIR spectra of OMC, MOMC-NP, MOMC-NP after Pb(II) adsorption.

Pb^{2+} Adsorption on MOMC-NP

Adsorption Kinetics. The effects of contact time and initial concentration on the adsorption of lead onto the MOMC-NP was studied as shown in **Fig. 3**. The adsorption capacity of Pb^{2+} increases with an increase of shaking time and reaches the equilibrium around 25 mins during the adsorption process for 60 mg L^{-1}, 80 mg L^{-1}, and 100 mg $L^{-1}$of Pb^{2+}. The corresponding equilibrium adsorption capacity was 56.06 mg g^{-1}, 68.82 mg g^{-1}, and 77.8 mg g^{-1}. The initial Pb^{2+} concentration has a strong effect on the adsorption capacity of the absorbent. From Fig. 3, the adsorption capacity of MOMC-NP for Pb^{2+} removal significantly increased from 56.06 mg g^{-1} to 77.8 mg g^{-1} when the initial concentration increased from 60 mg L^{-1} to 100 mg L^{-1}. This observation could be attributed to that higher Pb^{2+} concentration has the stronger driving force to combine with the active sites presenting on the surface of MOMC-NP. A similar result has been reported that the acceleration of adsorption capacity increased due to the stronger driving force by adsorbate (Shi et al., 2018).

Adsorption Isotherm. In the figure (**Fig. 4**) of Langmuir isotherm modeling, the R^2 is 0.99, which indicates that the Langmuir isotherm equation best fit the experimental data. Q_{max} represents the maximum adsorption capacity which is 47.89 mg g^{-1} and it is close to the experimental result of 46.96 mg g^{-1}. The dimensionless separation factor (R_L) is used to evaluate an essential feature of the Langmuir isotherm, for the test system, $R_L = 0.0116 < 1$ for MOMC-NP, indicating that the adsorption of Pb^{2+} was a favorable process. Freundlich model gives the parameters, n, indicative of bond energies between metal ions and the adsorbent and K_f, related to bond strength. The R^2 of the Freundlich equation is equal to 0.69 which indicated it doesn't fit reasonably well with the experimental data. Significantly, Langmuir model provides a great better fit than the Freundlich model. Freundlich model is characterized by 1/n heterogeneity factor; hence, it is applicable for sorption on heterogeneous surfaces. Smaller value of 1/n implies stronger interaction between the adsorbent and heavy metal, while 1/n values exist between 0 and 1 indicating the identical adsorption process and adsorption energies for all sites (Zhang et al., 2014). These results are similar to the ones described by Lalhmunsiama et al. (Lalhmunsiama et al., 2015).

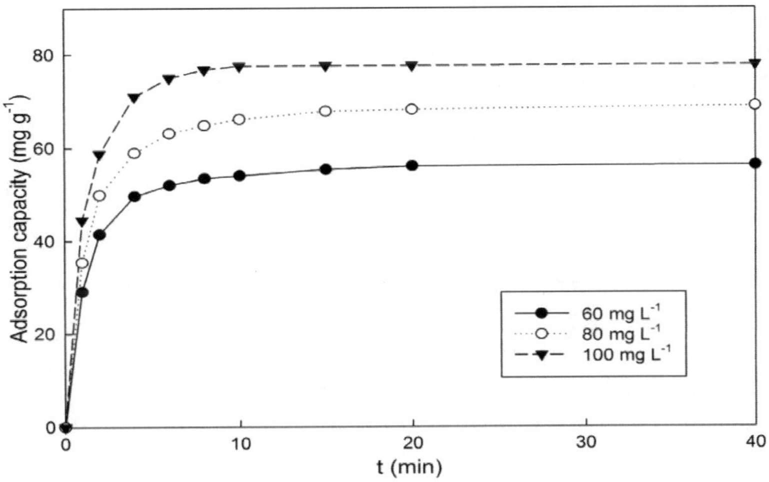

FIGURE. 3 Adsorption of lead on MOMC-NP at C_0 = 60 mg L^{-1}, 80 mg L^{-1}, and 100 mg L^{-1}, respectively. (T = 25 °C, S/V = 1 g L^{-1}, pH = 5)

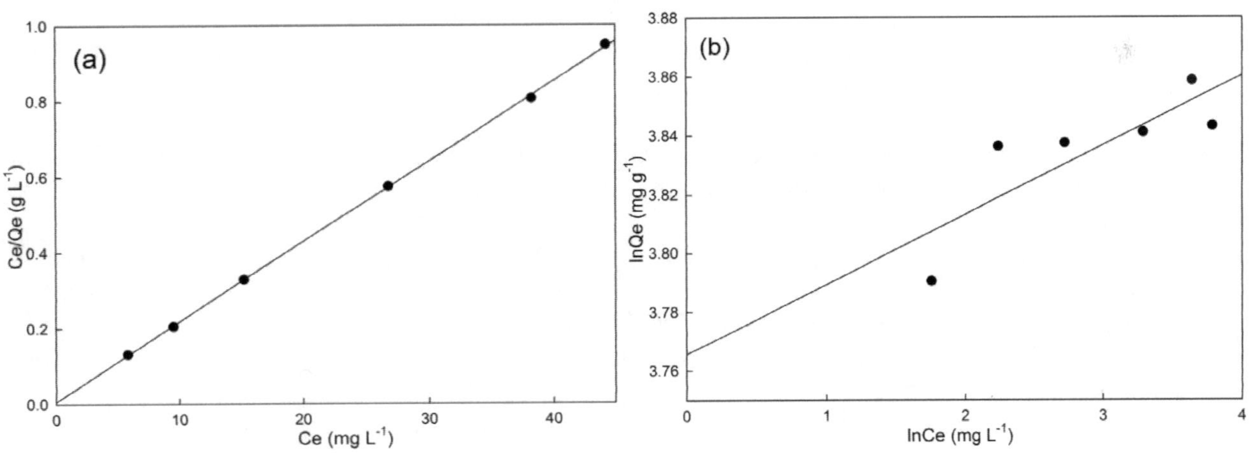

FIGURE. 4 (a) Langmuir isotherm; (b) Freundlich isotherm.

CONCLUSION

The OMC was synthesized via hard template method using the SBA-15 template. The MOMC-NP was successfully synthesized by the nitric acid and phosphoric acid treatments and characterized to study its application towards Pb^{2+} adsorption from aqueous solution. The maxmuium adsorption capacity of 77.80 mg g^{-1} for Pb^{2+} adsorption onto MOMC-NP was found at the initial Pb^{2+} concentration of 100 mg L^{-1}. The Pb^{2+} adsorption depends on the initial Pb^{2+} concentration, temperature, and pH. The adsorption capacity increased with the increase of Pb^{2+} initial concentration and temperature. The optimum pH value is 5.0 for Pb^{2+} adsorption onto MOMC-NP. The adsorption kinetics follows Pseudo-Second-Order kinetics model. The adsorption process was endothermic in nature. The equilibrium data fitted very well to Langmuir isotherm model. The mechanism of Pb^{2+} adsorption was determined that chemical adsorption and monolayer adsorption are the main adsorption but physical adsorption and heterogeneous adsorption still occurred in the adsorption process.

ACKNOWLEDGEMENTS

This work was supported by the University of Louisiana at Lafayette and the Lousisana Board of Regents (LEQSF(2016–17)-RD-C-15). Appreciation to the staffs and the assistance provided from Department of Chemical Engineering, Department of Chemistry and Department of Biology of University of Louisiana at Lafayette is also acknowledged.

REFERENCES

Zhang, Y., Zhao, J., Jiang, Z., Shan, D., & Lu, Y. (2014). Biosorption of Fe(II) and Mn(II) ions from aqueous solution by rice husk ash. *BioMed Research International*, 10.

Bai, R., Zhang, Y., Zhao, Z., Liao, Q., Chen, P., Zhao, P., & Li, L. (2018). Rapid and highly selective removal of lead in simulated wastewater of rare-earth industry using diglycolamic-acid functionalized magnetic chitosan adsorbents. *Journal of Industrial and Engineering Chemistry, 59*, 416–424.

Chen, F., Hong, M., You, W., Li, C., & Yu, Y. (2015). Applied Surface Science Simultaneous efficient adsorption of Pb^{2+} and MnO_4^- ions by MCM-41 functionalized with amine and nitrilotriacetic acid anhydride. *Applied Surface Science, 357*, 856–865.

Guo, R., Guo, J., Yu, F., & Gang, D. D. (2013). Microporous and Mesoporous Materials Synthesis and surface functional group modifications of ordered mesoporous carbons for resorcinol removal. *Microporous and Mesoporous Materials, 175*, 141–146.

Guyo, U., Mhonyera, J., & Moyo, M. (2014). Pb (II) adsorption from aqueous solutions by raw and treated biomass of maize stover–A comparative study. *Process Safety and Environmental Protection, 93*(6), 192–200.

Kadirvelu, K., & Namasivayam, C. (2010). Agricutural By-Product as Metal Adsorbent : Sorption of Lead(II) from Aqueous Solution onto Coirpith Carbon. *Environmental Technology, 21*(January 2014), 1091–1097.

Konggidinata, M. I., Chao, B., Lian, Q., Subramaniam, R., Zappi, M., & Gang, D. D. (2017). Equilibrium, kinetic and thermodynamic studies for adsorption of BTEX onto Ordered Mesoporous Carbon (OMC). *Journal of Hazardous Materials, 336*, 249–259.

Lalhmunsiama, T. D., & Lee, S. M. (2015). Physico-chemical studies in the removal of Sr(II) from aqueous solutions usingactivated sericite. *Journal of Environmental Radioactivity, 147*, 76–84.

Lead in Your Drinking Water; Actions You Take To Reduce Lead In Drinking Water (Report), EPA. (1993). EPA 810-F-93-001.

Liu, M.-X., Deng, X.-X., Zhu, D.-Z., Duan, H., & Xiong, W. (2016). Magnetically separated and N , S co-doped mesoporous carbon microspheres for the removal of mercury ions. *Chinese Chemical Letters, 27*(5), 795–800.

Ma, J., Qin, G., Zhang, Y., Sun, J., Wang, S., & Jiang, L. (2018). Heavy metal removal from aqueous solutions by calcium silicate powder from waste coal fly-ash. *Journal of Cleaner Production, 182*, 776–782.

Mahmoud, M. (1999). Selective solid phase extraction of mercury(II) by silica gel-immobilized-dithiocarbamate derivatives. *Analytica Chimica Acta, 398*, 297–304.

Momcilovic, M., Purenovic, M., Bojic, A., Zarubica, A., & Randelovic, M. (2011). Removal of lead(II) ions from aqueous solutions by adsorption onto pine cone activated carbon. *Desalination, 276*(1-3), 53–59.

Peng, X., & Fu, D. (2014). Preparation of ordered mesoporous carbons with ammonia modification for Orange II adsorption. *Desalination and Water Treatment, 54*(1), 255–264.

Shi, Z., Xu, C., Guan, H., Li, L., Fan, L., Wang, Y., & Zhang, R. (2018). Magnetic metal organic frameworks (MOFs) composite for removal of lead and malachite green in wastewater. *Colloids and Surfaces A: Physicochemical and Engineering Aspects, 539*, 382–390.

Removal of Glyphosate Using Commercial Grade Biochar and Activated Carbon: Statistical Optimization of Adsorption Parameters Using Response Surface Methodology

Gayana Anjali Dissanayake Herath[1], Richard D. Webster[1, 2], Wun Jern Ng[3, 4]

[1] Nanyang Environmental and Water Research Center, Nanyang Technological University, Singapore
[2] Division of Chemistry and Biological Chemistry, School of Physical and Mathematical Sciences, Nanyang Technological University, Singapore
[3] School of Civil and Environmental Engineering, Nanyang Technological University, Singapore
[4] Environmental Bio-innovations Group (EBiG), Nanyang Technological University, Singapore

ABSTRACT: The introduction of glyphosate, found in herbicides, to waterbodies is of concern due to its toxicity and hence a potential threat to public health and ecological systems. The present study compares commercially available activated carbon and biochar for glyphosate removal from aqueous solutions. Box–Behnken design and percent contribution with Pareto analysis techniques were used respectively in surface response and efficiency calculations to model the process conditions. The adsorption data better fitted the Freundlich isotherm model than Langmuir and BET models. The rate of glyphosate adsorption was found to follow a pseudo second order model. pH of the solutions was regulated by buffering during the adsorption process. Higher efficacy of glyphosate removal was obtained by optimizing parameters such as operating pH, initial glyphosate concentration, temperature, adsorbent dose, and contact time. The results showed that the optimal conditions yielding the best removals for the mentioned parameters were 8.0, 0.2 mg/L, 50.0°C, 11.4 g/L, 1.7 h for activated carbon and 5.0, 0.7 mg/L, 50.0°C, 12.3 g/L, 1.9 h for biochar, respectively. At the optimum adsorption conditions, the maximum removal capacity and efficiency were 0.0173 mg g^{-1} and 98.45% for activated carbon, and 0.0569 mg g^{-1} and 100.00% for biochar.

INTRODUCTION

Glyphosate (N-(phosphonomethyl) glycine) is a systematic, broad spectrum, non-selective, post emergence organophosphate herbicide, which has been widely used in agricultural and non-agricultural areas over the world (Woodburn, 2000). Recent studies have recorded that it is toxic to humans at concentrations below agricultural applied levels (Buffin & Jewell, 2001), and linked with cardiac and respiratory system issues causing respiratory, contact dermatitis and eczema (Hu, Zhao, & Sorohan, 2011). International Agency for Research on Cancer (IARC) has named glyphosate as a "probable human carcinogen" due to its mutagenic, genotoxic, and carcinogenic effects (Yamaguchi, Bergamasco, & Hamoudi, 2016). Application of the glyphosate has been increased by 15 million kilograms in the past years (Benbrook, 2016), that these increased amounts of application has been found to be phytotoxic to the crops (Williams, Kroes, & Munro, 2000). Thus, it is crucial to remove the excess amount of glyphosate in the soil and in groundwaters for public health and ecological systems.

The published literature shows adsorption processes may be suitable for removal of glyphosate (De Ridder et al., 2010; Sheals, Sjöberg, & Persson, 2002). A lot of commercial grade biochar and activated carbon are available in the market for contaminant adsorption. Powdered activated carbon (PAC) was a promising adsorbent for the removal of organic micropollutants (Mattson & Mark, 1971). The associated high carbon footprint and high energy requirement make PAC expensive for non-commercial applications. Biochar, which is a by-product of pyrolysis process can cost less and the biochar's made from paper sludge

(Mendez, Paz-Ferreiro, Araujo, & Gasco, 2014), wood waste (Hagner et al., 2015) and industrial residues (Hu et al., 2011) are comparatively high in moisture and has a negative carbon footprint.

The present study compares commercial grade activated carbon with biochar for glyphosate removal from aqueous solutions under various experimental conditions. The traditional method of varying one variable keeping the rest constant can be time consuming and unreliable as it doesn't explain the synergistic effect of different variables (Aleboyeh, Daneshvar, & Kasiri, 2008). Statistical optimisation using response surface methodology can be a better choice in building the model, designing the experimental plan and in evaluating and optimising the experimental data (Zaroual, Chaair, Essadki, El Ass, & Azzi, 2009) to overcome above issues. Response surface methodology is a collection of statistical and mathematical techniques. Over the widely used central composite design (CCD) and 3k full factorial design and Box-Behnken method, three factorial Box-Behnken method is applied in determining the treatment configuration, which is potentially low-cost with high efficiency for the parameters, operating pH, initial glyphosate concentration, temperature, adsorbent dose, and contact time. The data is fitted to a second order polynomial model, which is statistically validated by analysing the percent contribution with Pareto analysis techniques and performing an analysis of variance test.

MATERIALS AND METHODS

Chemicals. Glyphosate (N-(phosphonomethyl) glycine) (purity 96%) was purchased from Sigma-Aldrich, Singapore. Commercial grade coconut shell activated carbon was purchased from Puritas (Pvt) Ltd, Sri Lanka and wood biochar was purchased from Sigma-Aldrich, Singapore. All other reagents used were of analytic grade and used without further purification. MilliQ ultra-pure (18.2 MΩ cm) water was used in preparing the samples.

Adsorbent Characterisation. Adsorbent characterisation was done by means of qualitative, quantitative and spectroscopic analysis. Elemental composition of the carbons was analysed by Vario EL cube, Elementar Company, USA. Total ash content of the adsorbent was determined by heating 1 g samples to 900^0C for 2 h in a muffle furnace and then by weighing the remaining ash content. The Brunauer–Emmett–Teller (BET) specific surface area and pore size were determined by QUADRASORB™, USA equipment in micropore analysis mode. Scanning Electron Microscope Merlin with Cryo Work Station (Fision, USA) and Image Analyzer (Q500 Quantimax, USA) units were used to observe surface structure and morphology of the biochar and activated carbon.

Batch Adsorption Experiments. Glyphosate was dissolved in the milliQ water, sonicated and mixed with respective amounts of 0.01 M NaCl to prepare the stock solution. 1 M NaOH and 1 M HCl solutions were used in regulating the pH of the solutions. Prepared 40 ml of the glyphosate stock solution samples were added to 50 ml centrifuge tubes and the measured activated carbon or biochar were added to it. All experiments were conducted under 150 rpm mixing conditions in an incubating shaker under regulated temperature. After the adsorption test, the bulk fluid phase of sample was filtered using 0.22 µm syringe filters and the filtered volumes were transferred to brown vials. These solutions in vials were tested using LC-MS/MS (Agilent 6460 LC/MS QQQ, Agilent Technologies, Singapore) with Zorbax SB-C18 (2.1 x 100 mm, 3.5 µm particle size) (Agilent Technologies, Singapore) column to quantify the remaining glyphosate amounts in the water. Each experiment was triplicated, and each reading was duplicated to minimise the human and instrumental error.

Experimental Design. During the study, the Box-Behnken method was used to optimise the five operating parameters, operating pH, initial glyphosate concentration (mg/L), temperature (^{0}C), adsorbent dose (g/L), and contact time (h) to maximise the removal capacity and efficiency. Over central composite design (CCD) and 3k full factorial design, the Box-Behnken method was chosen because it's an independent quadratic design which doesn't contain any embedded factorials or fractional factorials. 3-factor Box-Behnken method represent a box. Each factor is varied at three levels, midpoints of the edges and at center as -1, 0 and +1, indicating low, medium and high values. The sample solutions which were used in adsorption tests were prepared to match the conditions at these each level points and the final concentration of glyphosate remaining after adsorption test for each level point was recorded. The recorded data representing each point are fitted using an empirical second- order polynomial model which is as follows:

$$Y = \beta_0 + \sum_{i=1}^{n} \beta_i x_i + \sum_{i=1}^{n} \sum_{i=1}^{n} \beta_{ij} x_i x_j + \sum_{i=1}^{k} \beta_{ii} x_{ii}^2 + \varepsilon \qquad (1)$$

Where, β is the regression coefficient for the intercept. The emphatical models found for activated carbon and biochar were plotted and analysed. The operation parameters were ranked according to the significance of each factor on the overall adsorption process using the perturbation analysis. In order to evaluate the accuracy of the model, pareto analysis and analysis of variance (ANOVA) tests were performed.

RESULTS AND DISCUSSION

Characterization of Adsorbent Material. The data on surface structure and pore dispersion (see Table 1) show that surface area and calculated pore volume of activated carbon was double the size of biochar. The SEM images of activated carbon and biochar are shown in Figure 1. The surface of the activated carbon was smooth, and its pores were very small of micropore range. Biochar had an irregular multilayer lamellar structure with the size of 5 μm.

TABLE 1. Surface characteristics and elemental composition of adsorption materials

	BET micro pore analysis							
	Surface Area (m^2/g)		**Pore Volume (cc/g)**			**Pore Radius $D_v(r)$ (Å)**		
	MultiPoint BET	DFT Cumulative	HK (Horvath-Kawazoe method)	SF (Saito and Foley method)	DFT Cumulative	HK (Horvath-Kawazoe method)	SF (Saito and Foley method)	DFT Cumulative
Activated carbon	1108.427	972.9	0.4323	0.3874	0.4181	2.913	5.076	4.297
Biochar	594.204	561.9	0.2555	0.2129	0.3427	2.963	1.754	4.297

	Elemental analysis					
	C % (wt./wt.)	**H % (wt./wt.)**	**N % (wt./wt.)**	**S % (wt./wt.)**	**O % (wt./wt.)**	**Ash Content % (wt./wt.)**
Activated carbon	90.800 ± 0.870	1.914 ± 0.280	0.224 ± 0.002	0.9886 ± 0.000	4.5804	1.492 ± 0.256
Biochar	72.710 ± 0.500	2.106 ± 0.170	0.515 ± 0.100	0.9886 ± 0.000	15.971	8.698 ± 0.177

Adsorption Behaviour

Kinetics of Adsorption. The adsorption rate curves for glyphosate with activated carbon and biochar are shown in Figure 2. The adsorption of biochar was spontaneous that it reached apparent equilibrium in less than 5 h and for activated carbon, it took 15 h. Then slow adsorption was observed to achieve equilibrium, taking 15 h and 24 h for biochar and activated carbon respectively. Similar adsorption behaviour has been reported in rice husk derived biochar (Herath et al., 2016) previously. It can believe the saturation of available adsorption sites with time leads to this slow adsorption behaviour, whereas initial rapid adsorption was due to the availability of active sites in the surface sufficiently and electrostatic attractions (Jiang et al., 2018).

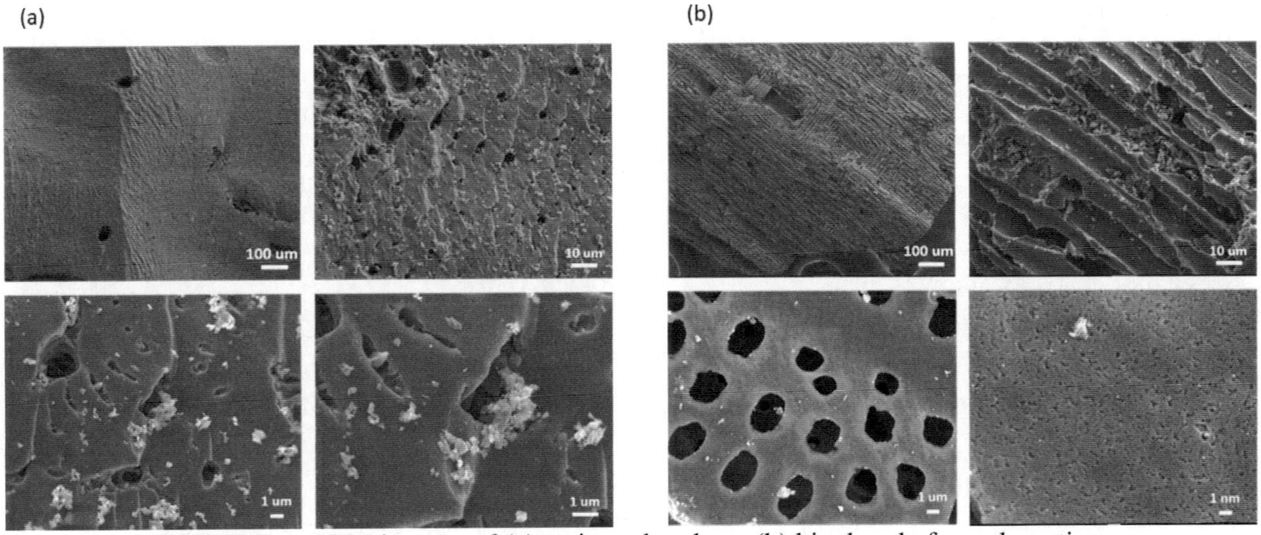

FIGURE 1. SEM images of (a) activated carbon; (b) biochar; before adsorption

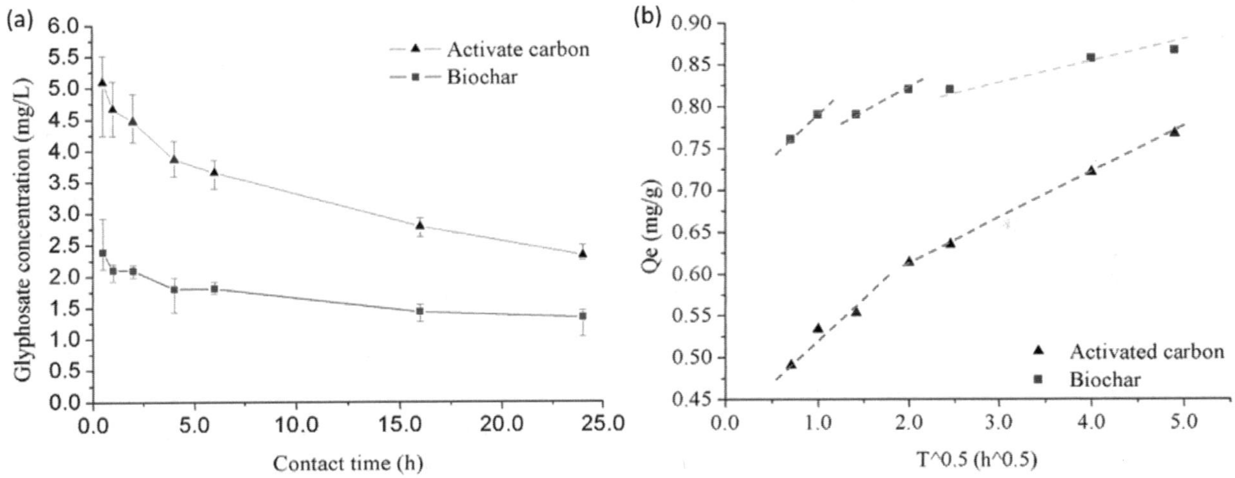

FIGURE 2. (a) Effect of contact time plot; (b) Weber-Morris intraparticle diffusion plot for the adsorption of glyphosate

Kinetic data were fitted to pseudo first order model, pseudo second order model and Weber-Morris internal particle diffusion model, to understand the adsorption process and rate limiting step. The equations used were as follows;

pseudo first order model; $\log(q_e - q_t) = \log(q_e) - \frac{k_1 \times t}{2.303}$ (2)

pseudo second order model; $\frac{t}{q_t} = \frac{1}{k_2 \times q_e^2} + \frac{t}{q_e}$ (3)

Weber-Morris internal particle diffusion model; $q_t = k_{id} \times t^{0.5} + c_i$ (4)

Where k_1, k_2 and k_{id} are pseudo first-order kinetic (min^{-1}), pseudo second-order kinetic (g/mg x min) and internal particle diffusion (mg/g * min ^ (-0.5)) constants. q_t *and* q_e are adsorption capacities at time t and at equilibrium.

TABLE 2. Adsorption kinetic parameters of glyphosate onto activated carbon and biochar

Kinetic model	Carbon type	Parameters			
		Co (mg/L)	**q_e (mg/g)**	**k_1(1/min)**	**R^2**
Pseudo-first-order model	Activated carbon	10	0.7669	0.1114	0.9897
	Biochar	10	0.8664	0.1471	0.9782
		Co (mg/L)	**q_e (mg/g)**	**k_2 (g/mg*min)**	**R^2**
Pseudo-second-order model	Activated carbon	10	0.7669	0.5812	0.9976
	Biochar	10	0.8664	0.1581	0.9997

Pseudo second order model, regression coefficient was better ($R^2 > 0.99$) than pseudo first order model (see Table 2). This indicates that adsorption involved a chemical reaction (Yamaguchi et al., 2016). It was also noticed that for activated carbon, k_2 (pseudo second-order kinetic constant) was inversely proportional to initial glyphosate concentration (C_o). This can happen due to intermolecular collisions between glyphosate molecules in the solution due to high concentrations that it delays the molecules from reaching the active sites in the adsorbent (Chakraborty, Karmakar, Mukherjee, & Kumar, 2014). For activated carbon, considering the linear segments, particle diffusion curve was divided into two different sections (see Figure 2). None of the lines passes through the origin which indicates that there are two rate limiting steps. In the first stage, the rate was limited by external film diffusion where molecules cross the boundary layer to reach the adsorbent surface. At the second step, the rate was dropped drastically due to interparticle resistance at the adsorption interface. In biochar, three rate controlling steps were observed. It involved a third step which further drops the removal rate which is governed by the pore diffusion. This further indicates that biochar pore structure is smaller in size compared to activated carbon (Zhao et al., 2014). The increase of c_i with time indicates the increase of boundary layer thickness with time. This suggested the importance in taking consideration of that external film diffusion at high glyphosate concentrations.

TABLE 3. Langmuir, Freundlich and BET isotherm constants

Adsorbent	Langmuir			Freundlich			BET		
	K_L (L/mg)	Q_m (mg/g)	R^2	K_F (mg^{n+1}* g*L^n)	n	R^2	K_b	Q^o (L/g)	R^2
Activated carbon	0.0861	1.0549	0.061	0.0941	1.0873	0.811	1055.889	1.0523	0.0611
Biochar	0.0451	1.1645	0.1238	0.2558	0.9772	0.9567	551	6.0496	0.1242

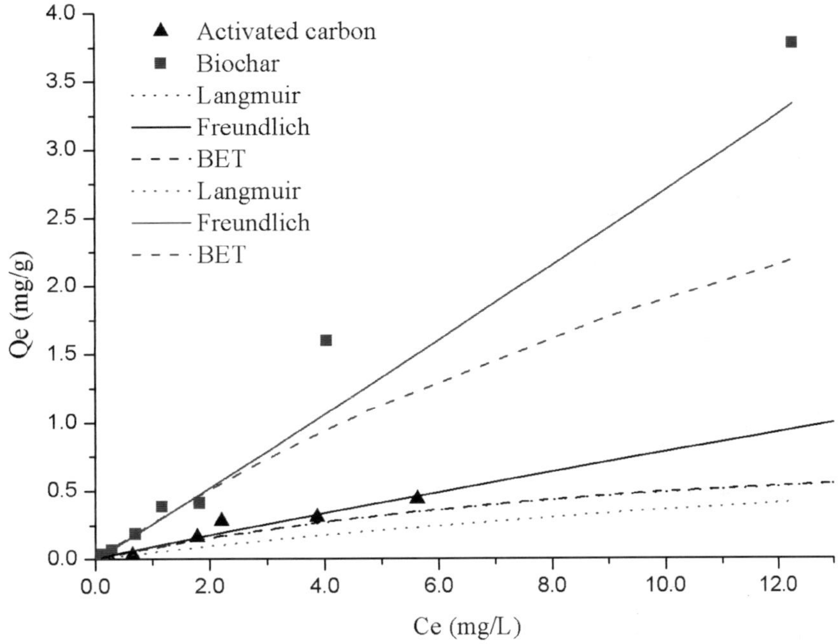

FIGURE 3. Adsorption isotherms of activated carbon and biochar

Adsorption Isotherms. Langmuir, Freundlich and BET (Brunauer, Emmett and Teller) isotherm models were used and data was fitted to those models. The equations used were as follows:

Langmuir model; $Q_e = \frac{Q_m \times K_l \times C_e}{1+ K_l \times C_e}$ (5)

Freundlich model; $Q_e = K_f \times C_e^{\frac{1}{n}}$ (6)

BET model; $Q_m = \frac{K_b \times C_e \times Q^0}{(C_s - C_e)\,[1 + (K_b - 1)(^{C_e}/_{C_s})]}$ (7)

Where K_l, K_f and K_b are Langmuir constant (L/mg), Freundlich constant ($(mg^{n+2}{*}L^n)$) and a parameter related to the binding intensity for all layers. $\frac{1}{n}$ represent the adsorption intensity. C_e and C_s are equilibrium concentration (mg/L) and solubility limit (mg/L) respectively. The adsorption isotherms for two adsorbents were plotted in Figure 3. Adsorption of glyphosate on both biochar and activated carbon was significantly linear and matched well with Freundlich isotherm ($R^2 > 0.8$). This indicated that adsorption was a multilayer process confirming the reusability of the adsorbent. Biochar showed better performance than activated carbon with a maximum capacity of 0.2558 $mg^{n+2}{*}L^n$ and with $\frac{1}{n} > 1$. Isotherm constants values calculated are as follows (see Table 3).

Response Surface Methodology

Effect of Initial Glyphosate Concentration. The combined effects of initial glyphosate concentration and contact time when adsorbent dosage, pH, and temperature were fixed at 25.5 g/L, 6.5, and 35.0^{0}C for biochar and activated carbon is shown in Figure 4. The percentage removal of glyphosate had increased when the initial glyphosate concentration was increased from 0.2 mg/L to 20.0 mg/L. This has been known as the capacity effect. When the glyphosate molecules concentration in the solution is increased, the number of effective collisions of glyphosate molecules with the adsorbent increases, which lead to the increase in uptake. However, when the initial concentration was further increased to 50.0 mg/L, the removal percentage dropped. This indicated the exhaustion of the surface of the carbon with time. Similar behaviour when glyphosate was adsorbing to forest soil has been observed by (Sen, Mondal, Chattoraj, & Datta, 2017), with a conforming decrease in $\mathbf{k_{id}}$ (distribution coefficient) suggesting a depletion of active sites at higher concentrations limiting the number of sites available for adsorption.

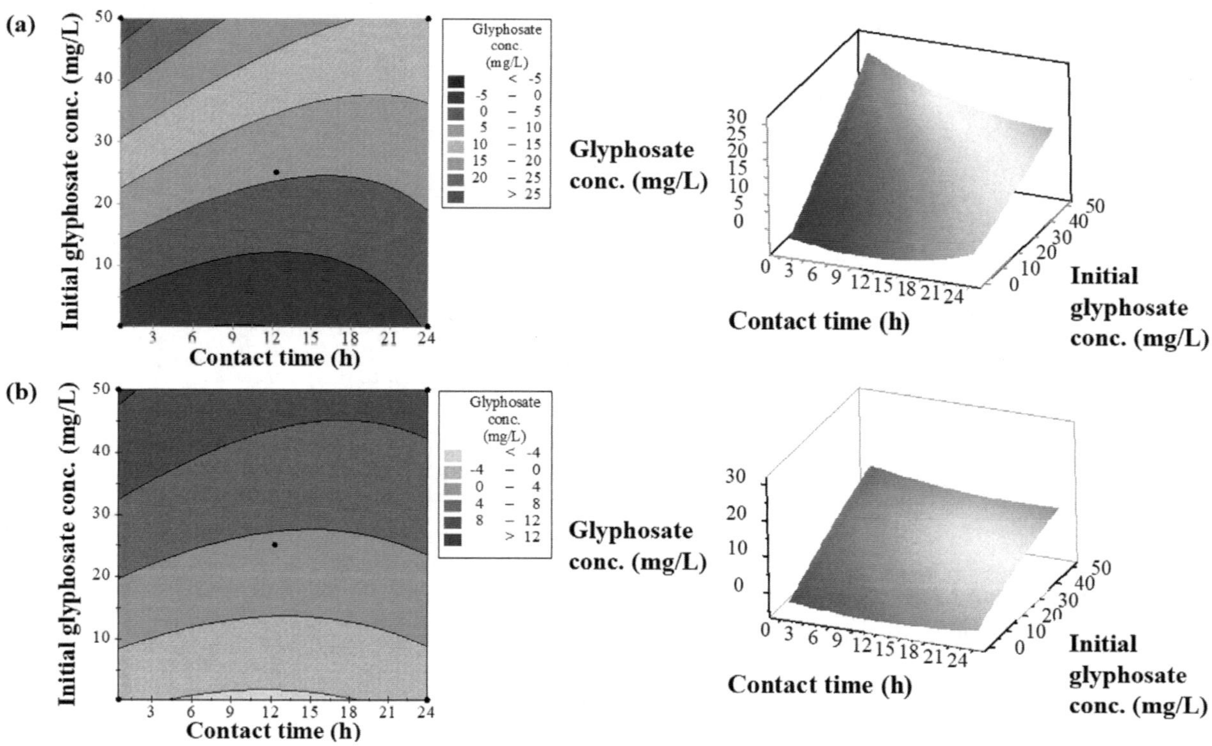

FIGURE 4. RSM plots. Contour plot and 3D surface graph of glyphosate removal (a) activated carbon; (b) biochar; showing the effect of contact time and initial glyphosate concentration

Effect of Dose. Synergistic effect of contact time and adsorbent dosage on adsorption of glyphosate when other parameters such as initial glyphosate concentration, pH, and temperature were fixed at 25.1 mg/L, 6.5 and 35.0^{0}C is shown in Figure 5. The percentage removal of glyphosate increased with the increase of dosage until 30.0 g/L. This is due to the enrichment of surface area and active sites with the increase of dosage (Bhaumik & Mondal, 2015). However, the final remaining glyphosate concentration wasn't reduced when the dosage is further increased. This can happen due to two reasons. The restraint of access to active sites due to the hindrance by the adsorbent particles (Rahmanifar & Dehaghi, 2014) and the increase in the interactions between the adsorbent-adsorbent than that of adsorbent-adsorbate (Akhtar, Hasany, Bhanger, & Iqbal, 2007).

Effect of pH. pH plays an important role in the removal process of micro organic compounds (Reddy, Lakshmipathy, & Sarada, 2014). In Figure 6, angularly shaped contours reveal that pH has affected the adsorption process of glyphosate by both biochar and activated carbon. Here, the synergistic effect of pH and contact time when initial glyphosate concentration, dosage and temperature fixed at 25.1 mg/L, 25.5 g/L and 35.0^{0}C is depicted. In biochar, maximum removal was recorded at lowest pH of 5.0, and the uptake is reduced with a further increase in pH. A similar observation was reported by (Herath et al., 2016) for adsorption of glyphosate with rice husk derived biochar, which also reported that adsorption capacity significantly dropped with the increase of pH. Glyphosate has four main acid dissociation constants, pK_a 2.0, 2.6, 5.6, and 10.6 and is negatively charged at most of the pH range. When pH is 5.0, the negativity of the molecule decreases.

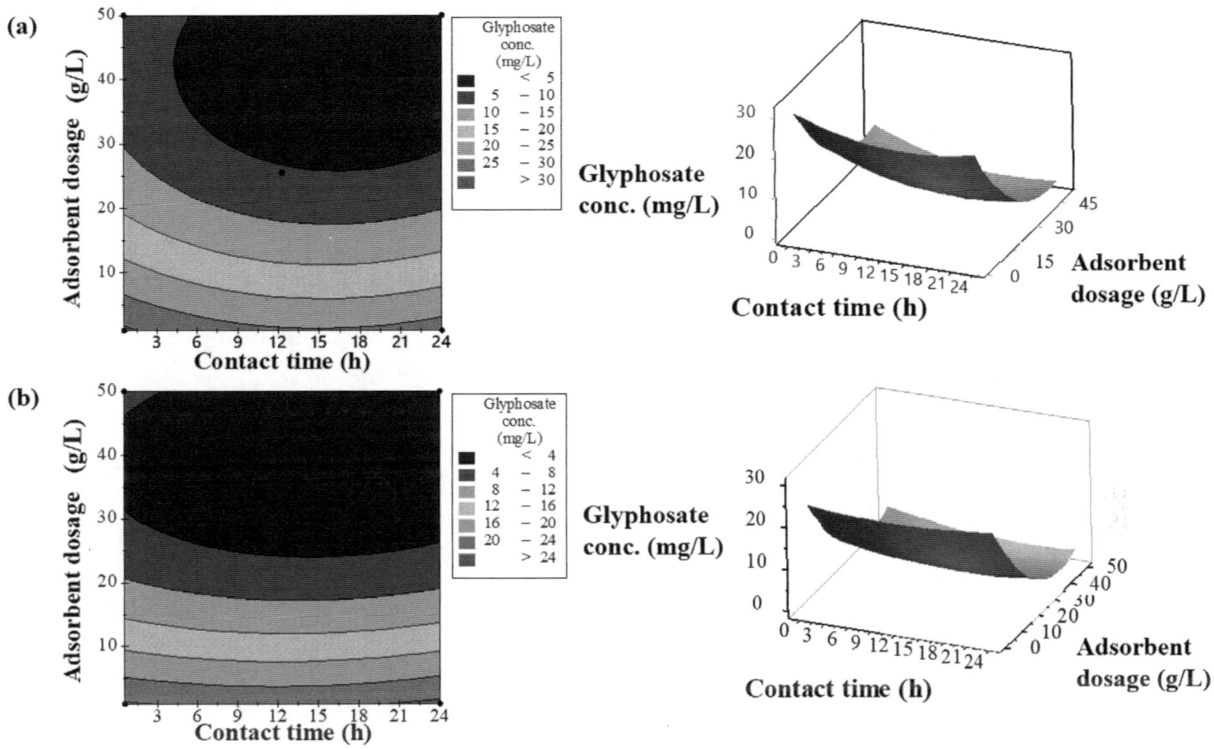

FIGURE 5. RSM plots. Contour plot and 3D surface graph of glyphosate removal (a) activated carbon; (b) biochar; showing the effect of contact time and adsorbent dosage

On the contrary to biochar, activated carbon performed well in higher pH. Best removal of glyphosate can be observed in the pH range from 7.0-8.0. Previous research highlighted a deportation of the acidic sites transforming the active sites to ionic species when increasing the pH which attracts the glyphosate molecules through electrostatic interactions (Sen et al., 2017). As the pH of the solution increases, the negative charge in the glyphosate molecule and the ionic species in the surface of activated carbon increases. This induces the attractions between the adsorbent and the adsorbate which result in a rise of removal of glyphosate.

Effect of Temperature. Temperature is one of the crucial factors which affect the adsorption process (Ghosh, Bhaumik, & Mondal, 2016). In Figure 7 with angular contours, it is clearly demonstrated the summed-up effect of temperature and time on the uptake of glyphosate at fixed initial glyphosate concentration, adsorbent dosage and pH of 25.1 mg/L, 25.5 g/L and 6.5. In biochar, comparing with activated carbon, the effect of temperature is the dominant factor. When observing the 3D plots, biochar seems to have a maximum whereas activated carbon seems to have a minimum which depicts that temperature requirement is higher for biochar than for activated carbon. Yet, both perform better at temperatures above 40^0C. However, research on thermodynamics and kinetic behaviour of the glyphosate adsorption has confirmed that glyphosate adsorption is exothermic and the adsorption capacity increases with the increase in temperature (Chen, Zhou, Li, & Peng, 2016). A better removal at higher temperature can expect if an endothermic reaction is taking place at the adsorbent interface during adsorption. The adsorption kinetics of the carbons also confirms a chemical adsorption at the interface than a physical adsorption which further implies that something other than the surface area is involved in the adsorption process especially at higher temperature levels.

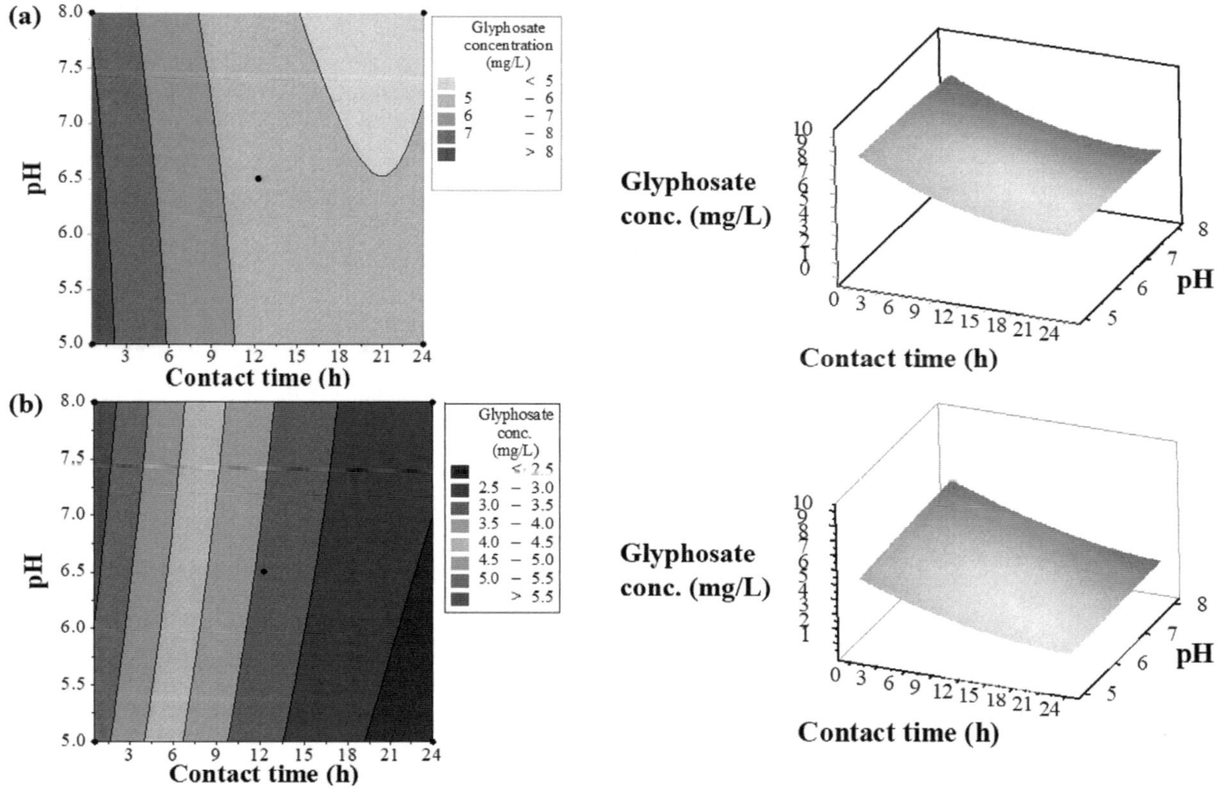

FIGURE 6. RSM plots. Contour plot and 3D surface graph of glyphosate removal (a) activated carbon; (b) biochar; showing the effect of contact time and operating pH

TABLE 4. Optimised conditions for the process for the maximum efficiency

Process parameter	Contact time (h)	Initial glyphosate concentration (mg/L)	Adsorbent dosage (g/L)	Operating pH	Operating temperature (°C)	Maximum removal capacity (mg/g)	Adsorption efficiency (%)
Activated carbon	1.7	0.2	11.4	8.0	50.0	0.0173	98.45
Biochar	1.9	0.7	12.3	5.0	50.0	0.0569	100.0

The Collective Effect of All Variables on the Removal Capacity. The optimisation using graphical user interface (GUI) for response surface methodology provided five plots demonstrating the effect of individual factors on adsorption for biochar and activated carbon. That graphs are summarised in Table 4. Biochar is more effective at high initial concentrations where activated carbon at low initial concentrations. According to the perturbation plot, the sequence of parameters influence on glyphosate removal was as follows:

Initial glyphosate concentration > adsorbent dose > contact time > temperature > pH

In the environment the initial glyphosate concentration and contact time is often fixed for the situation. At the treatment of effluent flows with high glyphosate concentrations, it is suggested to use biochar at the rough agent and activated carbon as the secondary polishing agent. That way it is possible to control the excessive use of adsorbents. This study explains the importance of controlling the temperature and pH at a limited dosage to enhance the removal in such instances.

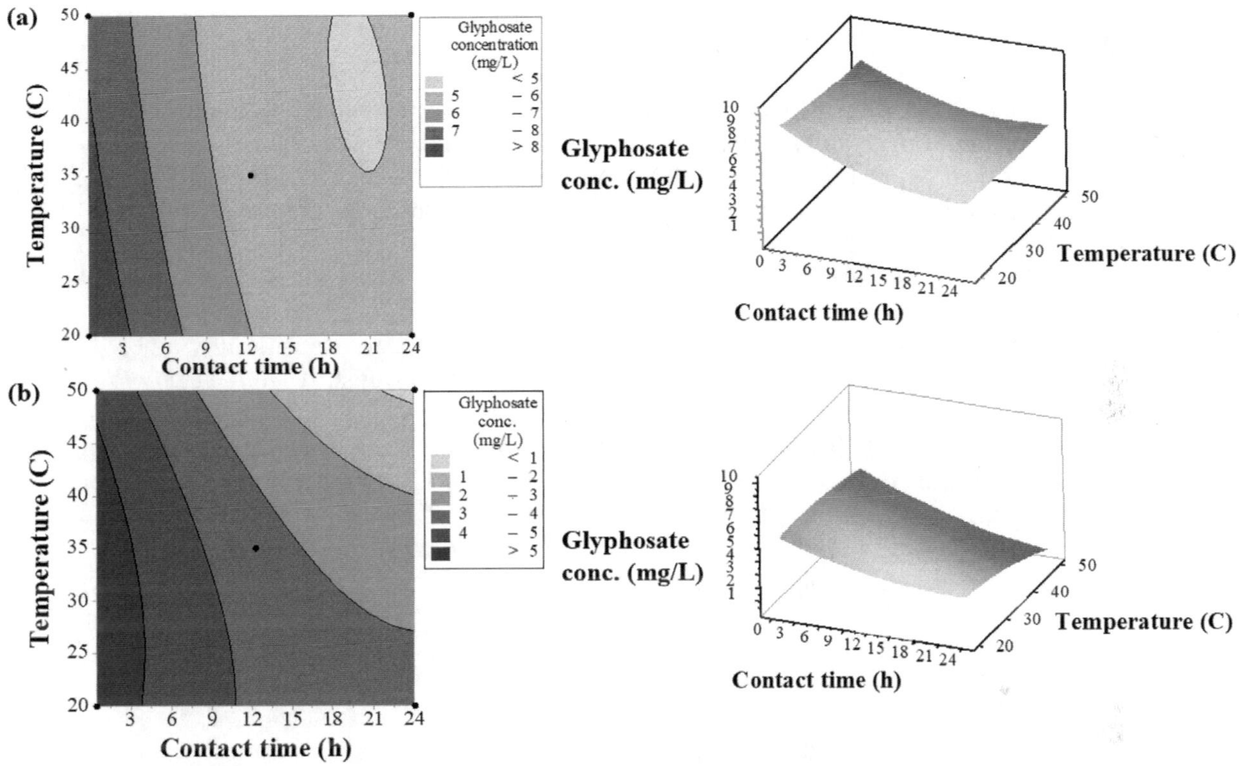

FIGURE 7. RSM plots. Contour plot and 3D surface graph of glyphosate removal (a) activated carbon; (b) biochar; showing the effect of contact time and operating temperature

CONCLUSIONS

In the current study, commercial grade biochar and activated carbon were studied to evaluate the removal of glyphosate from aqueous solutions. Freundlich isotherm was found to be a better fit for glyphosate removal for both biochar and activated carbon indicating multilayer adsorption of glyphosate. Pseudo second order model matched well for the adsorption behaviour. When the initial glyphosate concentration was increased, the removal was increased up to a limit and started decreasing due to the exhaustion of the active sites. The percentage removal was increased with the increase of adsorbent dosage. This paper also contributes to the debate on the effect of operating pH and temperature on the efficiency of the adsorption process. The rise of removal efficiency with the rise of temperatures suggests a chemical adsorption at the adsorbent interface than a physisorption at increased temperatures. Biochar performed well at low pH levels while activated carbon performed well at high pH. According to the perturbation plot, initial glyphosate concentration had the most influence on the uptake of glyphosate followed by adsorption dosage, contact time, temperature and pH respectively. The findings confirm that both activated carbon and biochar are effective although biochar may be the low-cost alternative for the removal of glyphosate from aqueous solutions. Thus, in this paper we argue biochar is the better candidate for the treatment wastewater in agricultural areas polluted with organic pollutants compared to activated carbon considering its strong absorbability, easy obtainability and inexpensive characteristics.

ACKNOWLEDGEMENTS

This work was supported in part by the Interdisciplinary Graduate School, Nanyang Technological University, Singapore and by Residue, Resource and Reclamation Centre, Nanyang Environment and Water Research Institute, Singapore.

REFERENCES

Akhtar, M., Hasany, S. M., Bhanger, M., & Iqbal, S. (2007). Low cost sorbents for the removal of methyl parathion pesticide from aqueous solutions. *Chemosphere, 66*(10), 1829-1838.

Aleboyeh, A., Daneshvar, N., & Kasiri, M. (2008). Optimization of CI Acid Red 14 azo dye removal by electrocoagulation batch process with response surface methodology. *Chemical Engineering and Processing: Process Intensification, 47*(5), 827-832.

Benbrook, C. M. (2016). Trends in glyphosate herbicide use in the United States and globally. *Environmental Sciences Europe, 28*(1), 3.

Bhaumik, R., & Mondal, N. K. (2015). Adsorption of fluoride from aqueous solution by a new low-cost adsorbent: thermally and chemically activated coconut fibre dust. *Clean Technologies and Environmental Policy, 17*(8), 2157-2172.

Buffin, D., & Jewell, T. (2001). *Health and environmental impacts of glyphosate: The implications of increased use of glyphosate in association with genetically modified crops*: Friends of the earth.

Chakraborty, R., Karmakar, S., Mukherjee, S., & Kumar, S. (2014). Kinetic evaluation of chromium (VI) sorption by water lettuce (Pistia). *Water Science and Technology, 69*(1), 195-201.

Chen, F.-x., Zhou, C.-r., Li, G.-p., & Peng, F.-f. (2016). Thermodynamics and kinetics of glyphosate adsorption on resin D301. *Arabian Journal of Chemistry, 9*, S1665-S1669.

De Ridder, D. J., Villacorte, L., Verliefde, A. R., Verberk, J. Q., Heijman, S., Amy, G. L., & Van Dijk, J. C. (2010). Modeling equilibrium adsorption of organic micropollutants onto activated carbon. *Water research, 44*(10), 3077-3086.

Ghosh, S. B., Bhaumik, R., & Mondal, N. K. (2016). Optimization study of adsorption parameters for removal of fluoride using aluminium-impregnated potato plant ash by response surface methodology. *Clean Technologies and Environmental Policy, 18*(4), 1069-1083.

Hagner, M., Hallman, S., Jauhiainen, L., Kemppainen, R., Rämö, S., Tiilikkala, K., & Setälä, H. (2015). Birch (Betula spp.) wood biochar is a potential soil amendment to reduce glyphosate leaching in agricultural soils. *Journal of environmental management, 164*, 46-52.

Herath, I., Kumarathilaka, P., Al-Wabel, M. I., Abduljabbar, A., Ahmad, M., Usman, A. R., & Vithanage, M. (2016). Mechanistic modeling of glyphosate interaction with rice husk derived engineered biochar. *Microporous and mesoporous materials, 225*, 280-288.

Hu, Y., Zhao, Y., & Sorohan, B. (2011). Removal of glyphosate from aqueous environment by adsorption using water industrial residual. *Desalination, 271*(1-3), 150-156.

Jiang, X., Ouyang, Z., Zhang, Z., Yang, C., Li, X., Dang, Z., & Wu, P. (2018). Mechanism of glyphosate removal by biochar supported nano-zero-valent iron in aqueous solutions. *Colloids and Surfaces A: Physicochemical and Engineering Aspects, 547*, 64-72.

Mattson, J. S., & Mark, H. B. (1971). *Activated carbon: surface chemistry and adsorption from solution*: M. Dekker.

Mendez, A., Paz-Ferreiro, J., Araujo, F., & Gasco, G. (2014). Biochar from pyrolysis of deinking paper sludge and its use in the treatment of a nickel polluted soil. *Journal of Analytical and Applied Pyrolysis, 107*, 46-52.

Rahmanifar, B., & Dehaghi, S. M. (2014). Removal of organochlorine pesticides by chitosan loaded with silver oxide nanoparticles from water. *Clean Technologies and Environmental Policy, 16*(8), 1781-1786.

Reddy, N. A., Lakshmipathy, R., & Sarada, N. (2014). Application of Citrullus lanatus rind as biosorbent for removal of trivalent chromium from aqueous solution. *Alexandria Engineering Journal, 53*(4), 969-975.

Sen, K., Mondal, N. K., Chattoraj, S., & Datta, J. K. (2017). Statistical optimization study of adsorption parameters for the removal of glyphosate on forest soil using the response surface methodology. *Environmental Earth Sciences, 76*(1), 22.

Sheals, J., Sjöberg, S., & Persson, P. (2002). Adsorption of glyphosate on goethite: molecular characterization of surface complexes. *Environmental science & technology, 36*(14), 3090-3095.

Williams, G. M., Kroes, R., & Munro, I. C. (2000). Safety evaluation and risk assessment of the herbicide Roundup and its active ingredient, glyphosate, for humans. *Regulatory Toxicology and Pharmacology, 31*(2), 117-165.

Woodburn, A. T. (2000). Glyphosate: production, pricing and use worldwide. *Pest Management Science, 56*(4), 309-312.

Yamaguchi, N. U., Bergamasco, R., & Hamoudi, S. (2016). Magnetic MnFe2O4–graphene hybrid composite for efficient removal of glyphosate from water. *Chemical Engineering Journal, 295*, 391-402.

Zaroual, Z., Chaair, H., Essadki, A., El Ass, K., & Azzi, M. (2009). Optimizing the removal of trivalent chromium by electrocoagulation using experimental design. *Chemical Engineering Journal, 148*(2-3), 488-495.

Zhao, F., Tang, W. Z., Zhao, D., Meng, Y., Yin, D., & Sillanpää, M. (2014). Adsorption kinetics, isotherms and mechanisms of Cd (II), Pb (II), Co (II) and Ni (II) by a modified magnetic polyacrylamide microcomposite adsorbent. *Journal of Water Process Engineering, 4*, 47-57.

External Phosphorus Adsorption and Immobility with the Addition of Amend Calcium Peroxide Material

Jing Zhou, Dapeng Li*, Shutong Chen
(Suzhou University of Science and Technology, Suzhou, China)

ABSTRACT: Calcium peroxide (CaO_2) are being widely applied as Phosphorous (P) inactivation agents (PIAs) for lake eutrophic management. However, Delaying the release of CaO_2 is important to immobilize phosphorus from wastewaters and sediment. In this study, compared phosphorus removal of material with calcium peroxide by using batch column experiments, and by means of capping. Besides, Rhizon samplers and Unisense micro sensor system were used to explore the sediment of micro-environment. The cumulative adsorption of external P was 9.65 mg, 7.32 mg, respectively for CaO_2 and ACPM. The result of Unisense micro sensor system indicated oxygen delivery capacity of CaO_2 is stronger than material, but shorter. Meanwhile, CaO_2 caused the DIP concentration in the pore water was considerably higher than the material, this increases the possibility of a large release in the later period. The study indicated that that material has the potential to be used as a P-inactivation agent for lake eutrophication control and by means of capping.

INTRODUCTION

At present, the method of in situ capping by P inactivation agents has been a promising measure to immobile the secondary P due to operational ease, efficiency, high controllability (Yin H. et al., 2015) resulting in retarding sedimentary P release and controlling lake eutrophication(Antunes E. et al., 2018).Because its oxygen releasing and oxidation properties, calcium peroxide (CaO_2) is widely used as a capping material(Zhou Y. et al., 2016). But it is impossible to change the micro-environment in the deep sediments, indicating that no improvement on the transformation of sedimentary P forms appeared due to the capping. Noticeably, the best method to inhibit the sedimentary P release is to improve the transformation of P from mobile forms to inert forms by reducing the mobility of sedimentary P(Jiménez-Cárceles F. J. et al., 2008). It is reported that the injection of the CaO_2 into the sediments is favor of the change of the micro-environment, resulting in the transformation of sedimentary P form(Xu Y. et al., 2018). Water purification sludge consists of various clay minerals, calcium carbonate particles and metal oxides or hydroxides. (Zhu P. et al.,2016) found that the adsorption capacity of the water purification sludge (WPS) increased obviously after ignition. This is beneficial to the specific adsorption on P. Therefore, the WPS can be mixed with CaO_2 by solid-state reactions to make amended calcium peroxide material (ACPM) and then to make the small particles using the ACPM as the capping materials. In this study, the small particles with CaO_2 and ignited WPS were made and used as the capping materials. Black and odorous sediments obtained from a channel in Suzhou city in China is capped with ACPM and CaO_2. 5racterize the transformation of P forms by the redistribution of sedimentary P forms.

MATERIALS AND METHODS

Material Preparation and Site Description. Calcium peroxide (75% CaO_2, 25% $Ca(OH)_2$) were purchased from Aladdin Reagent Co. Ltd. (Shanghai, China). Water purification sludge was acquired from the horizontal sedimentation tank of Suzhou water supply Co., Ltd. Xu Jiang water plant. Dry preservation after desiccated, manually ground and sieved through a 100-mesh sieve save. The properties of sludge are exhibited in Table 1. Sediments were sampled from the canal (N31°35′30.18″, E120°31′15.7″; depth: 1-2 m) which located in Suzhou City of China, classified as a typical black-odour and heavily eutrophic canal. Sediments were collected with a large diameter gravity sampler (Rigo Co., 11 cm i.d., 50 cm long), stored in capped glass flasks and immediately transported to the laboratory. After passed through a 100-mesh sieve, the sediments were homogenized and stored in airtight flasks at 4°C pending analysis. Overlying water (75 L) from the sampling points was collected simultaneously. The physicochemical properties of the overlying water and sediments were analyzed (Table. 2).

TABLE 5. Physicochemical properties of water purification sludge

Sample	SiO_2	Al_2O_3	Fe_2O_3	MgO	CaO	Na_2O	K_2O
Water purification Sludge	46.33	26.26	5.16	1.13	0.82	0.6	1.87

TABLE 6. Physic□ch□mic□l □r□□rti□s □f s□dim□nts □nd □v□rlying w□t□r.

Overlying water						Sediment		
DO	pH	TP	DIP	NH^{4+}-N	TOC	Water content	LOI	Tot-P
$mg\cdot L^{-1}$		$mg\cdot L^{-1}$	$mg\cdot L^{-1}$	$mg\cdot L^{-1}$	$mg\cdot L^{-1}$	%	%	$g\cdot kg^{-1}$
2.1	7.1	0.509	0.286	10.4	6.1	69.2	24.9	2309.5

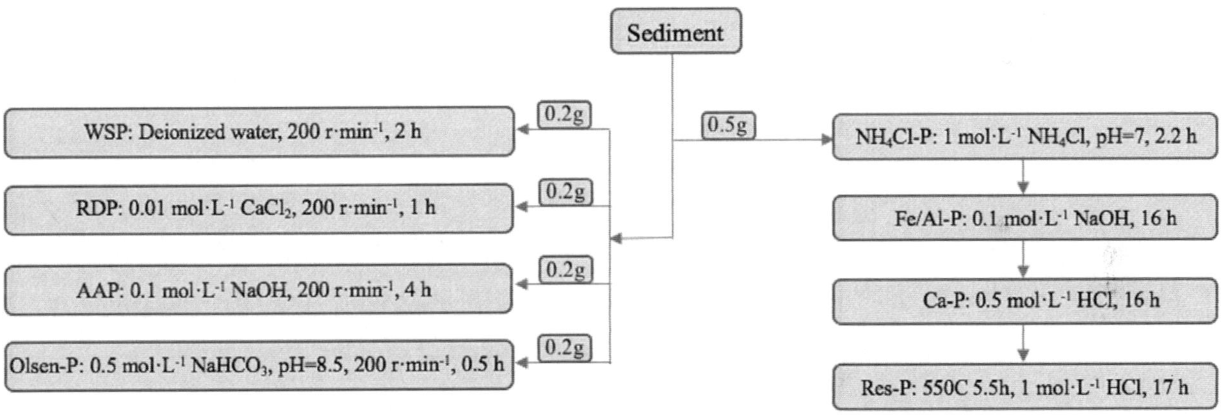

FIGURE 1. Chemical extraction methods for sediment P fractions

Capping Experimental Set-Up and Operation. The height of the wet sediments reached 15 cm in the containers. Three sets of cylinders (designated E0, E1 and E2) were used as the laboratory units, and each set included three parallel containers. For E0, the control, without ACPMs or CaO_2. For E1, the sediments were capped by 8 g ACPMs, which contained 4 g CaO_2. For E2, the sediments were capped by 4 g CaO_2. Overlying water (500 mL) containing 0.286 mg P L^{-1} was carefully added into the cylinders by siphoning, avoiding any disturbance of the sediments.

Analytical Methods. DO and pH were measured using a WTW Multi 9420 online apparatus. TP and DIP concentrations were measured by the molybdenum blue method(Franson M. A. H., 1995) with a spectrophotometer at 700 nm (UV-2550). NH_4^+-N concentrations were measured by the indophenol blue method. Sediment P fractions (NH_4Cl-P, Fe/Al-P, Ca-P, Res-P) were extracted in this study according to the method proposed by Hieltjes and Lijklema (Hieltjes A. H. M. et al., 1980). Bioavailable phosphorus was assessed by WSP, RDP, AAP, and Olsen-P, which can be measured by different chemical extraction methods shown in the Fig. **Error! Reference source not found.** (Zhou Q. et al., 2001).

Statistical Analysis. Variations in the results are reported as standard errors of the mean (± SEM) (n = 3). The data shown in this study used One-way analysis of variance (ANOVA) to determine the differences of NH_4^+-N, DIP, NH_4Cl-P, Fe/Al-P, Ca-P, Res-P etc. All statistical analyses of the data obtained were conducted by SPSS Data Editor.

RESULTS

The pH and DO in the Overlying Water. Differences in pH and DO between control and different capping treatments are presented in Fig.2. The pH values kept at higher level by capping ACPM and CaO_2, compared with control (P< 0.05). The pH values were sharply increased, up to 11.38, and then they kept stable (the mean, 10.53) under the capping of CaO_2. The trends of the DO are similar as the pH during the experiments. The DO values in the overlying water under the ACPM and CaO_2 capping were higher than that in the control (P< 0.01) (Fig. 2). The sharp increase of DO under the capping of CaO_2 was observed from 2.54 mg L^{-1} to 12.28 mg L^{-1} at day 3 and then the DO decreased obviously. From day 12, the DO value under the CaO_2 was lower than that under the ACPM (P< 0.01).

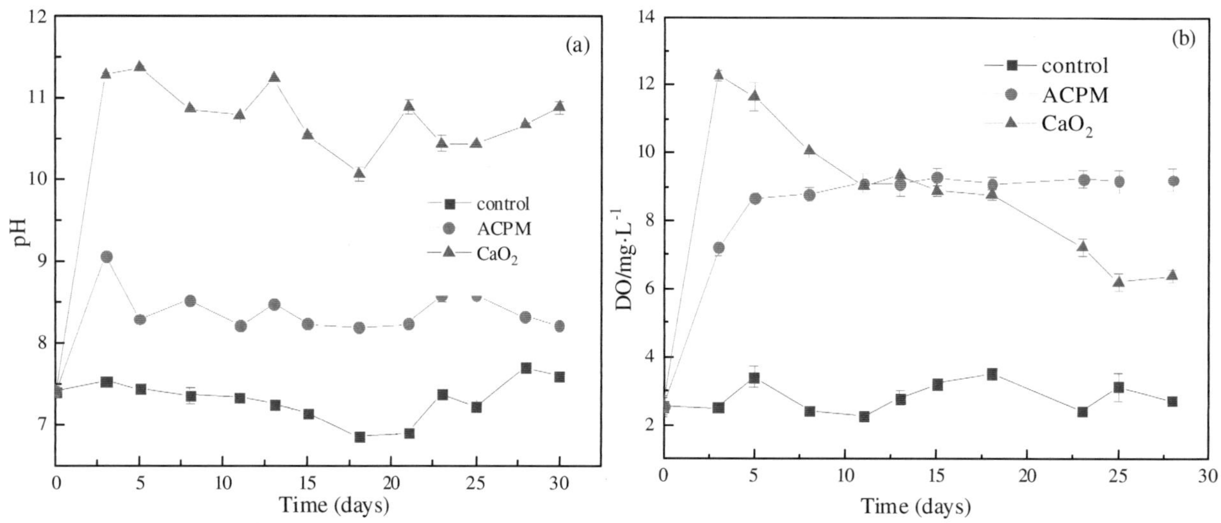

FIGURE 2. The pH (a) and Do (b) in overlying water with capping ACPM and CaO_2 compared with the control.

DIP in the Overlying Water and DIP in the Pore Water. The DIP concentrations in the overlying water under the capping of ACPM and CaO_2 were lower than that in the control (P < 0.01). Noticeably, the concentration of DIP kept at stable level under the capping of ACPM and CaO_2, around 0.160 mg L^{-1} and 0.080 mg L^{-1}, respectively. However, the values under the CaO_2 are slightly lower than that under the ACPM (P < 0.05). With the addition of the external P, the DIP in the pore water showed the obvious increase under the capping of CaO_2, up to 1.691 mg P L^{-1}, and the DIP was higher than that in the control (P < 0.01) (Fig. 3b). On the other hand, the concentration of DIP in the pore water under the capping of ACPM was lower than that in the control (P < 0.01) and the DIP kept constant during the experiment (the mean, 0.725 mg P L^{-1}).

Changes of P Fractions in Sediment. Concentrations of P fractions in layer 0-1 cm and layer 1-2 cm of sediments undergoing different change during 30 days are presented in Fig. 4. It is observed, under the capping of ACPM and CaO_2, NH_4CL-P and Ca-P increased obviously ($P<0.05$ and $P<0.05$, respectively), compared with the initial, While the Fe/Al-P and Res-P decreased marked ($P<0.05$ and $P<0.05$, respectively). The biggest increase is Ca-P under the CaO_2 at 0-1 cm and under the ACPM at 1-2 cm, up to 820.84 mg P kg^{-1} DW and 868.84 mg P kg^{-1} DW, respectively. The biggest reduction is Fe/Al-P under the ACPM at 0-1 cm and under CaO_2 the at 1-2 cm, up to 662.83 mg P kg^{-1} DW and 529.50 mg P kg^{-1} DW, respectively. After 30 days, the percentage of P fractions to the Tot-P changed obviously. The percentage of Ca-P to the Tot-P is the biggest, up to 64% and 66%(the mean under ACPM and CaO_2). The percentage of Fe/Al-P is the secondary.

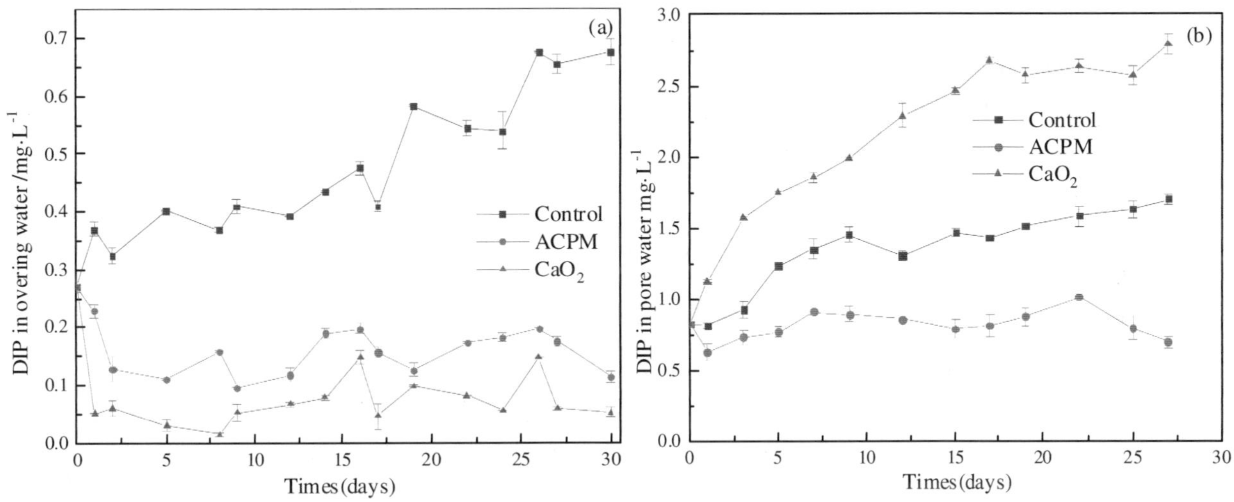

FIGURE 3. The DIP in overlying water (a) and pore water (b) with capping ACPM and CaO_2 compared with the control.

DISCUSSION

The pH and DO in the Overlying Water and Sediment. The CaO_2 and ACPM as the capping materials have a significant effect on the pH value and the DO content. The sharp increase of pH is attributed to the OH^- during the formation of calcium hydroxide ($Ca(OH)_2$) according to Eq. 1 and Eq. 2. However, the pH value of ACPM exhibited a comparatively lower variation, compared with the capping of CaO_2. The explanation is that the wrapping by the water purification sludge decreased the contacting probability between CaO_2 with the water, retarding the reaction of CaO_2 with the water. It is confirmed by the DO content in the Fig.2b. With the increase of reaction time, the DO under the CaO_2 is less than that under the ACPM. The most important is that the DO concentration under the ACPM kept constant during the reaction time. It is indicating that the mixing of WPS and CaO_2 hindered the quick release of oxygen from the CaO_2. This is favor of the maintaining of the oxic conditions in the natural waterbodies and beneficial to improve the oxic conditions for the sediments. However, under the interface, the DO content is opposite. The changes of sedimentary microenvironment from anaerobic conditions to oxic conditions is favor of the decrease of the reduction materials in the sediments.

$$2CaO_2 + 2H_2O \rightarrow 2Ca(OH)_2 + O_2 \uparrow \qquad \text{(Eq 1)}$$
$$CaO_2 + 2H_2O \rightarrow Ca(OH)_2 + H_2O_2 \qquad \text{(Eq 2)}$$

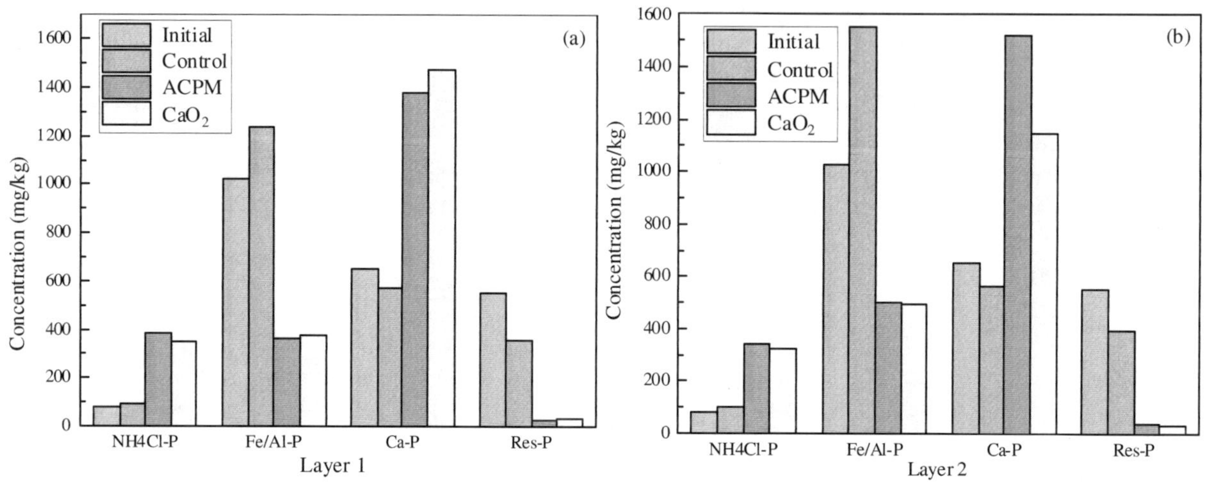

FIGURE 4. Sedimentary P fractions concentration at layer 0-1 cm (a) and layer 1-2 cm (b) under the capping of ACPM and CaO_2. Bars indicates SEMs (n=3).

Variables of DIP in the Overlying Water and in the Pore Water. The lower DIP in the overlying water under the capping of CaO_2 is could be attributed to the higher contacting probability between DIP and Ca^{2+} due to the quick reaction of CaO_2 with water. It is obvious that the disappearance of DIP is different completely from that under the capping of the ACPM because the reaction speed is slower due to the wrapping of the WPS. However, the aerobic conditions (Fig. 2b) in the overlying water is the key factor to impel the DIP disappearance. Under the aerobic conditions in the overlying water, the metal ions are in the oxidation state, and it is favor of the adsorption on P. It is expected that the capping materials has enough adsorption ability on the added external P and the released internal P during the incubation time according to Fig.3a, indicating that no increase of DIP in the overlying is observed under the capping of CaO_2 and ACPM. However, the DIP in the pore water increased under the capping of CaO_2, suggesting that not all P is fixed by the CaO_2. It is failed to differentiate the course of the P in the pore water. Considering the strong fixation ability of Ca^{2+} and isolation materials ($Ca(OH)_2$) between the water and the sediments, the DIP in the pore water mainly comes from the sedimentary P.

The Changes of Sedimentary P Fractions. Under the capping of ACPM and CaO_2, the change of different P fractions was observed according to Fig. 4. NH_4Cl-P is referred as the most available P for phytoplankton and algae growth (Pan G. et al., 2012), which is readily formed under the sorption of P on the surface of clays and the small particles. The apparent increase suggesting that the microenvironment under the ACPM is oxicer than that under the CaO_2, which caused the sedimentary P released into the pore water. Theoretically, the release of the sedimentary P depends on the concentration of Fe/Al-P (LI D. et al., 2011), which supplies the internal loading and considered as potentially mobile and algal-available P. According to the Fig.4, the sharp reduction of Fe/Al-P is similar under the capping of the ACPM and the CaO_2. It is indicate that the both microenvironment is similar, anaerobic conditions. However, the deduction shows the microenvironment under the ACPM is aerobic and it is not anaerobic under the capping of CaO_2. Ca-P is refractory and contributes to a permanent burial of P in the sediments. The Ca-P increased sharply under the both capping.

CONCLUSION

The DIP in the pore water shows greater reduction than CaO_2. The DO kept constant and higher than that under the CaO_2 from day 12 and played an important role to keep oxic microenvironment. The retarding release of oxygen from ACPM due to the addition of WPS maybe the main reason. Consequently, under both capping, the sedimentary P was redistributed in the different P fractions under the external P addition and the internal P release. However, the value at 1-2 cm under the CaO_2 is up to 15.15%, higher than that under the ACPM and the control, suggesting that the change of CaO_2 on the microenvironment is less than the ACPM. The implication is that the application of CaO_2 is not suitable for the direct uses, while combination with other materials (e.g. WPS) is the best selection on the sedimentary P immobilization by improving the transformation of P from bioavailable forms to nonbioavailable forms.

ACKNOWLEDGEMENTS

This work was supported by the National Natural Science Foundation of China (grant Nos. 51778393, 51278523). Financial support was also provided by the Collaborative Innovation Center of Water Treatment Technology and Material of Jiangsu Province.

REFERENCES

Antunes E., Jacob M. V., Brodie G. 2018. Isotherms, kinetics and mechanism analysis of phosphorus recovery from aqueous solution by calcium-rich biochar produced from biosolids via microwave pyrolysis, J. Journal of environmental chemical engineering. 6(1): 395-403.

Franson M. A. H. 1995. American public health association American water works association water environment federation, J. Methods. 6.

Hieltjes A. H. M., Lijklema L. 1980. Fractionation of Inorganic Phosphates in Calcareous Sediments1, J. Journal of Environmental Quality. 9(3):405-407.

Jiménez-Cárceles F. J., Álvarez-Rogel J. 2008. Phosphorus fractionation and distribution in salt marsh soils affected by mine wastes and eutrophicated water: A case study in SE Spain, J. Geoderma, 144(1):299-309.

LI D., Huang Y., Fan C. X. 2011. Phosphorus Bioavailability in a Phosphorus-Abundant System under Repeated Sediment Resuspension Conditions, J. Journal of Donghua University(English Edition). 28(5):523-527.

Pan G., Dai L., Li L. 2012. Reducing the Recruitment of Sedimented Algae and Nutrient Release into the Overlying Water Using Modified Soil/Sand Flocculation-Capping in Eutrophic Lakes, J. Environmental Science & Technology, 46(9):5077-5084

Xu Y., Han F., Li D. 2018. Transformation of internal sedimentary phosphorus fractions by point injection of CaO2, J. Chemical Engineering Journal, 343: 408-415.

Yin H., & Kong M. 2015. Reduction of sediment internal P-loading from eutrophic lakes using thermally modified calcium-rich attapulgite-based thin-layer cap, J. Journal of environmental management. 151: 178-185.

Zhou Q., Gibson C. E., Zhu Y. 2001. Evaluation of phosphorus bioavailability in sediments of three contrasting lakes in China and the UK, J. Chemosphere. 42(2):221-225.

Zhou Y., Fang X., Wang T. 2016. Chelating Agents Enhanced CaO_2, Oxidation of Bisphenol A Catalyzed by Fe^{3+}, and Reuse of Ferric Sludge as a Source of Catalyst, J. Chemical Engineering Journal. 313.

Zhu P. Y., Li D. P., Huang Y. 2017. Influence of ignited sediments on external phosphorus adsorption and sedimentary phosphorus forms, J. Environmental Science and Pollution Research. 24(28): 22622-22630.

The Effect of Conditioning on Improving Ammonium Removal Capacity of Clinoptilolites

Bilsen BELER-BAYKAL, Ahu YURTOGLU
(Istanbul Technical University, Department of Environmental Engineering, Istanbul, TURKEY)

Nitrogen is one of the foremost constituents in water to be controlled due to eutrophication in various kinds of water bodies. Furthermore, one of its common forms, ammonia, has to be removed to avoid its further impacts such as toxic effects on fish and additional oxygen demand which it will abstract form the water environment. Ion exchange is a proven method of nitrogen removal and it had been demonstrated that it may be removed successfully in the form of ammonium from domestic wastewater using ammonium selective natural zeolite clinoptilolite. The capacity of clinoptilolite to remove ammonium is one of the major determinants of the effectiveness of the process and higher capacities are always preferable.

This work was undertaken to investigate and quantify the effect of conditioning applied to clinoptilolite through the use of NaCl to improve its ammonium removal capacity by changing the original surfaces to the Na-form. Experiments were performed to observe capacity changes using three different Turkish clinoptilolites from Gordes, Bigadic and Yalova regions, and two different liquid phases, namely pure solutions of ammonium and municipal wastewater, for the original samples and samples conditioned to generate the Na-form. Within the context of this work, an appraisal of the change in surface capacities upon conditioning will be presented in a comparative manner based on equilibrium data, and isotherms for each sample together with their representative models will be provided.

Conditioning was accomplished using 1 M NaCl solution which was introduced into a fixed bed column of clinoptilolite at the rate of 1 bed volume per hour until the influent and effluent concentrations of sodium were practically the same, indicating that no more Na could be loaded onto the surface due to equilibrium. Chemical analyses of all six clinoptilolite samples were performed, pointing at the distinct increase in the Na ions on the surface, while other original exchangeable surface ions, especially Ca decreased upon conditioning.

Ion exchange isotherms were generated in batch experiments for liquid phase concentrations representing domestic wastewater in the range of 0-100 mg NH_4^+/l and the data were tested for compatibility with Langmuir and Freundlich isotherms. The results revealed that while the original samples from Gordes and Bigadic could successfully be represented by the Freundlich model, while their conditioned counterparts and both forms of Yalova clinoptilolite gave better fits with the Langmuir model.

A comparison of the original form with the conditioned one indicates that there was a considerable amount of capacity increase with conditioned samples in the Na-form for clinoptilolites from all three regions. The improvement in surface capacities could go as high as nearly three times of that of the original form. As expected, ammonium removal was lower when wastewater was used as the liquid phase as compared to pure solutions. A comparison between clinoptilolite samples from the three regions revealed that the highest capacity could be attained with the Gordes clinoptilolite while Bigadic and Yalova ranked the second and the third respectively, both in the Na-form as well as the original.

In conclusion, conditioning does improve ammonium removal capacity of clinoptilolite considerably and the use of the Na-form is indeed recommended for more effective results regarding ammonium removal through processing with clinoptilolite.

Reduced Graphene Oxide as an Efficient Adsorbent in Waste Water Treatment

Meena Ejjada[1], Ravi K. Biroju[2], P. K. Giri[2, 3], Chandan Mahanta[1]
([1] School of Civil Engineering, KIIT to be deemed University, Orissa - 751024, India; [2]Centre for Nanotechnology, Indian Institute of Technology Guwahati, Guwahati 781039, India; [3]Department of Physics, Indian Institute of Technology Guwahati, Guwahati 781039, India)

Removal of organic impurities from the Industrial waste water is a challenging task to Environmental engineers using conventional methods. Graphene and its functional materials have gained significant interest, which possess high adsorption strength due to its strong π-π* interaction on sp^2 hybridised carbon atoms and presence of various oxygenated functional groups to enhance the efficiency for the removal of organic pollutants present in the waste water. Here in, rGO (reduced graphene oxide) synthesized from top down approach (Chemical exfoliation of graphite) to implement as an efficient adsorbent for the waste water treatment. The morphological features, crystalline quality and presence of different oxygenated functional groups in rGO were extensively investigated by FESEM, TEM, μ-Raman, TGA/DTG and FTIR, respectively. BET surface area analysis had been performed for rGO and found its surface area ~ 5.9 m^2/g. Removal percent of organic matter was 16 % by rGO at optimised pH (4) within the duration of 2 h. Based on experimental results, quantitative study on adsorption kinetics have been performed and found the rate of adsorption by rGO using Pseudo first order and second kinetic model with best linear fitting function. Reduction of 30 % organic matter by optimizing the dosage of rGO (5 g/L) in the process of waste water treatment was achieved. In addition, evaluation of equilibrium kinetics by fitting the Langmuir and Freundlich isotherms was done, in which higher correlation coefficient was found in Langmuir isotherm, which is an evidence for the efficient adsorption capacity of rGO in the removal of organic pollutants present in our industrial waste water specimen. Further, the adsorption of organic impurities on the rGO surface was characterized using high resolution TEM and FTIR analysis, which is consistent with quantitative study on removal of organic entities present in the waste water.

Kinetics, Isotherm Studies, and Uptake Mechanisms of Anionic and Cationic Heavy Metal Adsorption onto Polymer-based Graphene Oxide Nanocomposite Beads

Jem Valerie D. Perez and Maria Lourdes P. Dalida
(University of the Philippines, Quezon City, Philippines)
Debora F. Rodrigues
(University of Houston, Houston, TX, USA)

Nanocomposite beads with optimum composition of graphene oxide (GO), polyethyleneimine (PEI), and chitosan (CS) were synthesized for chromium (VI) and copper (II) removal from water. Effects of several adsorption parameters were investigated by performing batch adsorption experiments at different pH, dosage, contact time, initial metal concentration, and temperature values. Kinetic studies were performed and a pseudo-second order kinetic equation with high correlation (0.999) for both Cr (VI) and Cu (II) adsorption were obtained. Adsorption equilibrium data for Cr (VI) and Cu (II) were best described by Freundlich isotherm model, suggesting multilayer adsorption on heterogeneous adsorption sites. Maximum adsorption capacities at 25 °C of 160 mg/g and 137 mg/g for Cr (VI) and Cu (II), respectively, were obtained.

FTIR and XPS analyses before and after adsorption revealed different uptake mechanisms by the CS-PEI-GO beads. Cr (VI) ions were reduced to less toxic Cr (III) on the surface of the beads, while Cu (II) ions reacted with surface amine groups. Finally, desorption studies were performed and high adsorption capacities were still obtained after 5 adsorption-desorption cycles.

Because of the excellent performance and multi-functionality of the synthesized CS-PEI-GO beads in this study, applicability to other contaminants can be explored, making the new adsorbent very promising for wastewater treatment.

Removal of Long and Short Chain Perfluoroalkyl Substances via Granular Activated Carbon Adsorption

Eric Forrester and Ralph Franco
(Calgon Carbon Corp., Pittsburgh, PA, USA)

Per- and polyfluoroalkyl substances (PFAS) and used in many industrial and commercial applications including non-stick cookware, stain resistant fabrics, food packaging, as well as fire fighting foam products, such as those used in civilian and military aviation firefighting. They are problematic because of their persistence in the environment and their long half-life in humans.

The two major PFAS of interest are perfluoro octanesulfonic acid (PFOS) and perfluoro octanoic acid (PFOA) due to their predominance in the above applications and associated USEPA Health Advisory Level of 70 ng/L (ppt). The USEPA recommends GAC adsorption as effective treatment technology for the removal of both PFOA and PFOS; as such, overall objective of this paper is to reinforce the EPA's guidelines that granular activated carbon is one of the best available treatment technologies for the removal of PFAS like PFOA and PFOS, but also show that GAC can be highly effective for removal of shorter chain perfluoroalkyl substances.

Using rapid small-scale column testing (RSSCT) and accelerated column testing (ACT) methodologies, this paper outlines test results relating to the removal of PFOA and PFOS via granular activated carbon (GAC) adsorption, as well as results exploring GAC efficacy for removal of other PFAS, specifically those with a shorter carbon chain length (e.g. PFBA, PFBS, PFHxA, and PFHxS). Additionally, column test results are compared to subsequent full-scale installation data to evaluate scalability of the bench-scale testing.

The results demonstrate the ability of domestic reagglomerated bituminous coal-based activated carbon to remove a variety of PFAS compounds and concentrations to non-detectable levels (< 2 ppt) in two groundwater well sites in PA, both in bench- and full-scale. Additionally, reagglomerated bituminous coal is shown to consistently be the preferred base material type for GAC used in PFAS removal.

Irreversible Adsorption of Sulfonamide Antibiotics on Luffa Cylindrica Fiber and Its Bioregeneration

Lan Zhang, ***Yun Liu**** and Yuan-hua Dong
(Institute of Soil Science, Chinese Academy of Sciences, Nanjing, China)

Luffa cylindrica fibers have been used in virtually all areas, and it is a valuable immobilization matrix for algal, bacterial and cells. The occurrence of sulfonamide antibiotics in aquatic environments has been recognized as a significant issue warranting focused attention. Special phenomenon in this paper is that adsorption of sulfonamides on Luffa is irreversible. It takes four days to reach equilibrium, and sulfonamides antibiotics adsorption is favored over the pH range of 2.0-11.0. The effects of anions (Cl^-, NO_3^- and $C_2O_4^{2-}$) and cations (Na^+, Ca^{2+}, Fe^{3+}) on the adsorption of sulfonamides antibiotics were studied. $C_2O_4^{2-}$ and Fe^{3+} enhance the removal of sulfonamides antibiotics from aqueous. The irreversible adsorption of sulfonamides antibiotics on Luffa is mainly due to the chemical reaction between carboxyl groups in Luffa and amide groups in sulfonamides. Bioregeneration is a method that renews the adsorbent by microorganisms for further adsorption, which involves desorption and biological removal of adsorbate. Luffa cylindrica fiber was successfully reused after bioregenerstion although its irreversible adsorption.

Reinvestigating the Role of Reactive Species in the Oxidation of Organic Co-Contaminants during Cr(VI) Reactions with Sulfite

Hongyu Dong, Xiaohong Guan
(Tongji University, Shanghai, China)
Timothy J. Strathmann
(Colorado School of Mines, Golden, USA)

Previous researchers have not considered the possible inhibition of scavenging agents (alcohol and aniline) on HSO_3^- consumption during Cr(VI) reactions with sulfite (Cr(VI)/HSO_3^- process) since alcohols and some organics could be effective inhibitors in the intrinsic oxidation of sulfite, which may lead to misleading conclusions. Consequently, it is essential to re-evaluate the importance of these radicals and other transient intermediates generated during Cr(VI)/HSO_3^- reactions for mediating the oxidation of co-contaminants.

4-chlorophenol (4-CP) was used as a model co-contaminant and experimental work was undertaken in this study to re-investigate the mechanisms and active species responsible for oxidation of co-contaminants in the Cr(VI)/HSO_3^- reaction system. Batch experiments showed that the degradation rates of 4-CP correlated well with the rates of Cr(VI) reduction by sulfite in the same solutions, and that $O_{2(aq)}$ was necessary for the oxidation of 4-CP. Multiple lines of evidences indicate that Cr(VI)/HSO_3^- reaction is a $SO_4^{\bullet-}$-based oxidation process. $SO_3^{\bullet-}$ was generated in Cr(VI)/HSO_3^- system based on the electron spin resonance spectra, which could be transformed to secondary radicals ($SO_4^{\bullet-}$, $SO_5^{\bullet-}$, and $HO^{\bullet}$). The contribution of $SO_5^{\bullet-}$ was ruled out through almost complete inhibition of methanol (MeOH) on 4-CP degradation. Considering the negligible inhibition of tert-butanol (TBA) on 4-CP degradation, $SO_4^{\bullet-}$ was identified to be reactive species in Cr(VI)/HSO_3^- process. This result was further verified by almost no degradation of nitrobenzene (NB) and the inhibiting effect of Cl^- in Cr(VI)/HSO_3^- process.

$SO_4^{\bullet-}$ was verified to be the major reactive oxidation species in Cr(VI)/HSO_3^- process combining the ESR spectra, radical quenching experiments, and Cl^- influence. The contribution of HO• was negligible for organic contaminant oxidation in Cr(VI)/HSO_3^- process, evidenced by the negligible NB degradation in Cr(VI)/HSO_3^- process. Clarification of the reactive oxidation species will lead to better industrial application of Cr(VI)/HSO_3^- system in wastewater treatment.

The Effect of Sodium Hypochlorite(NaOCl) on RO Membrane Surface During Backwashing in Seawater Desalination Process

Minjin Kim, Jayeong Seong, Nayoung Park, Hyungsoo Kim
(Sungkyunkwan University, Suwon, Republic of Korea)

The cleaning efficiency of NaOCl(Sodium hypochlorite) is higher than that of physical flushing and other chemical cleaning. And NaOCl(Sodium hypochlorite) is effective in cleaning foulants from NOM(natural organic matter), EPS(Extracellular Polymeric Substances) & microbes in desalination process. Also, unlike NaOH(sodium hydroxide) which is generally using in RO membrane cleaning, it is available to direct apply to feed because it doesn't make scale in process. Despite this excellent chlorine cleaning efficiency, there was no solution to the membrane damage, so it was not possible to apply the NaOCl(Sodium hypochlorite) cleaning. Because a major disadvantage of polyamide type RO membranes is their sensitivity to chemical disinfectants such as chlorine. Generally, foulant of RO membrane is compressed by high pressure but if the pressure of feed side is removed, permeated water will move to feed side by osmotic force. Such a cleaning method is called as FOB(forward osmotic backwashing). Such the basic principle of FOB makes foulants easy to remove. Using this principle, backwashing water layer between a membrane surface and a foulant accumulated layer minimizes direct contact between NaOCl and PA membrane and it is assumed that the direct chlorine cleaning is possible in the RO membrane process. In this study, therefore, we investigated relationship between FOB and chlorine cleaning and we checked membrane contamination before and after cleaning, efficiency of removal, analysis of organic matter, possibility of chemical damage of RO membrane. In the FOB test with chlorine cleaning, NaOCl(Sodium hypochlorite) solution was injected at the concentration of the raw water, and the membrane damage was checked with time. The RO membrane damage over time was confirmed by Fujiwara Test and Dye Test, and the analysis method was based on the Autopsy analysis method which is provided by the pharmaceutical company.

Removal of COD from Pigment Industrial Wastewater by Fenton Oxidation Process

Vishnu Tejani, Nibedita Pani, Dr. Anantha Singh TS
(Pandit Deendayal Petroleum University, Gandhinagar, Gujarat, India)

Various industries including chemical, intermediate manufacturing, textiles are responsible for highly concentrated COD. The textile industry generates large quantities of highly coloured wastewaters coming from dyeing/printing processes which contains prodigious amount of COD. Textile Wastewater is characterized with intense colour, TSS, COD and other pollutants. To overcome the disadvantages of the conventional methods there is need of a more efficient method which can remove COD efficiently in an economic manner. This paper primarily attempts to remove COD from Dye Manufacturing Industrial wastewater by the means of Chemical Fenton Oxidation process. In this process Fenton's reagent (H_2O_2 and Fe^{2+}) was used. H_2O_2 was used to generate Hydroxyl radicals to oxidize organic compounds present in wastewater using Fe^{2+} as catalyst. The effects of pH value, H_2O_2 dosage, the molar ratio of Fe^{2+} to H_2O_2, oxidation time on COD removal were investigated. Experiments were done at various pH. The results showed that COD removal efficiency first increased then decreased with the increase of solution pH. Different molar ratios of H_2O_2 and FeSO4 were combined in acidic pH value of 2.5. Molar ratio of Fe^{2+} to H_2O_2 (1:3) was found to be most efficient with the removal efficiency of 74.26%. The oxidation time was kept 45 - 60 minutes at room temperature.

Efficient Degradation of Organic Pollutants by a Dual-Reaction-Center Fenton-Like Process over 4-Phenoxyphenol-Functionalized Reduced Graphene Oxide Nanosheets

Lai Lyu, Yumeng Wang, Wenrui Cao and Chun Hu
(Guangzhou University, Guangzhou, China)

Triggered by growing water pollution issues, research aiming at a more efficient water purification technology has been thriving. Fenton reaction as a typical advanced oxidation process is especially powerful, since it can rapidly and nonselectively degrade various organic pollutants via •OH. However, whether homogeneous or heterogeneous Fenton process is used, the reaction always depends on the redox of the metal ions in the single-metal center. This intrinsic property of metal-containing Fenton catalysts often results in some problems, including the need for acidic reaction conditions (pH = 2–4), occurrence of a rate-limiting step due to the low reaction rate constant for the reduction of $M^{(n+m)+}$ to M^{n+}, excessive consumption of H_2O_2, and secondary pollution owing to the production of iron-containing sludge or metal leaching, which narrow the application of Fenton reactions for environmental remediation.

Here, we report for the first time a highly effective stable metal free Fenton-like catalyst with dual reaction centers consisting of 4-phenoxyphenol-functionalized reduced graphene oxide nanosheets (POP-rGO NSs) prepared through surface complexation and copolymerization. X-ray photoelectron spectroscopy (XPS) and Fourier-transform infrared spectroscopy (FTIR) analyses revealed that the connection between POP and graphene is achieved through C-O-C bonding, resulting from the deprotonated phenolic OH group of POP bonding with the C atoms in the graphitic rings. Electron paramagnetic resonance (EPR) analysis and density functional theory (DFT) calculations verified that dual reaction centers are formed on the C-O-C bridge of POP-rGO NSs. The electron-rich center around O is responsible for the efficient reduction of H_2O_2 to •OH, while the electron-poor center around C captures electrons from the adsorbed pollutants and diverts them to the electron-rich area *via* the C-O-C bridge. By these processes, pollutants are degraded and mineralized quickly in a wide pH range, and a higher H_2O_2 utilization efficiency is achieved. Our findings address the problems of the classical Fenton reaction and are useful for the development of efficient Fenton-like catalysts using organic polymers for different fields.

Enhanced Catalytic Ozonation Oxidation Using Magnetic Carbon Nanotubes Composites for Removal of Bisphenol A With Coexistence of Humic Acid in Aqueous Solution

Yajing Huang, Qing Zhang, Wenjun Xu, Lingling Hu, Chun He
(Sun Yat-sen University, Guangzhou, China)

Endocrine disrupting compounds (EDCs), including bisphenol A, has drawn lots of attention recently because of their adverse health impacts on human beings and ecological environment. Till now, numerous effective methods and strategies have been developed to mineralize BPA at low concentration in aqueous system. Among the methods, the application of advanced oxidation processes, for example, catalytic ozonation, is an attractive option for the removal of BPA. However, the low solubility and utilization of ozone hinders the practical application of catalytic ozonation. In this work, MWCNTs/Fe_3O_4 magnetic composites, combined adsorption and catalytic ozonation behavior, were synthesized by a hydrothermal process. The prepared MWCNTs/Fe_3O_4 magnetic composites were characterized by X-ray diffraction (XRD), scanning electronic microscopy (SEM), transmission electron microscopy (TEM), Fourier transform infrared spectroscopy (FTIR), X-ray photoelectron spectroscopy (XPS), element analysis and Brunauer-Emmett-Teller (BET) surface area methods. The composites were applied for degrade BPA in aqueous solution, and the results showed that the synthesized MWCNTs/Fe_3O_4 exhibited an excellent adsorption ability and catalytic ozonation activity owning to its large specific surface areas, hollow tube channels and abundant surface oxygen-containing groups, leading to high adsorption for BPA and O_3, high retention and utilization of O_3 molecules on the composites. The highest BPA removal efficiency has reached 99% under the condition of pH 9.6 at the BPA concentration of 50 mg L^{-1}. In addition, the fluorescence spectra and •OH radicals measurement displayed that HA could greatly increase •OH radicals via inducing more surface oxygen-containing groups on the catalyst, thus, promoting BPA removal. Moreover, MWCNTs/Fe_3O_4 showed the great stability and durability during several reaction recycles. Therefore, MWCNTs/Fe_3O_4 magnetic composites have great application potential to remove endocrine disrupting compounds because of its excellent adsorption ability, catalytic activity, simple separation and good stability.

Efficient Catalytic Ozonation of Bisphenol-A with Three-Dimensional MnO_2 Porous Hollow Microspheres and an Insight into Probable Catalytic Mechanism

Lingling Hu, Qing Zhang, Wenjun Xu, Yajing Huang, Chun He
(Sun Yat-sen University, Guangzhou, China)

With growing environmental awareness among the public, bisphenol A (BPA), considered as an endocrine disrupting compound (EDC), has trigged much social concern due to its adverse impact to human beings and ecosystem. Till now, numerous effective methods and strategies has been developed to remove BPA in water and wastewater. But BPA are relatively hydrophilic and therefore less vulnerable to traditional physical/chemical treatment processes. Thus, the application of advanced oxidation processes based on catalytic ozonation is an attractive option for the removal of BPA. However, the catalytic activity in ozonation process was still limited to its low solubility and utilization of ozone. In this work, three-dimensional (3D) MnO_2 porous hollow microspheres (δ- and α- MnO_2 PHMSs), with high adsorption and catalytic ozonation performance, were synthesized by a self-template ($MnCO_3$ microspheres) process at room temperature. The synthesized MnO_2 PHMSs were characterized by X-ray diffraction (XRD), scanning electronic microscopy (SEM), transmission electron microscopy (TEM), X-ray photoelectron spectroscopy (XPS) and Brunauer-Emmett-Teller (BET) surface area. The results showed that PHMSs exhibit the excellent adsorption ability and catalytic activity owning to their hollow spherical structure, mesoporous shell and well-defined interior voids, leading to the strong adsorption for bisphenol A (BPA) and the retention of O_3 molecules on catalyst. Moreover, the catalytic performance of α-MnO_2 PHMSs was better than that of δ-MnO_2 PHMSs which was attributed to the richer lattice oxygen of α-MnO_2 PHMSs to accelerate O_3 decomposition by producing more reactive oxidative species. The degradation efficiency of BPA using 3D α-MnO_2 PHMSs was more than 90% in the presence of ozone within 30 min reaction time. The probe tests for reactive oxidative species (ROSs) displayed that BPA degradation by catalytic ozonation is dominated by $\bullet O^{2-}$ and •OH in our present study. Furthermore, the organic compounds as intermediates of the degradation process were identified by LC/MS. This study may expand the further development of other porous hollow microspheres for the removal of organic pollutants using catalytic ozonation methods.

Influence of Polyelectrolyte Modification on Nanoscale Zero-Valent Iron (nZVI): Aggregation, Sedimentation, and Reactivity with Ni(II) in Water

Liu Jing, ***Liu Airong***, Zhang Wei-xian
(Tongji University, Shanghai, P.R. China)

Polyelectrolyte modification on the surface of nanoscale zero-valent Iron (nZVI) particles could affect colloidal stability, aggregation, mobility, oxidation as well as reactivity. In this article, two anionic polyelectrolytes with different molecular weight (MW), anionic polyacrylamide (APAM, MW=3 million) and carboxymethylcellulose sodium (CMC, MW=300-800) were modified on the surface of nZVI. Fourier transform infrared spectroscopy (FT-IR), scanning electron microscope (SEM), and X-ray diffraction (XRD) were used to characterize the polyelectrolyte-modified nZVI. Results from UV-Vis spectrophotometer (UV) and sedimentation tests showed that APAM modification led to the aggregation of nZVI in suspension, CMC modification made nZVI disperse well. XRD results indicated that polyelectrolyte modification made the slow oxidation of nZVI under ambient conditions. APAM-modified nZVI had much lower Ni^{2+} removal efficiency than that of bare nZVI while CMC-modified nZVI had higher Ni^{2+} removal efficiency after 1.5 hours (h). The study provided the information on the modification effect of nZVI on the colloidal stability and aggregation, sedimentation, reduction and sorption of heavy metal ions in the environment.

Micropollutant Removal Efficiency in an Activated Sludge System Operated at Low Sludge Retention Time

Maggie Green
(United States Environmental Protection Agency, NPDES, 1650 Arch Street Philadelphia, PA, USA)
Emel Topuz, Gülten Yüksek, Emine Ubay Cokgor, Didem Okutman Tas
(Istanbul Technical University, Environmental Engineering Department, 34469 Maslak, Istanbul, Turkey)

Micropollutants has been an increasing concern due to their presence in receiving environments as a result of rising global consumption. Identifying the occurrence of pharmaceuticals in the environment is becoming increasingly important because over the past decade, the discovery of various pharmaceutical residues have been found in the environment. Procedures for implementing preventative entry into the environment, as well as removal mechanisms are constantly being updated and performed as new knowledge is acquired. A semi continuously fed activated sludge reactor (working volume of 10L) with a sludge retention time of 5 days was operated for two months. The culture was fed daily with a synthetic wastewater (ISO 8192) (600 mgCOD/L) and 10 μg/L micropollutant mixture (Naproxen, Diclofenac, Ketoprofen, Ibuprofen and Indomethacin) in methanol (MeOH) (40 μL). The culture was kept in a 22ºC constant temperature room and aerated with constant air flow (> 2 mg O_2/L). Micropollutants in the samples was concentrated with OASIS HLB SPE cartridges (200 mg, 6 cc) (Waters, Millford, MA, USA) and measured with LC-MS/MS (Thermo Accela UPLC coupled with Thermo Quantum Access tandem MS, USA) according to the method explained in Topuz et al. (2014). In order to determine biodegradation potential of the anti-inflammatory pharmaceuticals and to determine their removal efficiency, liquid samples were taken from the reactor before the addition of daily pharmaceutical feeding as well as directly after feeding. The removal efficiencies of the anti-inflammatory micropollutants observed in the reactor is given in Figure 1. Diclofenac showed a negative removal following 40 days of biomass acclimation, going from 60% to 30% removal efficiency. Ibuprofen showed only a slightly lower removal efficiency whereas Indomethacin also had a decreasing removal efficiency from 80% to 60%. In the case of Ketoprofen and Naproxen, both showed an improved impact on removal efficiency (55% to 85% and 30% to 90%, respectively). Whitacre (2010) also reported <40% removal efficiency for DCF, IBU AND NAP while a 50-89% removal for INDO and KET. Whitacre (2010) also indicates that with the addition of UV or a biofilter, removal efficiency greatly increases for all aforementioned anti-inflammatories.

Remediation of Polluted Wetlands through Phytoremediation

S Mohan and Abhishek Tippa
(Indian Institute of Technology Madras, Chennai, India)

The wetlands are unique ecological sites which play a substantial role in the environment, relating to its unique characteristics of biology, hydrology and ecology. The buffer zone provides high carrying capacity for all the organisms, because of its unique soil characteristics which make habitat for benthic organisms. In the recent decade, the surface water quality and soil characteristics are getting deteriorated because of anthropogenic activities and land encroachment in the urban and coastal zones affecting the inland and coastal wetlands. The untreated wastewater and excessive dumping of solid waste increase the concentration of heavy metals and the excessive nutrients (nitrogen and phosphorous) which become a major pollution contributor to the coastal and urban wetlands. Therefore, it becomes imperative to remove the nutrients and heavy metals from the wetland without altering its ecosystem. Among many biological methods phytoremediation is imaging out to be the most effective method in removing the heavy metals and nutrients from the wetlands.

Worldwide, work has been done on remediation of heavy metals and excessive nutrients by phytoremediation mostly on lab scale, but the same is to be implemented on a large scale. The lab scale working condition is a miniature version which lacks the various on-field conditions like the inhibition factors for nitrogen fixation and bioavailability of heavy metals to the plants. To overcome these factors effective microorganisms especially bacteria will be used along with hyperaccumulator plants. The hyperaccumulator plants have capacity to store higher amounts of heavy metals in their aerial tissues and is defined as the plants that accumulate more than 1gm/kg of heavy metals. But, when the physiology of plants has been looked into, it is found that the accumulation is mainly dependent on molecular interaction of plants and heavy metals. The plants which have better heavy metal transporters can accumulate higher heavy metals in their vacuoles.

Most known proteins that facilitate the transport of heavy metals in plants belong to the following families: the zinc-regulated transporter, iron-regulated transporter of protein (ZIP) family, the cation diffusion facilitator (CDF) family, the P1B-type subfamily of P-type ATPs, the natural resistance associated macrophage protein (NRAMP) family and the yellow-stripe 1-like (YSL) family. Among these, P1 type ATPs play important role in transporting the heavy metals to all the parts of the plants. Among them P1-ATPs play a role in plant nutrition by contributing to micronutrient delivery throughout the plant. Thus, these could serve in xylem loading of heavy metals for long-distance transport to the shoot and leaves. P1 type also contributes to metal detoxification and sequestration of heavy metals in the vacuole. The P1-ATPases can be provided by the endophytic bacteria which resides inside the plant tissues. Thus, a proper development of symbiotic relationship between endophytic bacteria and plants is necessary to achieve this. The symbiotic nature of endophytic bacteria and plants helps in transporting the heavy metals to all the parts of the plants and also the bacteria promotes plant growth by decreasing the ethylene levels in the plants which is said to be a stress or ageing hormone. The expected outcomes from the study may help in increasing the accumulation capacities of the plant with incoluation of specific endophytic bacteria which were also able to provide the lacking heavy metal transporters to the plants.

Accelerated Photocatalytic Degradation of Organic Pollutant over Metal-organic Framework MIL-53(Fe) under Visible LED Light Mediated by Persulfate

Chun Hu and ***Yaowen Gao***
(Guangzhou University, Guangzhou, Guangdong, P.R. China)
Hui Zhang
(Wuhan University, Wuhan, Hubei, P.R. China)

Photocatalysis based on metal-organic frameworks (MOFs) is being actively investigated as a potential technology in visible light harvesting and utilizing processes. Herein we report that MIL-53(Fe), an earth-abundant Fe-containing MOF material, shows photocatalytic activity for the degradation of Acid Orange 7 (AO7) from aqueous solution under visible LED light irradiation, yet the photocatalytic performance of bare MIL-53(Fe) was not satisfactory due to the fast recombination of photoinduced electron-hole pairs. This can be effectively overcome by adding the external electron acceptor (e.g., persulfate) to the catalytic process. The accelerated photocatalytic degradation of AO7 is demonstrated by the result that degradation efficiency of AO7 in the MIL-53(Fe)/PS/Vis process reached almost 100% within 90 min as compared to only 24% under the identical experimental conditions for the MIL-53(Fe)/Vis process. To investigate the mechanism of the MIL-53(Fe)/PS/Vis process, photoluminescence (PL) spectra, electrochemical measurements and electron paramagnetic resonance (EPR) analysis were performed. It was concluded that the efficient separation of photogenerated electrons and holes by the introduced PS and the subsequent formation of reactive radicals resulting from the activation of PS by photogenerated electrons accounted for the accelerated photocatalytic degradation of AO7 in the MIL-53(Fe)/PS/Vis process. Furthermore, the applicability of MIL-53(Fe) used in persulfate-mediated photocatalytic process was systematically investigated in terms of the identification of reactive radicals, the reusability and stability of the photocatalyst, as well as the effect of operating parameters. The findings of this work highlighted the great potential of MOFs as photocatalysts and elucidated a new opportunity for persulfate remediation of contaminated water.

Continuous Treatment of Emulsion Effluents of Steel-Rolling Mills Using Magnetite (Fe_3O_4) Nanoparticles

Parsanta and Ashok N. Bhaskarwar*
(Chemical Engineering Department, Indian Institute of Technology Delhi, Hauz Khas, New Delhi-110016, INDIA)
*Corresponding Author. E-mail: ashoknbhaskarwar@yahoo.co.in, parsantaiitd@gmail.com

ABSTRACT: The kinetically stable oil-in-water emulsions are widely used in steel industries for cooling and lubrication of contacting metal surfaces, like rollers and steel plates. Such types of emulsions have high specific heats and heat transfer coefficients, which make them operable under high temperature and pressure conditions. Besides enhancing the shelf life of the tools, it also brings down the materials' cost. A major disadvantage of such emulsions is that these are difficult to dispose of because of their high stability, and thus become hazardous to the environment. Although there is a number of conventional techniques available for the treatment (or breaking) of such emulsions, such as acidification, centrifugation, electrolytic method, UV radiation, and membrane separation, these are energy-intensive, and hence expensive, time–consuming, or environmentally harmful. Thus, there is a need for a new cost-effective, eco-friendly, and less energy-intensive treatment. In the present work, we introduce a new method of breaking these emulsions into water and oil, by using magnetic nanoparticles (MNPs), with almost a full recovery of the nanoparticles. We synthesized magnetic nanoparticles by co-precipitation method and characterized them using XRD, SEM, and TEM. We designed and fabricated a continuous setup for the treatment of emulsion effluents (1%, v/v), and separated the oil-loaded (Fe_3O_4) nanoparticles from the treated water. Effect of the conveyor belt's linear velocity on the recovery of oil was studied while keeping the flowrates of emulsion and, a slurry of nanoparticles constant. It could be concluded that with an increase in the linear velocity of the conveyor belt, there was an increased separation. The magnetic nanoparticles could be reused three times, with approximately 96.03 % (w/w) cumulative recovery, in the treatment of 1 % (v/v) emulsion effluents.

Keywords: metal-working industries, steel-rolling mills, emulsion effluents, oil-in-water emulsions, Fe_3O_4 nanoparticles.

INTRODUCTION

The oil-in-water emulsions have potential applications in all the metalworking industries. These kinetically stable emulsions and can be applied under high temperature and pressure conditions of the industries. In steel industries such types of emulsions are used in huge amount under thinning process. It helps in cooling and lubrication of contacting metal surfaces and thus, not only increases the shelf life of the tools but also reduces the economy of the whole process (Mang, 2017). However, these are difficult to beak and are carcinogenic by nature, thereby causing tremendous hazardous effects in the environment. It is recommended by the government to treat such emulsions up to at least till 10 mg/l concentration before being disposed off in the environment. The conventional techniques to break such emulsion are acidification, centrifugation, electrolysis, membrane filtration, gravity separation, and UV- radiation (Hupka, 1992). The operational costs of such processes are very high than the actual product or the process and are also not eco-friendly. All these factors motivate us to explore a novel method which should be less energy intensive, eco-friendly, quick, and economic. In the present work, we introduce a new method for continuous treatment of emulsion effluent of steel –rolling mills.

MATERIALS AND METHODS

Synthesis of Magnetite Nanoparticles. The magnetite nanoparticles were synthesized by a cost-effective co-precipitation method using ferric and ferrous chloride salt solution in 1:1 M ratio and ammonium hydroxide solution as precipitating agent (Berger et al., 1999) (Zhou et al., 2015). The reaction as such, resulted into the formation of black coloured precipitates of magnetite. These were further dried in an oven at 70ºC for 6 hours and stored in an airtight container (Talekar et al., 2012).

Co-precipitation Method.

$$2FeCl_3\,(s) + 6H_2O \rightarrow 2Fe(OH)_3 + 6HCl \quad [1]$$

$$FeCl_2 + 2H_2O \rightarrow Fe(OH)_2 + 2HCl \quad [2]$$

$$2FeCl_3\,(s) + FeCl_2\,(s) + 8NH_3 + 4H_2O \rightarrow FeO(Fe_2O_3) + 8NH_4Cl \quad [3]$$

The synthesized nanoparticles were then characterized by XRD confirmation of Fe_3O_4, and for crystal phase determinations, SEM for analysis of the surface morphology and TEM for the confirmation of the Fe_3O_4 nanoparticles.

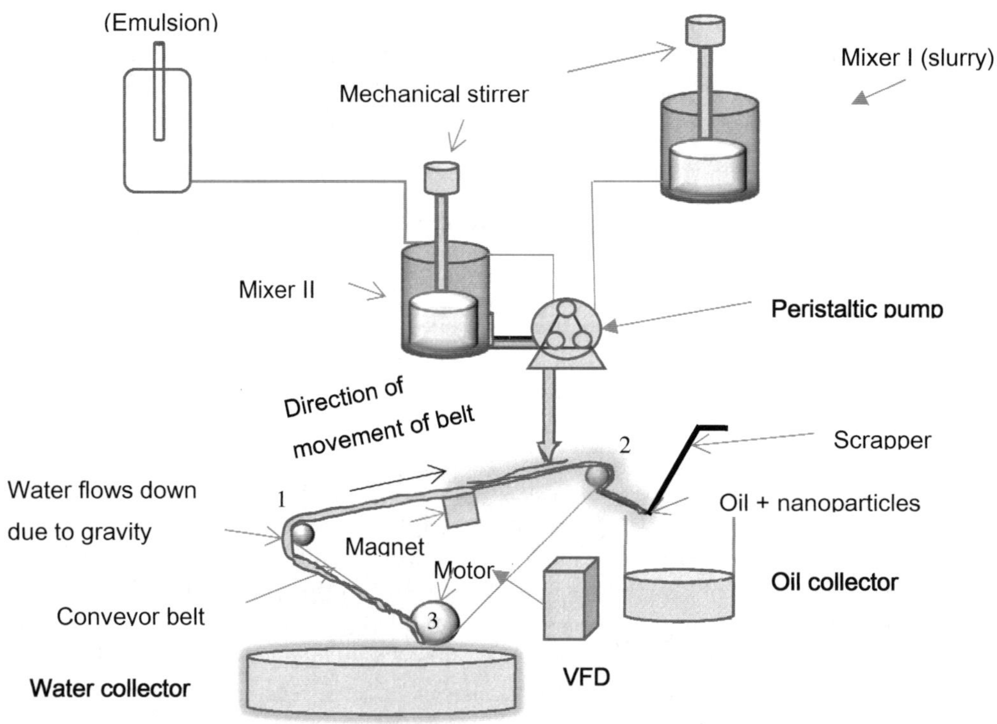

Figure 1. Schematic Representation Of The Continuous Setup (Built-In-House) For Treatment Of Emulsion Effluent Using Fe_3o_4 Nanoparticles And Separation Of Oil Loaded Nanoparticles (Thick Paste) From The Treated Water.

Experimental Setup. It consists of two feed streams, an emulsion and a slurry of magnetite nanoparticles in water (Figure 1). An aspirator bottle provides a constant flow of emulsion and a mixer I attached with a mechanical stirrer, and a peristaltic pump provides a constant flow rate of the slurry. Both the constant feed streams of emulsion and slurry allowed to mix in a mixer II attached with a mechanical stirrer and peristaltic pump. A conveyor belt, rotating on three pullies, is placed on an inclination underneath the mixer II. The lower pulley (3) is powered by a variable frequency drive (VFD) which gives a constant linear velocity to the belt. A permanent magnet is placed below the conveyor belt at the site where the mixture from the mixer II falls on the conveyor belt in order to separate magnetic and nonmagnetic phases of the mixture. The materials were collected from the two ends of the conveyor belt.

Working. A stream of emulsion and slurry was allowed to mix in mixer II. Then this mixture was allowed to pass through the separation process on the conveyor belt. The separated products were collected in the two collectors which were placed at the upper and lower end of the conveyor belt. The products were again placed on a permanent magnet for certain length of time to removes the excess of water from the supernatant. The following effects were studied on this setup.

The Effect of an Increase in Linear Velocity of the Conveyor Belt in the Separation Process. It was studied on specific set of conditions using slurry (slurry 5.6 % (w/w), slurry volume 1000 ml, slurry flow rate 120 ml/min.) and vary the linear velocity of belt from 34 to 90 cm/s. The treated water was collected from the water and oil collector ends for the full duration of the experiment. The magnetic nanoparticles were separated from water by placing the sample collecting beakers on the permanent magnet. The separated MNPs were dried in oven till we get the constant weight of it. A relationship between amount of MNPs (g) collected from both water and oil collector vs. linear velocity of belt was studied (Figure 3). Similarly experiments were conducted using mixture of emulsion and slurry at specific set of conditions on setup (slurry 5.6 %(w/v), 1 % (v/v) oil-in-water emulsion 1000 ml, stirring at 800 RPM, stirring time 20 min., flow rate of mixture 300 ml/min.). The linear velocity of belt was varied from 34 to 90 cm/s to study the relation between volume of water collected from both the collectors (water and oil collectors) vs. linear velocity of the conveyor belt (Figure 4 a). The remaining oil loaded nanoparticles washed with acetone to remove the oil. The washed MNPs were further dried in oven till we achieved the constant weight. A plot between amount washed dried MNPs Vs linear velocity of the belt was studied (Figure 4 b).

The Re-Use of Magnetite Fe_3O_4 Nanoparticles in De-Emulsification of Emulsion Effluent. On the specific set of conditions of 1000 ml of 1 % (v/v) oil-in-water emulsion, 105.5 g of MNPs dry powder, while keeping the rest of parameters same as above, experiments were conducted to analyse the reusability of the recovered amount of MNPs in the treatment of a fresh batch of emulsion. This step was repeated till no more complete emulsion breaking was observed.

RESULTS AND DISCUSSION

The black coloured MNPs characterized by XRD techniques (Figure 2 a.) confirms the synthesis of Fe_3O_4 nanoparticles. The six characteristic peaks at $2\theta \approx 30°, 35°, 43°, 53°, 57°$, and $62°$ were corresponding to the (220), (311), (400), (422), (511), and (440) crystal planes of Fe_3O_4 with a spinal structure (JCPDS file PDF no. 65-3107). No characteristic peaks of impurities were detected in the XRD pattern (Hariani et al., 2013). TEM analysis of Fe_3O_4 nanoparticles confirms the synthesis of nanoparticles of 10 nm size range (Figure 2 b.) (Mahdavi et al., 2013). SEM analysis of Fe_3O_4 nanoparticles shows the agglomeration due to lack of capping agent (Figure 2 c.) (Shen et al., 2009).

During the study of effect of linear velocity of conveyor belt on the separation process, it was found that the volume of water and MNPs in water collector decrease with a linear velocity of belt, whereas the carry over water to oil collector increases. The similar trend was observed for water and MNPs separation in both (I. slurry and II. the mixture of emulsion and slurry) the cases. It was also found that the separation increases with the increase in the linear velocity of the conveyor belt which was caused due to a decrease in the residence time.

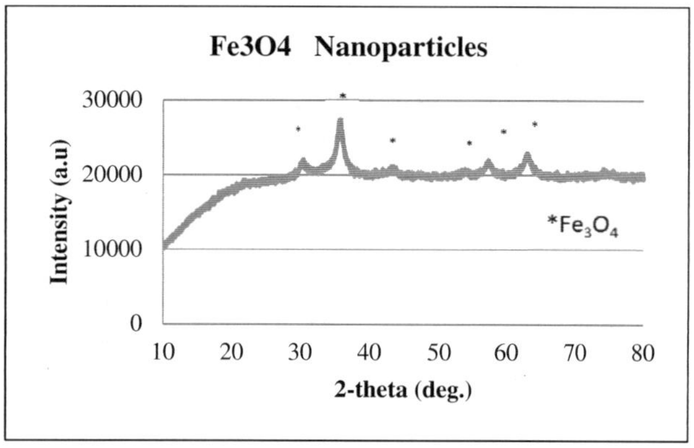

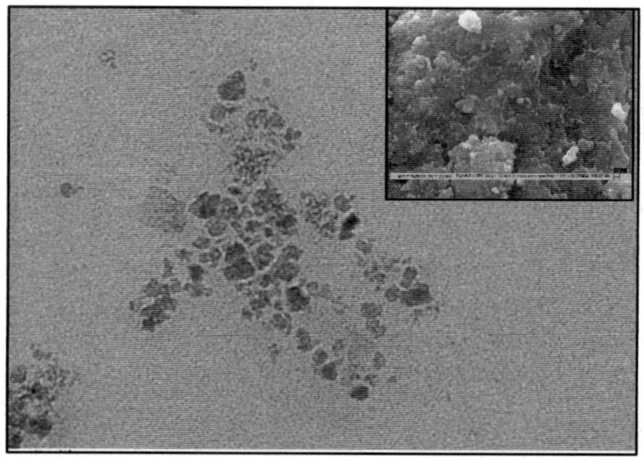

Figure 2. (**a**) XRD analysis of Fe_3O_4 nanoparticles; (**b**) TEM image of Fe_3O_4 nanoparticles; (**c**) SEM image of Fe_3O_4 nanoparticles (inset)

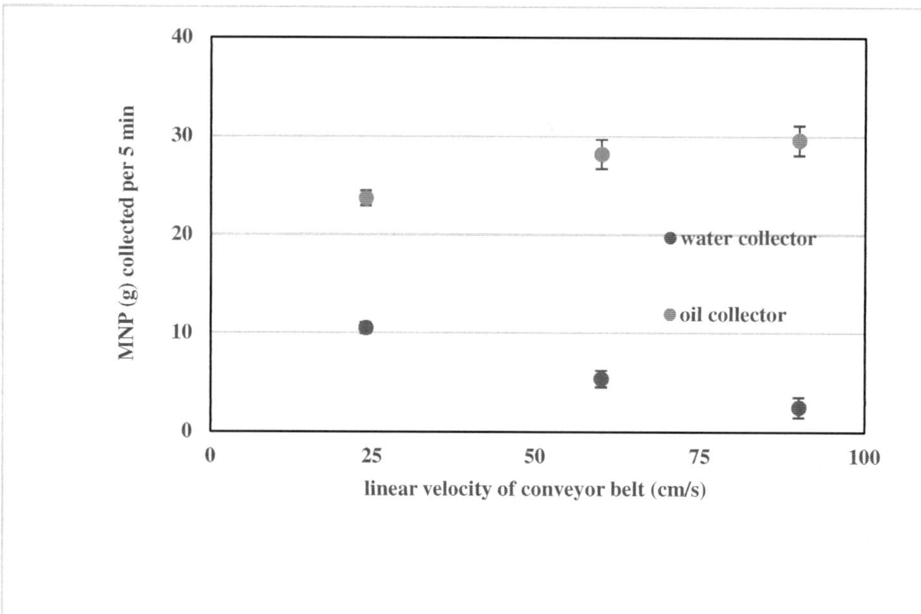

Figure 3. The mass of MNPs collected in water and oil collector per 5 minutes at different linear velocities of the belt

During the study of reusability of MNPs from the first batch of 1000 ml of emulsion treatment, 97 % (v/v) of water, 97.09 % (w/w) of MNPs, and 55% (v/v) of water was recovered. The recovered MNPs (102.44 g) from the first batch was used in the treatment of second fresh batch of emulsion (1000 ml). Here, we got recovery of water 98 % (v/v), of MNPs (92.8%) and of oil 20 % (v/v). Similarly, when experiment was repeated with third batch of fresh emulsion, it was found that the recovered MNPs (100.63 g) were not able to break the emulsion completely. So, to that, fresh 10 g MNPs were added to break the emulsion completely. It was observed that MNPs could be reused thrice with a little amount of fresh addition of MNPs (TABLE 1).

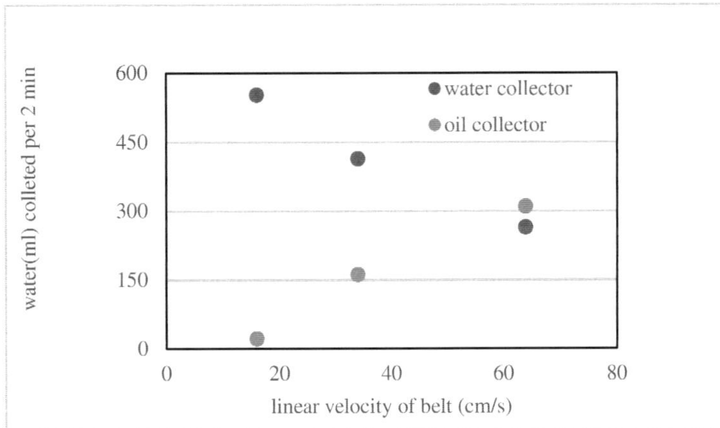

(a)

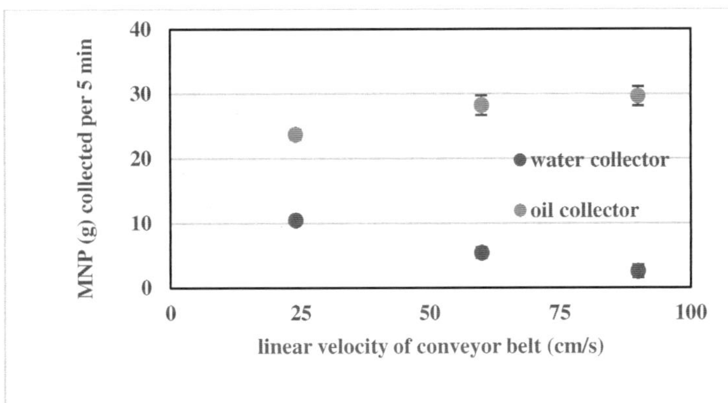

(b)

Figure 4. (**a**). Volume of water collected from water and oil collector per 2 minutes at different linear velocities of the belt; (**b**). mass of MNPs collected from water and oil collector per 2 minutes at different linear velocities of the belt.

TABLE 1. Re-use of Magnetite nanoparticles in de-emulsification of emulsion effluents

S.NO	CONTENTS	FIRST TIME TOTAL RECOVERY	SECOND TIME TOTAL RECOVERY	THIRD TIME TOTAL RECOVERY	AVERAGE VALUE AND PERCENTAGE RECOVERY
1	Water	970 ml, 97%	980, ml 98%	990 ml ,99%	980 ml **98.98 %** (v/v)
2	MNP's	102.44 g, **97.09 %w/w**	100.63 g, **98.2% w/w**	102.08 g, **92.8% w/w**	105.63 g, **96.03%** (w/w)
3	0il	5.5 ml , 55% (v/v)	2 ml, 20 % (v/v)	1 ml, 10 % (v/v)	8.5 ml i.e **28.3 %** (v/v)

CONCLUSIONS

The magnetite nanoparticles could be successfully synthesized in bulk amount by co-precipitation method. XRD, SEM and TEM analysis confirmed the synthesis of Fe_3O_4 nanoparticles, effectively treat the emulsion effluents of steel rolling mills. The studies on the effect of an increase in the linear velocity of the conveyor belt show that the carryover-amount of nanoparticles and treated water increases with the linear velocity of the belt. The studies on reusability of nanoparticles showed that these could be reused trice for the treatment of 1% (v/v) oil-in-water effluents. This process is economic, eco-friendly, quick and less energy-intensive.

ACKNOWLEDGEMENTS

The authors would like to thank the Department of Chemical Engineering, Indian Institute of Technology Delhi (IITD), for providing us with labs and support to carryout the work. Ms Parsanta would also like to thank Petrotech for providing her with their research fellowship. The authors thank undergraduate student Mr Rishab Gupta, Somesh Gupta and Mr. Sandeep Meena, the experimental engineer and research scholar, and late Mr. Vinod Taneja for fabricating and modifying the basic experimental set-up, and Professor A. Sethuramaiah of Center for Industrial Tribology, Machine Dynamics and Maintenance engineering Centre (ITMEC) of IITD for suggesting the difficult and challenging practical problem to one of us (Professor Ashok N. Bhaskarwar).

REFERENCES

Berger, P., Adelman, N.B., Beckman, K.J., Campbell, D.J., Ellis, A.B., Lisensky, G.C., 1999. Preparation and Properties of an Aqueous Ferrofluid 76.

Hariani, P.L., Faizal, M., Setiabudidaya, D., 2013. Synthesis and Properties of Fe 3 O 4 Nanoparticles by Co-precipitation Method to Removal Procion Dye 4. https://doi.org/10.7763/IJESD.2013.V4.366

Hupka, J., 1992. Method and apparatus for treatment of waste emulsions, especially from washing of machine parts. PL 157058 B1 19920430.

Mahdavi, M., Ahmad, M., Haron, M., Namvar, F., Nadi, B., Rahman, M., Amin, J., 2013. Synthesis, Surface Modification and Characterisation of Biocompatible Magnetic Iron Oxide Nanoparticles for Biomedical Applications. Molecules 18, 7533–7548. https://doi.org/10.3390/molecules18077533

Mang, T., 2017. Lubricants and Lubrication, 3rd ed. Wiley-VCH Verlag GmbH & Co. KGaA.

Shen, Y.F., Tang, J., Nie, Z.H., Wang, Y.D., Ren, Y., Zuo, L., 2009. Preparation and application of magnetic Fe3O4 nanoparticles for wastewater purification. Sep. Purif. Technol. 68, 312–319. https://doi.org/10.1016/j.seppur.2009.05.020

Talekar, S., Ghodake, V., Ghotage, T., Rathod, P., Deshmukh, P., Nadar, S., Mulla, M., Ladole, M., 2012. Novel magnetic cross-linked enzyme aggregates (magnetic CLEAs) of alpha amylase. Bioresour. Technol. 123, 542–7. https://doi.org/10.1016/j.biortech.2012.07.044

Zhou, W., Yang, M., Song, Z., Xing, J., 2015. Magnetic Nanoparticles. Biotechnol. Bioprocess Eng. 20, 117–123. https://doi.org/10.1007/s12257-014-0504-8

Effect of HCO_3^- on Physicochemical Properties of Silver Nanoparticles and Toxicity to *E. Coli.*

Bojie Yuan and Minghao Sui
(Tongji University, Shanghai, China)

With the widespread use of silver nanoparticles (Ag NPs) in medical and commercial products, their potential risk to human health and ecological balance, once released into environment, are of concern. Understanding the interaction of Ag NPs with different ions and organic materials presented in natural water is crucial to determine their fate and toxicity in the environment. HCO_3^-, ubiquitously present in aquatic environment, may have an effect on the behavior of Ag NPs in environment. Thus, this study investigated the evolution of Ag NPs physicochemical properties, like optical property, surface charge and the release of Ag^+, as well as its toxicity to E.coli during long-term incubation with HCO_3^-. Our results suggested that an ultrathin Ag_2CO_3 layer may formed on the surface of Ag NPs after treating with HCO_3^-, which resulted in the red-shifting of UV-vis spectra, increasing of zeta potential values (more negative) and reducing of Ag^+ concentration. Due to the protection of Ag_2CO_3 layer, the dissolution rate of Ag NPs was completely inhibited or significantly decreased even in the presence of strong oxidant, H_2O_2. Ambient temperature played an important role in the interaction of Ag NPs with HCO_3^-. With the increasing of temperature, the change of Ag NP physicochemical properties were obviously promoted. We further investigated the toxicity of Ag NPs to E.coli and found that the antibacterial efficiency of Ag NPs was significantly inhibited after treating with HCO_3^-. Finally, we proposed a reaction schematic to describe the interaction of Ag NPs with HCO_3^-, which could provide a reference for better understanding the effect of aquatic compounds on Ag NPs.

Kinetics, Mechanisms and Pathways of Sulfadiazine Photodegradation by $ZnO/ZnFe_2O_4/g\text{-}C_3N_4$ Composite

Guanglan Di, Zhiliang Zhu
(Tongji University, Shanghai 200092, China)

Sulfonamides, as one kind of the most commonly used antibiotics, have been frequently detected in different water bodies, which have posed potential threat to human health and ecological environment. In this study, novel $ZnO/ZnFe_2O_4/g\text{-}C_3N_4$ composites were synthesized by a facile hydrothermal-calcination method for the enhanced degradation of sulfadiazine (SDZ). Results showed that the composite exhibited a higher photoactivity compared to the bulk $g\text{-}C_3N_4$. Under the condition of no pH adjustment, more than 98% removal of SDZ was achieved within 4 hours. The promoted activities were mainly ascribed to the increased specific surface, higher light absorption capacity and efficient charge separation. Active species trapping experiments demonstrated that hydroxyl radicals (·OH) dominated the degradation process. A Z-scheme mechanism was proposed for the efficient charge separation in the system of $ZnO/ZnFe_2O_4/g\text{-}C_3N_4$ composite, together with $ZnFe_2O_4$ functioning as a light sensitizer. By virtue of UPLC-ESI-QToF-MS technology, a total of 11 intermediates were identified and the degradation pathways for SDZ involving hydroxylation, desulfonation and bond cleavage pathways were established. This study might be enlightening for rationally designing of potential $g\text{-}C_3N_4$ based photocatalysts for the decontamination of pharmaceuticals in water.

Physico-chemical Approach of Harvesting Chlorella Sorokiniana and Scenedesmus Obliquus with Magnetic Ferrite Nanoparticles Containing Mn

Louie Lapenas and Mark Daniel De Luna
(University of the Philippines, Diliman, QC, Philippines)
Debora Rodrigues
(University of Houston, Houston, Texas, USA)

One of the drawbacks of microalgal technologies, specifically in the field of biofuel production is the expensive cell concentration procedures. In this work, magnetic ferrite nanoparticles (Mn Ferrite and Mn doped with magnetite) containing manganese were synthesized and used as a new agent to harvest *Chlorella sorokiniana* and *Scendesmus obliquus*. Harvesting and recovery efficiencies of both strains to the two nanomaterials were tested under a model environment (KCl) at different pH, revealing the dependency not only on the pH and the amount of material but as well as the strain itself. The separation efficiency of *Chlorella sorokiniana* was over 90% on both materials at 1: 0.150 g (algae: nanomaterial) and 1: 0.100 g for Mn Ferrite and Mn doped with magnetite respectively. On the other hand, *Scenedesmus obliquus* showed poor affinity to both of the two nanomaterials which resulted to a much lower harvesting efficiency, around 50 - 60%. Both of the microalgal cells however were successfully recovered at > 80% (Chlorella) and > 90% (Scenedesmus) efficiency. Using the predictions made by colloidal adhesion (XDLVO) model, the interactions and mechanisms leading to the cell – nanomaterial attachment and detachment were also studied.

Fabrication of Cu-Ni/graphene Catalysts for Eletrochemical Degradation of Chloramphenicol

Xiaozhe Song and Hui Wang
(Beijing Forestry University, Beijing, China)
Zhaoyong Bian
(Beijing Normal University, Beijing, China)

The chlorinated nitroaromatic antibiotic, chloramphenicol (CAP), is frequently detected in water environments, which may cause adverse effects in ecosystems and organisms due to its lethal toxicity to aplastic anemia, genotoxicity and potential carcinogenicity. Graphene decorated with bimetallic nanoparticles (NPs) was considered as efficient cathode materials for their outstanding electrocatlytic activities, such as excellent electronic conductivity and significant electrochemical production of H_2O_2. Moreover, bimetal Pd-Fe NPs is considered as the preferred modified material in hydrodechlorination and oxygen reduction reaction (ORR) processes. Cu-Ni NPs were decorated on graphene (Cu_xNi_y/G, X, Y stand for the weight ratio) via modified chemical reduction methods. Higher peak currents (0.242 mA) for H-adsorption, a lower onset potential and higher peak current were achieved in $Cu_{0.5}Ni_{0.5}/G$ (-0.045 mA at -0.136 V) for ORR than other Cu-Ni/graphene catalysts with different metal loading. The number of electrons transferred calculated in the ORR using Koutecky-Levich plots was 1.92, which indicated a two-electron pathway for producing H_2O_2. Then, the electrochemical reduction-oxidation process was designed to degrade chloramphenicol by using $Ti/IrO_2/RuO_2$ anode and selected $Cu_{0.5}Ni_{0.5}/G$ cathodes. The $Cu_{0.5}Ni_{0.5}/G$ cathode showed higher electrochemical activities in the degradation procedure under the alkaline conditions. The main influence factors on the degradation of CAP including electrolyte concentration, current density and initial CAP degradation were exported by the removal efficiency of CAP and TOC. The complete removal of CAP was occurred within 30 min in cathodic chambers for the rapid nitro-reducing. The removal rates of TOC were reached 98.92%, 96.63% and 96.11% respectively in anodic and cathodic chambers under the optimized conditions after 4 h electrolysis in 0.05 mol L^{-1} Na_2SO_4 electrolytes at 30 mA cm^{-2}. The dechlorination process has taken place within 1 h by feeding H_2 in cathodic chambers and then the oxygen reduction reactions in cathode and anodic oxidation processes jointly promoted the mineralization of CAP during electrolysis. The degradation mechanism and possible pathways including reduction of nitro group, dechlorination and mineralization in cathodic chambers as well as a consistent mineralization in anodic chamber were exhibited.

Effect of Environmental Factors on Interaction between Tetracycline and Copper Oxide Nanoparticle

Jun Ma and G. Daniel Sheng
(Tongji University, Shanghai, China)

Antibiotic is a widely used by the whole world for treating human diseases, preventing animals' disease and promoting their growth. Tetracycline is one of the most common antibiotic, which as an efficient ligand with cations, and has many functional groups. Metallic oxide nanoparticle is an emerging contaminant, with the widespread use and randomly discharge, antibiotics and Metallic oxide nanoparticle will inevitably interact with each other in the water and soil environment. In order to investigate the effect of environmental factors on the interaction between antibiotics and Metallic oxide nanoparticle, tetracycline and copper oxide nanoparticle were chose as research objects, and the change of pH, ionic strength were used to explore the effect on the interaction with each other. The results revealed that with the decline of pH, CuO nanoparticle could degrade tetracycline effectively with the Cu^{2+} gradually was dissolved from CuO nanoparticle in the first six hours after their interaction, especially when pH < 6.0, and the interaction attained balance in the forty-eight hours. On the contrary, with the increasing of ionic strength from 1 mmol/L to 50 mmol/L and the valence states of metal ions enhance from NaCl to $CdCl_2$, the ability of Cu^{2+} dissolved from CuO nanoparticle was enhanced rapidly, accordingly, the time of CuO nanoparticle degrade tetracycline effectively was shortened. These results indicate that antibiotics may be degrade effectively by CuO nanoparticle when the environment is acid and full of higher valence states metal ions.

Plasmonic Photocatalysts for Environmental Applications in Detection and Treatment of Organic Pollutants

Jai Prakash
(Department of Energy, Materials and Telecommunications, INRS-EMT, Quebec, J3X 1S2, Canada; University of the Free State, Bloemfontein 9300, South Africa)

A great deal of research in nanotechnology has been dedicated to the design and the development of new nanomaterials with tunable physico-chemical properties for several environmental issues. Recent advances and an improved fundamental understanding of the coupling of different nanomaterials has shown great potential in interdisciplinary research with enhanced fundamental properties owing to their synergistic effects. Plasmonic (Au/Ag) and metal-oxide (TiO_2/ZnO) nanostructures based plasmonic photocatalysts have been emerged as novel functional nanomaterials for various environmental and engineering applications. These plasmonic photocatalysts show enhanced optical and transport properties resulting in visible light activity due to the Shottkey and plasmonic effects in presence of noble metals. Engineering of their shapes and sizes cause significant variation in the physical and chemical properties adding flavor to the tunability. This not only enhances the photocatalytic photo-degradation activities of metal oxide semiconductors but also extend its utilization in the photovoltaic industry. Similarly, plasmonic nanostructures on semiconductor's surface allow them to use as recyclable surface enhance Raman scattering (SERS) substrate for ultra-detection of bio-organic molecules. In this presentation, interdisciplinary aspect of plasmonic photocatalyst nanocomposites with special emphasis on their environmental applications in detection and treatment of organic pollutants in waste water treatment will be discussed.

Effects of Fabrication and Modification Method of Magnetic Nanomaterials on Green Algae Capture Performance

Pei-rui Liu and Yu Hong
(Beijing Forestry University, Beijing, China)

Introduction: The explosive growth of microalgae in eutrophic water leads to ecological crisis. In bloom-forming algae, green algae are needed to become more concerned about. Compared with traditional technologies, recently, capture green algae by magnetic nanomaterials has come into the spotlight because of the advantages of higher capture efficiency, lower cost and convenient operation.

Objectives: Fe_3O_4 NPs were fabricated and modified with amino acid by different methods in this study, and utilized to capture green algae efficiently.

Methods: Fe_3O_4 NPs were synthesized by chemical coprecipitation and pyrolysis method, and then modified with amino acid by ultrasonic, long time mixing, and one-step method. The green algae used in this study was *Chlorella* sp. HQ (Collection No. GCMCC7601 in the China General Microbiological Culture Center). The magnetic algae capture was conducted by mixing algal suspension with magnetic nanomaterial suspension of a certain concentration on a rotary shaker and the nanomaterial coated algae cells were separated from the solution by a permanent magnet, and the density of algal cells in the supernatant was measured after the magnetic separation using a UV spectrophotometer. The effects of fabrication and modification method on capture efficiency were investigated. Then Fe_3O_4@PAMAM magnetic nanomaterial was fabricated by optimal method and used to capture green algae at different dosage and pH value. The regeneration performance of Fe_3O_4@PAMAM was also evaluated.

Results and Conclusions: Considering capture efficiency and fabrication cost, Fe_3O_4 NPs synthesized by chemical coprecipitation method presented better capture efficiency. Modifying by one-step method is more suitable for magnetic green algae capture, with the capture efficiency of 95%. The nanomaterial Fe_3O_4@PAMAM fabricated by optimal method could achieve more than 95% capture efficiency within 5 minutes at 80mg L^{-1} dosage, and the high capture efficiency was not affected by pH values. Also, Fe_3O_4@PAMAM presented good regeneration performance during 5 times cycle. The results showed that chemical coprecipitation and one-step method were suitable to fabricate magnetic nanomaterials for green algae capture.

AIR POLLUTION AND AIR QUALITY CONTROL

Trend Analysis of Air Pollutants from Non-Conventional Emission Sources in the North Texas Region

Guo Quan Lim, Kuruvilla John
(Department of Mechanical and Energy Engineering, University of North Texas, Denton, Texas)

Growth in shale gas production at the Barnett Shale region introduced non-conventional air pollutant sources to the Dallas-Fort Worth (DFW) region. Gas well count and shale gas production volume spiked, with the production volume growing from 216 million-cubic-feet-per-day (mmcf) in 2000 to 4928 MMcf-per-day in 2014. A trend analysis was conducted on volatile organic compound (VOC), oxides of nitrogen (NO_x) and ozone (O_3), which were collected from the Texas Commission on Environmental Quality (TCEQ) monitoring sites at Dallas Hinton, Fort Worth Northwest and Denton Airport South. The Dallas site had 5 wells within a 10-mile radius while the Denton site had 1,119. The Dallas site is located within a dense urban region with 10 times the population when compared to the Denton site. The Dallas and Denton sites serve as perfect counterparts for this evaluation. The Fort Worth site is in a suburban location and has higher number of gas wells than the Dallas site. NO_x concentrations dropped at all sites from 2005 to 2014. At Dallas, mean and maximum NO_x dropped by 50% and 18%, respectively, while it was about five times higher than in Denton. However, the maximum VOC monitored at Denton was ten times greater than Dallas. Over the study period, the mean VOC concentration decreased by 21% at Dallas while the peak concentration increased 46%. At Denton, the mean value increased 620% while the peak grew by 1960%. The mean ethane-to-TNMOC (total non-methane organic carbon) ratios increased from 0.1 to 0.15 at Denton and from 0.05 to 0.09 in Dallas. On the other hand, acetylene-to-TNMOC ratio dropped from 0.012 to 0.005 at Denton and 0.034 to 0.018 at Dallas. This suggests traffic sources had a decreasing contribution to the total VOC distribution, while natural gas sources were becoming much more dominant.

Episodic Air Pollution Events in Delhi during Crop Residue Burning: A Comprehensive Study of Multiple Pollutants

Pavan Kumar Nagar, ***Mukesh Sharma***
(Indian Institute of Technology Kanpur, Kanpur, Uttar Pradesh, India)

In the months of October-November, a large scale (area about 30,000 sq-km) crop residue burning (CRB) of paddy crop in the upwind (North-West) of Delhi is an annual practice. The emissions from CRB are believed to severely impact the air quality of Delhi to an episodic level, causing wide-spread panic. To develop a causal linkage between the emissions and the impacts, a comprehensive study of multiple pollutants (NO_2, CO, $PM_{2.5}$) including detailed chemical composition of $PM_{2.5}$ (25 metals, organic carbon (OC), elemental carbon (EC), ions (NO_3^-, SO_4^{-2}, Cl^-, NH_4^+, K^+)) was undertaken at two locations (i) Rohini (November 5 – 23, 2013; CRB Period) and (ii) Dwarka (December 2 – 22, 2013; Post-CRB period) in Delhi. To estimate the contribution of sources, chemical mass balance model (of USEPA) was employed. $PM_{2.5}$ pollution levels were critically high during CRB (November 5 - 23, 2013) as the concentrations exceeded the national air quality standard (60 µg/m^3) by a factor of 5-7. The mean concentrations were 428±119 µg/m^3 during CRB period (at Rohini) and 348±150 µg/m^3 during post-CRB period (at Dwarka). It was estimated that OC, EC and K (potassium) contributed 75, 38 and 15 µg/m^3 to PM2.5 during CRB period. The K can be considered as marker for CRB. It was observed that K/EC (CRB: 0.40 and post-CRB: 0.32) was higher during CRB than post-CRB which indicates impact of CRB on $PM_{2.5}$ levels in Delhi. A validated WRF-Chem model showed an increase of 7 mg/m^3 in CO levels during CRB period. Based on chemical mass balance model, the CRB contribution was estimated 32 % (137 µg/m^3) during CRB and 27% (95 µg/m^3) in post-CRB periods. The other major sources (during CRB and post-CRB periods) were vehicles (22% and 18%), inorganic secondary aerosol (19% and 35%), municipal solid waste burning (11% and 4%) and coal and flyash (5% and 3%). It was concluded that CRB cause significant increase in air pollution in the Delhi. There is an immediate need to find alternatives to CRB to prevent episodic air pollution levels not only in Delhi but for a larger area.

Analysis of Land Use on Air Quality in Pennsylvania Using Geospatial Technologies

Alex Criswell and ***Rachel Hodgins***
(Wilkes University, Wilkes Barre, PA, USA 18766)

Land use in Pennsylvania varies highly from cropland to expansive forestland and pastureland to densely populated urban land. According to the United States Census Bureau's most recent Decennial Census of 2010, nearly 78 percent of the state's population resides on urban land. As of US EPA's air quality report published in February 2017, twenty counties in the Commonwealth of Pennsylvania have been designated as "nonattainment", *i.e.* counties that do not meet the National Ambient Air Quality Standards (NAAQS) for at least one of the six criteria air pollutants published in the Clean Air Act Amendments (CAAA) of 1990 (the most recent amendment). The criteria air pollutants are carbon monoxide, nitrogen oxides, sulfur oxides, lead, particulate matter (PM_{10} and $PM_{2.5}$), and ozone. Sources for most of these air pollutants are natural (generally wildland fires, volcanoes, and dust storms) and anthropogenic (industries, automobiles, and agricultural operations). Except for ozone, all criteria air pollutants are primary in nature, meaning that they are released from a source such as a stack or an automobile tailpipe among others; ozone is a secondary air pollutant, meaning it is not released from a specific source but is formed due to chemical reactions between various constituents in the atmosphere. While many studies (epidemiological, clinical and toxicological) have been conducted over the last 25 years (the CAAA of 1990 was passed by the U.S. Congress in November 1989) to study the effects of air pollution on human health and human welfare, there is a need to better understand the impact of land management on the air quality at local and regional levels. The goal of this preliminary study was to investigate the relationship between land use and air quality in Pennsylvania using geospatial technologies. Levels of $PM_{2.5}$, PM_{10}, ozone and nitrous oxide were obtained from the EPA for the year of 2010. A raster map showing land usage for the state of PA was obtained from the Multi-Resolution Land Characteristics Consortium. The average level for each pollutant for the year was mapped against the land cover map to show where the highest levels of each pollutant occurred. Other factors such as seasonal trends, and population were examined also at the counties that showed the highest levels for the year. For PM_{10} and nitrous oxides, the highest levels were observed in areas with the most urban development. Ozone and $PM_{2.5}$ occurred in counties with less urban development.

Evaluation of PM2.5 Emissions and PAHs Concentration in $PM_{2.5}$ from the Exhausts of Diesel Vehicles

Kassian T.T. Amesho, Ya-Ching Li, Feng-Chih Chou, Syu-Ruei Jhang, Yu-Chieh Lin, Yuan-Chung Lin
(Institute of Environmental Engineering, National Sun Yat-sen University, Kaohsiung 804, Taiwan)

In order to reach the pollution and renewable energy targets for 2020, considerable research and development efforts are required. Because new technologies require a significant amount of time to cover a large enough market share, fuel improvements are a viable solution that has the capability of affecting the whole vehicle park. Urban areas can be subject not only to poor air quality, but also to contamination of other environmental media by air pollutants. To that end, our study intends to estimate the PM emissions from diesel vehicles using diesel fuels and to analyze the PM2.5 emissions and PAHs concentration in PM2.5. The study examined the data on the emission factors of PM2.5, PM emissions through the diesel fuel combustion in diesel vehicles engines [g/GJ]. In this study, particulate matter (PM2.5) were characterized from a place impacted by diesel vehicles fueled with diesel in Kaohsiung City, Taiwan. Sixteen priority polycyclic aromatic hydrocarbons (PAH) concentrations were analyzed in the samples by their association with atmospheric PM2.5. We estimated that the total of NOx and CO emissions at 31.5% and 23.6% respectively. This would also result in approximately 5% increase in estimation of national anthropogenic NOx emissions. Thus, more effective control measures (such as promotion of CNG buses and a new in-use compliance testing program) are urged to secure the goal of total NOx mitigation for the diesel vehicles fleet in the future. Our results further indicate that the major PAHs compounds contribution (%) to PM2.5 emissions (ng/m^3) are as follows: Pyrene contributed the highest concentration levels of 32.4%, Naphthalene contributed 18.5%, while Fluorene and Phenanthrene contributed 17.6%, and 16.8% respectively. Anthracene was accounted for 4.28% and Fluoranthene was responsible for 3.82%.

Reducing Emissions from Gasoline Engines Using Hydrogen

Yu-Chieh Lin[1], Syu-Ruei Jhang[1], Kang-Shin Chen [1], Sheng-Lun Lin[2] , Yuan-Chung Lin [1]*
([1] Institute of Environmental Engineering, National Sun Yat-sen University, Kaohsiung 804, Taiwan;
[2]Super Micro Mass Research and Technology Center, Cheng Shiu University, Kaohsiung)

Since most people in Taiwan own car or motorcycles, the air pollution has become more and more serious of recent. There also has been a strong impact on climate change and this air pollution impacts account for a great part of emissions of carbon monoxide (CO), hydrocarbons (HC) and nitrogen oxide (NOx) within the urban environment in Taiwan. But due to increase in environmental awareness, researchers have developed more sustainable energies for vehicles to both effectively reduce the fossil fuel demands and air toxic emissions. EU also promote that we can use replaceable fuels for vehicles to reduce the emission of exhaust. At the same time, it can also handle the growing demand of energy for transport use. We know clearly carbonyl compounds are everywhere and they have bad effects on human health. The cleanest fuel is hydrogen (H2) which has more advantage because of its high ignition temperature and combustion properties that can be used for improving combustion and emissions performance of gasoline-fueled spark ignition (SI) engines. For the past years, H2 has been researched in term of production, transportation, storage, and combustion around the world in order to finding alternatives for fossil fuels. This study was modified to be precisely focused on gasoline vehicle fueled with the mixture of gasoline and hydrogen injected into the intake ports simultaneously. The engine was tested at four different condition: Idling, 40 km/h, 70 km/h, 100 km/h, using pure gasoline and four H2 mixtures: 4-16 L/min, interval 4 L/min. Nevertheless, the major carbonyl compounds detected in the collected samples were formaldehyde, acetaldehyde, acrolein, acetone and benzaldehyde and the total carbonyl compound emissions shows the exciting results with 4 L/min and 8 L/min H2 intake, while they reduced 6.2~78.1% and 7.0~78.3% at various car speed. Consequently, this study partially solved the CO-NOx traditional pollutants problem and effectively reduced the toxic emissions of a real gasoline vehicle, while the petro-energy could also be saved, especially at the idling operation.

Occurrence and Health Risk Assessment of Phthalate Esters in Road Dust in Parks of Tianjin City during Spring and Winter

Yaqin Ji, ***Lei Zhang***, ***Shibao Wang***
(College of Environmental Science and Engineering, Nankai University, Tianjin 300350, China; jiyaqin@nankai.edu.cn)

This study reports the concentrations and distribution of phthalate esters (PAEs) in road dust from 15 parks of Tianjin city in winter and spring. Through ultrasonic extraction, nitrogen blowing instrument, GC-MS for analysis, $\sum_6$PAEs concentrations from road dust in winter and spring ranged from 0.016 to 6.858 mg kg^{-1} and 0.008 to 1.482 mg kg^{-1}, respectively. The temperature plays a key role in the change of PAEs concentration. DEHP and DnBP were the predominant PAEs detected in samples. Spearman correlation analysis and principal component analysis (PCA) were used to qualitatively analyze sources of PAEs, the results showed that PAEs mainly come from PVC plastic products in winter, PVC plastic products, cosmetics, personal care products in spring. Furthermore, the noncancer and carcinogenic risks of six PAEs were estimated. The hazard indexes (HQ) of PAEs in all samples were below 1. The carcinogenic risk of DEHP and BBP via non-dietary routes were also in a low risk category.

Adsorptive Removal of Formaldehyde from Indoor Air Using Mixed-Metal Oxides

Fateme Rezaei, Anirudh Krishnamurthy
(Missouri University of Science & Technology, Rolla, MO, USA)

ABSTRACT: Volatile organic air pollutants such as aldehyde compounds have been identified as progressively damaging chemicals impacting human health at small albeit dangerous quantities. This study focuses on evaluating the dynamic adsorption of formaldehyde from indoor air over binary mixed-metal oxides (MMOs) such as ZrO_2/SiO_2 and TiO_2/SiO_2 with different metal ratios. In addition, a metal-organic framework (MOF), namely, MIL-101(Cr) was synthesized and used as a base adsorbent to which the performance of MMOs was compared. The formaldehyde dynamic adsorption capacity of the materials was determined by performing breakthrough experiments. Our results indicated that zironia-based materials exhibit a comparatively higher affinity toward formaldehyde than their titania-based counterparts at very dilute concentrations. The ZrO_2/SiO_2 with weight ratio of 25/75 exhibited a dynamic adsorption capacity of 2.9 mmol g^{-1} at room temperature using a formaldehyde concentration of 166 ppm_v, which was only slightly lower than that of MIL-101(Cr). Characterization of the materials before and after formaldehyde exposure indicated that formaldehyde was chemically adsorbed on the MMOs.

INTRODUCTION

Volatile organic compounds (VOCs) are a diverse set of compounds with low vapor pressures that are frequently found in indoor and outdoor emissions. They can be easy inhaled and cause irreversible damage to human beings when present in concentrations as low as 10 ppm_v in air, while slightly higher concentrations of 30–50 ppm_v on continuous inhalation and prolonged exposure can cause life-threatening conditions.[1] The health effects of VOCs range from stress, irritation to mutagenicity and cancer, depending on the compound emitted.[2] VOCs also have consequential atmospheric effects as well. For example, when common VOCs such as benzene or isoprene react with SO_x and NO_x emissions from vehicular transport, a tropospheric ozone smog is formed, which effects visibility and is highly carcinogenic.[3] In terms of oral dosage levels, toxic effects of VOCs can be determined by reference dose (RfD), which is a time-based measurement of their inhalation and consequent ill effects.[4,5] Formaldehyde (HCHO) is a notorious carcinogen at low concentrations along with a low RfD of 0.2 mg/kg-day and is listed in EPA's Hazardous Air Pollutant (HAP) list..[6] In enclosed environments such as commercial or residential buildings airborne formaldehyde is among the most toxic VOC compounds. While identifying emissions from individual products may be challenging, the largest sources of formaldehyde in the indoor environment have been identified to be wood floor finishes, wood-based furniture, carpets and paints.[7,8] The concentration levels are typically dependent on a various factors including number of formaldehyde sources, temperature, humidity, and air exchange rate.

Although catalytic oxidation of formaldehyde into CO_2 is highly efficient,[9] it requires the use of expensive noble metals and external thermal energy. The abatement of airborne pollutants in enclosed environments by adsorption has been shown to be the most effective method for air quality control.[10–13]

This method is particularly efficient given the dilute concentration of formaldehyde. To date, various materials such as activated carbons,[8,14–18] amine-modified mesoporous silica,[19–21] metal-organic frameworks (MOFs),[22] mesoporous alumina,[23] and mixed-metal oxides (MMOs)[24] have been investigated for formaldehyde removal from indoor air with activated carbon being the most widely used adsorbent. One of the early experiments performed on the adsorption of formaldehyde, although not specifically so, was performed by Eriksson et al.[25] on different adsorbents like activated carbon, aluminum oxide in a household environment. Although the experiment showed positive results, the accurate adsorption potential of HCHO wasn't determined due to its interactions with other pollutants. Due to the polar nature of formaldehyde, the adsorption on activated carbon is not very efficient and thus surface modifications by primarily nitrogen and amines have been widely practiced to address this issue. Boonamnuayvitaya et al.[14] evaluated a series of activated carbons for formaldehyde adsorption and demonstrated that $ZnCl_2$-impregnated carbon activated with nitrogen shows highest affinity toward formaldehyde due to the hydrophilic functional groups (e.g. O–H, C=O, C–O) on the surface. It was concluded that surface chemistry and physical properties play an important role in formaldehyde abatement. Similar findings were reported by Rong et al.[15,16] In another study, amine functionalization proved to be a facile strategy to significantly improve adsorptive performance of activated carbon in formaldehyde abatement.[18] Passive control of formaldehyde by air-cleaning materials by which formaldehyde gas is decomposed into CO_2 at room temperature has also been reported.[26] The air-cleaning materials consisting of activated carbon particles and manganese oxides reduced formaldehyde concentration from 0.21 to 0.04 ppm_v.

Saeung and Boonamnuayvitaya[20] tested amine-modified mesoporous silica for the adsorption of formaldehyde vapor and results showed a much higher adsorption capacity as compared to activated carbon owing to their amine groups and large pore size of the mesoporous silica support. Moreover, the characterization results confirmed the reaction between formaldehyde molecules and amine groups on pore surface of the adsorbents. In another investigation,[21] the use of amine-functionalized MCM-41 and SBA-15 as adsorbents for formaldehyde vapor from contaminated air was studied and it was concluded that NH_2-MCM-41 is capable of reducing formaldehyde in indoor air by 80%. Furthermore, the performance of amine-functionalized hierarchically hollow SiO_2 microtubes was evaluated for indoor air purification and a maximum formaldehyde adsorption capacity of 20.65 mg/g was reported for these materials.[19] In recent works by Nomura and Jones,[27–29] amine-functionalized silicates were used for formaldehyde abatement. The authors reported an adsorption capacity of 5.7 mmol g^{-1} for silica-supported poly(allyl amine) amine when formaldehyde gas concentration was kept at 100 ppm_v.[28]

MOF materials have been largely studied as formaldehyde sensors,[30] however, their capability in adsorptive removal of formaldehyde has been rarely explored. In a recent investigation by Wang et al.,[22] the formaldehyde capture of diamine-modified MIL-101 was studied and an adsorption capacity of 5.49 mmol g^{-1} was reported for this material using an initial HCHO concentration of 150 ppm_v, whereas the unmodified MIL-101 exhibited a capacity of 3.34 mmol g^{-1}. Although functionalization of porous adsorbents with amine moieties contributes to enhanced formaldehyde capture, there is a potential health risk associated with the amines leaching from the adsorbents in enclosed environments, especially for aminopolymers that have weak bonds with the support.

Various MMOs, in particular TiO_2/SiO_2 and TiO_2/ZrO_2, have been commonly evaluated for the capture of aromatic hydrocarbons such as benzene, toluene, and xylene (BTX compounds).[31–35] However, their use as formaldehyde abatement adsorbents has been scarcely discussed. In a study done by Carlos-Cuellar et al.,[36] a range of metal oxides were studied for the adsorption of formaldehyde, among other carbonyl compounds. It was shown that SiO_2 has the lowest uptake when compared to Fe and Al based oxides. Further analysis also showed that a weak hydrogen bonding was responsible for the interaction between the adsorbent and adsorbate. The acidic character of the material was revealed to be an important factor contributing to a higher uptake potential. Typically, in their elemental state, the use of metal oxides for adsorption faces many challenges including low surface area or low adsorption capacity while utilization with other metal oxides or porous materials broadens their potential in adsorption. Xu et al.[24] synthesized a series of hierarchical porous $Ni(OH)_2/SiO_2$ composites with flake-like nanostructure and evaluated their performance in the removal of formaldehyde from air. A maximum adsorption capacity of 8.30 mg/g was obtained for the composites with a better stability than with the parent $Ni(OH)_2$ over six adsorption-desorption cycles. The authors attributed the improved performance to the specific framework and synergistic effect between $Ni(OH)_2$ and SiO_2. Most recently, Chen et al.[37] used hierarchical TiO_2 nanospheres functionalized with diethylenetriamine (DETA) for airborne formaldehyde removal at room temperature. The materials exhibited ~ 100% removal efficiency with good degree of recyclability when exposed to 200 ppm_v formaldehyde.

As noted above, although formaldehyde abatement by adsorption has been the subject of several studies, the majority of the materials investigated so far suffer from low adsorption capacity. Furthermore, the removal of formaldehyde from indoor air has not been investigated over MMO materials that have been shown efficient in capturing other VOCs. In this investigation, we synthesized two different MMO materials, TiO_2/SiO_2 and ZrO_2/SiO_2, with varying metal oxide ratios and evaluated their formaldehyde adsorption characteristics by performing dynamic breakthrough experiments in a fixed-bed system. In addition, the performance of MIL-10(Cr), as a base adsorbent that has been proven efficient in abating formaldehyde, was evaluated and compared with that of MMOs. The adsorbents were systematically characterized and their textural and physical properties were correlated with their adsorption performance in abating airborne formaldehyde.

EXPERIMENTAL SECTION

Materials. Tetraethyl orthosilicate (TEOS, 99 %), zirconium (IV) propoxide solution (70 % in 1-propanol), titanium (IV)butoxide (TBT, reagent grade, 97 %), , hexadecyltrimethylammonium bromide (CTAB), triethanolamine (TEAH), formaldehyde solution (ACS reagent, 37 wt.% in H_2O with methanol stabilizer), chromium nitrate nonahydrate ($Cr(NO_3)_3{\cdot}9H_2O$), terephthalic acid (H_2BDC) and nitric acid (HNO_3) were all purchased from Sigma-Aldrich. Moreover, all solvents were of analytical grade and purchased from Sigma-Aldrich. Ultra high purity nitrogen gas was purchased from Airgas.

Adsorbents Synthesis.

Preparation of Mixed-Metal Oxides. Binary metal oxides were prepared according to the modified atrane synthesis route described by Zarate et al. Typically, TEOS was added dropwise to 17.7 ml of TEAH. The required amount of metal alkoxide was added very slowly dropwise to the above solution and was subsequently heated to 150 °C to obtain atrane complexes. The solution is then cooled to 90 °C and 4.5 g of CTAB is added to the solution. The mixture was then cooled to 60 °C and 75 ml of DI was added and the obtained white colored liquid was set to gel for 24 h at room temperature. The samples were then washed with DI and ethanol multiple times before drying at 80 °C overnight. The white powders were then calcined at 550 °C for 6 h. The samples obtained were named as Zr/Si (x) and Ti/Si (x), where x is the metal ratio.

Preparation of MIL-101(Cr). The synthesis method followed was a slightly modified procedure used by Zhao et al.[38] Briefly, about 800 mg of chromium nitrate nonahydrate ($Cr (NO_3)_3{\cdot}9H_2O$) and 328 mg of H_2BDC were mixed together. As an acid additive, HNO_3 was selected and 180 mg was added to the mixture. The reactants were mixed and sonicated until a fairly homogeneous solution was obtained. The contents were then transferred to an autoclave and kept at 220 °C for 8 h. The material was then cooled, filtered and solvent-exchanged. The obtained solid was washed with DMF, water and then refluxed in hot ethanol at 70 °C for 5 h.

Adsorbents Characterization. Low angle powder X-Ray diffraction (XRD) was performed using a PANanalytical X'Pert multipurpose X-Ray diffractometer. Diffraction patterns were recorded for all prepared samples to identify their crystal structure. Using crystal structure file obtained from the open crystal database, a simulated pattern for MIL-101(Cr) was generated and compared with the synthesized material. Scanning electron microscopy (SEM) was used to assess structural morphology of the materials. This was performed using a Hitachi model S4700 field-emission SEM system. N_2 physisorption isotherms were obtained using a Micromeritics 3Flex gas analyzer at 77 K. Degassing in a Micromeritics Prevac was performed for various temperatures at elongated time frames, depending on the material. Surface area was consequently determined using Brunauer-Emmett-Teller (BET) while pore volume was estimated using the Barrett-Joyner-Halenda (BJH) method, obtained from the same device. ^{13}C solid-state nuclear magnetic resonance (NMR) measurements were conducted on a Bruker DSX-300 spectrometer. The materials were spun at a frequency of 10 kHz, and 15000 scans were obtained. X-ray photoelectron spectroscopy (XPS) was performed using Kratos Axis 165 Photoelectron Spectrometer using an aluminum X-ray source to excite the samples.

Formaldehyde Breakthrough Experiments. Dynamic adsorption capacities of the synthesized materials were measured by performing breakthrough experiments using a borosilicate packed column and a mass spectrometer (BELMass) with a setup used in previous works. The samples were degassed in nitrogen prior to tests. Formaldehyde solution (37 wt. %) was filled into the saturator and a pure N_2 stream at a flow rate of 40 mL/min was used to carry a vapor concentration of 166 ppm to the column. Experiments were performed at 25 °C and atmospheric pressure. It should be noted here that this concentration is much higher than the realistic formaldehyde levels at an indoor environment (< 5 ppm), however, to accurately quantify the uptake capacities of the materials, the experiments were conducted under these conditions. The schematic of the setup is presented in Figure **1.**

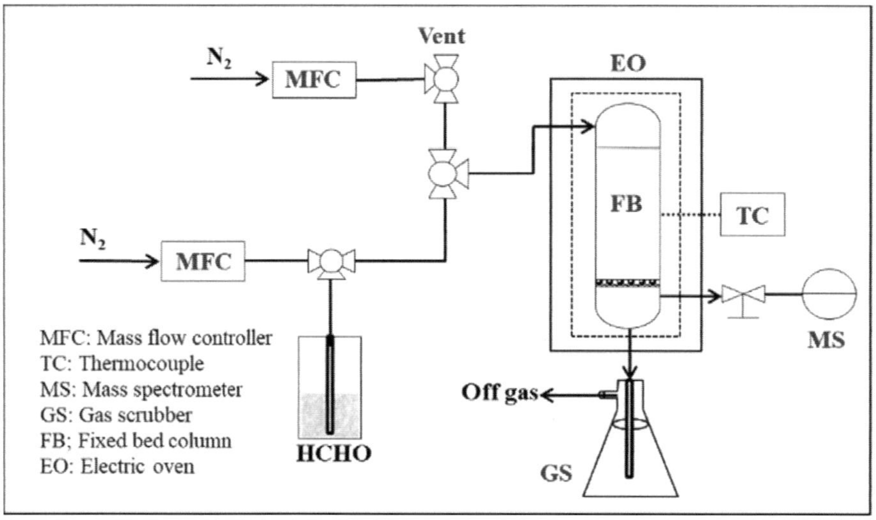

FIGURE 1. General scheme of the breakthrough setup.

RESULTS AND DISCUSSION

Adsorbents Characterization. The low angle XRD patterns for the prepared adsorbents are presented in Figure **2**. Low angle XRD was considered in this study due to the nature of the absorbent particles, which show low angle diffraction patterns, typical of interplanar spacing in mesoporous materials.[39]. The nano-sized particles did not seem to show any peak intensities in regular angle XRD (2θ = 10 – 50° approximately). The obtained patterns were characteristic of interparticular short range hexagonal ordered mesoporous materials, similar to the case of MCM-41[40]. This corresponds to the peaks displayed at 2θ = 2.5. This is however seen to a lesser degree in the higher loaded samples, with peaks being detected at low intensities. It is also observed that with the increase of Zr or Ti loading, there is an observable change in peak intensities. Considering the structure of the adsorbent, this may be due to a larger number of Zr or Ti molecules depositing throughout the walls of the silica support[41]. The degree of mesoporosity increased when Zr/Ti was deposited on the silica surface, evident by the low intensity peak observed with pure SiO_2. Standard XRD profiles were obtained for MIL-101(Cr). By using crystallographic parameters obtained from the crystallographic open database in project mercury, a simulated pattern for MIL-101(Cr) was obtained and was shown to correspond with the obtained pattern in the MIL-101(Cr) material synthesized in this work, as evident from Figure **2c**.

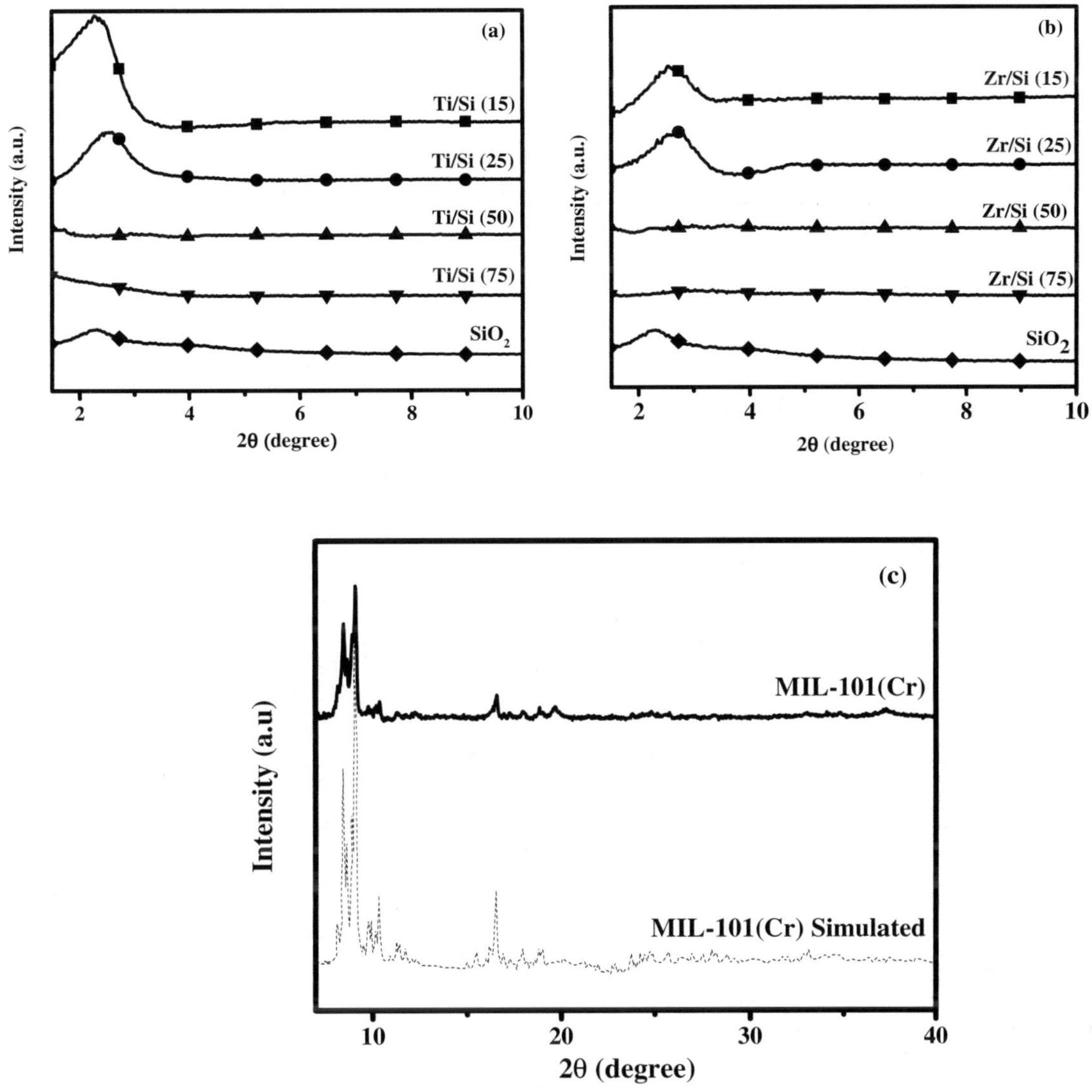

FIGURE 2. XRD patterns for (a) Ti/Si, (b) Zr/Si, and (c) MIL-101(Cr) adsorbents.

N_2 physisorption isotherms obtained at 77 K along with pore size distribution profiles are displayed in Figure **3**. Pore size distributions were obtained using the NLDFT method. As evident from this figure, all MMOs showed a type IV isotherm with a H4 hysteresis indicative of their mesoporous nature. The figure also illustrates the bimodal pore system that is characteristic of UVM-7 silica. The loading of Zr/Ti was shown to sharply decrease the BET surface area and pore volume. While the surface properties of lower loaded samples were shown to be comparable to the bare support, there is a sharp decrease in these properties with increased loading percentages (50 and 75). The higher loaded samples also revealed some macropores in the range of 100 nm or possibly above. Textural characteristics of the adsorbents obtained from N_2 physisorption isotherms are quantitatively represented in Table **1**.

All the loaded samples were shown to have lower BET surface area and pore volume compared to bare silica. This is expected, as the the doped oxides occupy the surface and the interparticular ordered mesopores. The bare silica showed two distinct adsorption steps[42]. The first, at lower P/P_0 indicated the filling of mesopores. This process seemed to be gradual, unlike the curve at higher P/P_0 which indicated the rapid filling of very large macro pores (possibly >100 nm), implying a larger concentration of mesopores, which can be confirmed in the pore size distribution.

Table 7. Textural Properties of the Adsorbents

TABLE Sample	**TABLE** S_{BET} (m^2/g)	**TABLE** V_{tot} (cm^3/g)	**TABLE** d_p (nm)
SiO_2	1483	1.35	3.2
Zr/Si 15	768	0.56	4.3
Zr/Si 25	635	0.4	6.0
Zr/Si 50	226	0.18	4.2
Zr/Si 75	3.5	0.01	26
Ti/Si 15	1115	0.64	2.3
Ti/Si 25	1030	0.60	2.8
Ti/Si 50	271	0.18	3.6
Ti/Si 75	73	0.08	4.3
MIL-101(Cr)	2200	1.16	2.0

The best surface characteristics from the mixed metal oxide samples was obtained from Ti/Si (15) at 1115 m^2/g with a pore volume of 0.64 cm^3/g, while the lowest was Zr/Si (75) at 3.5 m^2/g with a pore volume of 0.01 cm^3/g. Another observable trend is the comparatively better surface properties of Ti/Si (x) when compared to Zr/Si (x). The larger size of ZrO_2 explains this trend, with a greater portion of the surface being occupied when compared to its TiO_2 counterpart. For MIL-101(Cr) sample, a typical type I isotherm indicative of microporous nature of this MOF was observed (Figure **3f**) with uniform pores having sizes on the order of 2 nm, as can be observed in Figure **3g**. MIL-101(Cr) was shown to have superior surface characteristics when compared to the metal oxide samples, as expected, with a BET surface area of 2200 m^2/g and a pore volume of 1.16 cm^3/g.

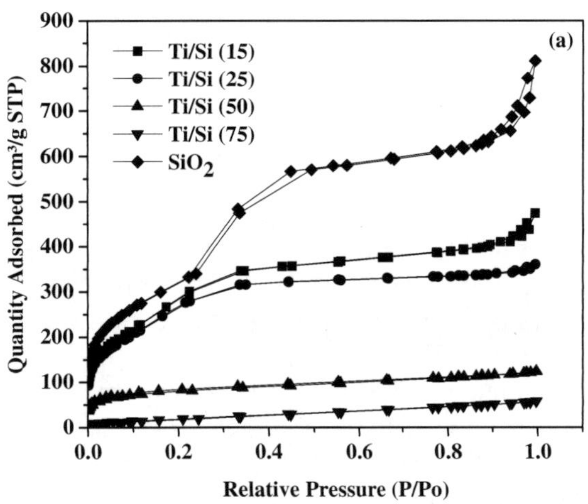

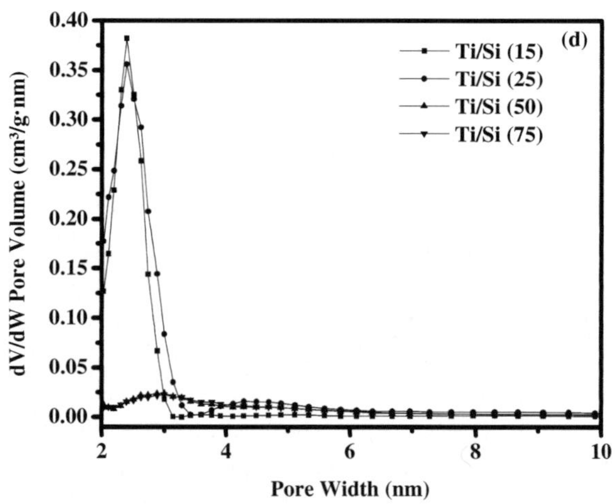

FIGURE 3(a,b). N_2 physisorption isotherms and pore size distributions for (a-b) Ti/Si materials

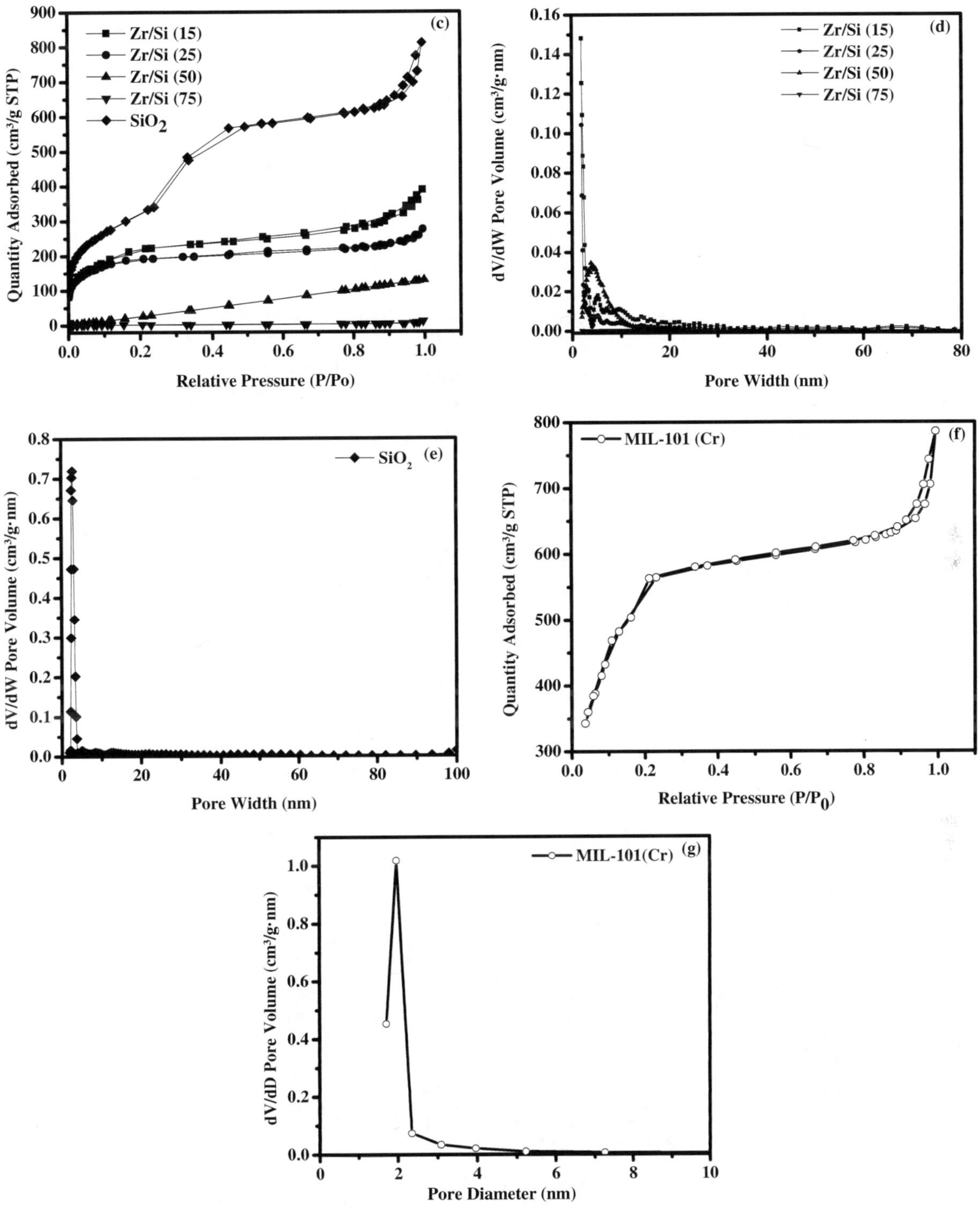

FIGURE 3. N_2 physisorption isotherms and pore size distributions for (c-d) Zr/Si materials, (e) Bare SiO_2 support and (f-g) MIL-101(Cr).

Figure 4 illustrates the SEM images of all mixed metal oxide samples. What can be observed is spherical nanoparticles similar to that of UVM-7 silica[42]. The MMOs appear to be fairly spherical in shape and seem to be aggregated together. They are also arranged in an inter-particular hexagonal array, which can be confirmed by surface characteristics earlier and is commonly detected in such materials[40].

The lower doped materials seemed to have a more pronounced spherical particle shape, while the higher doped materials had more irregular particle shapes. High level magnification of Zr/Si (75), revealed large cracks and voids through the material, as seen in Figure **4e**, which could explain the trends obtained in the pore size distribution, which showed large mesopores and some macropore volume. When comparing the higher doped Ti/Si and Zr/Si materials, it is evident that the adsorbent somewhat retains a fairly spherical shape, while the Zr/Si sample appears to be agglomerated. This can be a factor in the large difference in surface characteristics of higher doped Zr/Si materials when compared to their Ti/Si counterparts. In fact, almost all sorbents showed some small macropore volume between 60 – 100 nm. Considering the microporous MIL-101(Cr), Figures **4i, j** confirms the formation of well-intergrown crystals with octahedron shape and sizes in the range of 20 – 300 nm, as expected.

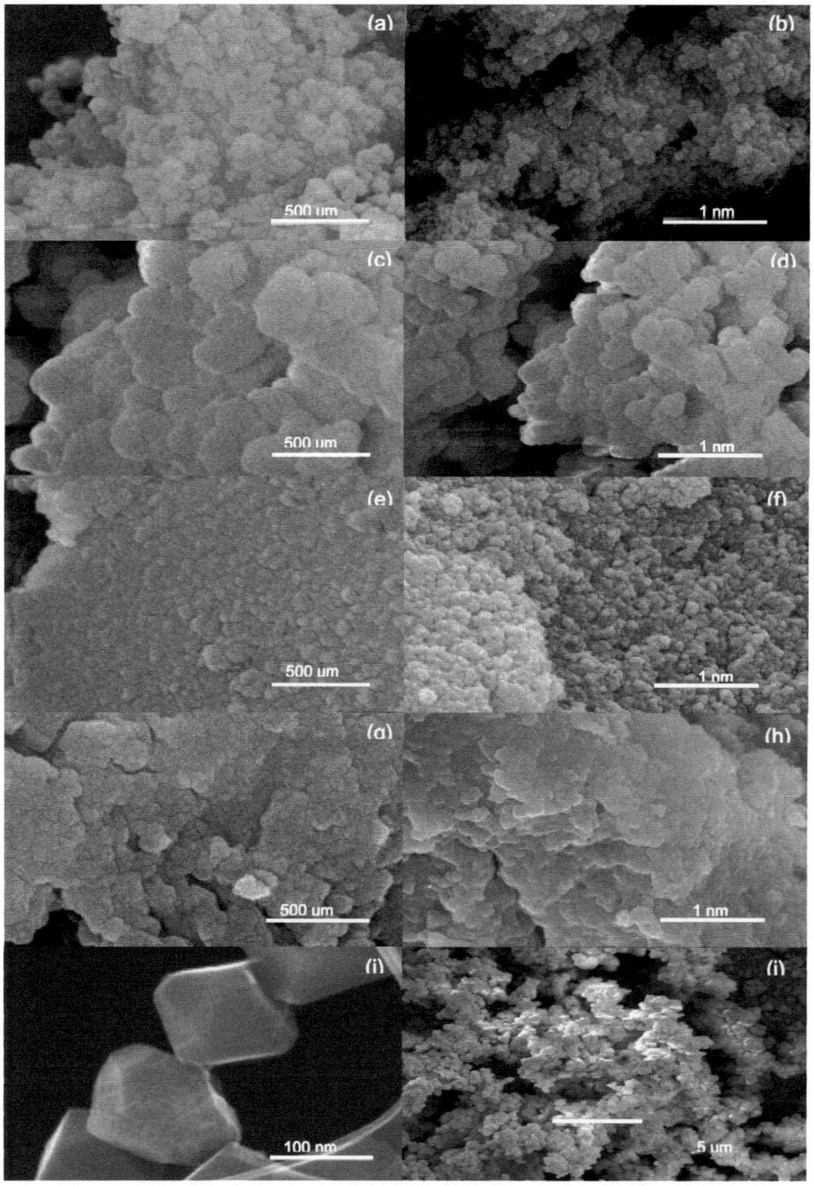

FIGURE 4. High- and low-magnification SEM images of (a,b) Ti/Si (15), (c,d) Ti/Si (75) and (e,f) Zr/Si (15), (g,h) Zr/Si (75), (i,j) MIL-101(Cr)

Figure 5 illustrates the XPS spectra obtained for the MMO samples. From Figure **5a**, two distinct Si peaks were detected at 103 and 154 eV which correspond to SiO_2 in the 2p and 2s atomic levels. This confirms the existence of Si-OH groups within the materials prepared. What is also noticeable is the decrease in intensity of these peaks as the metal oxide doping increased. The most prominent oxygen peak, which was detected in all samples at 530 eV (Figures **5a, c**) corresponded to oxygen bonded with silicon (Si-O). This peak also had the highest intensity, which can be seen in the results derived from XPS, listed in Table **2**, and is the atomic level of oxygen present in metal oxides in general.

Table 8. Metal Loading Calculated from XPS

TABLE Sample	TABLE Metal content (Atomic %)	TABLE Si content (Atomic %)	TABLE O content (Atomic %)	TABLE Metal loading (%)
Ti/Si (15)	2	24	36	8.3
Ti/Si (25)	5	34	56	14.7
Ti/Si (50)	13	24	45	54
Ti/Si (75)	19	23	49	82
Zr/Si (15)	3	23	51	13
Zr/Si (25)	5	22	49	22
Zr/Si (50)	7.3	17	41	43
Zr/Si (75)	10	16	40	62

In Figure **5a**, the Ti spectra obtained is displayed. Ti/Si (15) and (25) showed a lack of Ti 3s and 3p atomic levels. This could be due to the smaller loading of the TiO_2 in these samples. All samples however, showed a spin orbital splitting at Ti 2p and Ti $2p_{1/2}$ atomic levels. This corresponded to the atomic levels commonly detected in TiO_2 which confirms the formation of TiO_2 on the SiO_2 support. With the increase in TiO_2 percentage, the expected reduction in peak intensities were detected. In case of Zr/Si materials, in Figure **5b**, we can observe a larger number of atomic levels detected when compared to titania samples, possibly implying a larger degree of bonding between the two metal oxides when compared to Ti/Si materials. Zr 4s and 4p atomic levels weren't detected in Zr/Si (15) and (25) samples. Zr 3s, 3p and 3d levels were detected in all samples, with increasing intensities with increase in ZrO_2 content. Figure **5e**, depicts the results obtained for the control experiment. There were no peaks detected in the Ti and Zr range, but the Si and O peaks were shown to be similar to the MMOs with lowest loadings. Table **2** illustrates the atomic concentrations calculated from the XPS results. C 1s was also detected in the spectra and was detected due to the carbon tape used in the measurement.

Dynamic Adsorption Performance. Fixed-bed adsorption experiments were performed to determine dynamic adsorption capacity of the adsorbent materials. Formaldehyde vapor concentration in the inlet gas stream was calculated to be 166 ppm at room temperature and atmospheric pressure. Moreover, we used **eq. 1** to estimate the amount of formaldehyde adsorbed onto the adsorbents:

$$q_d = \frac{Q_F C_0 t_s}{W} \tag{1}$$

where Q_F is the inlet feed flow rate, C_0, formaldehyde feed concentration, W is the weight of the adsorbent, and t_s is the stoichiometric time, which can be estimated from the breakthrough profile using **eq. 2**:

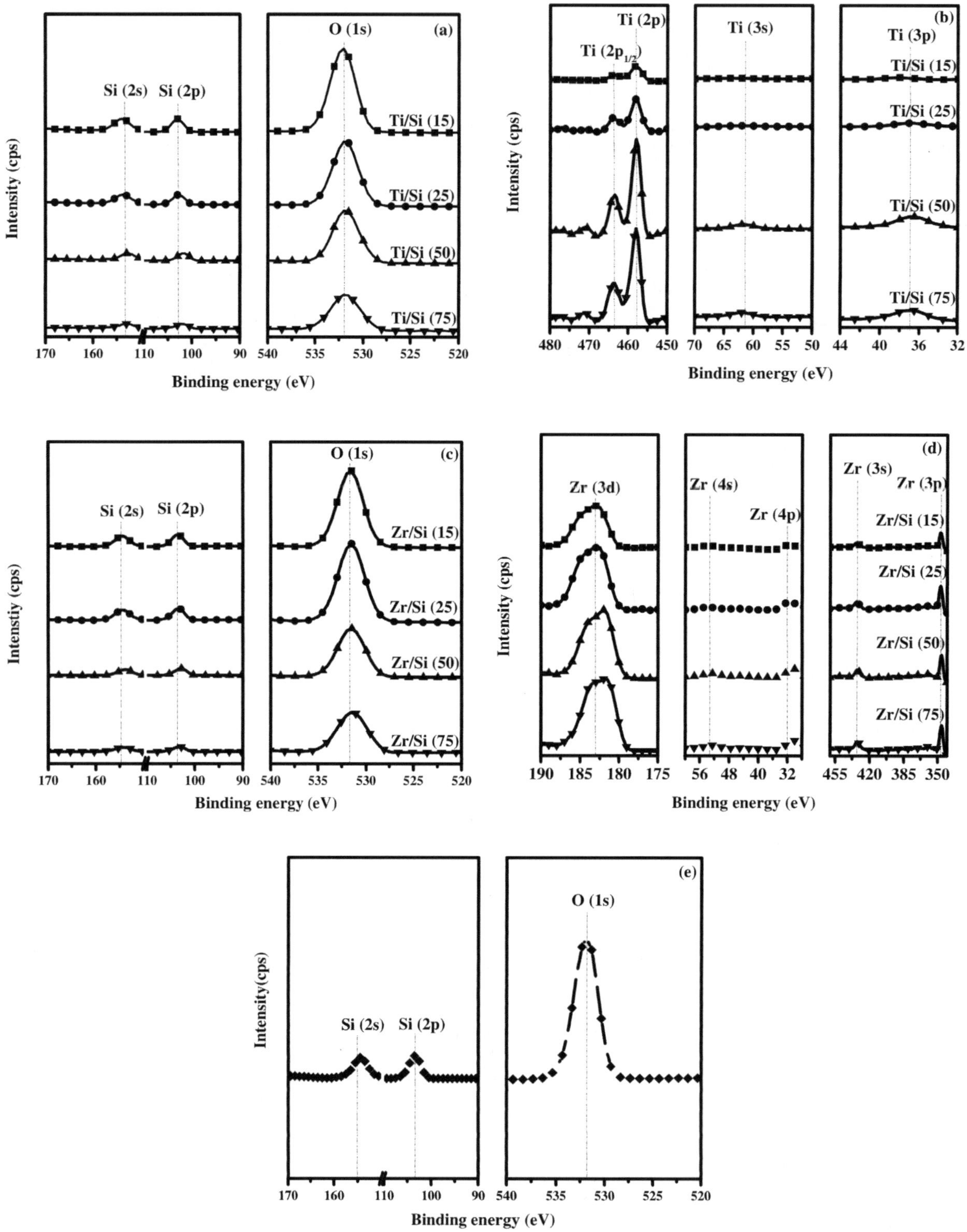

FIGURE 5. XPS spectra for all samples. (a) Si and O spectra for Ti/Si (b) Ti spectra for Ti/Si (c) Si and O spectra for Zr/Si (d) Zr spectra for Zr/Si materials (e) Si and O spectra for bare SiO_2

$$t_s = \int_0^\infty \left(1 - \frac{C_A}{C_0}\right)\partial t \qquad (2)$$

In the above equation, C_A is the formaldehyde concentration at the column outlet at time t.

Figure 5 illustrates the formaldehyde breakthrough profiles over the adsorbents investigated and **6** shows the corresponding dynamic capacities (q_d) estimated from **eq. 1**. Overall, the formaldehyde adsorption on MIL-101(Cr) was higher than on other MMOs, as characterized by a longer breakthrough time with an uptake capacity of 3.3 mmol g^{-1}. The capacity obtained for MIL-101(Cr) was comparable with the literature data reported before[43].

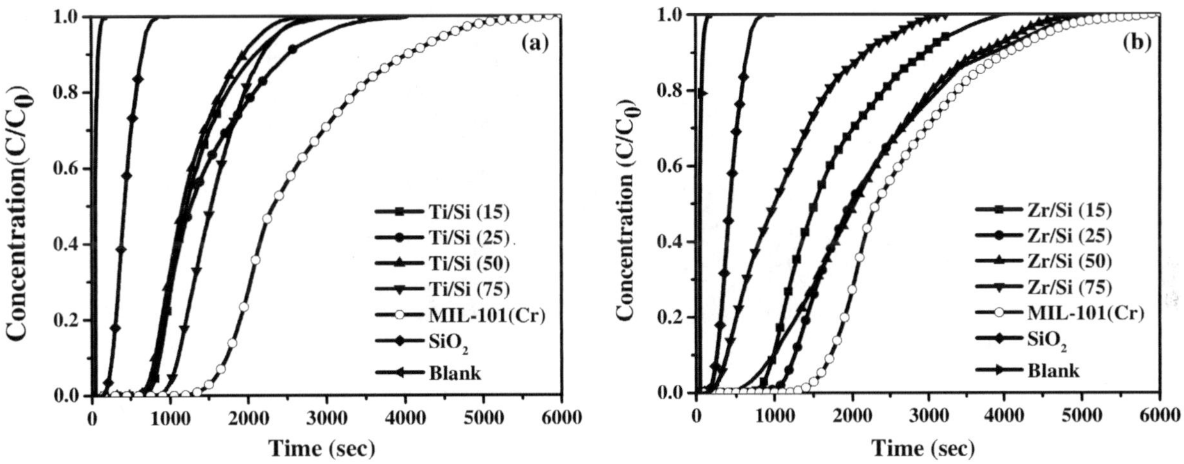

FIGURE 5. Formaldehyde breakthrough profiles over (a) Ti/Si and (b) Zr/Si adsorbents with comparison with Mil-101(Cr) profile.

For the titania class of materials, 25 wt. % titania supported by silica was observed to have the highest capacity at q_d = 1.7 mmol g^{-1}, while other samples were in a slightly lower range (c.a. 1.1 – 1.2 mmol g^{-1}). Hydroxyl groups have been identified to be the active site during the adsorption of formaldehyde on metal oxides[44]. Despite the surface properties of the 15 and 25 wt. % could be considered fairly similar, we see a large increase in capacity due to a larger number of hydroxyl groups present in the latter, which was shown by the XPS results earlier. It is also important to note that adsorption kinetics were shown to be slightly slower in the case of Ti/Si (25) when compared to other materials in its family, indicated by the sharpness of its curve. As the loading of titania increased, the adsorption kinetics also got faster, albeit with reduced semi-equilibrium capacities. This could be attributed to increased diffusion resistance by the successively lower pore volume as the loading of the samples was increased. A small increase in capacity is observed when the loading is increased from 50 to 75, due to the presence of macropores, as suggested by the surface characterization earlier.

In the case of zirconia samples, similar to the titania family, the 25 wt. % loading showed the highest semi-equilibrium adsorption capacity at q_d = 2.9 mmol g^{-1}, which was almost 30 % of an increase in formaldehyde uptake and was close to MIL-101(Cr). However, when increasing the loading from 25 to 50, there was only a slight drop in capacity (q_d = 2.9 – 2.7 mmol g^{-1}). From **5b**, it was noticed that the adsorption profiles of zirconia based materials was broad when compared to the tiania class of materials. This revealed that the process was highly diffusion controlled, and the broader profiles for zirconia based materials could be attributed to the larger size of ZrO_2 in general. Loading percentages above 25 % resulted in a downward trend in capacity. The consistent decrease in case of zirconia based materials further confirms the control the size of the molecule has in the diffusion kinetics of formaldehyde on MMOs, despite the presence of largely microporous cracks in case of the 75 wt. % loaded sample. The higher capacities obtained from the Zr/Si(x) samples compared to the Ti samples were explained by surface acidity. It has been proven that Zr/Si materials show a higher acidity than Ti/Si materials, which could be an influential factor in the uptake values recorded.[45] Furthermore, the level of hydroxyl bonding in case of zirconia materials has been shown to be much higher[46] than its titania counter-part.

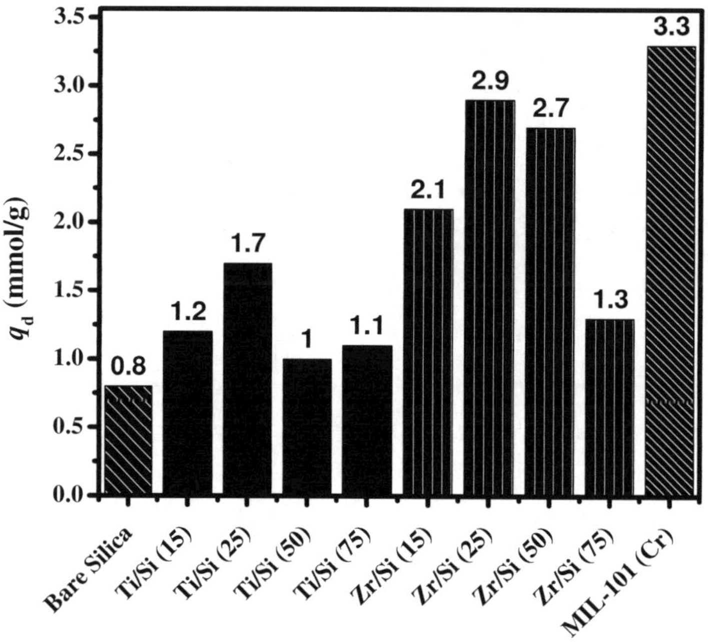

FIGURE 6. Dynamic adsorption capacity of formaldehyde estimated from breakthrough profiles obtained at room temperature and atmospheric pressure using 166 ppm_v formaldehyde vapor.

As a control experiment, bare silica (SiO_2) was also tested for HCHO adsorption, as evident from Figures **5 & 6**. Despite very high surface characteristics, it showed the lowest semi-equilibrium capacity at 0.8 mmol g^{-1} when compared to all the MMO samples. These results showed us that in general, Zr-Si materials were shown to have a higher affinity towards HCHO when compared to titania based materials for the airborne abatement of formaldehyde. Considering the low costs to prepare this material and its comparable performance to MIL-101, a MOF commonly used for VOC abatement which is expensive to synthesize, zirconia based materials could potentially be an optimum physisorption strategy to control and abate formaldehyde and other VOCs.

Figure 7 illustrates the FT-IR spectra obtained for the samples prepared. A band detected in all samples, between 998 and 1234 cm^{-1}, corresponded to the fundamental component of these materials. This range depicted the SiO_4 and Si-O-Si absorption bands, and was indicative of the role of silica in these materials as a support and as a building block for these MMO materials[47]. An absorption band was observed at around 920 cm^{-1} for both classes of MMO samples, which corresponded to Si-O-M bond[48], where M = Si/Ti. The broad peaks beyond 3000 cm^{-1} were assigned to O-H stretching vibrations. These bands depict the hydroxyl group present in the materials. From Figure **7a,** in case of the titania samples, multiple smaller peaks were also observed in the 500 – 700 cm^{-1} range, which suggests that titania was present in its anatase phase within the MMO system[49]. Peaks at and around 920 cm^{-1} were detected, which corresponded to the Ti-O-Si bond[50]. Higher loadings of titania displayed a shift in SiO_4 vibration modes, which suggests that the loading of titania effected the primary silica structure. Peaks around 1639 cm^{-1} referred to the bending modes of H-O-H, which was an indication of chemisorbed water on the surface of the material[51]. Peaks between 3235 – 3554 cm^{-1} represented the surface hydroxyl groups present in the samples. Ti/Si (25) showed the largest concentration of surface hydroxyl groups quantitatively, which falls in line with semi-equilibrium adsorption measurements. As the loading starts to increase, we see a drop in the number of hydroxyl groups.

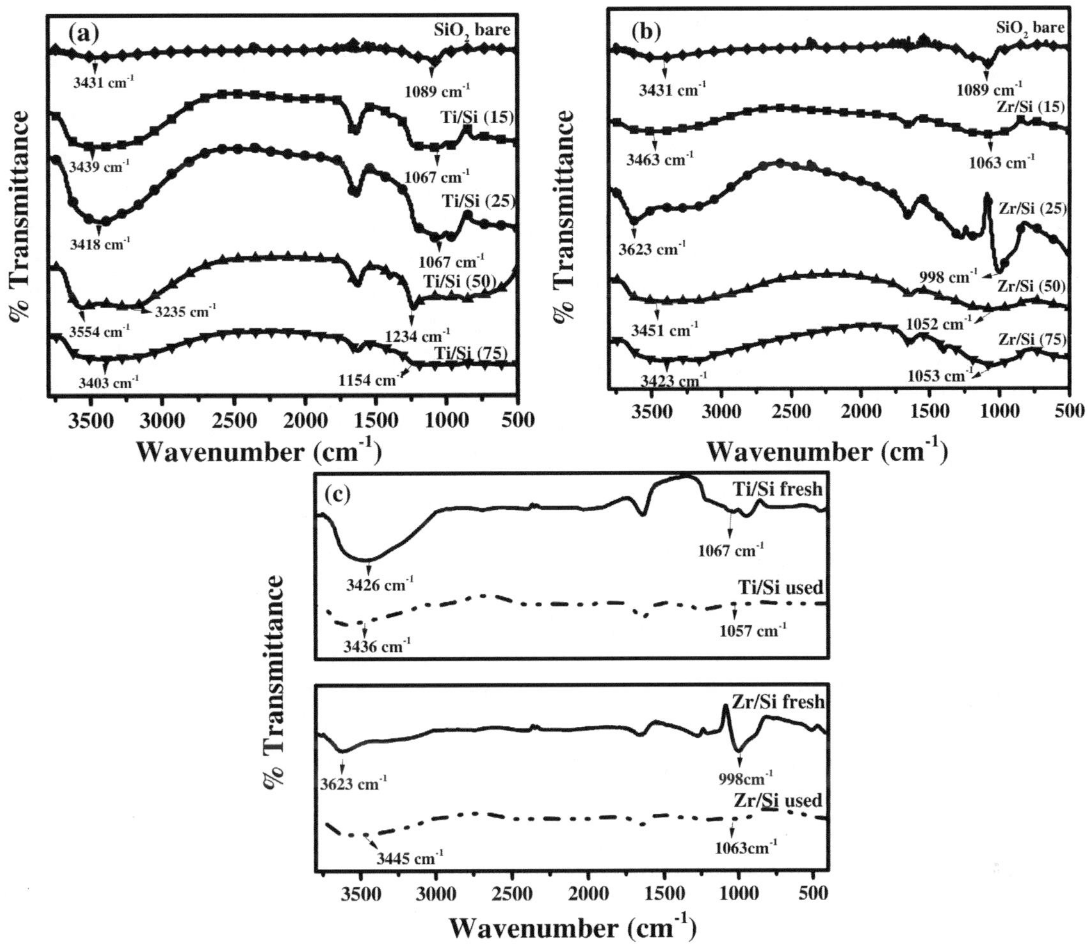

FIGURE 7. (a) FT-IR spectra obtained for Ti samples. (b) FT-IR spectra obtained for Zr samples. (c) FT-IR spectra obtained used and fresh adsorbent

Figure **7b** illustrates the spectra obtained for zirconia samples. Peaks between 998 - 1063 cm^{-1} were due to asymmetric stretching of the SiO_4 unit[48]. Peaks observed in the zirconia samples show a more similar absorption band when compared to bare SiO_2, which shows that the silica network was fairly maintained for the zirconia sample. This also shows the nature of the zirconia incorporation within the silica matrix, i.e. each Zr atom is coordinated with the SiO_4 molecule through an edge and corner connection[52]. While hydroxyl group peaks are observed for the titania samples as well, it is smaller quantitatively when compared to zirconia samples. This can also be the reason for zirconia's superior adsorption characteristics when compared to titania samples. The spectrum obtained around 1050 - 1063 cm^{-1} indicated the presence of SiO_4^- tetrahedra. In contrast to their titania counter-parts, zirconia samples showed either very minimal or no peaks at the 1640 cm^{-1} range, which suggests that the effect of humidity could potentially be insignificant. As observed with titania samples, the number of surface hydroxyl groups reaches a maximum for the 25/75 loading, which confirms the theory that the hydroxyl groups are the active sites for HCHO adsorption. **7c** compares the FT-IR spectra obtained for fresh and used MMO samples. The range of peaks below 1200 cm^{-1} seem to be less evident in the used adsorbent samples. This could suggest the deformation of the MMO matrix due to the adsorption of formate ions on the surface hydroxyl groups. Furthermore, the different types of charged ions being adsorbed onto the surface[44] could have caused structural changes. From 13 C NMR experiments performed, it is revealed that a host of ions are adsorbed onto the surface (from figure 8). This also confirms the fact that deformation can be caused due to the repulsions caused by the adsorbed ions. The adsorbed HCHO, therefore, effects the structure of the adsorbent. Beyond 3000 cm^{-1}, we see a shift and deformation in the hydroxyl group spectra, which depicts the reduction in the number of available surface hydroxyl groups after the adsorption process.

To determine if the formaldehyde adsorption on the zirconia and titania adsorbents was reversible, the NMR experiments on the fresh, used and regenerated adsorbents were performed and the corresponding spectra are displayed in Figures **8a, b**. On conclusion of adsorption experiments, the samples were consecutively subjected to a regeneration method, which involved heating the sample to 100 °C for approximately 5 hours to facilitate the removal of any adsorbed species on the adsorbent. As can be observed, the fresh materials showed no signs of carbon as observed by lack of peaks, whereas several peaks were observed on the spectra of both adsorbents after breakthrough tests, indicating the change in their chemical structure due to the formaldehyde adsorption. For Ti/Si sample, several peaks were detected after breakthrough experiments at 29 ppm, 49 ppm and 54 ppm, which corresponded to lower level hydrocarbons attached to the surface. A peak at 89 ppm signifies formaldehyde hydrate which could have formed due to the reaction with humidity, while the peak at 97 ppm denotes paraformaldehyde, which was formed due to the polymerization of surface formaldehyde groups.[53] After sample regeneration, the lack of prominent C-H bonds was observed with only a small peak at 49 ppm. A small peak was detected at 64 ppm which could signify a C=O. A significant peak was detected at 72 ppm which is attributed to surface adsorbed formaldehyde groups.[54] This can be explained by the severe reduction in peak intensity of paraformaldehyde, which breaks down to formaldehyde groups, observed at a small shift to 104 ppm. A larger portion of the paraformaldehyde was retained on the surface due to its higher stability when compared to formaldehyde hydrate. With a smaller peak intensity similar to the scenario with paraformaldehyde, at 86 ppm, formaldehyde hydrate is thought to have evaporated during the regeneration process. For Zr/Si sample after breakthrough test, a peak at 64 ppm was detected which was ascribed to C=O bond. Similar to the titania sample, the most significant peak related to formaldehyde was detected at 71 - 74 ppm. Upon regeneration, the peaks were retained, indicating a strong chemical interaction between HCHO and the adsorbent. These results highlight that formaldehyde chemically adsorbed on the MMOs, which is beneficial if the materials are used for permanent storage of airborne formaldehyde that is present in ultra-dilute concentration in indoor air. This strong chemisorption could potentially be utilized in the catalytic destruction of formaldehyde on the surface of these adsorbents.

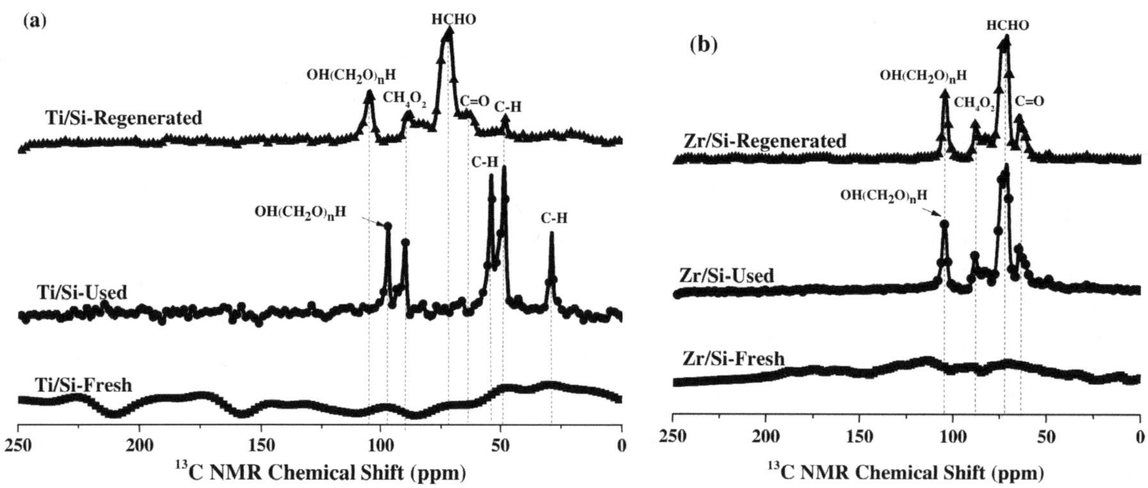

FIGURE 8. ^{13}C NMR spectra of the (a) Ti/Si (25) and (b) Zr/Si (50) for fresh, after breakthrough and regenerated materials.

CONCLUSIONS

This paper provides insight into the formaldehyde removal from indoor air using mixed-metal oxide adsorbents. A series of titania- and zirconia-doped silica materials were synthesized and their formaldehyde adsorption performance was investigated by dynamic breakthrough experiments. In addition, MIL-101(Cr) material was synthesized and used as a base adsorbent for comparing the adsorptive characteristics of MMOs. The obtained capacities were correlated with the titania/zirconia loading and it was demonstrated that zirconia samples are more efficient in abating airborne formaldehyde than their titania analogues. A maximum dynamic capacity of 2.9 mmol g^{-1} was observed for Zr/Si (25) which was comparable to that of MIL-101(Cr) with a capacity of 3.3 mmol g^{-1} when exposed to 166 ppm_v formaldehyde vapor. Furthermore, the characterization of the materials after experiments highlighted chemisorption of formaldehyde on the surface of MMOs. The findings reported in this investigation reveal the potential of zirconia-based mixed-metal oxides as efficient solid adsorbents in the abatement of formaldehyde.

ACKNOWLEDGEMENTS

The funding from *Innovation at Missouri S&T* is acknowledged. The authors also thank Materials Research Center (MRC) of Missouri S&T for SEM and XRD measurements.

REFERENCES

[1] US EPA, Interim acute exposure guideline levels (AEGLs) for formaldehyde (CAS Reg. No. 50-00-0), (2008).

[2] C.S. Mitchell, J. Zhang, T. Sigsgaard, M. Jantunen, P.J. Lioy, R. Samson, M.H. Karol, Current state of the science: Health effects and indoor environmental quality, Environ. Health Perspect. 115 (2007) 958–964. doi:10.1289/ehp.8987.

[3] R. Atkinson, Atmospheric chemistry of VOCs and NO(x), Atmos. Environ. 34 (2000) 2063–2101. doi:10.1016/S1352-2310(99)00460-4.

[4] D. U.S. Environmental Protection Agency, Risk Assessment Forum, Washington, U.S. EPA. A review of the reference dose and reference concentration processes, 2002.

[5] M.D. Barnes, D., Reference Dose (RfD): Description and Use in Health Risk Assessments, 1989.

[6] G.D. Nielsen, P. Wolkoff, Cancer effects of formaldehyde: A proposal for an indoor air guideline value, Arch. Toxicol. 84 (2010) 423–446. doi:10.1007/s00204-010-0549-1.

[7] An Update on Formaldehyde, U.S. Consum. Prod. Saf. Comm. (2016) 4. doi:10.1177/016235329101400308.

[8] E.M. Carter, L.E. Katz, G.E. Speitel, D. Ramirez, Gas-phase formaldehyde adsorption isotherm studies on activated carbon: Correlations of adsorption capacity to surface functional group density, Environ. Sci. Technol. 45 (2011) 6498–6503. doi:10.1021/es104286d.

[9] J.Q. Torres, S. Royer, J.P. Bellat, J.M. Giraudon, J.F. Lamonier, Formaldehyde: Catalytic oxidation as a promising soft way of elimination, ChemSusChem. 6 (2013) 578–592. doi:10.1002/cssc.201200809.

[10] N.A. Khan, Z. Hasan, S.H. Jhung, Adsorptive removal of hazardous materials using metal-organic frameworks (MOFs): a review., J. Hazard. Mater. 244–245 (2013) 444–56. doi:10.1016/j.jhazmat.2012.11.011.

[11] P.E. Rajan, A. Krishnamurthy, G. Morrison, F. Rezaei, Advanced buffer materials for indoor air CO2 control in commercial buildings, Indoor Air. 27 (2017) 1213–1223. doi:10.1111/ina.12386.

[12] P.M. Green, G. D.; Swedo, R. J.; Tomlinson, I. A.; Whetten, A. R.; Coburn, C. E.; Henning, M. A.; Novy, Methods for reducing airborne formaldehyde, 8236263 B2, 2012.

[13] M.A. Sidheswaran, H. Destaillats, D.P. Sullivan, S. Cohn, W.J. Fisk, Energy efficient indoor VOC air cleaning with activated carbon fiber (ACF) filters, Build. Environ. 47 (2012) 357–367. doi:10.1016/j.buildenv.2011.07.002.

[14] V. Boonamnuayvitaya, S. Sae-Ung, W. Tanthapanichakoon, Preparation of activated carbons from coffee residue for the adsorption of formaldehyde, Sep. Purif. Technol. 42 (2005) 159–168.

doi:10.1016/j.seppur.2004.07.007.
[15] H. Rong, Z. Ryu, J. Zheng, Y. Zhang, E ffect of air oxidation of Rayon-based activated carbon fibers on the adsorption behavior for formaldehyde, 40 (2002) 2291–2300.
[16] L.I. Jing, L.I. Zhong, L.I.U. Bing, X.I.A. Qibin, X.I. Hongxia, Effect of Relative Humidity on Adsorption of Formaldehyde on Modified Activated Carbons, Chinese J. Chem. Eng. 16 (2008) 871–875. doi:10.1016/S1004-9541(09)60008-2.
[17] D. Das, V. Gaur, N. Verma, Removal of volatile organic compound by activated carbon fiber, Carbon N. Y. 42 (2004) 2949–2962. doi:10.1016/j.carbon.2004.07.008.
[18] C. Ma, X. Li, T. Zhu, Removal of low-concentration formaldehyde in air by adsorption on activated carbon modified by hexamethylene diamine, Carbon N. Y. 49 (2011) 2873–2875. doi:10.1016/j.carbon.2011.02.058.
[19] Y. Le, D. Guo, B. Cheng, J. Yu, Bio-template-assisted synthesis of hierarchically hollow SiO_2 microtubes and their enhanced formaldehyde adsorption performance, Appl. Surf. Sci. 274 (2013) 110–116. doi:10.1016/j.apsusc.2013.02.123.
[20] S. Srisuda, B. Virote, Adsorption of formaldehyde vapor by amine-functionalized mesoporous silica materials, J. Environ. Sci. 20 (2008) 379–384. doi:10.1016/S1001-0742(08)60059-5.
[21] A.M. Ewlad-Ahmed, M.A. Morris, S. V. Patwardhan, L.T. Gibson, Removal of formaldehyde from air using functionalized silica supports, Environ. Sci. Technol. 46 (2012) 13354–13360. doi:10.1021/es303886q.
[22] Z. Wang, W. Wang, D. Jiang, L. Zhang, Y. Zheng, Diamine-appended metal–organic frameworks: enhanced formaldehyde-vapor adsorption capacity, superior recyclability and water resistibility, Dalt. Trans. 45 (2016) 11306–11311. doi:10.1039/C6DT01696K.
[23] D. Chen, Z. Qu, Y. Sun, Y. Wang, Adsorption-desorption behavior of gaseous formaldehyde on different porous Al_2O_3 materials, Colloids Surfaces A Physicochem. Eng. Asp. 441 (2014) 433–440. doi:10.1016/j.colsurfa.2013.10.006.
[24] Z. Xu, J. Yu, G. Liu, B. Cheng, P. Zhou, X. Li, Microemulsion-assisted synthesis of hierarchical porous $Ni(OH)_2/SiO_2$ composites toward efficient removal of formaldehyde in air, Dalt. Trans. 42 (2013) 10190–10197. doi:10.1039/c3dt51067k.
[25] and I.S. Eriksson, B., L. Johanssin, The Nordest Symposium on Air Pollution Abatement by Filtration and Respiratory Protection, in: 1980.
[26] Y. Sekine, A. Nishimura, Removal of formaldehyde from indoor air by passive type air-cleaning materials, Atmos. Environ. 35 (2001) 2001–2007. doi:10.1016/S1352-2310(00)00465-9.
[27] A. Nomura, C.W. Jones, Amine-Functionalized Porous Silicas as Adsorbents for Aldehyde Abatement, (2013). doi:10.1021/am400810s.
[28] A. Nomura, C. W.Jones, Enhanced Formaldehyde-Vapor Adsorption Capacity of Polymeric Amine-Incorporated Aminosilicas, (n.d.).
[29] A. Nomura, C.W. Jones, Airborne aldehyde abatement by latex coatings containing amine-functionalized porous silicas, Ind. Eng. Chem. Res. 54 (2015) 263–271. doi:10.1021/ie504165d.
[30] Y. Yu, X. Zhang, Q. Liu, P. Wang, Y. Dong, Cu (I)-MOF: naked-eye colorimetric sensor for humidity and formaldehyde in single-crystal-to-single-crystal fashion., Chem. Commun. 50 (2014) 1444–1446. doi:10.1039/c3cc47723a.
[31] L. Zou, Y. Luo, M. Hooper, E. Hu, Removal of VOCs by photocatalysis process using adsorption enhanced TiO_2-SiO_2 catalyst, Chem. Eng. Process. Process Intensif. 45 (2006) 959–964. doi:10.1016/j.cep.2006.01.014.
[32] H.O. Seo, D.H. Kim, K.-D. Kim, E.J. Park, C.W. Sim, Y.D. Kim, Adsorption and desorption of toluene on nanoporous TiO_2/SiO_2 prepared by atomic layer deposition (ALD): influence of TiO_2 thin film thickness and humidity, Adsorption. 19 (2013) 1181–1187. doi:10.1007/s10450-013-9550-3.
[33] S. Morales-Torres, F. Carrasco-Marín, A. Pérez-Cadenas, F. Maldonado-Hódar, Coupling Noble Metals and Carbon Supports in the Development of Combustion Catalysts for the Abatement of BTX Compounds in Air Streams, Catalysts. 5 (2015) 774–799. doi:10.3390/catal5020774.
[34] K. Kosuge, S. Kubo, N. Kikukawa, M. Takemori, Effect of pore structure in mesoporous silicas on

VOC dynamic adsorption/desorption performance, Langmuir. 23 (2007) 3095–3102. doi:10.1021/la062616t.

[35] H.L. Tidahy, S. Siffert, J.F. Lamonier, E. a. Zhilinskaya, A. Aboukaïs, Z.Y. Yuan, A. Vantomme, B.L. Su, X. Canet, G. De Weireld, M. Frère, T.B. N'Guyen, J.M. Giraudon, G. Leclercq, New Pd/hierarchical macro-mesoporous ZrO_2, TiO_2 and ZrO_2-TiO_2 catalysts for VOCs total oxidation, Appl. Catal. A Gen. 310 (2006) 61–69. doi:10.1016/j.apcata.2006.05.020.

[36] S. Carlos-Cuellar, P. Li, A.P. Christensen, B.J. Krueger, C. Burrichter, V.H. Grassian, Heterogeneous uptake kinetics of volatile organic compounds on oxide surfaces using a Knudsen cell reactor: Adsorption of acetic acid, formaldehyde, and methanol on α-Fe2O3, α-Al2O3, and SiO2, J. Phys. Chem. A. 107 (2003) 4250–4261. doi:10.1021/jp0267609.

[37] F. Chen, S. Liu, J. Yu, Efficient removal of gaseous formaldehyde in air using hierarchical titanate nanospheres with in situ amine functionalization., Phys. Chem. Chem. Phys. 18 (2016) 18161–18168. doi:10.1039/c6cp03037h.

[38] T. Zhao, F. Jeremias, I. Boldog, B. Nguyen, S.K. Henninger, C. Janiak, High-yield, fluoride-free and large-scale synthesis of MIL-101(Cr)., Dalton Trans. 44 (2015) 16791–801. doi:10.1039/c5dt02625c.

[39] R. T.Yang, Adsorbents: Fundamentals and Applications, 2003.

[40] V.B. Fenelonov, V.N. Romannikov, A.Y. Derevyankin, Mesopore size and surface area calculations for hexagonal mesophases (types MCM-41, FSM-16, etc.) using low-angle XRD and adsorption data, Microporous Mesoporous Mater. 28 (1999) 57–72. doi:10.1016/S1387-1811(98)00280-7.

[41] D. Ortiz De Zarate, A. Gomez-Moratalla, C. Guillem, A. Beltran, J. Latorre, D. Beltran, P. Amoros, High-zirconium-content nano-sized bimodal mesoporous silicas, Eur. J. Inorg. Chem. (2006) 2572–2581. doi:10.1002/ejic.200501140.

[42] M. Pérez-Cabero, A.B. Hungría, J.M. Morales, M. Tortajada, D. Ramón, A. Moragues, J. El Haskouri, A. Beltrán, D. Beltrán, P. Amorós, Interconnected mesopores and high accessibility in UVM-7-like silicas, J. Nanoparticle Res. 14 (2012). doi:10.1007/s11051-012-1045-8.

[43] Z. Wang, W. Wang, D. Jiang, L. Zhang, Y. Zheng, Diamine-Appended Metal-Organic Framework: Enhanced Formaldehyde-Vapor Adsorption Capacity, Superior Recyclability and Water Resistibility, Dalt. Trans. (2016) 11306–11311. doi:10.1039/C6DT01696K.

[44] G. Busca, J. Lamotte, J. ciaude Lavalley, V. Lorenzelli, FT-IR Study of the Adsorption and Transformation of Formaldehyde on Oxide Surfaces, J. Am. Chem. Soc. 109 (1987) 5197–5202. doi:10.1021/ja00251a025.

[45] K. Shibata, T. Kiyoura, J. Kitagawa, T. Sumiyoshi, K. Tanabe, Acidic properties of Binary Metal Oxides, Bull. Chem. Soc. Jpn. 46 (1973) 2985–2988. doi:10.1246/bcsj.46.2985.

[46] F.U. Xianzhi, L.A. Clark, Q. Yang, M.A. Anderson, Enhanced photocatalytic performance of titania-based binary metal oxides: TiO2/SiO2 and TiO2/ZrO2, Environ. Sci. Technol. 30 (1996) 647–653. doi:10.1021/es950391v.

[47] a. Zecchina, S. Bordiga, C. Lamberti, G. Ricchiardi, C. Lamberti, G. Ricchiardi, D. Scarano, G. Petrini, G. Leofanti, M. Mantegazza, Structural characterization of Ti centres in Ti-silicalite and reaction mechanisms in cyclohexanone ammoximation, Catal. Today. 32 (1996) 97–106. doi:10.1016/S0920-5861(96)00075-2.

[48] D.A. Neumayer, E. Cartier, Materials characterization of ZrO2–SiO2 and HfO2–SiO2 binary oxides deposited by chemical solution deposition, J. Appl. Phys. 90 (2001) 1801–1808. doi:10.1063/1.1382851.

[49] A. Adamczyk, E. Długoń, The FTIR studies of gels and thin films of Al 2O 3-TiO 2 and Al 2O 3-TiO 2-SiO 2 systems, Spectrochim. Acta - Part A Mol. Biomol. Spectrosc. 89 (2012) 11–17. doi:10.1016/j.saa.2011.12.018.

[50] V.A. Zeitler, C.A. Brown, The infrared spectra of some Ti-O-Si, Ti-O-Ti and Si-O-Si compounds, J. Phys. Chem. 61 (1957) 1174–1177. doi:10.1021/j150555a010.

[51] J. Yu, X. Zhao, J. Yu, G. Zhong, J. Han, Q. Zhao, The Grain Size and Surface Hydroxyl Content of Super-Hydrophilic TiO2/SiO2 Composite Nanomeer Thin Films, J. Mater. Sci. Lett. 20 (2001) 1745–1748. doi:10.1023/A.

[52] G. Lucovsky, G.B. Rayner, Microscopic model for enhanced dielectric constants in low concentration SiO2-rich noncrystalline Zr and Hf silicate alloys, Appl. Phys. Lett. 77 (2000) 2912–2914. doi:10.1063/1.1320860.

[53] T. Okachi, M. Onaka, Formaldehyde Encapsulated in Zeolite : A Long-Lived , Highly Activated One-Carbon Electrophile to Carbonyl-Ene Reactions, J. Am. Chem. Soc. 126 (2004) 2306–2307. doi:10.1021/ja039737p.

[54] H. Nakayama, A. Hayashi, T. Eguchi, N. Nakamura, M. Tsuhako, Adsorption of formaldehyde by polyamine-intercalated α-zirconium phosphate, Solid State Sci. 4 (2002) 1067–1070. doi:10.1016/S1293-2558(02)01367-5.

The Dechlorination Effect of Chlorobenzene on Vanadium-based Catalysts for Low-temperature NH_3-SCR

Dong Wang, Yue Peng , Jun-hua Li and Ji-ming Hao
(Tsinghua University, Beijing, 100084, China)

Both the low temperature selective catalytic reduction (SCR) and the catalytic oxidation of chlorinated volatile organic compounds (CVOCs) require catalysts with strong redox property. The possibility of simultaneous removal of nitrogen oxides and CVOCs using the SCR convertor is a scientific topic. Compared with C-H bonds, the C-Cl bonds in CVOCs are more facile to be attacked by nucleophiles. Thus, the dechlorination of CVOCs may occur during chlorobenzene catalytic oxidation. Dissociative adsorption of Cl on the active sites would happen via Cl abstraction in the chlorobenzene molecular combustion. And the absorbed Cl cannot be removed until the dynamic equilibrium of adsorption and desorption rates was achieved or the combination of the Deacon Reaction and active oxygen species at higher temperature (> 350 °C). When SCR reaction is carried out, due to the strong electronegativity of Cl, the adsorption of ammonia would be inevitably affect at low temperature. The dechlorination effect of chlorobenzene on SCR reaction at low temperature were studied using DRIFTS, ion chromatography method and etc.

Chlorobenzene (CB) inhibits SCR activity of V-based catalysts at low temperature, though the adsorption amount is small. Ammonia adsorption increases due to the break in the C-Cl bond during CB oxidation. The dissociated Cl ions provide extra Brønsted acid sites, leading to the formation of inactive ammonium chloride. Ce and Mn improve dechlorination, forming more ammonium chloride on the catalyst surface.

Pb and SO_2 Poisoning Effect on CeO_2-WO_3/TiO_2-SiO_2 Catalyst for SCR

Yue Peng, Dong Wang, Jun-hua Li and Ji-ming Hao
(Tsinghua University, Beijing, 100084, China)
John Crittenden
(Georgia Institute of Technology, Atlanta, 30332-0595, United States)

In the flue gas from stationary sources, SO_2 and Pb are the most common poisons and, in most cases, exist simultaneously. SO_2 is generated by the oxidation of sulfur contained in fossil fuels. The SO_2 concentrations in flue gas can be less than 100 ppm after desulfurization to 5000 ppm for high-sulfur coal sources in Southwest China. Pb is generated from the combustion of municipal solid waste and the flue gas from smelting operations. The Pb concentration can exceed 3000 ppm in an SCR catalytic convertor that is used in a waste incinerator after approximately 2000 h of operation.20 Our recent works examined the Pb poisoning mechanisms of CeO_2-WO_3 catalysts via in situ infrared (IR) and Raman spectra. We found that Pb preferentially bonded with WO_3 (acid sites) over CeO_2 (redox sites), whereas SO_2 preferentially bonded with CeO_2 to form $Ce(SO_4)_2$ and did not form a bond with WO_3. In other words, SO_2 and Pb poison the redox sites and acid sites of CeO_2-WO_3 catalysts, respectively.

A CeO_2-WO_3/TiO_2-SiO_2 catalyst was employed to investigate the poisoning mechanisms of Pb and SO_2 during selective catalytic reduction (SCR). The introduction of Pb and SO_2 suppressed the catalytic performance by decreasing the numbers of surface acid and redox sites. Specifically, Pb preferentially bonded with amorphous WO_3 species rather than with CeO_2, decreasing the numbers of both Lewis and Brønsted acid sites but exerting less influence on the reducibility. SO_2 preferentially bonded with CeO_2 as sulfate species rather than with WO_3, leading to a significant decrease in reducibility and the loss of surface active oxygen groups. Although SO_2 provided additional Brønsted acid sites via the interaction of SO_4^{2-} and CeO_2, it had little positive effect on catalytic activity. A synergistic deactivation effect of Pb and SO_4^{2-} on CeO_2 was found. Pb covered portions of the weakly bonded catalyst sites poisoned by SO_4^{2-}, which increased the decomposition temperature of the sulfate species on the catalyst.

Characterizatıon of Volatile Organic Compound Emissions from a Wastewater Treatment Plant

Kadir Ulutaş and ***Hakan Pekey*** (Kocaeli University, Kocaeli, Turkey)
Selami Demir (Yıldız Technical University, Istanbul, Turkey)
Faruk Dinçer (TUBITAK Marmara Research Center, Kocaeli, Turkey)

ABSTRACT: Odor problem is one of the most common problems in wastewater treatment processes. Thus, determination of volatile organic compounds (VOCs) has vital importance to implement measures for proper abatement. In this study, VOC samples were collected in a biological wastewater treatment plant with 250,000 m^3/d capacity and analyzed for VOC species in a GC-MS. Samples were collected within the treatment plant that represents winter (2015), spring and summer (2016) concentrations in nine units. A total of 41 VOC species were identified of monoaromatics, alkanes, halogenated VOCs, ketones, alcohols, acetates, and sulfur containing VOCs. For all samples, average contributions of VOC groups to total concentrations were determined as 35.9% from alcohols, 35.5% from sulfur containing VOCs, 11.2% from alkanes, 10.3% from monoaromatics, 5.5% from halogenated VOCs, 1.3% from ketones, and 0.3% from acetates.

Keywords: Biological wastewater treatment, odor emissions, treatment operations, volatile organic compounds

INTRODUCTION

A wastewater treatment plant is a complex system of interconnected units that are responsible for achieving various steps of treatment (Lewkowska et al, 2016), and is used for reducing pollutant loads by carefully engineered physical, chemical, and biological processes prior to discharge of wastewaters to the receiving water bodies (Lebrero et al, 2011). The sole purpose of a wastewater treatment plant is to protect receiving water bodies, public health and neighbors' quality of life (Thomas, 2007).

Odor emissions are inevitable during the transportation as well as biochemical processes under aerobic and anaerobic treatment (Shao et al, 2014), and characterization of volatile organic compounds could help implement better abatement strategies. Odorous emissions from wastewater treatment plants primarily contain hydrogen sulfide and ammonia as inorganic forms as well as alcohols, volatile fatty acids, aldehydes, ketones, thioethers, and mercaptans as organic forms (Ölmez, 2008). For instance, sludge from primary sedimentation, sludge thickening and other operations such as anaerobic digestion contains high concentrations of (especially sulfur containing) volatile organic compounds (Liu et al, 2012). In addition, various odorous species can be detected in emissions from wastewater treatment plants that industrial wastewaters are handled (DEFRA, 2006)

A person inhales an average of 20 cubic meters of air daily. Thus, the quality of the atmospheric air the person lives in is of vital importance (Marotta et al, 2015). In this aspect, identification and quantification of odorous VOCs in the vicinity of wastewater treatment plants are a priori topic of air quality investigations because of the fact that trace concentrations of these odorous compounds can deteriorate environmental conditions (Giungato et al, 2016) as well as quality of life, and can result in several health effects (Dinçer and Müezzinoğlu, 2007). Environmental pollution from odorous VOCs has adverse effects on humans, economy, social life, and the whole ecological system (Zarra et al, 2011; Giungato et al, 2016).

The aim of this study is to collect ambient air samples within a wastewater treatment plant that has the highest capacity in the region and that was at the center of odor complaints. Samples were collected from nine units within the plant and analyzed for volatile organic compounds. This paper presents a thorough discussion of VOC concentrations within the plant.

MATERIALS AND METHODS

Wastewater Treatment Plant and Sampling. The wastewater treatment plant chosen for this study is one of the largest wastewater treatment plants with 250,000 m^3/d capacity. The system employs biological processes for carbon, nitrogen, and phosphorous removal, and uses a co-generation system for electricity generation within the plant. Waste heat is used for the purpose of drying the sludge without any other inputs of energy to produce a useful product as a supplementary fuel for electricity generation. Besides, biogas is produced in anaerobic digestion units. The odorous components in emissions are removed in an ozone-enhanced wet scrubber. Flue gas streams are collected from the units shown in Fig. 2 by a draft fan with 140,000 m^3/d capacity, and forced into the ozone-enhanced scrubber. The wastewater treatment plant of concern includes several steps of process tanks to accomplish its ultimate goal of wastewater purification prior to discharge. Thus, samples that represent three seasons (2015 winter, 2016 spring, and 2016 summer) were collected at nine points within the plant: (1) coarse screens and pumping station, (2) fine screens, (3) grit chambers, (4) Primary settlers, (5) Aerobic process tanks, (6) Secondary settlers, (7) sludge thickeners, (8) sludge drying, and (9) ozonation-wet scrubber. The sampling points are shown in Fig. 1.

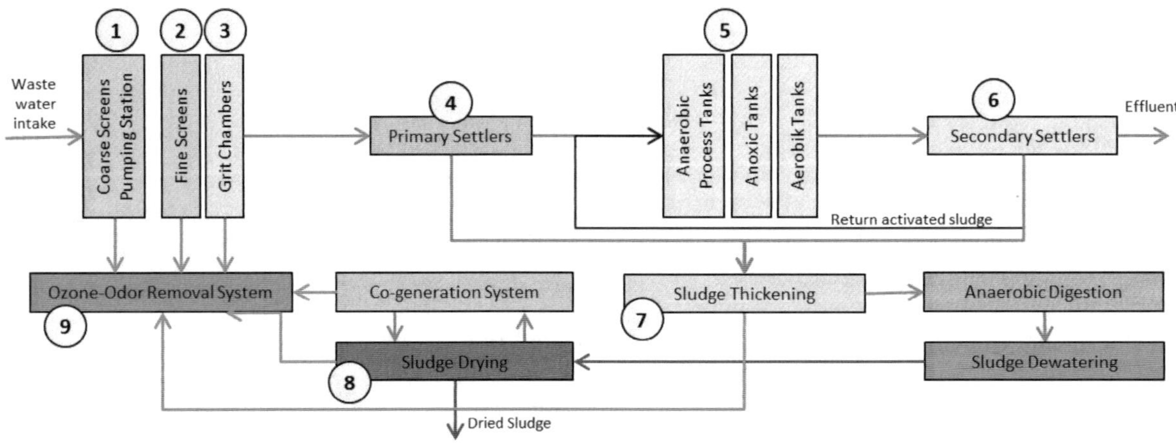

FIGURE 1. Sampling Points within the Wastewater Treatment Plant

Analytical Methods. Specially designed techniques allows identification of one or two specific VOC species whereas there are generalized techniques of VOC analyses that allow a wide matrix of VOCs (Lorngage et al, 2004). Aldehydes, amins, organic acids, and sulfur-containing VOCs are the most commonly identified pollutants in odor samples (Giungato et al, 2016; Kim et al, 2014; Agus et al, 2012; Krach et al 2008). TS EN 13649:2003 and ASTM D5504-98 methods were used in this study. Agilent 6890N GC-MS system was used for analyses. Method detection limits (MDLs) are shown in Table 1. Carbon disulfide (CS_2) was used for desorbing the sample. A known volume of the CS_2 extract was injected on activated carbon and analyzed in GC-MS. Calibration was performed with standard solutions. A sample chromatogram is shown in Fig. 2. Calibration curves for o-xylene and toluene are shown in Fig. 3.

RESULTS AND DISCUSSION

Concentrations of VOCs. Samples were collected at nine different points within a biological wastewater treatment plant and analyzed for 41 VOC species. Monoaromatics, alkanes, halogenated VOCs, ketones, alcohols, acetates, and sulfur-containing compounds were detected in the samples. Measurement results are shown in Table 2.

TABLE 1. VOC Detection Limits

Compound	Detection limit (μg)	Compound	Detection limit (μg)	Compound	Detection limit (μg)
1,2,3-trichlorobenzene	0.66	Ethylbenzene	0.71	n-octane	0.73
1,2,4-trichlorobenzene	0.81	Ethylacetate	0.81	n-propane	1.35
1,2,4-trimethylbenzene	1.16	Isopropylalchol	0.66	n-tetradecane	2.3
1,2-dichlorobenzene	0.89	m,p-xylene	0.76	n-tridecane	1.15
1,3,5-trimethylbenzene	0.78	Methanol	0.89	n-undecane	1.62
1,4-dichlorobenzene	0.91	n-buthylbenzene	0.97	o-xylene	0.67
4-chlorotoluene	0.87	n-butane	1.12	n-propylbenzene	0.73
Methylisobutylketone	0.5	n-decane	0.58	Styrene	0.96
Acetone	0.43	n-dodecane	2.5	tert-buthylbenzene	0.55
Benzene	0.5	n-heptane	0.57	tetrachloroethene	0.91
Butylacetate	0.65	n-hexane	0.73	toluene	0.98
Ethanol	0.73	n-nonane	0.48	Trichloroethylene	0.97

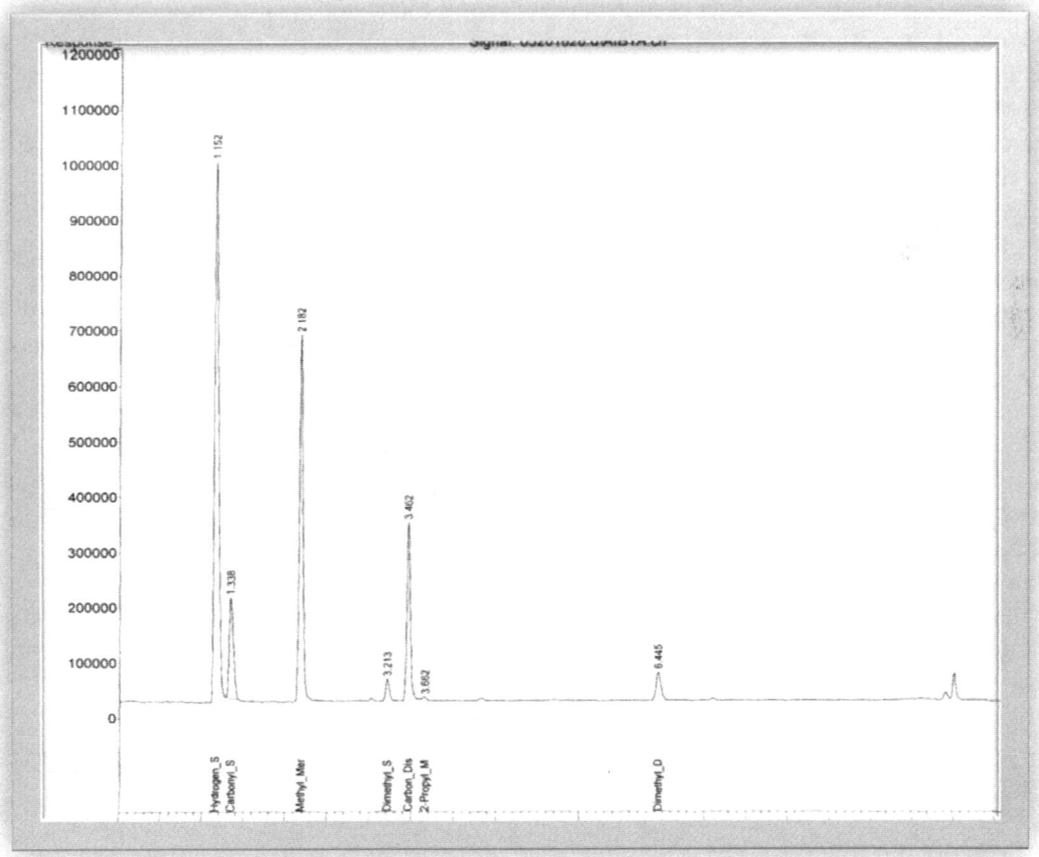

FIGURE 2. Sample chromatogram

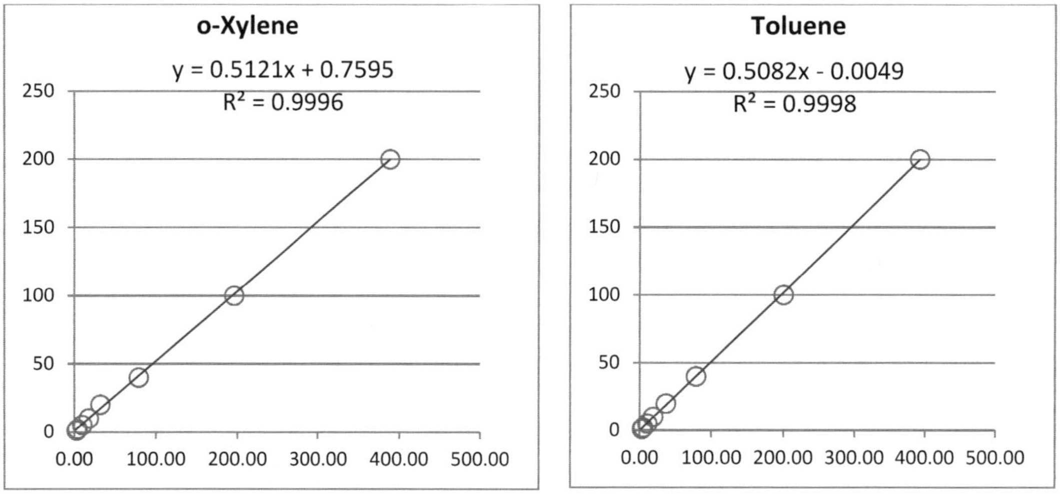

FIGURE 3. Sample Calibration Curves

The only two species in the alcohol group makes up a considerable fraction of all species. In this group, the highest concentration was observed for methanol. Isopropyl alcohol was the second. As expected, the highest concentration was observed for hydrogen sulfide in sulfur-containing group. In this group, most of the compounds were not detected in winter and summer samples. In spring samples, respectively methyl mercaptan, carbonyl sulfide, dimethyl sulfide, dimethyl disulfide, and isopropyl mercaptan followed hydrogen sulfide concentrations. Of the alkanes group, on the other hand, the concentration of n-hexane was the highest followed by n-tridecane, n-propane, n-butane, n-dodecane, n-undecane, n-decane, n-nonane, n-heptane, and n-octane. Similar to all other gorups, the highest concentrations of monoaromatic compounds were observed in spring samples, too, with the highest average concentration of n-propyle benzene. The order of concentrations were determined as m,p-xylene, toluene, o-xylene, ethyl benzene, 1,2,4-trimethyl benzene, 1,3,5-trimethyl benzene, styrene, isopropyl toluene, tert-butyl benzene, and n-butyl benzene. Highest concentrations of halogenated compounds were observed in spring samples. Of these, concentration of tetrochloro ethylene was the highest, followed by trichloro ethylene, methyl chloride, 1,2,4-trichloro benzene, 1,2,3-trochloro benzene, 4-chloro toluene, 1,4-dichloro benzene, and 1,2-dichloro benzene. Two ketones were detected in all samples: acetone and methyl isobutyl ketone. Of these, the summer concentrations of methyl isobutyl ketone were higher than acetone. Butyl acetate was the only compound detected during the whole study (only in spring samples).

The highest contribution to total concentrations of odorous compounds was 35.9% from the alcohol group. Moreover sulfur-containing compounds were close to alcohol groups and their contributions share nearly the same ratio 35.5%. Alkanes and monoaromatics compounds followed them (11.2% and 10.3%, respectively). Halogenated, acetate and ketones contributions were the smallest (0.3% and 1.3%, respectively). Fig. 5 shows average percent contributioned to total concentrations by all groups of compounds.

CONCLUSIONS

Volatile organic compound emissions are the main air pollutant emissions from processes that take place in wastewater treatment plants, and are usually the main source of odor pollution in neighboring locations. This study involves sampling of odorous emissions at nine different points in a wastewater treatment plant and evaluation of measured concentrations.

TABLE 2. Measurement Results

VOC species (mg/m^3)	Winter season average	Spring season average	Summer season average	Minimum and maximum concentrations	Overall average and standard deviation
Total monoaromatics	**0.835**	**2.108**	**0.501**	**0.016–2.293**	**1.148±0.263**
1,2,4-trimethylbenzene	0.114	0.104	0.037	0.026–0.202	0.076±0.061
1,3,5-trimethylbenzene	0.086	0.053	0.044	0.018–0.088	0.058±0.025
Ethylbenzene	0.106	0.157	0.043	0.016–0.255	0.096±0.075
Isopropyltoluene	–	0.049	–	0.032–0.070	0.049±0.017
m,p-xylene	0.239	0.466	0.065	0.050–0.723	0.234±0.224
n-butylbenzene	–	0.023	0.052	0.023–0.052	0.029±0.013
n-propylbenzene	0.035	0.528	–	0.035–1.559	0.380±0.572
o-xylene	0.121	0.146	0.041	0.039–0.216	0.122±0.059
Styrene	–	–	0.058	0.022–0.071	0.058±0.024
Tert-butylbenzene	–	0.037	–	–	0.037
Toluene	0.134	0.545	0.161	0.032–2.293	0.225±0.385
Total alkanes	**1.422**	**0.700**	**1.634**	**0.013–2.696**	**1.252±0.602**
n-butane	–	–	0.282	0.273–0.291	0.282±0.013
n-decane	–	0.036	0.140	0.014–0.149	0.070±0.056
n-dodecane	0.180	0.059	–	0.059–0.240	0.150±0.079
n-hexane	0.968	–	0.761	0.089–2.696	0.904±1.092
n-heptane	0.055	0.013	–	0.013–0.055	0.044±0.021
n-nonane	0.063	0.032	0.068	0.023–0.087	0.052±0.022
n-octane	–	0.017	–	–	0.017
n-propane	–	–	0.383	0.348–0.418	0.383±0.050
n-tridecane	–	0.445	–	0.027–1.040	0.445±0.402
n-undecane	0.156	0.098	–	0.038–0.277	0.123±0.090
Total halogenated	**0.074**	**1.437**	**0.318**	**0,021–1.838**	**0.610±0.352**
1,2,3-trichlorobenzene	0.032	0.098	–	0.032–0.161	0.058±0.058
1,2,4-trichlorobenzene	–	0.248	–	–	0.248
1,2-dichlorobenzene	–	0.031	–	0.021–0.051	0.031±0.017
1,4-dichlorobenzene	–	0.043	–	–	0.043
4-chlorotoluene	0.042	0.047	–	0.021–0.073	0.045±0.015
Methylenechloride	–	–	0.254	0.188–0.315	0.254±0.064
Tetrachloroethylene	–	0.659	0.064	0.054–1.838	0.489±0.630
Trichloroethylene	–	0.311	–	0.052–0.982	0.311±0.352
Total ketones	**–**	**0.148**	**–**	**0.025–0.123**	**0.148±0.069**
Acetone	–	0.123	–	–	0.123
Methylisobutylketone	–	0.025	–	–	0.025
Total alcohol	**–**	**–**	**4.002**	**0.214–3.723**	**4.002±1.579**
Isopropylalcohol	–	–	0.484	0.214–0.608	0.484±0.182
Methanol	–	–	3.518	3.313–3.723	3.518±0.289
Total acetates	**–**	**0.031**	**–**	**0.015–0.057**	**0.031±0.020**
Butylacetate	–	0.031	–	0.015–0.057	0.031±0.020
Sulfur compounds	**8.604**	**3.033**	**0.229**	**0.011–8.500**	**3.955±1.752**
Hydrogensulfide	8.500	2.302	0.229	0.046–8.500	1.525±2.850
Carbonylsulfide	–	0.206	–	0.061–0.350	0.206±0.204
Methylmercaptan	0.092	0.294	–	0.025–0.930	0.294±0.426
Dimethylsulfide	–	0.048	–	0.016–0.079	0.048±0.045
Carbondisulfide	0.012	0.107	–	0.011–0.350	0.107±0.163
Isopropylmercaptan	–	0.016	–	–	0.016
Dimethylsulfide	–	0.060	–	0.020–0.099	0.060±0.056

The concentrations of detected groups of odorous compounds from highest to lowest were ranked as alcohols, sulfur compounds, alkanes, monoaromatics, halogenated compounds, ketones, and acetates. Specifically alcohol group emissions were dominant in ozone-enhanced scrubber while their emissions were not detected in other units. In contrast to what is expected, alcohol concentrations were higher in effluent of ozone-enhanced scrubber. This shows that an accumulation of alcohol group emissions take place in the scrubber in contrast to a reduction by absorption-oxidation. Although identified in a very limited number of samples, the high concentrations of alcohol group emissions accounted for the highest percent contribution to total concentrations.

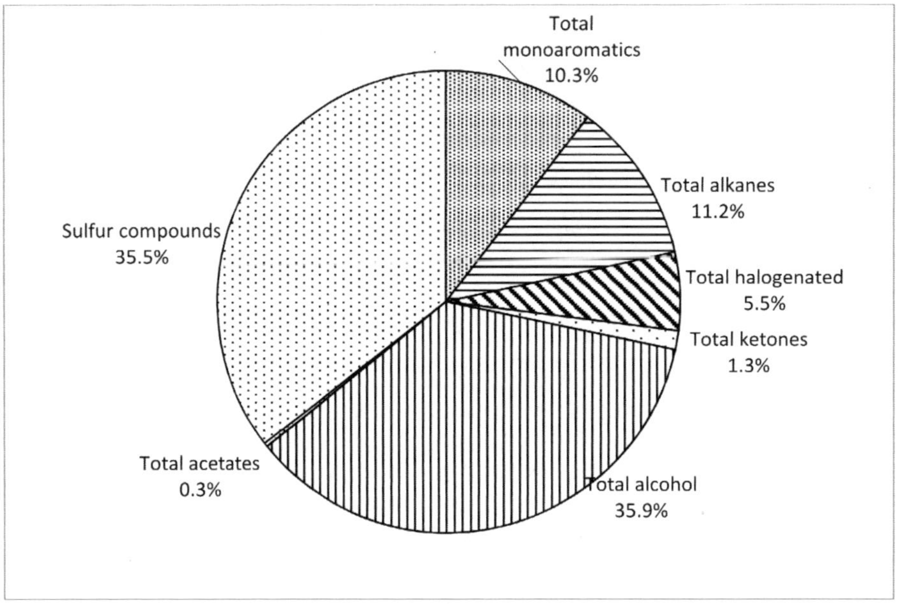

FIGURE 5. Percent contributions by all groups of compounds

ACKNOWLEDGEMENT

This research was supported by Kocaeli University Scientific Research Projects Coordination Department under grant no. 2015/23. The authors would like to express their deepest gratitude to the personnel of the wastewater treatment plant for their help and understanding during the study.

REFERENCES

Agus, E., Sedlak L.D., and L. Zhang. 2012. "A Frame Work for Identifying Characteristic Odor Compounds in Municipal Wastewater Effluent." *Water Research*, 46:5970-5980.

DEFRA. 2006. *Code of Practice on Odour Nuisance from Sewage Treatment Works*. Department for Environment Food & Rural Affairs UK.

Dinçer, F., and A. Muezzinoğlu. 2007. "Odor Determination at Wastewater Collection Systems: Olfactometry versus H_2S Analyses." *Clean Journal*, 35:565 – 570.

Giungato, P., Gennaro, G., Barbieri, P., Briguglio, S., Amodio, M., Gennaro, L., and F. Lasigna. 2016. "Improving Recognition of Odors in a Waste Management Plantby Using Electronic Noses with Different Technologies, Gas Chromatography Mass Spectrometry/Olfactometry and Dynamic Olfactometry." *Journal of Cleaner Production*, 133:1395-1402.

Kim H., Lee H., Choi E., Shin T., Im H., Ahn S., and I. Choi. 2014. "Characterization of Odor Emission from Alternating Aerobic and Anoxic Activated Sludge Systems Using Real-Time Total Reduced Sulfur Analyzer." *Chemosphere,* 117:394–401.

Krach, K.R., Li,B., Burns,B.R., and J. Mangus. 2008. "Benchandfull-Scalestudies for Odor Control from Lime Stabilized Biosolids: the Effect of Mixing on Odor Generation." *Bioresource Technology*, 99:6446–6455.

Lebrero, R., Bouchy, L., Stuetz, R., and R. Munoz. 2011. "Odor Assessment and Management in Wastewater Treatment Plants: a Review." *Critical Reviews in Environmental Science and Technology*, 41(10):915-950.

Lewkowska, P., Cieślik, B., Dymerski, T., Konieczka, P., and J. Namieśnik. 2016. "Characteristics of Odors Emitted from Municipal Wastewater Treatment Plantand Methods for Their Identification and Deodorization Techniques." *Environmental Research*, 151:573–586.

Liu, H., Luo, G.Q., Hu, H.Y., Zhang, Q., Yang, J.K., and H. Yao. 2012. "Emission Characteristics of Nitrogen and Sulfur Containing Odorous Compounds During Different Sewage Sludge Chemical Conditioning Processes." *Journal of Hazardous Materials*, 15:298–306.

Lorngage R., Fine L., Chiraiac R., and T. Lagier. 2004. "Investigation on Municipal Solid Waste Gases Composition and the Olfactory Nuisances Generated." *Environmental Odour Management, International Conference*, Cologne, Germany.

Marotta L., Kwoka T., Snow M., and S. Varsico. 2015. "Analysis of Volatite Organic Compounds (VOCs) in Air Using USEPA Method TO-17, Application Note." PerkinElmer Inc. Waltham, (03.03.2017, https://www.perkinelmer.com/lab-solutions/resources/docs/APP_Analysis-of-VOCs-in- Air-Using-EPA-Method-TO-17-011909_01.pdf)

Ölmez, S.S. 2008. *Odour Control in Wastewater Treatment Plants Using Ozonation and Chemical Scrubbing*. Marmara University, Institute for Graduate Studies in Pure and Applied Sciences, Master Thesis, İstanbul.

Shao, L.M., Zhang, C.Y., Wu, D., Lü, F., Li, T.S., and P.J. He. 2014. "Effects of Bulking Agent Addition on Odorous Compounds Emissions During Composting of OFMSW", *Waste Management*, 34:1381–1390.

Thomas, D.S. 2007. *Reducing Hydrogen Sulfide (H_2S) Concentrations at Wastewater Collection Systems and Treatment Facilities Using Chemical Oxidation*. Florida State University, Department of Civil and Environmental Engineering, master Thesis, Florida.

Zarra, T., Giuliani. S., Naddeo. V., and V. Belgiorno. 2011. "Odour Impact Evaluation from Wastewater Treatment Plant." *Proceeding of the 12th International Conference on Environmental Science and Technology*, 8-10 September, Rhodes, Greece, 2065-2072.

Characteristics of Peroxyacyl Nitrate Pollution during a 2015 Winter Haze Episode in Beijing

Boya Zhang, Ximeng Zhao, Jianbo Zhang*
(State Key Joint Laboratory of Environmental Simulation and Pollution Control, College of Environmental Sciences and Engineering, Peking University, Beijing 100871, China)

ABSTRACT: Peroxyacyl nitrates (PANs) are effective indicators of photochemical pollution, and also play an important role in regional oxidant balance. Surprisingly, in recent years, PANs have also been detected under conditions that do not favor the photochemical processes thought to form them. To obtain a better understanding of the mechanisms of formation of atmospheric compound pollution, this study examined the relationships between concentrations of PANs and other pollutants (e.g., ozone [O_3] and $PM_{2.5}$) during a winter haze episode. The observation periods were December 31, 2015, to February 2, 2016, and February 19, 2016, to March 4, 2016. The maximum daily concentration of PANs during haze episodes was 4 to 10 times higher than during non-haze episodes. The continuous cumulative increase in PAN concentrations was the result of a combination of photochemical production during the daytime and accumulation during the nighttime. During the focal haze episode, the correlation between concentrations of PANs and O_3 was weak, while a significant correlation was observed between PAN and $PM_{2.5}$ concentrations ($R^2 = 0.82$). This may have been due to higher concentrations of particulate matter impairing illumination, which can then inhibit the photochemical reactions that produce PAN and O_3. OH radicals can replace the role of light in PAN formation, which can cause concentrations of PAN and O_3 to vary independently. During the haze episode, the ratio of PAN/O_3 was around 0.3, which was much higher than that during the clean period.

INTRODUCTION

Peroxyacyl nitrates (PANs) are secondary products formed by a series of chemical reactions of volatile organic compounds (VOCs) and oxidants (e.g., OH, O_2, or NO_2). PAN compounds are not only an effective indicator of photochemical pollution, but also play an important role in regional oxidant balance.

Although the atmospheric concentration of PANs is usually about an order of magnitude lower than that of O_3, the phytotoxic effect of PANs is about one to two orders of magnitude greater than that of O_3. PANs can have toxic effects on humans, plants, and animals (Temple and Taylor, 1983), including tearing of the human eye (Altshuller, 1993), phytotoxicity (Taylor, 1969), and mutagenicity (Shepson et al., 1986), which can even change DNA bases (Peak et al., 1969) and cause skin cancer (Lovelock, 1977). Therefore, both the US Environmental Protection Agency and the World Health Organization have established concentration limit standards for PANs. The harmful effects of PANs are due to the following properties of the molecular structure: (1) oxidation, similar to peroxides and peroxynitrates; (2) nitrification and hydroxylation, similar to peroxynitrates; and (3) acetylation, similar to acetic anhydride (Lin et al., 2000).

Research on photochemical smog began with outbreaks of smog in Los Angeles, USA, and has since been conducted across the globe. PAN concentrations are important indicators of photochemical pollution, and researchers have conducted extensive PAN observations and simulations. Observational studies have focused on urban areas and suburbs affected by urban air masses (Duane et al., 2002; Grosjean et al., 2002; Lee et al., 2008). Researchers have also surveyed remote villages, mountains, oceans, and Polar Regions (Bottenheim and Gallant, 1989; Beine and Krognes 2000; Dassau et al., 2004; Zhang et al., 2009). Western developed countries rely on advanced technologies to study the spatial distribution and large-scale transmission of PANs through aerial surveys performed by aircraft and satellites (Glatthor et al., 2007; Tereszchuk et al., 2013).

PAN concentrations vary from the parts per trillion (ppt) levels in distant areas (Jacobi et al., 2000) to several parts per billion by volume (ppbv) in urban areas. As a photochemical secondary product, PANs typically display a unimodal diurnal pattern characterized by a continuous increase in PAN concentration during the morning, a peak in the afternoon, and a decrease in the evening (Jacobi et al., 1999; LaFranchi et al., 2009). Under special conditions, PAN concentration may also increase at night or present several concentration peaks in one day. These unusual patterns generally indicate a complex pollution process.

PANs and O_3 are both secondary products that are co-produced in photochemical reactions from similar precursors. However, PAN and O_3 concentrations are not always highly correlated (Lee et al., 2008). There are several differences between PANs and O_3 that lead to these patterns. First, from the point of view of production, almost all reactive VOCs can participate in the generation of O_3. However, only VOCs that produce peroxyacetyl radicals can generate PANs (Grosjean et al., 2002). In addition, the pathways of removal differ. PANs are primarily removed by thermal decomposition, while O_3 is removed by reactions with NO or unsaturated VOCs (Tuazon EC, 1991). Therefore, changes in the VOC species composition, the NO_2/NO ratio, and the ambient temperature can lead to a decrease in the correlation between PANs and O_3. In addition, heterogeneous reactions among aerosols on the surface may also affect the correlation between PANs and O_3.

With the rapid development of China's economy, the prevalence of harmful photochemical pollution has caused widespread concern. Atmospheric pollution leads to reduced visibility and enhanced atmospheric oxidization in the short term, and it endangers human health and causes climate change in the long term. In recent years, the government and scientific research institutions have focused on $PM_{2.5}$ control and reduction of VOC emissions. The annual $PM_{2.5}$ concentrations in the PRD (Pearl River Delta) and Yangtze River Delta have declined over time, but nevertheless photochemical pollution has continued to intensify. In 2015, the days with O_3 as the primary pollutant accounted for 40.5% of the days in whole year in Guangdong, and it is the first time that O_3 exceeded $PM_{2.5}$ (38.6%). In 2016, the proportion of O_3 as the primary pollutant rose to 42%. In field campaign, during severe haze days, concentrations of O_3 and $PM_{2.5}$ showed opposite trends, and the correlation between O_3 and PAN concentrations became insignificant, despite being formed by similar photochemical reactions. To understand the formation of winter haze episodes, more data are needed on the relationships between PANs and other pollutants (e.g. O_3 and $PM_{2.5}$). This study reports the concentration of PANs and its relationship with O_3 and $PM_{2.5}$ during a winter haze episode in Beijing, and explores the atmospheric chemical reactions affecting PAN concentrations in the haze formation process.

MATERIALS AND METHODS

Sampling Information. The monitoring site was located on the Yanqi Lake Campus of the University of Chinese Academy of Sciences (Huairou District, Beijing; 116.68°E, 40.41°N). The PAN analyzer was in the laboratory on the top floor of the test building, and the height of the PAN sampling inlet was consistent with the sampling inlet for conventional gas and particulate matter. Beijing Yanqi Lake is located in the northern suburbs of Beijing, where pollutant concentrations are relatively low. This site is affected by both polluted air masses from the urban areas and clean air masses from the northwest. Therefore, this site captures both the formation and the removal processes of pollution, which allows observations of the mechanisms of pollution formation. The monitoring periods were December 31, 2015, to February 2, 2016, and February 19, 2016, to March 4, 2016. The pause in monitoring in early February was due to the Chinese Spring Festival holiday.

PAN Online Detection System. A PAN online detection system developed by Peking University was used to monitor PAN concentrations. It consists of three main parts: 1) the sample and calibration unit (including the auto-sampler system and the instrument calibration system), 2) the GC-ECD analysis unit, and 3) the computer control unit.

The sample and calibration unit quantitatively collects air samples, and performs auto-sampling and calibration. Its main components are as follows: 1) the six-way injection valve, 2) loop, 3) mass flow controller, 4) data acquisition card, 5) solenoid valve, 6) pyrolysis furnace, and 7) photolysis reaction chamber. A vacuum pump at the end of the air circuit provides the sampling power. The air sample is filtered by a filter in the sampling inlet, which is then introduced into the equipment through the PFA tube. The sampling status is shown in Fig. 1a. The air sample is filled using a quantitation tube. The injection process is shown in Fig. 1b. Carrier gas brings the air sample from the quantitation tube to the capillary column, where it is analyzed by the GC-ECD analysis system.

The GC-ECD analysis unit consists of a Shimadzu GC-2010 GC and a cryogenic column temperature control system. High-purity He is used as a carrier gas and high-purity N_2 is used as a makeup gas. This analysis unit separates and quantitatively detects PAN in the atmosphere.

The computer control unit includes computers, data acquisition cards, and control software. An independently developed computer control program uses AD/DA conversion and IO output of the data acquisition card to control the switch and flow of the electromagnetic valve, pyrolysis furnace, sampling pump, and UV lamp in the system. This allows automatic sampling, automatic calibration, data recording, and other functions of the system.

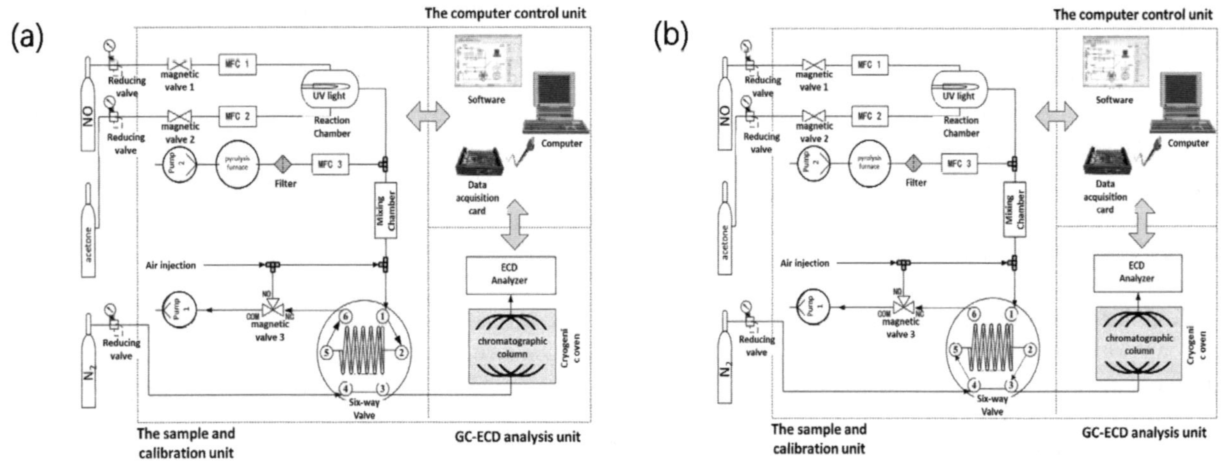

Figure 1. The structure of the PAN analyzer system
(a. The sampling status; b. The injection process)

Instrument Calibration and Data Processing. The analyzer calibrates the instrument by directly synthesizing PAN standard gas online, using NO standard gas (1 ppm), acetone standard gas (20 ppm), and a built-in UV light source. The online calibration system consists of two sections. One section is the PAN standard gas source composed of NO, acetone gas, and the reaction chamber UV lamp. The other section is a zero air source, which is used to dilute and adjust the concentration of the PAN standard gas; this section consists of a vacuum pump and a pyrolysis furnace. The basic principle of the synthesis of the standard gas is as follows: under ultraviolet rays of 285 nm, acetone gas is photolyzed to generate PA radical [$CH_3C(O)O_2$]. The excessive PA radical reacts with the quantitative NO gas to produce a certain concentration of PAN standard gas. The reaction equation is as follows:

$$CH_3C(O)CH_3 + hv \rightarrow CH_3C(O) + CH_3$$
$$CH_3C(O) + O_2 \rightarrow CH_3C(O)OO$$
$$NO + O_2 \rightarrow 2\ NO_2$$
$$CH_3C(O)OO + NO_2 \rightarrow CH_3C(O)OONO_2$$

The frequency of calibration depends on the state of the instrument and the length of the contamination process. The slope of the fitted line before and after calibration should differ by less than 10%. Figure 2 shows a calibration curve made during this field campaign.

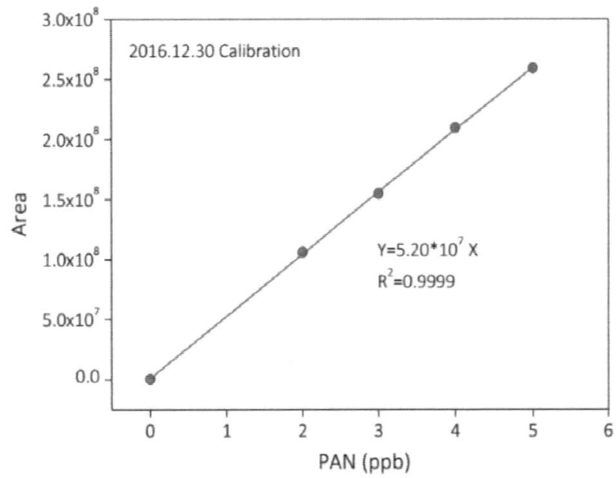

Figure 2. Calibration curve during Beijing winter field campaign

Shimadzu's software was used to acquire and process PAN data. Data processing included the following steps: 1) Set the appropriate time range in the batch method file and export editable data files according to the retention time of the PANs in the chromatogram. 2) Check the spectrum after batch processing. 3) Remove or manually integrate data in cases of abnormal integration. There may be time misalignment or software failures during the monitoring process, which need to be manually corrected. 4) Finally, the calibration results were used to convert peak area to concentration, resulting in a complete PAN time series. To facilitate model calculations and comparison with the existing literature, parts per billion by volume (ppbv) was used as the unit for reporting the atmospheric concentration of PANs.

RESULTS AND DISCUSSION

PAN Concentration Patterns during the Field Campaign. Figure 3 shows PAN concentrations during the field campaign as a time series (resolution: 5 min). Table 1 shows summary statistics of PAN concentrations. In contrast to the unimodal daily variations in summer, PAN concentrations during winter pollution periods showed continuous high values during day and night, while the non-pollution periods showed unimodal daily variations. PAN concentrations in the two states (pollution and non-pollution) are significantly different. A total of six high-concentration PAN pollution periods were captured during the monitoring process, each lasting about 3–4 days. The first and last pollution periods were the most severe (December 31 to January 3, February 29 to March 4), and were characterized by a longer duration than the other pollution periods, higher mean PAN concentrations, and maximum PAN concentrations exceeding 5 ppbv. It is noteworthy that the maximum and mean PAN concentrations observed during the photochemical period (daytime; 9:00–18:00) and non-photochemical period (nighttime; 18:00–9:00) are relatively similar (as shown in Table 1) during pollution periods. This indicates that nighttime chemistry also contributes greatly to high concentrations of PAN pollution. The maximum and average PAN concentrations throughout the monitoring period were 5.96 and 1.04 ppbv, respectively.

Owing to the peculiar patterns of PAN variations during winter, non-pollution and pollution periods are discussed separately. Pollution periods were characterized by high PAN levels, long duration, and maximum concentrations of 5.96 ppbv. Non-pollution periods were characterized by single-peak daily variation and low concentration of PAN (average daily value of 0.53 ppbv), where the maximum daily concentration of PAN does not exceed 1 ppbv. PAN concentrations during the two conditions differed significantly, and the maximum daily concentration of PAN during pollution episodes was 4–10 times higher than that during non-pollution episodes.

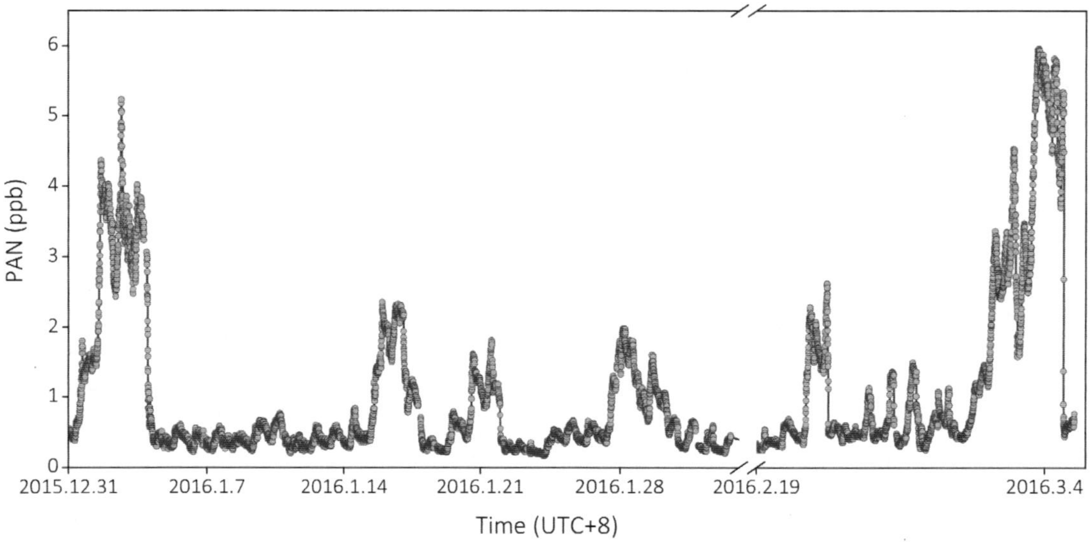

Figure 3. Time series of PAN concentrations during field campaign (time resolution: 5 min)

Table 1. Summary Statistics of PAN Concentrations during Winter Monitoring

Time period	Sample size	Maximum (ppbv)	Minimum (ppbv)	Mean (ppbv)
Dec	285	1.8	0.38	0.85
Jan	8475	5.23	0.16	0.88
Feb	3469	2.62	0.2	0.65
Mar	1269	5.96	0.44	3.23
Total	13498	5.96	0.16	1.04
9:00–18:00	5667	5.96	0.19	1.16
Non-photochemical period	7831	5.87	0.16	0.95

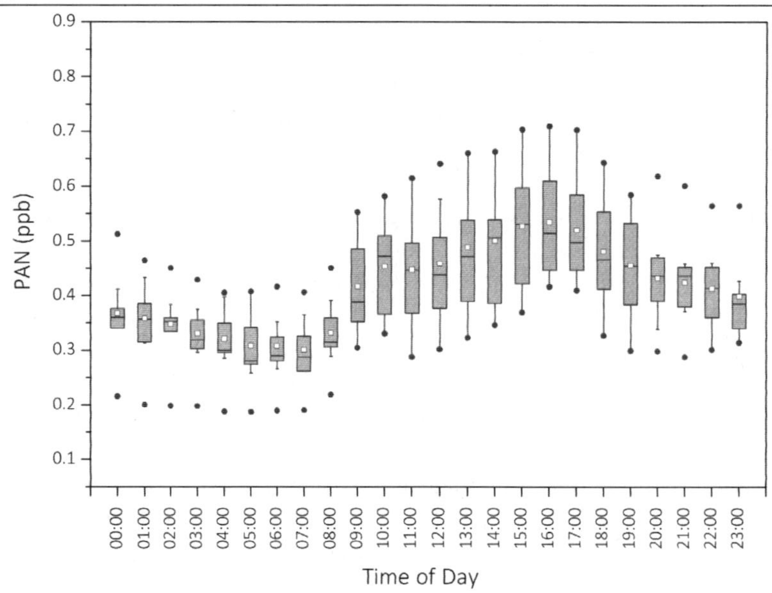

Figure 4. Diurnal change of PAN concentration during clean days (Data time range from January 5th to 13th and from January 24th to 26th)

Characteristics of PAN Concentration during Clean Period. During the monitoring period, January 5th to 13th and January 24th to 26th consisted of typical clean days. Observations during this period represent the PAN concentrations in winter in the Yanqi Lake area. As shown in Fig.4, the PAN concentrations showed a typical unimodal diurnal pattern, with the concentration rising from 9:00 AM and peaking at 15:00–16:00. The maximum daily average was 0.53 ppbv. The daily minimum average was 0.30 ppbv (05:00–07:00). The timing of these daily patterns in Beijing in winter was very similar to the daily patterns of PAN concentrations in summer in Chongqing. This indicates that the photochemical reaction in the suburbs of Beijing is highly active.

Atmospheric conditions during clean days were characterized using meteorological parameters from January 7 to January 13 (shown in Fig.5). The photochemical period was defined as 9:00–18:00. During this time, the ambient temperature during the photochemical period ranged from −5°C to 3°C, whereas nighttime lows reached −10°C. The air was relatively dry: humidity ranged from 30% to 50%. No obvious patterns were found in wind speed or wind direction. The proportion of southerly winds was relatively small. Wind speeds were relatively low during the photochemical period (less than 4 m/s). During three nights in this subset, stronger northerly winds were observed (greater than 6 m/s). Two days in this subset (January 9 and 10) showed slight increases in PAN concentration, with higher concentrations at night. These days were preceded by nights (January 8 and 9) lacking strong northerly winds. PAN concentrations were significantly lower during the nights than during the days, and PANs did not accumulate during the nighttime. PAN has a long half-life, and in low-temperature conditions, pyrolysis is not the main removal mechanism; therefore, at this monitoring site, the north wind is an important force for clearing pollution and ensuring clean days.

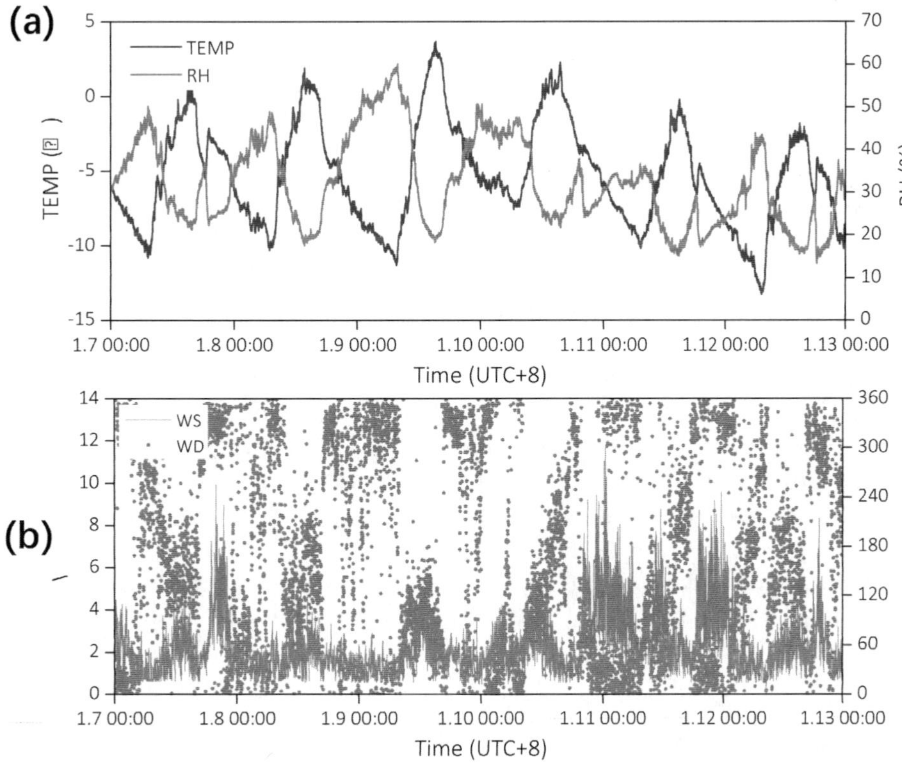

Figure 5. Meteorological conditions during a clean episode. (a) Temperature and humidity. (b) Wind speed and direction.

During the non-pollution periods, the maximum daily average concentration of PANs in Beijing in winter was only 0.53 ppbv, which is very low compared to other cities in China and even other countries. However, when pollution periods are included, the maximum and average PAN values were 5.96 ppbv and

1.04 ppbv, respectively. These concentrations are significantly higher than those observed in 2010 in Beijing, in 2011 at a Hong Kong background site, and in 2012 in Ziyang during winter; haze was not reported during the 2010 Beijing or 2011 Hong Kong observations. This average concentration is higher than summer PAN concentrations in many areas. These comparisons indicate that the haze pollution process in winter contributed to a high level of PAN pollution.

PAN Concentration during the Winter Haze Process and Its Relationship with the Concentrations of other Pollutants. During the fall and winter, compound atmospheric pollution occurs frequently in the Beijing-Tianjin-Hebei region. Particulate matter and photochemical oxidants coexist to form compound atmospheric pollution.

A current focus and challenge for atmospheric chemistry research is to understand the mechanisms of formation of pollution. Although PANs are important photochemical secondary products, changes in their concentration and behavioral characteristics during the haze process have not been reported in the literature. Therefore, this section examines a typical haze process during the winter monitoring period, and analyzes changes in PAN concentration in relation to causes of pollution. These analyses aim to elucidate the role of PANs in compound atmospheric pollution, and provide a basis for studying mechanisms of pollution formation.

During this winter monitoring period, six pollution episodes with high levels of both PANs and $PM_{2.5}$ were observed. The last accumulation process was the most persistent, with the highest concentration of pollutants. Therefore, the data from February 29 to March 5 were selected as the focal pollution process.

Meteorological Conditions during the Focal Pollution Episode. **Fig**. 6 shows the meteorological conditions during the focal pollution episode. The ambient temperature increased over the episode, gradually increasing from around 0°C at the beginning to 5–15°C later in the episode. On March 2nd, the temperature rise was most pronounced. An increase in temperature increases the efficiency of the photochemical reaction, but it may also accelerate the pyrolysis of PAN, which could slow PAN accumulation. Early in the pollution episode, humidity was similar to that observed on clean days (30% to 50%). However, humidity rose to about 80% two days after the pollution episode began. On the last day, strong northerly winds removed much of the pollution, and humidity dropped sharply to 10%.

In the first four days of the pollution episode, wind speed and direction showed clear patterns, as shown in Fig. 6(b). During the day (09:00 to 17:00), southerly winds dominated, with speeds greater than 4 m/s. At other times (18:00 to 08:00), northerly winds dominated and wind speed was low (less than 2 m/s). The southerly wind usually brings a polluted air mass (possibly the precursor or the generated secondary product) from the city. Southerly winds happened to dominate during the daytime period of photochemical pollution, which likely exacerbated the accumulation of pollutants. In addition, the lower atmospheric boundary layer in winter is also conducive to the accumulation of pollutants.

Pollution Characteristics of PAN, O_3, and $PM_{2.5}$. Fig.7 shows the patterns of PAN and conventional gas concentrations during the focal pollution episode. Raw data for PANs and conventional gas were collected at time resolutions of 5 and 1 min, respectively. The data points in the figure represent hourly averages.

Fine particulate matter ($PM_{2.5}$) is not a photochemical product, so generally the concentration of $PM_{2.5}$ does not have a regular diurnal pattern, regardless of season. During summer, concentrations of PANs and O_3 generally follow a similar pattern of unimodal diurnal variation. PAN and O_3 concentrations are usually highly correlated, but the correlation between PAN and PM2.5 concentrations is weak.

Surprisingly, during this haze episode, PAN concentrations experienced a similar cumulative increase as $PM_{2.5}$ concentrations. In contrast, O_3 concentrations maintained their normal daily variation, and were only weakly correlated with PAN concentrations. Therefore, at this site, PANs, which are an index of photochemical pollution, can also be used as an indicator for the formation of haze.

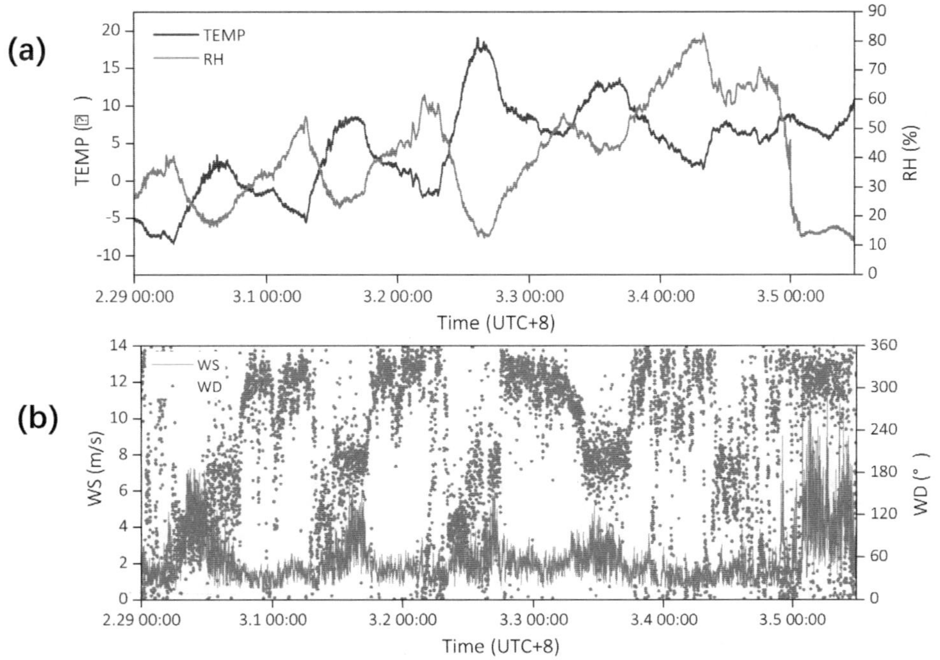

Figure 6. Meteorological conditions during the pollution episode. (a) Temperature and humidity. (b) Wind speed and direction

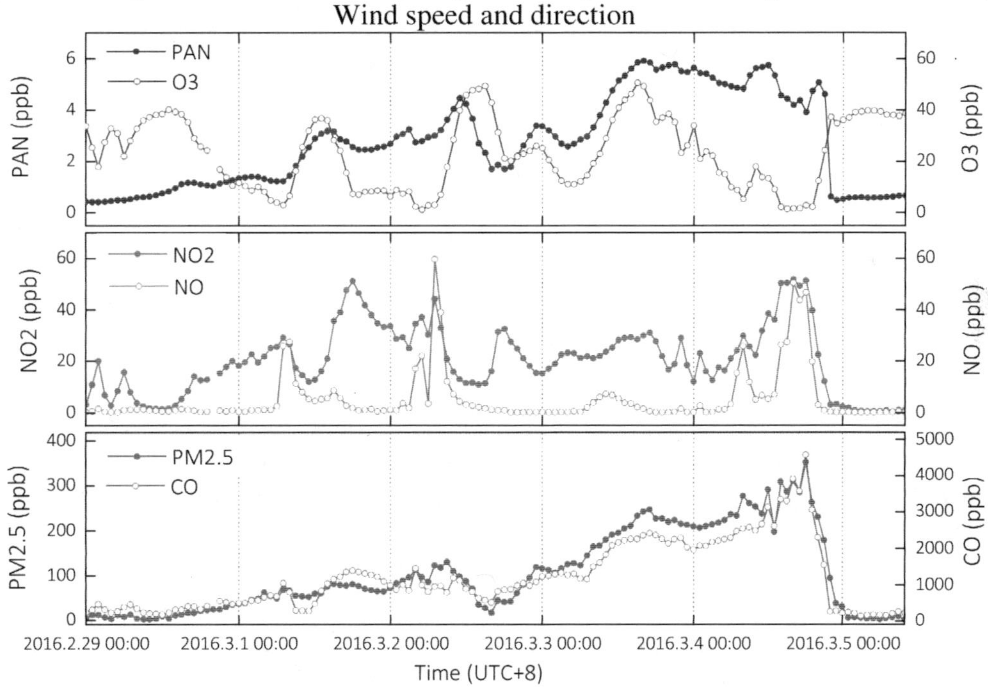

Figure 7. Change in concentrations of PANs and conventional gases during pollution episode (hourly average data)

During haze episodes, high concentrations of particulates are present in the atmosphere, which affect atmospheric visibility and impair light intensity. This is detrimental to atmospheric photochemistry, which depends on light to trigger a chain reaction, and therefore PAN formation may be inhibited during haze episodes. In addition, rapid declines in PAN concentrations have been observed in the afternoon; one hypothesis to explain this is that heterogeneous reactions of PANs on the surface of black carbon particles may accelerate PAN decomposition (Gaffney et al., 1999). This hypothesis depends on high-density particulate matter already present in the atmosphere, which in turn impacts the generation and removal of

PANs. However, VOCs, NO_x, and other substances that form particulate matter are also precursors of PANs, which means that concentrations of PANs and $PM_{2.5}$ are likely to be positively correlated. In the present study, high concentrations of particulate matter were not associated with inhibition of the generation of PANs, or with decreased concentrations of PANs. Instead, PAN concentrations were significantly positively correlated with concentrations of $PM_{2.5}$ ($R^2 = 0.82$).

In this five-day pollution episode, the cumulative increase in PAN concentrations includes two processes: photochemical production during the daytime and accumulation of PANs at night. Although the peak height of PAN concentration is not as significant as that of O_3, the trend of PAN concentration is still similar to that of O_3 during this period. Since the photochemical period during this pollution episode happens to be dominated by the southerly wind, the transmission process may also contribute to the accumulation of PANs. The second part is the accumulation and growth of concentrations at night. Temperatures are low during winter nights, so the pyrolysis rate of PANs is very low. Therefore, nighttime PAN concentrations are only slightly lower than in the afternoon, and sometimes there is even a significant increase in concentration at night.

During the clean period observed in winter, PAN concentrations generally began to increase at 09:00. However, during the focal heavy pollution episode, the fast-acting photochemical reaction period started earlier, with PAN concentrations beginning to rise even in the absence of light. Meteorological observations indicate that urban transmission is unlikely to contribute during this period of time, so the formation of PANs in the early morning may be dominated by free radical chemistry.

The 5-day pollution episode ended quickly over the course of 20 min on the evening of March 4th with a northerly wind (wind speed: 3 to 7 m/s). The concentration of PAN dropped from 5.24 ppbv (21:40) to 1 ppbv (22:00). At the same time, concentrations of PM2.5, VOCs, NOx, and CO all dropped rapidly to their lowest values. Only O3 quickly rebounded from low concentrations.

There are reports of the coexistence of high concentrations of PANs and particulate matter in the literature. For example, Zhang et al. (2014) detected high concentrations of PAN during a haze episode in February 2011 (average daily PAN concentration of 5.71 ppbv on February 23), but no detailed time series was given. During a monitoring period in Ziyang, Sichuan Province, in 2012 (Gao Tianyu, 2014), a transient particulate pollution period also occurred: the maximum concentration of PM2.5 was about 200 ppbv when the ambient temperature was 10 to 15°C. However, those observations of particulate pollution are lower than those observed in this study. Compared with the daily changes in PAN concentrations and the PAN/O3 ratio during the non-pollution period, PAN concentrations remained higher during the transient haze process, and the production of O3 was significantly inhibited.

Generation and Removal of PANs and O_3, and Correlations between PAN and O_3 Concentrations. The increased moisture absorption of aerosols significantly reduces the penetration of UV and visible wavelengths (Erlick and Frederick, 1998). Since PANs can only be generated by secondary atmospheric reactions, OH radicals play an important role in PAN formation when precursors cannot efficiently generate PANs by photolysis. Reactions R1 to R6 are the major free radical sources in the troposphere. Light during winter haze is weak; therefore, less OH is generated from the photolysis of O_3 and reactions between O_3 and unsaturated olefins. The photolysis rate of H_2O_2 is 10^{-5}, and the concentration of H_2O_2 is usually 1–2 times lower than that of O_3; therefore, photolysis of H_2O_2 is less important near the ground. HONO photolysis occurs in the presence of longer wavelengths, and the photolysis rate is 0.001, so photolysis of HONO may be an important source of OH radicals in winter. During a haze episode, there are many particles in the air. Heterogeneous reactions of NO_2 on the surface of these particles can generate HONO, which may be an important source of OH radicals (Zhang et al., 2014). Moreover, high concentrations of VOCs are inevitable during a haze period. These conditions can contribute to the high concentrations of PAN in winter.

The half-life of PAN depends on the ambient temperature and the NO/NO_2 ratio (Roberts et al., 2007). During the pollution period observed in this study, the ambient temperature was low (0–10°C), so the pyrolysis reaction rate of PAN was very slow (R17). Even if the NO concentration is high, it cannot have a significant impact on the PAN concentration. In summary, the OH radical generated by the photolysis of HONO during haze periods may replace the role of light and play an important role in PAN generation chemistry. Low-temperature conditions allow PAN to accumulate.

O_3 is generated by the photolysis of NO_2 (R7), and the effect of light on O_3 formation is more important than that of PAN. Therefore, when haze reduces light intensity, O_3 production will be hindered. In the first few days of the focal pollution episode, the concentration of particulate matter was still relatively low and did not affect the ability of O_3 to increase in concentration during the day. On March 4th, the $PM_{2.5}$ concentration reached more than 250 ppbv (humidity up to 80%), and the maximum concentration of O_3 on this day was less than half that of the previous days. This was likely due to the inhibition of O_3 formation by the high concentration of particulate matter. NO_2 that is not photolyzed may be used for PAN production, resulting in the appearance of higher concentrations of PAN. In addition, the O_3 concentration does not depend on temperature, and O_3 does not accumulate in a low-temperature environment. The presence of a large amount of NO will lead to O_3 removal (R9). These are the primary explanations for the inconsistent trends of PANs and O_3 in the non-photochemical period.

The sources of tropospheric atmospheric free radicals

$O_3 + hv \rightarrow O(^1D) + O_2$ (< 325 nm)	R1
$O(^1D) + H_2O \rightarrow OH + OH$	R2
$HONO + hv \rightarrow OH + NO$ (< 400 nm)	R3
$H_2O_2 + hv \rightarrow OH + OH$ (< 350 nm)	R4
$HCHO + hv \rightarrow H + HCO$ (< 340 nm)	R5
$OVOCs + hv \rightarrow RO_2 + HO_2 + prod.$ (< 800 nm)	R6

Basic photochemical cycles of NO_2, NO, and O_3

$NO_2 + hv \rightarrow NO + O(^3P)$ (< 420 nm, 对流层)	R7
$O(^3P) + O_2 + M \rightarrow O_3 + M$	R8
$O_3 + NO \rightarrow NO_2 + O_2$	R9

PAN formation chemistry

$RH + OH/O_3/NO_3 \rightarrow R\cdot + H_2O$	R10
$R\cdot + O_2 \rightarrow RO_2\cdot$	R11
$RO_2\cdot + NO \rightarrow RO\cdot + NO_2$	R12
$RO\cdot + O_2 \rightarrow R'(C{=}O) + HO_2$	R13
$R'(C{=}O) + OH/hv \rightarrow CH_3C(O)\cdot$	R14
$CH_3C(O)\cdot + O_2 \rightarrow CH_3C(O)O_2\cdot$	R15
$CH_3C(O)O_2\cdot + NO_2 \rightarrow PAN$	R16
$PAN \rightarrow CH_3C(O)O_2\cdot + NO_2$	R17
$CH_3C(O)O_2\cdot + NO \rightarrow CH_3C(O)O\cdot + NO_2$	R18
$CH_3C(O)O\cdot \rightarrow CH_3\cdot + CO_2$	R19

The day-to-day variation in the concentration of O_3 during the clean period was not obvious, and there were often multiple peaks in a day. The daily maximum of O_3 concentration was about 35 ppbv. The concentration changes in PAN and O_3 were not correlated, which is very different from the typical summer pattern. The ratio of PAN and O_3 daily maxima, that is, daily max. PAN/daily max. O_3, is used to evaluate the relative generation efficiency of both. The average value of the ratio on clean days was 0.017, while the PAN/O_3 ratio during the pollution episode was maintained at about 0.3 most of the time. The higher ratio during the pollution episode indicates that polluted air masses are conducive to the generation of PANs.

The two highest ratios appeared on March 2 and 4. From 03:00 to 08:00 on March 2, the concentration of PANs increased in the early morning, while O_3 was still at a lower concentration. The maximum ratio of PANs/O_3 was 2.5. From noon to evening on March 4, the production of O_3 was significantly inhibited, while PAN concentration remained at a high level, leading to a maximum PAN/O_3 ratio of 3.22. Previous studies found that PAN production is more efficient than O_3 production in highly contaminated air masses (Roberts et al., 1995). This contamination process confirms this view, but observations of PAN/O_3 ratios exceeding 1 are exceedingly rare.

In the literature, it is generally concluded that, when the correlation between PANs and O_3 is strong, this indicates that the generation of both is dominated by local chemistry. On the other hand, when the correlation between the two is weak, this indicates that the sources of PANs are complex, and may come from local photochemical generation, long-distance transmission, and diffusion from high-level atmosphere. The high concentration of particulates during winter haze episodes changes the photochemical process by affecting light, making the reaction conditions more favorable for PAN production and unfavorable for the formation of O_3, resulting in a weak correlation between the two.

Night Generation of PAN. During this monitoring period, PAN concentrations demonstrated many nighttime increases. One such increase was on March 2. The increase in PAN concentration during nighttime began at 24:00. After experiencing a small drop, the concentration continued to rise at 05:00 in the morning, despite this being well before dawn. Concentration peaked at 4.52 ppbv at 11:00. This time period was dominated by slow northerly winds, so transmission is unlikely to have contributed to this increase in concentration.

The second nighttime increase of PAN concentrations began on March 2 at 19:00, peaking at 24:00 (3.47 ppbv) after a sustained increase over 5 hours from 1 ppbv. During this time, the concentration of NO_2, a precursor of PAN, decreased significantly. $PM_{2.5}$ concentration increased similarly to that of PANs, while O_3 concentration increased only slightly. At 18:00 on March 2, the wind direction changed from southerly to northwesterly, and it continued at low speeds until the early morning of March 3. This wind direction indicates that urban air pollution was unlikely to contribute to PANs, and instead the nighttime increase in PAN concentration was likely due to local chemical generation.

Existing literature provides only limited data on PAN generation at night. Doussin et al. (2003) conducted smoke box experiments and model calculations, and proposed that free radicals contribute to PAN generation through VOCs. Rappengluck (2004) also noticed a nighttime increase in PAN concentrations. During the nighttime, PANs increased from 30 ppt to 2.5 ppbv. However, the meteorological conditions (increased temperature and wind speed) were not conducive to an increase in PAN concentration. Within a few hours, the PAN/O_3 ratio rose from 0.0067 to 0.0286.

NO_3 is formed by the oxidation of NO_2 at night (R20 to R22), and NO_3 can exist in the form of N_2O_5 in the equilibrium reaction. At high concentrations of NO_3, the oxidation of acetaldehyde at night can lead to PAN production (Cantrell et al., 1986), as shown by R23 to R26. At the same time, NO_3 reacts with PA radicals, which may be one of the nocturnal sources of OH (R27).

NO_3 generation

$NO_2 + O_3 \rightarrow NO_3 + O_2$ R20

$NO_2 + NO_3 \rightarrow N_2O_5$ R21

$N_2O_5 \rightarrow NO_2 + NO_3$ R22

NO_3 reacts with VOCs to generate PANs

$CH_3CHO + NO_3 \rightarrow HNO_3 + CH_3C(O)\cdot$ R23

$CH_3C(O)\cdot + O_2 \rightarrow CH_3C(O)O_2\cdot$ R24

$CH_3C(O)O_2\cdot + NO_2 \rightarrow CH_3C(O)O_2NO_2$ R25

$CH_3C(O)O_2NO_2 \rightarrow CH_3C(O)O_2\cdot + NO_2$ R26

NO3 reacts with PA radicals

$NO_3 + CH_3C(O)O_2 \rightarrow CH_3C(O)O + NO_2 + O_2$ R27

CONCLUSION

PAN concentrations showed a cumulative growth pattern, consisting of two parts: daytime photochemical production and nighttime accumulation. The OH radical generated by the photolysis of HONO during haze episodes may replace the effect of light and play an important role in PAN generation. Low temperatures allow PANs to accumulate. Multiple nighttime PAN concentration increases were observed, which were likely generated chemically by the precursors after they had been transported locally. When particulate matter reaches a sufficiently high concentration, it inhibits the formation of O_3. Because O_3 does not accumulate at low temperatures, the correlation between PANs and O_3 was inconsistent during the non-photochemical period. There was no correlation between the changes of PAN and O_3 concentrations during the non-pollution episodes. Local VOCs in Beijing contributed less to PAN generation. The ratio of PAN/O_3 was maintained at about 0.3 for the majority of the pollution episode, with maxima exceeding 1. This ratio is much larger than ratios observed during the non-pollution period or in other regions. In a highly polluted air mass, the production of PANs is more efficient than that of O_3.

REFERENCES

Altshuller, A. (1983). "Review: Natural volatile organic substances and their effect on air quality in the United States." Atmospheric Environment (1967) 17(11): 2131-2165.

Beine, H. J. and T. Krognes (2000). "The seasonal cycle of peroxyacetyl nitrate (PAN) in the European Arctic." Atmospheric Environment 34(6): 933-940.

Bottenheim, J. W. and A. J. Gallant (1989). "PAN over the Arctic; observations during AGASP-2 in April 1986." Journal of Atmospheric Chemistry 9(1): 301-316

Cantrell, C. A., J. A. Davidson, K. L. Busarow and J. G. Calvert (1986). "The CH_3CHO-NO_3 reaction and possible nighttime PAN generation." Journal of Geophysical Research: Atmospheres **91**(D5): 5347-5353

Dassau, T. M., P. B. Shepson, J. W. Bottenheim and K. M. Ford (2004). "Peroxyacetyl nitrate photochemistry and interactions with the Arctic surface." Journal of Geophysical Research: Atmospheres 109(D18).

Doussin, J., B. Picquet-Varrault, R. Durand-Jolibois, H. Loirat and P. Carlier (2003). "A visible and FTIR spectrometric study of the nighttime chemistry of acetaldehyde and PAN under simulated atmospheric conditions." Journal of Photochemistry and Photobiology A: Chemistry **157**(2): 283-293.

Duane, M., B. Poma, D. Rembges, C. Astorga and B. Larsen (2002). "Isoprene and its degradation products as strong ozone precursors in Insubria, Northern Italy." Atmospheric Environment 36(24):3867-3879.

Erlick, C. and J. E. Frederick (1998). "Effects of aerosols on the wavelength dependence of atmospheric transmission in the ultraviolet and visible: 1. A "single-scattering-separate" delta-Eddington model." Journal of Geophysical Research: Atmospheres **103**(D10): 11465-11472.

Gao Tianyu. Study on the Distribution of Atmospheric Concentration of PAN and the Influence of Heterogeneous Reaction on It[D]. Peking University, 2014.

Gao Tianyu, Han Li, Wang Bin, Yang Guang, Xu Zhenqiang, Zeng Limin, Zhang Jianbo (2014). Peroxyacetyl nitrate observed in Beijing in August from 2005 to 2009; Journal of Environmental Sciences 26:2007-1017

Gaffney, J., N. Marley, M. Cunningham and P. Doskey (1999). "Measurements of peroxyacyl nitrates (PANS) in Mexico City: implications for megacity air quality impacts on regional scales." Atmospheric Environment **33**(30): 5003-5012.

Glatthor, N., T. v. Clarmann, H. Fischer, B. Funke, U. Grabowski, M. Höpfner, S. Kellmann, M. Kiefer, A. Linden and M. Milz (2007). "Global peroxyacetyl nitrate (PAN) retrieval in the upper troposphere from limb emission spectra of the Michelson Interferometer for Passive Atmospheric Sounding (MIPAS)." Atmospheric Chemistry and Physics 7(11): 2775-2787.

Grosjean, E., D. Grosjean, L. F. Woodhouse and Y.-J. Yang (2002). "Peroxyacetyl nitrate and peroxypropionyl nitrate in Porto Alegre, Brazil." Atmospheric Environment 36(14): 2405-2419.

Jacobi, H. W., R. Weller, T. Bluszcz and O. Schrems (1999). "Latitudinal distribution of peroxyacetylnitrate (PAN) over the Atlantic Ocean." Journal of Geophysical Research: Atmospheres 104(D21):26901-26912.

Jacobi, H.-W., R. Weller, A. Jones, P. Anderson and O. Schrems (2000). "Peroxyacetyl nitrate (PAN) concentrations in the Antarctic troposphere measured during the photochemical experiment at Neumayer (PEAN'99)." Atmospheric Environment 34(29): 5235-5247.

LaFranchi, B., G. Wolfe, J. Thornton, S. Harrold, E. Browne, K. Min, P. Wooldridge, J. Gilman, W. Kuster and P. Goldan (2009). "Closing the peroxy acetyl nitrate budget: observations of acylperoxy nitrates (PAN, PPN, and MPAN) during BEARPEX 2007." Atmospheric Chemistry and Physics 9(19): 7623-7641

Lee, G., Y. Jang, H. Lee, J.-S. Han, K.-R. Kim and M. Lee (2008). "Characteristic behavior of peroxyacetyl nitrate (PAN) in Seoul megacity, Korea." Chemosphere 73(4): 619-628.

Lin, J.-K., K.-J. Chen, G.-Y. Liu, Y.-R. Chu and S.-Y. Lin-Shiau (2000). "Nitration and hydroxylation of aromatic amino acid and guanine by the air pollutant peroxyacetyl nitrate." Chemico-biological Interactions 127(3): 219-236.

Lovelock, J. E. (1977). "PAN in the natural environment; its possible significance in the epidemiology of skin cancer." Ambio: 131-133.

Peak, M. and W. L. Belser (1969). "Some effects of the air pollutant, peroxyacetyl nitrate, upon deoxyribonucleic acid and upon nucleic acid bases." Atmospheric Environment (1967) 3(4):385IN1395-394397

Rappenglück, B., C. Forster, G. Jakobi and M. Pesch (2004). "Unusually high levels of PAN and ozone over Berlin, Germany, during nighttime on August 7, 1998." Atmospheric Environment **38**(36): 6125-6134

Roberts, J. M., M. Marchewka, S. Bertman, R. Sommariva, C. Warncke, J. De Gouw, W. Kuster, P. Goldan, E. Williams and B. Lerner (2007). "Measurements of PANs during the New England air quality study 2002." Journal of Geophysical Research: Atmospheres **112**(D20).

Roberts, J., R. Tanner, L. Newman, V. Bowersox, J. Bottenheim, K. Anlauf, K. Brice, D. Parrish, F. Fehsenfeld and M. Buhr (1995). "Relationships between PAN and ozone at sites in eastern North America." Journal of Geophysical Research: Atmospheres **100**(D11): 22821-22830

Shepson, P. B., T. E. Kleindienst, E. O. Edney, C. M. Nero, L. T. Cupitt and L. D. Claxton (1986)."Acetaldehyde: the mutagenic activity of its photooxidation products." Environmental Science & Technology 20(10): 1008-1013.

Taylor, O. (1969). "Importance of peroxyacetyl nitrate (PAN) as a phytotoxic air pollutant." Journal of the Air Pollution Control Association 19(5): 347-351.

Temple, P. and O. Taylor (1983). "World-wide ambient measurements of peroxyacetyl nitrate (PAN) and implications for plant injury." Atmospheric Environment (1967) 17(8): 1583-1587

Tereszchuk, K. A., G. González Abad, C. Clerbaux, J. Hadji-Lazaro, D. Hurtmans, P.-F. Coheur and P.F. Bernath (2013). "ACE-FTS observations of pyrogenic trace species in boreal biomass burning plumes during BORTAS." Atmospheric Chemistry and Physics 13(9): 4529-4541.

Tuazon, E., W. Carter and R. Atkinson (1991). "Thermal decomposition of peroxyacetyl nitrate and reactions of acetyl peroxy radicals with NO and NO_2 over the temperature range 283-313 K." Journal of Physical Chemistry 95(6): 2434-2437.

Zhang, J. (2009). Measurement of atmospheric peroxyacetyl nitrate (PAN) and the implications to photochemical pollution, The Hong Kong Polytechnic University.

Zhang, G., Y. Mu, J. Liu, C. Zhang, Y. Zhang, Y. Zhang and H. Zhang (2014). "Seasonal and diurnal variations of atmospheric peroxyacetyl nitrate, peroxypropionyl nitrate, and carbon tetrachloride in Beijing." Journal of Environmental Sciences **26**(1): 65-74.

Air Pollution Distribution, Modelling and Development of an Air Quality Index based on Land use Profiles, Trinidad, WI

Himawatee Baboolal* and Valerie Stoute
(University of Trinidad and Tobago, T&T)

Trinidad is the most industrialised of the Caribbean islands, with a large energy and petrochemical economic base. The island supports several world scale industries and has a transportation fleet that is mainly diesel fueled, and an inefficient public transport system. It also has one of the highest per capita GHG emission ratings in the world. In addition, Trinidad, like the rest of the Caribbean region is affected seasonal by Sahara dust (<PM2.5) across the Atlantic on the Saharan Air Layer during the African desert's winter months (June to October). These all have an impact on the ambient air quality and the public health risks. The industrial development of Trinidad has outpaced the management and legislative strategies for protection of the environment and public health and has placed a strain on the local environmental regulators to control and curb industrial pollution.

Objective: To provide baseline representative air quality profiles for different geographic areas and major land use activities in Trinidad to platform air pollution management.

Methodology: Measurements were made of meteorological parameters, levels of particulates (three size fractions in the fine and respirable range), heavy metals in these size fractions, criteria pollutant gases and air toxics at four sites over the heavily populated west coast (March '15-May '16). Stations were cited to represent rural, urban, mixed background and industrial land uses. Sampling took place once every sixth days simultaneously at each station using regulatory grade equipment. Trace metals were analysed using standard methods (microwave acid digestion for ICP-MS analyses) which were compared to standard reference materials, field and reagent blanks. Particulate fractions were measured gravimetrically. Gases were measured in real time using FTIR technology. Statistical analyses included factorial analysis (univariate and multivariate techniques) to elucidate a statistical model for the variability of pollutants and spatio-temporal trends. Pollutant data was also modelled using USEPA's AERMOD with some limitations on accuracy due to the unavailability of comprehensive point and area source data in the public records. The measurement data collected in this study was used to calculate and validate an air quality index that could be readily applied to ongoing data collection on air pollutants.

Results and Conclusions: Particulate levels exceeded the WHO guidelines for protection of public health over seventy percent of the time at the urban and industrial stations. Factor analysis indicated significant interaction effects determining the PM distribution and variability among fixed factors of the time of year, location of station (land use), the particulate size fraction. The PM measured at the industrial station was markedly different from the other three stations. Trace metals measured included (As, Ba, Be, Cd, Co, Cr, Cu, Fe, Hg, Li, Mn, Ni, Pb, Sr, V, Zn). Levels of Fe, Ni and Zn tended to be much higher the industrial station for most periods, coinciding with the closure of a major steel manufacturing plant near the end of the sampling period. Factor analysis affecting the variability of trace metals measured vary with trace metal. Gaseous pollutants (criterial pollutants and toxic gases) varied from station to station, with other significant main factors and interaction effects depending on the gas. The air quality index (incorporating coefficients for particulate (fine and respirable) and criteria pollutants) was developed using and validated using canonical discriminant function yielding four tiers for air quality (Good, normal, high, very high). The industrial station has the highest frequency of 'very high' pollutant levels, as well as the highest frequency of 'good' air quality days. The urban station had highest frequency of 'normal' to 'high' rankings. The rural station, as expected, had much better overall air quality.

Characterizing Spatial Distribution of Coarse Particle Concentrations Using Passive Aerosol Sampler in Kocaeli, Turkey

Beyhan Pekey (Kocaeli University, Kocaeli, Turkey)
Robert D. Willis (US EPA, North Carolina, USA)

Particulate matter, which has important health and environmental effects in terms of different physical properties and chemical composition, is one of the important pollutants in ambient air. For this reason, it is very important to monitor the amount of atmospheric particulate matter and take precautions accordingly.

A passive aerosol sampler (PAS) has been developed as an alternative to active sampling by Wagner and Leith (2001) the University of North Carolina at Chapel Hill. It is about 15 mm in diameter, weighs about 2 g, is inexpensive and energy-free sampler. The PAS combined with computer controlled scanning electron microscopy (CCSEM) provides ambient size distributions based on particle number, mass, and elemental concentrations. In this respect, the PAS may be especially useful in cases where active sampling is not feasible due to lack of power; where it is too expensive to deploy; or in situations involving long term sampling duration. The PAS has been used in numerous ambient air quality applications throughout the world over the past few years. In this manuscript provide an overview of one application in Kocaeli, the second important industrial city in Turkey with a population of over 1.5 million inhabitants.

Sixteen sites were selected for the passive aerosol sampling in the scope of this study. These sites were located around the Izmit Bay, where the majority of people are living and working. Two consecutive and identical sampling campaigns, each of which lasted for 3 weeks, were performed simultaneously on 16 sites. The first sampling campaign was conducted from 20 August to 10 September 2014 and the second from 10 September to 1 October 2014. The size and elemental composition of individual particles deposited on the passive sampler were determined by CCSEM-EDX.

Concurrent PAS measurements at 16 sites showed highest $PM_{10-2.5}$ concentrations in the northwest part of the city where most of the industrial plants are located and in the eastern part of Kocaeli which has less industrial activity.

Insight into the Sources of $PM_{2.5}$ at Roadside Micro-Environment in Tianjin, China

Shi-Bao Wang, ***Ya-Qin Ji***, Lei Zhang
(Nankai University, Tianjin, China)

As a micro-environment where people frequently activities, the roadside air quality is directly affected by traffic pollution sources. In order to explore the pollution characteristics and sources of $PM_{2.5}$ at roadside micro-environment, samples were collected from five different types of road in 2015 in Tianjin. The annual mean concentration of $PM_{2.5}$ at roadside was 166 ± 38 μg m^{-3}. The coefficient of divergence of $PM_{2.5}$ and road dust, construction dust, coal combustion dust, soil dust, steel dust, and vehicle emission source were 0.29, 0.41, 0.43, 0.53, 0.55, 0.56, respectively. The correlation coefficients of $PM_{2.5}$ and soil dust, steel dust, road dust, construction dust, coal combustion dust, and vehicle emission source were 0.78, 0.83, 0.85, 0.87, 0.89, 0.97, respectively, indicating that $PM_{2.5}$ was highly affected by these kinds of pollution sources. A PCA/MLR-PMF-CMB combined receptor model was used to analyze the sources of $PM_{2.5}$ at roadside. The results obtained by the combined models showed that the combined source apportionment technique is feasible. Four, six and eight main sources were identified by PCA-MLR, PMF and CMB, respectively. Secondary source, vehicle emission source, combustion source, and fugitive dust were identified by all the three receptor models, and totally contributed over 60% to $PM_{2.5}$ at roadside.

Hexabromocyclododecane (HBCDD), a brominated flame retardant, is used in heat insulation materials, which are expanded polystyrene and extruded polystyrene, and in furniture, textile and other industries for incombustibility. HBCDD threatens human health and the environment due to its potential for biomagnification and bioaccumulative properties. It is a persistent organic pollutant (POP) regulated by the Stockholm Convention on POPs, and listed under Annex-A Elimination List of the Convention in 2013. Only in the last decade, studies are reported on HBCDD degradation in the literature, and they show that the degradation of HBCDD in soil and sediment samples is more favorable in anaerobic environments. There are currently a very limited number of studies in the literature on HBCDD degradation, and no mesocosm-scale study (to the best of our knowledge) on evaluation of biodegradation in HBCDD contaminated sediments. In this study, anaerobic biodegradation of HBCDD was investigated in two laboratory sediment mesocosms (2L total glass tank volume with 685 g sediment mass and 550mL overlying liquid volume), set up as natural attenuation and biostimulation sets. While natural attenuation contains only distilled water and HBCDD contaminated sediment, biostimulation set is prepared by adding organic medium containing electron donor (ethanol) and carbon source (sodium formate) to sediment contaminated with HBCDD. There are also 2 control mesocosms: contaminant control and sterile mesocosms. Contaminant control contains only distilled water and clean sediments (without adding HBCDD). Sterile set contains distilled water and HBCDD contaminated sediment, and it is prepared by autoclaving and poisoning with $HgCl_2$. All reactors were prepared as duplicate, completely under N_2:CO_2:H_2 environment in an anaerobic glovebox (PlasLabs 818GB/Exp) and incubated in the dark at 25°C. The mesocosms are operated for 49 days and sampling is done in a 7-day period. During sampling, triplicate sediment samples were taken from each mesocosm and extracted using an ultrasonic bath (USEPA Method No: 3550C). In the analysis of HBCDD, GC-MS was used for evaluation of total-HBCDD and LC-MSMS is used for evaluation of α, β, γ-HBCDD at different sampling times of operation. As a result, percent reduction in total-HBCDD, and α, β, γ-HBCDD as well as degradation rates were obtained for natural attenuation and biostimulation, followed by comparison with the limited literature available on HBCDD. In natural attenuation set, 92-96% reduction was observed in 42 days, while HBCDD was below detection in the last sampling day. In biostimulation mesocosms, 96-98% reduction was achieved in 35 days, after which HBCDD was below detection. As expected, HBCDD was never detected in the contaminant control set, indicating absence of any cross-contamination in laboratory procedures. However, degradation was observed in sterile sets, at 95% reduction of HBCDD in 49 days. Observation of degradation in sterile sets has been cited in the literature, and results of this study and those are comparatively discussed. Overall, this mesocosm study indicate HBCDD anaerobic biodegradation can be stimulated with the addition of organic rich medium containing an electron donor and carbon source. All degradation results, both terms of total-HBCDD and stereoisomer-specific-HBCDD are discussed with relevant data from the literature for the purpose of improving our understanding while remediation of sediments contaminated with this persistent organic pollutant.

Degradation of Ternary Mixture of Trihalomethanes in a Biotrickling Filter Seeded with Biosurfactant and Fungi

Sanaiya Islam*, George Sorial, and Bineyam Mezgebe
(University of Cincinnati, Cincinnati, OH, USA)

Harmful trihalomethanes (THMs) are formed when disinfectants like chlorine react with natural organic matter in water. Chloroform (CF), bromodichloromethane (BDCM) and dibromochloromethane (DBCM) are among the different type of THMs. These compounds can also be found in chemical industry air emissions. US Environmental Protection Agency has set the maximum contaminant level (MCL) for THMs to be 0.08 mg/L in water. To achieve the MCL, aeration by diffused-air have been found effective in removing THMs from the water phase to the gaseous phase. In order to decontaminate the gaseous phase, a cost effective technology needs to be utilized. The use of aerobic biotrickling filter (BTF) with filamentous fungi as biodegrading organism, had proven to have significant impact on removal of chloroform in the presence of cometabolites from the gaseous phase. In this study, biosurfactant was added to an aerobic BTF seeded with filamentous fungal consortium. This system was used to observe the degradation of a ternary mixture of CF, BDCM and DBCM. The biosurfactant is synthesized from *Bacillus subtilis* and their attribute to accumulate at the interface of air and water due to their amphiphilic molecular structure. The biosurfactant enhances the interaction between the volatile THMs mixture and biofilm surface. A gas stream containing the ternary mixture is treated within the BTF system. The system is kept under acidic condition to ensure the growth of fungi colonies. The compounds were mixed in equal ratio (1:1:1) and loaded at the rate of 3.1, 3.02 and 2.94 ppmv for CF, BDCM and DBCM, respectively. The removal efficiencies observed at this loading rates were 85.6 %, 89.2% and 90.1% for CF, BDCM and DBCM, respectively. Further studies were conducted at different ratio of CF, BDCM and DBCM (10:7:3), which is commonly found in air emissions from chemical industries. In addition, the effect of different empty bed residence times in the biodegradation of the THMs are being observed.

Effects of Nanoparticle Based Fuel Additives on Diesel Engine Performance and Emissions

Asanga Wijesinghe, Fanxu Meng, Carolyn LaFleur, and Richard Haut
(Houston Advanced Research Center, The Woodlands, TX, USA)

ABSTRACT: Fuel Additives are chemical compounds added in very small quantities to diesel base fuel to modify or create certain fuel properties. They have played a major role in enhancing fuel performance for many decades. Fuel additives improve fuel efficiency, reduce soot emissions, and alter the emissions of carbon monoxide (CO), nitrogen oxides (NOx), and hydrocarbon (HC) species, including several hazardous air pollutants (HAPs). Fuel additives incorporating nano-sized cerium oxide have been increasingly used in diesel engines to promote combustion by decreasing the amount of soot in diesel exhaust and increasing the fuel efficiency. However, only few studies have evaluated the impact of these nanotechnology based fuel additives on pollutant emissions. To assess the impact of cerium oxide nanoparticles on the environment, a laboratory study was conducted to find how these nanoparticles affect diesel engine performance and exhaust emissions. The results show that the cerium oxide nanoparticles are efficient in improving the combustion properties of ultra-low sulfur diesel fuel that in turn improve the engine performance and reduce some hazardous air pollutants.

INTRODUCTION

Excessive use of fossil based fuels dissipates the petroleum reserves and increases air pollution leading to global climate change (Wijesinghe, et al. 2018). In addition to the environmental pollution, higher fuel cost and increase in demand of petroleum resources have forced researchers, engine manufacturers, and consumers to focus on finding efficient and less polluting alternatives and/or modifications to conventional diesel fuel. Various solutions have been proposed, including utilizing alternative fuels, fuel additives, engine technologies, and retrofit (Kannan, Karvembu, & Anand, 2011; Tiwari, 2015; Huang, et al., 2014). Among these various solutions, the current focus is on the use of fuel additives due to their advantage of increasing in fuel efficiency while reducing the harmful exhaust gas emissions.

Fuel additives are chemical compounds added in very small quantities (in the order of mg of additives per kg of fuel) to diesel base fuel to modify or create certain fuel properties, with the intention of improving the performances and reducing the emissions of a diesel engine. Some examples of additives include deposit removers, combustion improvers, lubricity improvers, and flow improvers. These fuel additives improve fuel efficiency, reduce diesel soot emissions, and alter the emissions of carbon monoxide (CO), nitrogen oxides (NOx), and hydrocarbon (HC) species, including several hazardous air pollutants (HAPs) (Mukhopadhyay & Chakraborty, 2015).

The diesel fuel additive industry has existed since the 1920's and has played a major role in improving fuel performance. In the beginning it mostly consisted of cetane boosters to improve fuel ignition quality, and thus shorten the ignition delay. Since then the need for additives has grown, for both economic as well as legislative reasons (ACEA, 2006).

Nano-sized cerium oxide (CeO_2) fuel additives have been extensively used in diesel engines to promote combustion and thereby decreasing the amount of soot in diesel exhaust and increasing the fuel efficiency (Zhang, et al., 2013). Cerium oxide acts as an oxygen buffer in the combustion process donating oxygen and assisting the oxidation of hydrocarbon fuel, thus reducing both harmful emissions and amount of fuel used (Jung, Kittelson, & Zachariah, 2005; Sajeevan & Sajith, 2013). However, only few studies have evaluated the impact of these nanotechnology based fuel additives on pollutant emissions. To assess the impact of cerium oxide nanoparticles on the environment, a laboratory study was conducted. The objective of this experimental study was to examine the effectiveness of the cerium oxide nanoparticles as fuel additive on ultra-low sulfur diesel fuel and its influence on the engine performance and exhaust emission characteristics of a medium-duty compression ignition engine.

MATERIALS AND METHODS

Test Engine. The subject engine was a 2008 model inline 6 cylinder, Cummins ISB diesel engine without aftertreatment control system. The rated power of the engine is 350 HP at 2600 rpm. The engine was equipped with a common rail injection operating with pilot, main, and post injection pulses for combustion control. It was connected to an engine dynamometer (dyno) from SuperFlow, model SF-901. It is a water brake dyno with a "pressure boost" option that allows for higher torque absorption in the low speed range in comparison to the standard SF-901. The project engine installed on the test bench is shown in Figure 1 and the basic specifications are given in Table 1.

FIGURE 8. Photograph of Cummins Engine Installed on the Test Bench

TABLE 9. Engine Specifications

Engine Make/Model	Cummins/ISB (2008)
Engine Configuration	Inline 6-Cylinder
Engine Displacement	6.7 Liters
Rated power	350 HP
Rated Speed	2600 RPM
Peak Torque	827 NM (610 ft-lb)
Compression Ratio	17.3:1
Aspiration	Turbocharged & Aftercooled

A schematic of the experimental set-up used in this research is shown in Figure 2. The fuel flow rate was measured on a volume basis using two KRAL volumeter series OME (accuracy 0.1% of rate) flow meters. The flow meters were installed (in-line) in the diesel fuel supply and return lines respectively to measure fuel consumption. SIERRA 620S precision (accuracy 1.0% of full scale) intake hotwire mass air flow meter was installed at the air intake pipe after the intake air filter, before the turbocharger compressor as shown in Figure 2. Soot measurement probe, gaseous emissions measurement probe, and temperature measurement probe were installed on the tail pipe. The gas emissions of the engine was directly sampled and sent through a temperature controlled (190°C) heated sampling line to the MKS 2030 HS gas analyzer. The soot emissions of the engine was directly sampled and sent through a temperature controlled (50°C) heated sampling line to the AVL-415S smoke meter. A K-type thermocouple was installed to monitor the exhaust gas temperature.

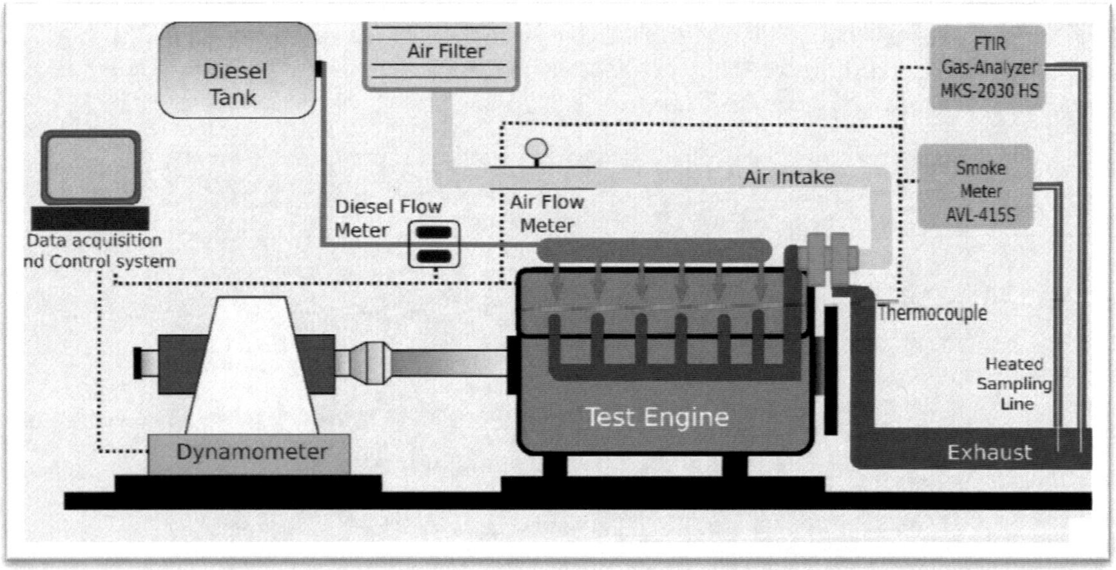

FIGURE 9. Schematic of the Experimental Set-up

Test Cycle. B-type ISO 8178 Test Modes with D-2 weighting factors as shown in Table 2 was selected to evaluate the emissions and performance of the fuels. Points at five engine loads were taken corresponding to 10%, 25%, 50%, 75%, and 100% of the 350 HP maximum load. The engine speed was maintained at 2600 rpm rated speed for all cases. Figure 4 shows the position of the test points in the speed load map. Experiments were carried out at steady state with intake manifold temperature of 45°C and block coolant temperature reaching 88°C.

TABLE 10. ISO 8178 Test Mode and Type D-2 Weighting Factors

Weighting Factors of B-Type ISO 8178 Test Cycles					
Mode Number	1	2	3	4	5
Torque, %	100	75	50	25	10
Speed	Rated Speed				
Constant Speed					
Type D2	0.05	0.25	0.3	0.3	0.1
Engine torque is expressed in percent of the maximum available torque at a given engine speed					
Rated speed is the speed at which the manufacturer specifies the rated engine power					

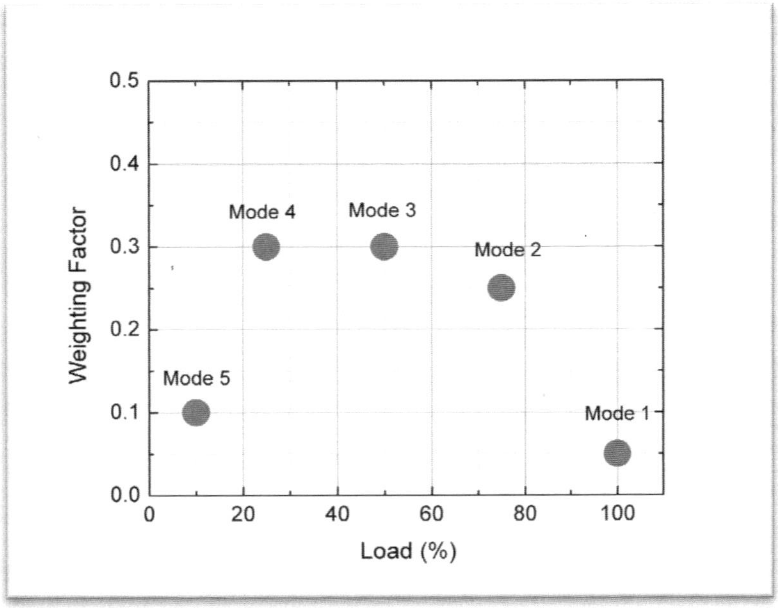

FIGURE 10. Position of the Test Points in the Speed Load Map

Test Fuel. Standard ultra-low sulfur diesel (ULSD) fuel was used as the base fuel. The fuel additive used in this experiment was cerium oxide in the form of commercially available nanoparticle of size 10 to 20 nanometers. Three different fuel additive samples ranging the cerium oxide composition from 10 ppm, 20 ppm and 30 ppm were prepared by adding the nanoparticles to the baseline diesel fuel and agitating for 5 min, sufficient to disperse the nanoparticles to prepare a uniform suspension.

RESULTS AND DISCUSSION

Experiments were performed to investigate the effect of cerium oxides nanoparticles in diesel fuel. The variation of fuel consumption and the emissions of CO, NOx, HC, and smoke were measured under all the desirable loads as prescribed by the ISO 8178 D-2 drive cycle. The results of cerium oxide nanoparticles blended diesel were compared with the results of standard diesel baseline fuel. The operation of the engine was found to be very smooth throughout the speed load map, without any operational problems with cerium oxide blended diesel fuel.

Fuel Consumption. The variation of fuel consumption with engine load at rated engine speed for baseline diesel and cerium oxide nanoparticles blended diesel is shown in Figure 4. From the data, the fuel savings at lower load levels were lower than the higher load level fuel savings. An average reduction of 1% to 4% in the fuel saving was observed for cerium oxide nanoparticles blended fuel when compared to diesel baseline.

Weighted Cycle Averaged Brake Specific Fuel Consumption. The composite weighted cycle averaged engine brake specific fuel consumption (BSFC) of the baseline fuel and the cerium oxide nanoparticles blended diesel are shown in Figure 5. On a cycle-weighted basis, an average reduction of 1% to 3% in the BSFC was observed for cerium oxide nanoparticles blended fuel when compared to diesel baseline. The optimum fuel additive mixture (Diesel+ CeO_2 -30 ppm) reduces BSFC by 3.0% when compared with the baseline fuel.

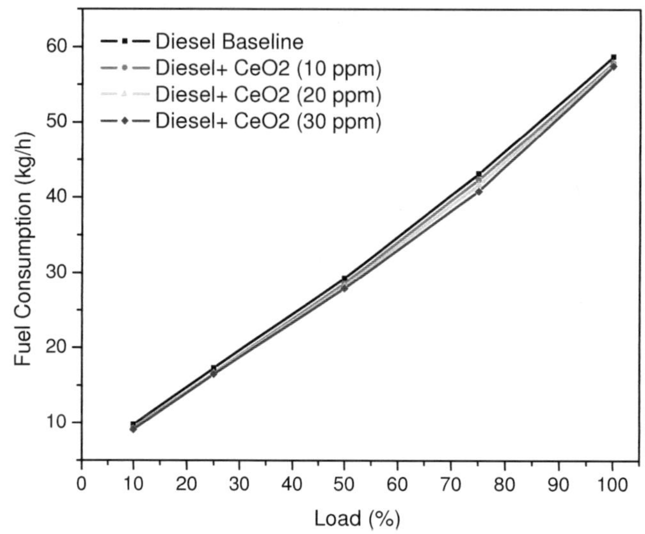

FIGURE 11. Variation of Fuel Consumption in Different Engine Loads at Rated Engine Speed (2600 rpm)

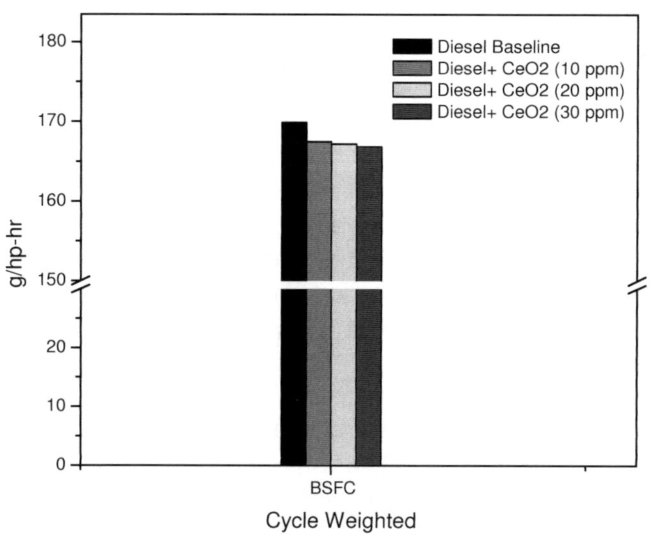

FIGURE 12. Weighted Cycle Averaged Fuel Economy (BSFC) Comparison of Baseline and Fuel with Additive

Exhaust Gas Temperature. The variation of the exhaust gas temperature with different engine load at rated engine speed for diesel fuel and different combination of cerium oxide nanoparticles blended diesel fuel is shown in Figure 6. The exhaust gas temperature is observed to be increased by addition of the cerium oxide in diesel fuel as compared to diesel fuel alone. The reason may be due to the catalytic action of the cerium oxide nanoparticles which improves the combustion process resulting in increased combustion gas temperature.

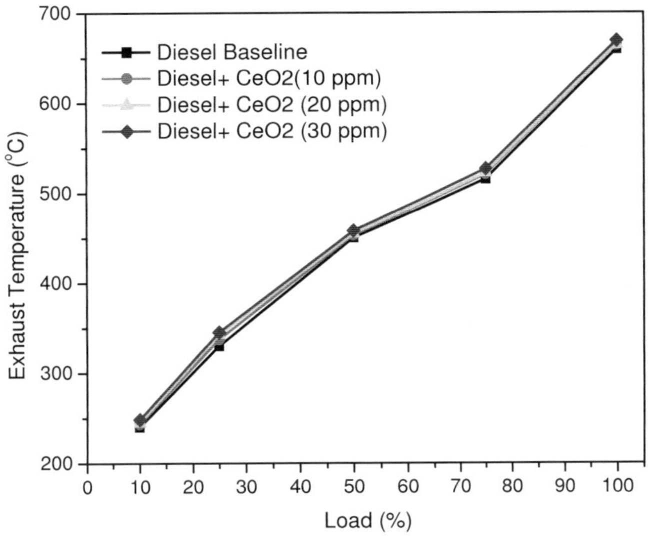

FIGURE 13 . Variation of Exhaust Gas Temperature in Different Engine Loads at Rated Engine Speed (2600 rpm)

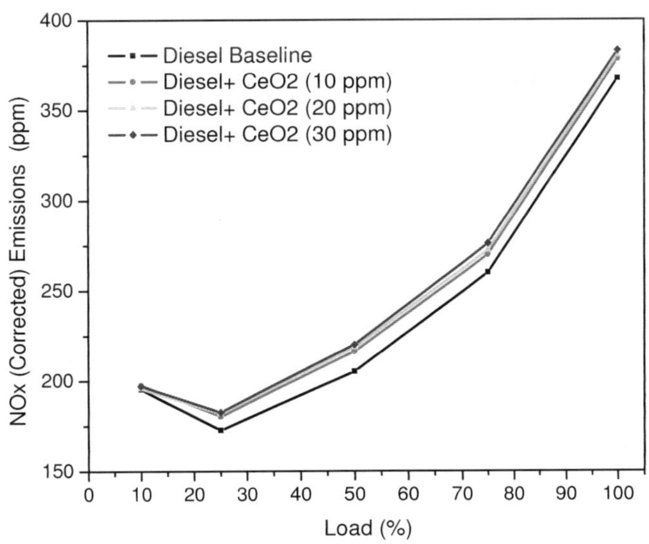

FIGURE 14. Variation of NO_x Emissions in Different Engine Loads at Rated Engine Speed (2600 rpm)

Oxides of Nitrogen (NO_x) Emission. Nitrogen oxides, NO_x, primarily nitric oxide (NO) and nitrogen dioxide (NO_2), are produced by oxidation of atmospheric nitrogen in combustion. Nitrogen compounds in diesel fuel also act as an additional source of NO_x. The formation of NO_x is propagated by higher combustion temperatures and excess oxygen. The variation of NO_x emission with engine load at rated speed for baseline diesel and cerium oxide nanoparticles blended diesel is shown in Figure 7. An average increment of 3% to 7% in the NO_x emissions was observed for cerium oxide nanoparticles blended fuel when compared to diesel baseline. The additional oxygen content in the cerium oxides nanoparticles blended fuel increases the in cylinder combustion temperature which in turn increases the NO_x emission.

Carbon Monoxide (CO) Emission. Carbon monoxide, CO, forms in internal combustion engines as a result of incomplete combustion when a carbon based fuel undergoes combustion with insufficient air or cold engine temperature. The lack of oxygen and/or low temperatures in the cylinder results a failure in the oxidization process of CO into CO_2. The variation of CO emission with engine load at rated speed for the baseline diesel and cerium oxide nanoparticles blended diesel fuel is shown in Figure 8. From these results it can be seen that cerium oxide nanoparticles blended diesel fuel clearly exhibits lower CO emissions. An average reduction of 10% to 20 % in the CO emissions was observed for cerium oxide nanoparticles blended fuel when compared to diesel baseline.

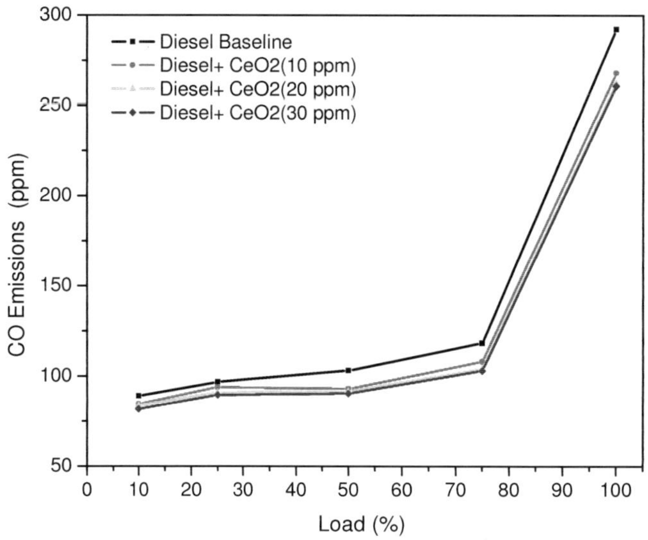

FIGURE 15. Variation of CO Emissions in Different Engine Loads at Rated Engine Speed (2600 rpm)

Hydrocarbon (HC) Emission. Hydrocarbon, HC, emissions are often a product of incomplete combustion of the hydrocarbon fuel. In compression ignition diesel engines, there are two major sources of hydrocarbon emissions. First, during the ignition delay period, the fuel mixture is too lean, and thus outside the flammability boundaries of the fuel form unburned hydrocarbon. Later in the combustion process when fuel leaves the injector at low velocity resulting in under-mixing of the fuel, and consequently form unburned hydrocarbon. The variation of HC emission with engine load at rated speed for the baseline diesel and cerium oxide nanoparticles blended diesel fuel is shown in Figure 9. A reduction of HC emissions was observed on cerium oxide nanoparticles blended diesel fuel. An average reduction of 10% to 25 % in the HC emissions was observed for cerium oxide nanoparticle blended fuel when compared to diesel baseline. The catalytic action of cerium oxide improves the combustion process leading to oxidation of unburnt HC resulting in decrease in HC emission.

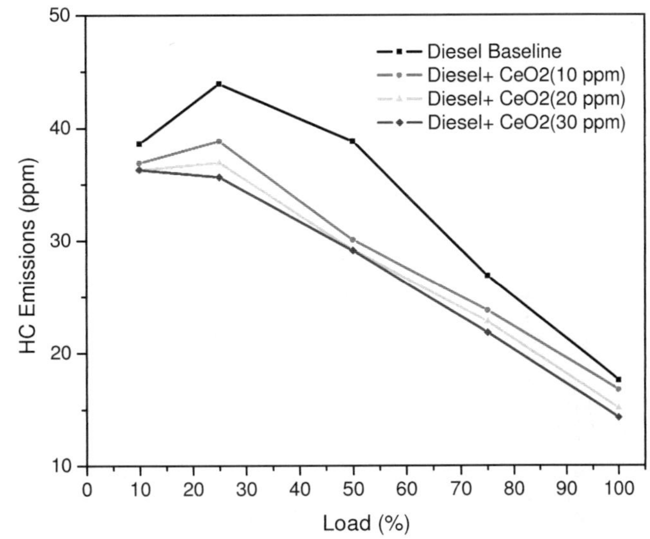

FIGURE 16. Variation of HC Emissions in Different Engine Loads at Rated Engine Speed (2600 rpm)

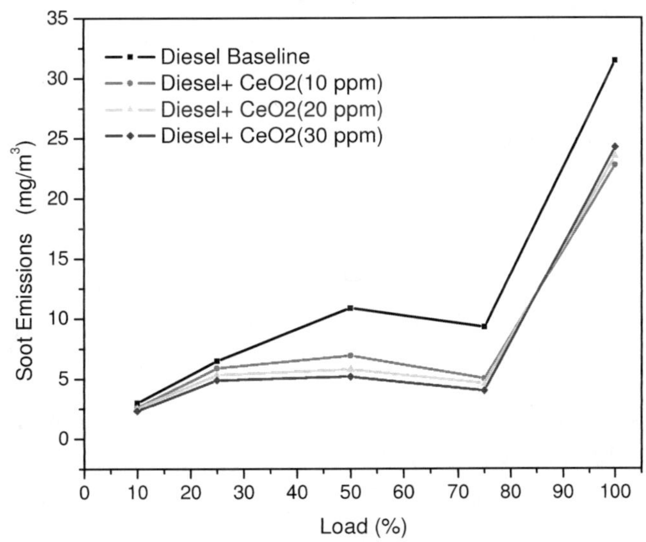

FIGURE 17. Variation of Soot Emissions in Different Engine Loads at Rated Engine Speed (2600 rpm).

Soot Emission. Soot consists of a complex aggregation of solid and liquid material that can be present in diesel engine exhaust. Generally, the occurrence and concentration of soot depends on the fuel composition, engine operating parameters (i.e. load, speed) and exhaust conditions. The variation of soot concentration in the exhaust with engine load at rated speed for the baseline diesel and cerium oxide nanoparticles blended diesel is shown in Figure 10. An average reduction of 30% to 60% in the soot emissions was observed for cerium oxide nanoparticles blended fuel when compared to diesel baseline. This is mainly due to improvement in the combustion process leading to complete oxidation of carbon in to CO_2.

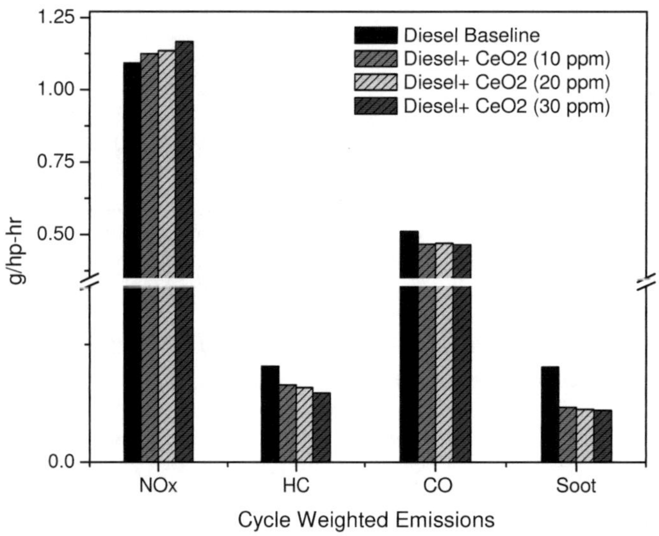

FIGURE 18. Weighted Cycle Averaged Emissions Comparison of Baseline and Fuel with Additive.

CONCLUSION

Fuel additive can vary significantly in its design and effects on improvement in fuel economy and potential emission reduction. Fuel additives incorporating of cerium oxide in nanoparticle form is found to be more effective due to higher surface to volume ratio, catalytic activity, and chemical properties that change the properties of fuel, which will affect the emissions and fuel economy through combustion. In this work, the performance and exhaust emission characteristics of the diesel engine were evaluated for diesel and diesel with different combination of cerium oxide nanoparticles and the results were compared with the standard diesel fuel. The major observations are listed below.

- Slight improvement in brake specific fuel consumption was observed with cerium oxide nanoparticles blended diesel fuel. The specific fuel consumption decreases with an increase in the dosing level of nanoparticles.
- Increase in NO_x emission was observed with cerium oxide nanoparticles blended diesel fuel.
- Significant reduction in HC, CO, and soot emission were observed with cerium oxide nanoparticles blended diesel fuel.

In conclusion, the cerium oxide nanoparticles are efficient in improving the properties of ULSD fuel that in turn improve the engine performance and reduce some hazardous air pollutants from diesel engine.

ACKNOWLEDGEMENTS

The authors gratefully acknowledge the assistance of Prof. Ron Matthews at University of Texas, Austin, and John Colvin at Houston Advanced Research Center (HARC). This project was supported by the HARC's Environmentally Friendly Drilling (EFD) Systems Program. The authors also gratefully acknowledge financial support from EFD program industry sponsors.

REFERENCE

ACEA, A. E. (2006). *Worldwide Fuel Charter* (4th ed. s.l.:s.n. ed.).

Huang, Y., Colvin, J., Wijesinghe, A., Wang, M., Hou, D., & Fang, Z. (2014). Dual Loop EGR in Retrofitted Heavy-Duty Diesel Application. *SAE Technical Paper*, (pp. 2014-01-1244).

Jung, H., Kittelson, D. B., & Zachariah, M. R. (2005). The influence of a cerium additive on ultrafine diesel particle emissions and kinetics of oxidation. *Combustion and Flame, 142*(3), 276-288.

Kannan, G., Karvembu, R., & Anand, R. (2011). Effect of metal based additive on performance emission and combustion characteristics of diesel engine fuelled with biodiesel. *Applied Energy, 88*(11), 3694-3703.

Mukhopadhyay, P., & Chakraborty, R. (2015). Effects of Bioglycerol Based Fuel Additives on Diesel Fuel Properties, Engine Performance and Emission Quality: A Review. *Energy Procedia, 79*, 671-676.

Sajeevan, A. C., & Sajith, V. (2013). Diesel Engine Emission Reduction Using Catalytic Nanoparticles: An Experimental Investigation. *Journal of Engineering, 2013-589382.*

Tiwari, A. (2015). Converting a Diesel Engine to Dual-Fuel Engine Using Natural Gas. *International Journal of Energy Science and Engineering, 1*(5), 163-169.

Wijesinghe, A., Lafleur, C., Meng, F., Colvin, J., & Haut, R. (2018). Fuel Economy and Emission Characteristics of a High Horsepower Natural Gas/Diesel Dual-Fuel Engine in Oil & Gas Operations. *IADC/SPE Drilling Conference and Exhibition.* Society of Petroleum Engineers.

Zhang, J., Nazarenko, Y., Zhang, L., Calderon, L., Lee, K.-B., Garfunkel, E., . . . Mainelis, G. (2013). Impacts of a Nanosized Ceria Additive on Diesel Engine Emissions of Particulate and Gaseous Pollutants. *Environmental science & technology, 47*(22), 13077-13085.

A Novel $Cu_wMg_{2-w}Mn_yAl_{1-y}O_x$ Catalyst from LDHs for NH_3-SCR Catalyst and Resistance to Its SO_2 and H_2O

Qinghua Yan and Qiang Wang
(Beijing Forestry University, Beijing, China)

Nitrogen oxides (NO_x) is an important part of waste incineration or coal-fired flue gas, which is the main pollutant of air pollution and can cause great harm to the ecological environment and human health. In the current De-NO_x technology, SCR due to high efficiency, maturity, and has been widely used. Selective Catalytic Reduction (SCR) is the use of ammonia, urea, hydrocarbons, and H_2 as a reducing agent, under a certain temperature and the role of the catalyst, the selective reduction of NO_x in the flue gas to N_2 and H_2O. The core of SCR technology lies in the development of catalysts. In this text, we report a novel NH_3-SCR catalyst ($Cu_wMg_{2-w}Mn_yAl_{1-y}O_x$) synthesized from layered double hydroxides with superior activity in a wide temperature range and improved SO_2 and H_2O resistance comparing to the conventional doped Mn/γ-Al_2O_3. The results show that the CuMgMnAl-CO_3 layered double hydroxides can be successfully synthesized when the M^{2+}/M^{3+} molar ratio is 2, and the properties of the catalyst can be measured. The optimal catalyst $Cu_{0.5}Mg_{1.5}Mn_{0.5}Al_{0.5}O_x$ resulted in a high NO_x removal efficiency of 87.0%–96.6% at a wide range of temperatures of 100–250 °C, which is much higher than that of Mn/γ-Al_2O_3 catalyst (35.0%–67.2%). Besides, taking into account the impact of SO_2 and H_2O poisoning in the low-temperature, the performance of the catalyst against SO_2 and H_2O poisoning was tested at 150 °C. The results show that the conversion of NO_x was 72.2% and 78.2% in the poisoning test of 100 ppm SO_2 and 5% H_2O, respectively. The conversion of NO_x was 68.2% in the 100 ppm SO_2 + 5% H_2O synergistic poisoning test. Thus, $Cu_{0.5}Mg_{1.5}Mn_{0.5}Al_{0.5}O_x$ exhibited significant resistance to SO_2 and H_2O due to the introduction of Cu and Mg to the Mn-based NH_3-SCR catalyst.

LAND (SOIL, SOLID WASTE) POLLUTION AND REMEDIATION

An Investigation into the Microbial Attenuation of Hydrocarbon Contaminated Soils

Grace Zvinowanda and ***Memory Tekere***
(Department of Environmental Sciences, Science Campus, University of South Africa Florida 1710, Johannesburg, South Africa)

Environmental contamination by hydrocarbons is of major concern as most of the component hydrocarbons are toxins with far reaching imparts to the living organisms in the ecosystem. Microorganisms are known to play a major role in the environmentally friendly remediation of hydrocarbon contaminated soils through biodegradation, bio-augmentation and bio-stimulation. This study was designed to investigate the biodegradation of petroleum hydrocarbons in hydrocarbon impacted soils, investigating the extent of soil contamination, physicochemical and microbial properties of soils as well as potential to enhance bioremediation using bio-stimulation. Hydrocarbon contamination levels were analysed and the identified hydrocarbon included decane, undecane, hexadecanal, 2-ethylcridine, octadecane and 1-iodo. Soil characterization for N, P and microbial content, and the supplementation of nitrogen and phosphorus to stimulate naturally adapting microorganisms were investigated. Naturally adapting microbial species capable of degrading hydrocarbons that were identified included *Pseudomonas, Bacillus, S. marcescens, Flavobacterium, Micrococcus, Streptomyces* and *yeasts.* The N: P nutrient ratio and moisture levels were identified as potential limiting factors in microbial degradation. Soils weighing 10kgs with hydrocarbons levels of about 265mg/kg were subjected to eight treatments with seven different combinations of N: P concentrations including the control. Moisture was added through irrigation, and tilling for aeration was done on a weekly basis. Changes in Total Petroleum Hydrocarbons, C: N: P, microbial mass and pH were evaluated over 111 consecutive days. The TPH concentration was reduced by 73% from the initial concentration within the first 74 days. Beyond 74 days there were no significant changes in the TPH concentration and this was attributed to the presence of less recalcitrant more complex insoluble hydrocarbons which need more time. It can be concluded however that bio-stimulation of natural microbial populations can be used in the bioremediation of petrochemical impacted soils with potential to reduce the likely environmental impacts of the pollution.

Interacting Effects of Salinity, Sewage Sludge, and Earthworms on the Fractionation of Heavy Metals in Soil around A Lead-Zinc Mine

Ghasem Rahimi, Fatereh Karimi
(Bu-Ali Sina University, Hamedan, Iran)

This study assessed the effects of salinity and sewage sludge on the fractionation of Zn and Cu in soils around a lead-zinc mine, as well as their uptake by earthworms (*Eisenia fetida*) in order to identify novel methods for employing earthworms in waste management techniques. *E. fetida* specimens were kept under laboratory conditions for 42 days. The first treatment involved the addition of 0%, 2%, 4%, or 8% (w/w) sewage sludge to contaminated soil. In the second treatment, NaCl was added to the soil at concentrations of 0, 1170, 2340, 3510, or 4680 mg L^{-1}. The mortality rate was assessed on days 3, 7, 14, 21, 28, 35, and 42. The Cu and Zn fractions in the contaminated soils and their uptake by earthworms were assessed only on day 42. The results indicated that sewage sludge and salinity had negative effects on the earthworms. The lowest pH (7.56) was found with the highest level of sewage sludge in the absence of earthworms. Comparisons of the sewage sludge treatments with and without earthworms showed that their activity decreased the amount of Zn bound to organic matter. As the salinity increased, the exchangeable and organic forms of Zn increased whereas the carbonate and residual forms decreased. After incubating the contaminated soils for 42 days, the Zn and Cu contents increased with the earthworms when exposed to higher levels of sludge. Increases in Zn and Cu were also observed at the maximum salinity application rate (electrical conductivity = 8 dS m^{-1}).

A New Concept for Utilization of Industrial Solid Wastes in Building Structures

Arunima Shukla and ***Ashok N. Bhaskarwar***[*]
(Department of Chemical Engineering, Indian Institute of Technology, Delhi, Hauz Khas, New Delhi 110 016, INDIA)
[*]Email: ashoknbhaskarwar@yahoo.co.in

ABSTRACT: The solid-waste management is one of the important issues of present times due to rapid industrialization. Research studies reveal that globally, a total estimated annual industrial solid-wastes generation was about 11 billion tons for 2002, which is expected to increase in the coming years (Chavan and Patil, 2013).

We have developed a new technology for making light-weight aerated concrete blocks, which allows the integration of industrial solid wastes. In the present study, solid waste of autoclaved aerated concrete (AAC) industries and a by-product of coal-fired power plants known as fly ash have been successfully integrated into colloidal-gas aphrons (CGA) based aerated slurry. Experimental results have shown that it is possible to utilize flyash up to 75 % and AAC solid waste up to 25 % with respect to total mass of solids in aerated concrete.

INTRODUCTION

Now-a-days, industrial solid waste is one of the major environmental issues. The construction materials can be established as secondary industry in the area of solid waste and recycling (Chavan and Patil, 2013). There is plenty of literature available on the generation and utilization of flyash. In USA, the annual production of flyash is 71 metric tons, of which only 31% is being utilized (Singh and Gupta, M Singh, 2013). In India, the percentage utilization of flyash is around 10 to 15%, where more than 88 million tons of fly ash is getting generated per year (Siddique, 2004). Due to high demand of energy, the quantity of fly ash, as a solid waste, is also increasing (Singh and Gupta, M Singh, 2013). In the autoclaved aerated concrete (AAC) manufacturing industries, a considerable amount of solid-waste, to the tune of at least about 10 to 12 % of total production, is generated because of breakage due to thermal stresses arising during autoclaving, and during handling and transportation.

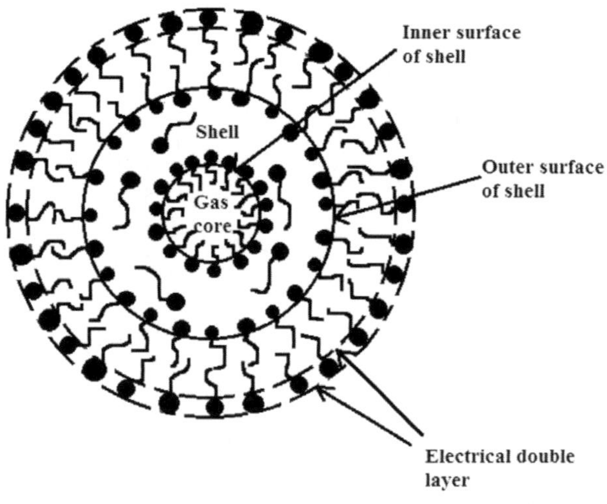

Figure 1. Conceptual Diagram of CGA, As Explained by Sebba.

Considering the above factors, we have developed a new technology which will simultaneously manufacture light weight aerated concrete blocks, and utilize considerable amount of solid wastes generated by various industries. Aerated concrete is known for its insulating properties, as well as for its structural properties (Narayanan and Ramamurthy, 2000). Its properties mainly depend on the method of aeration, as void in the matrix is the key ingredient of light weight aerated concrete. We are working with colloidal gas aphrons (CGAs) as an aerating medium for creating the voids in the cement matrix. CGAs are micro-bubbles, typically in the range of 10 to 100 μm diameter, having high interfacial area, flow properties similar to water, and high stability (Jauregi, Mitchell, and Varley, 2000; Sebba, 1987). The conceptual diagram of CGAs is shown in Figure 1. Sebba had proposed that CGAs have inner gas phase, surrounded bi-layer of surfactant and an electrical double layer.

MATERIALS AND METHODS

Raw Material. Sodium lauryl sulfate (Fisher scientific), ordinary portland cement (43 grade), flyash, autoclaved aerated concrete waste (Biltech Ltd.) of two sizes.

Equipment.
Accelerated curing tank (Ferrotek equipment)
Stirrer (Remi elecktrotechnik Ltd., 220/230 V, 1 phase, 50 Hz)
Hot air oven (Macro scientific works pvt. Ltd.)
Optical microscope (Motic microscope)
SEM (Hitachi, TM 3000, tabletop microscope)

Experimental Set-Up. The experimental set-up for CGA formation consists of a spinning disc CGA generator which is an open cylindrical vessel, attached with inverted L- shaped baffles. Stirring is done by a circular disc mounted on stirrer.

Methodology. Sodium lauryl sulfate was mixed with fixed volume of ordinary tap water for 15 minutes at 1500 rpm, in order to form a homogenous solution. This is followed by increasing the rpm of stirrer to 5500 rpm, which causes generation of waves at the surface of surfactant solution. The generated waves create thin film of gas, which being unstable, break up into small sized droplets, known as CGAs. Continuous recirculation of surfactant solution led to increase in the air hold up (up to 0.70 to 0.80) in CGAs. The CGAs generated were mixed with solid waste and cement. The samples were allowed to set for 24 hours, followed by de-moulding and steam curing at 65°C for 4 to 5 hours. The samples were then dried in oven for 24 hours, to measure oven dried weight at 110±10°C.

RESULTS AND DISCUSSION

Colloidal Gas Aphrons. The CGAs were prepared by using anionic surfactant, sodium lauryl sulfate ($NaC_{12}H_{25}SO_4$). Photomicrographs of CGAs were taken using optical microscope, as shown in Figure 2. Motic software was used to determine the sizes of the micro-bubbles.

Solid Waste. Fly ash was obtained from two different resources in order to study its effect on the properties of aerated concrete. The SEM images of fly ash are shown in Figure 3. It was observed that the type 1 flyash was finer as compared to type 2. The solid waste from AAC was utilized in the form of pebbles and granules, in order to obtain re-integrated blocks of aerated concrete. Higher densities of aerated concrete were targeted to get the intact blocks in case of AAC solid waste utilization. The SEM images of different AAC waste were observed (Figure 4) to study the surface morphology of the waste. The needle-like and plate-like structures were observed in pebbles, which indicate that ettringite and calcium silicate hydrate were already developed (Stutzman, 2001). However, in case of granules, such kind of structures could not be observed.

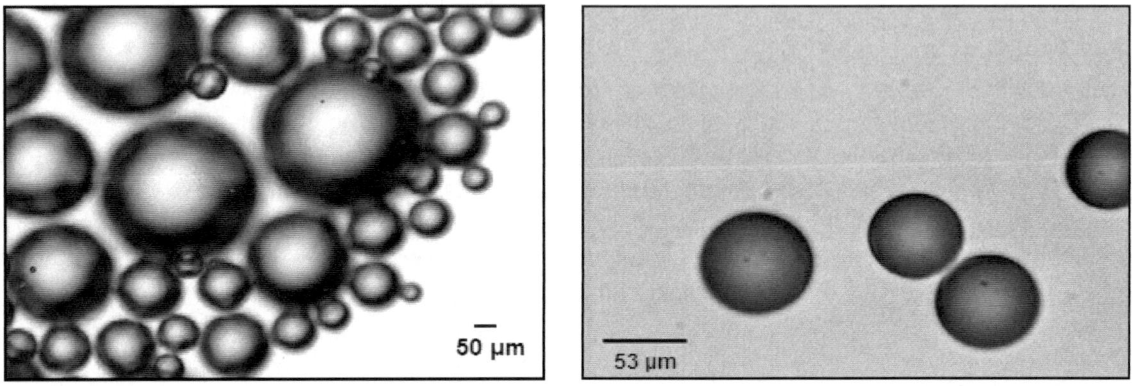

Figure 2. Photomicrographs of CGAs at different magnification: (a) At 10X; (b) At 40 X.

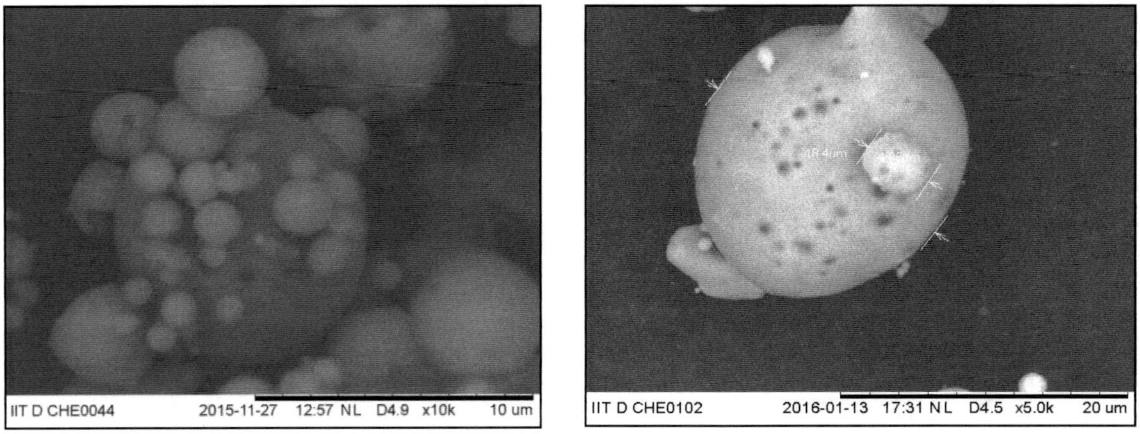

Figure 3. Scanning electron micrographs of fly ash type 1 (left) and type 2 (right).

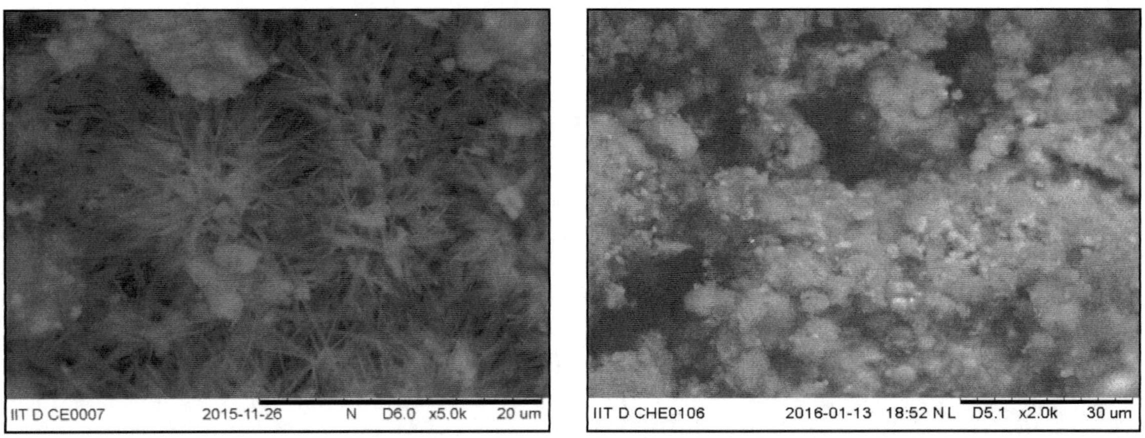

Figure 4. Scanning electron micrographs of autoclaved aerated concrete solid wastes in the form of pebbles (left) and granules (right).

Effect of Fly Ash on the Properties of Aerated Concrete. The aerated slurry has variations in the CGA distribution, which results in difference in the properties of concrete blocks. Thus, the samples, in each set of experiments, were characterized based on pouring of samples, the 1st poured was named as 1st cut, similarly, the 2nd and 3rd poured were named as 2nd and 3rd cut respectively. The weight percentages of cement, flyash and water were kept constant, within the range of ±0.15%. It was observed that the 1st cut was lighter, as compared to 2nd and 3rd cut, because of more air hold-up. Thus with increase in air weight % in slurry, dry density of the sample decreases, as shown in Figure 5. It was also noted that with the increase in air weight % in slurry, the compressive strength of the samples initially decreases till 0.1 air weight % w/w, and then increases slightly, as shown in Figure 6. The increase is more significant in type 2 flyash samples. This is due to the fact that porosity directly influences compressive strength (Erniati et al., 2015; Zhao et al., 2014) and the colloidal gas aphrons provide a well distributed pore size distribution in the cement matrix.

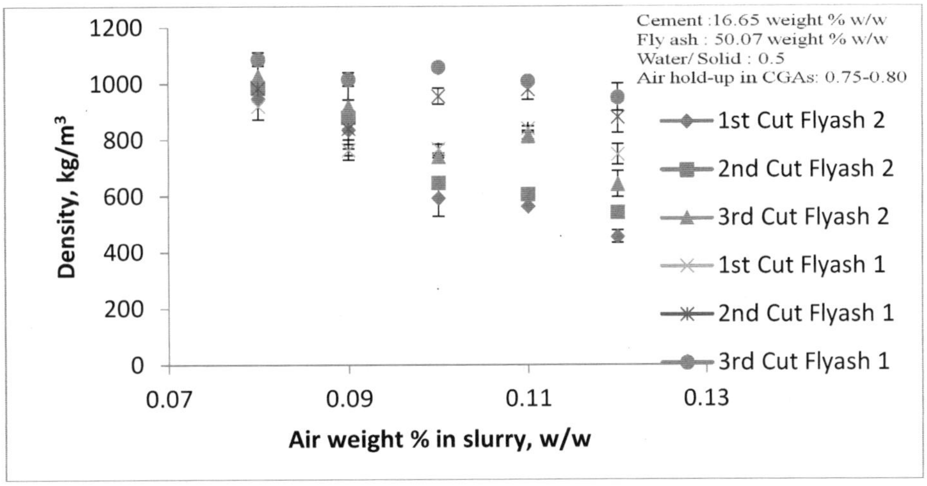

Figure 5. Density vs. Air Weight % w/w in Slurry.

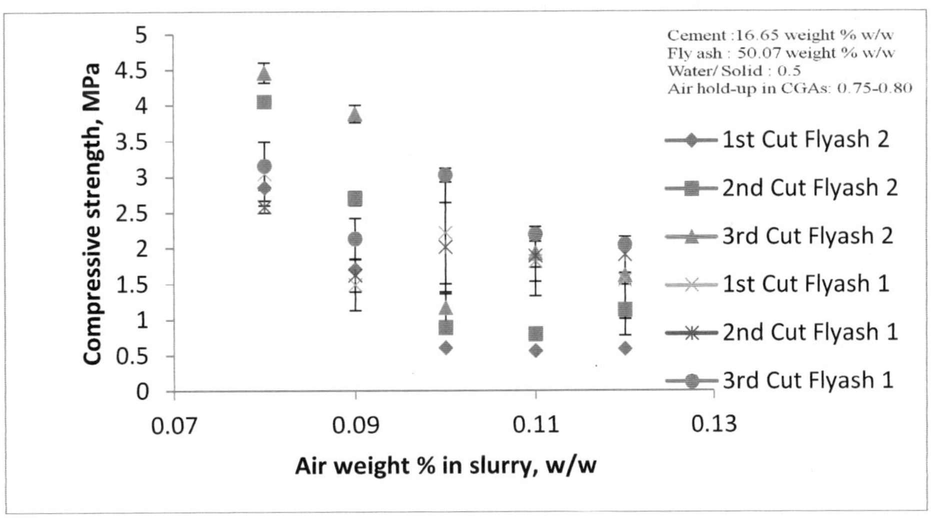

Figure 6. Compressive Strength vs. Air Weight % w/w in Slurry.

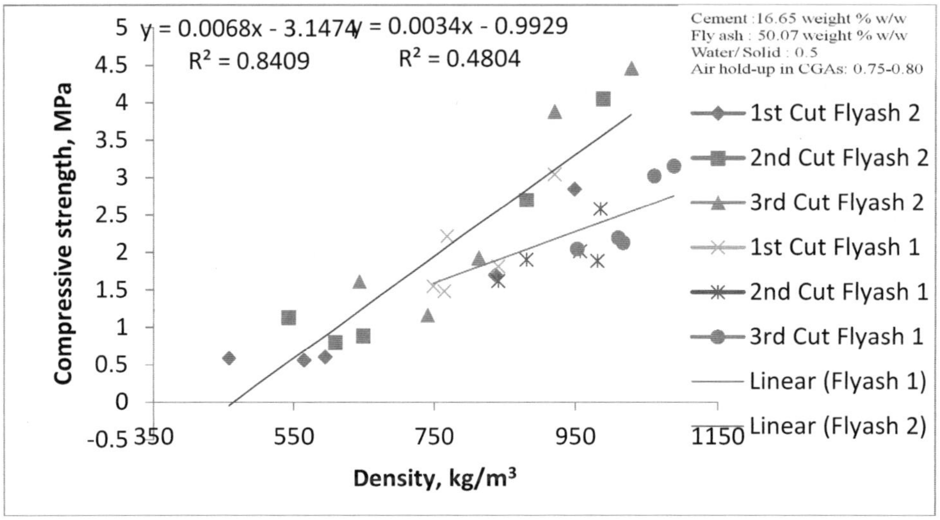

Figure 7. Variation in Compressive Strength vs. Density Based on Pouring of Samples.

All the data points were plotted (Figure 7), based on cuts, in order to check the correlation between density and compressive strength, as it was reported that compressive strength increases linearly with density (Narayanan and Ramamurthy, 2000).The range of densities from 712 to 1112 kg/m^3 and compressive strengths 1.01 MPa to 3.49 MPa were observed for the fly ash type 1, and densities range of 392 to 1072 kg/m^3 and compressive strengths 0.52 MPa to 4.6 MPa were observed for fly ash type 2, respectively. Type 2 flyash shown more linear trend as compared to type 1.

Autoclaved Aerated Concrete Solid Waste. The solid waste from autoclaved aerated concrete industries was incorporated into the CGAs based slurry, varying from 5 to 25 weight % of total solid mass. The water to solid ratio was fixed as 0.5. The variation in the composition is given in Table 1.

Table 1. Varying Percentages of Composition With Solid AAC Waste

S.No.	AAC waste (% of total solid mass)	Cement (weight %)	Flyash (weight %)	Water (weight %)	Air (weight %)
1	5	21.09	42.19	33.31	0.06
2	10	19.98	39.98	33.31	0.06
3	15	18.88	37.76	33.32	0.05
4	20	17.77	35.54	33.32	0.05
5	25	16.66	33.32	33.32	0.04

All the samples were intact without any deformation. A good bonding between solid waste and aerated slurry was observed without any crack formation, as shown in Figure 8. The color difference between AAC waste and slurry was observed, which was marked on the sample using a marker, in order to take the SEM image of interface.

The densities and compressive strengths of the samples were also measured and plotted in Figure 9 and Figure 10, respectively. The ranges of densities were from 652 to 896 kg/m^3 and compressive strengths 0.26 MPa to 0.75 MPa.

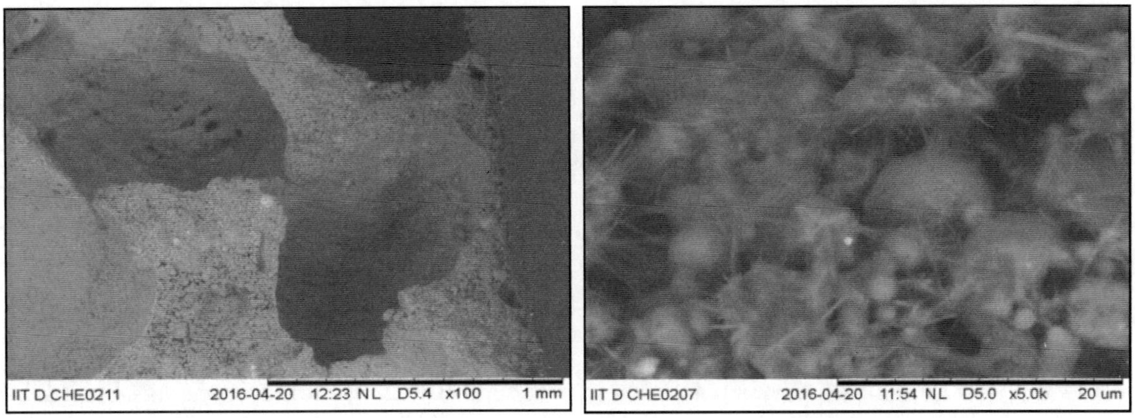

Figure 8. Scanning Electron Micrographs of Bonding between AAC Solid Waste and CGAs Based Aerated Slurry, at Lower Magnification (Left) and at Higher Magnification (Right) Respectively.

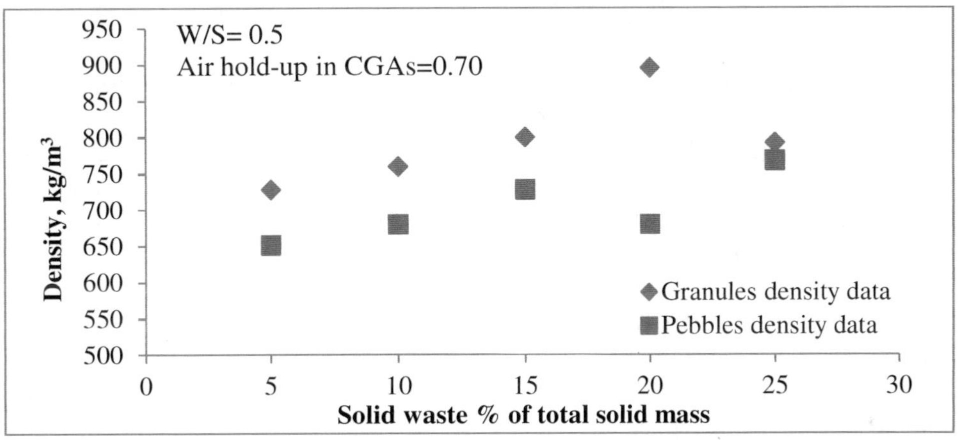

Figure 9. Density vs. Solid Waste Percentages of Autoclaved Aerated Concrete (AAC) sample.

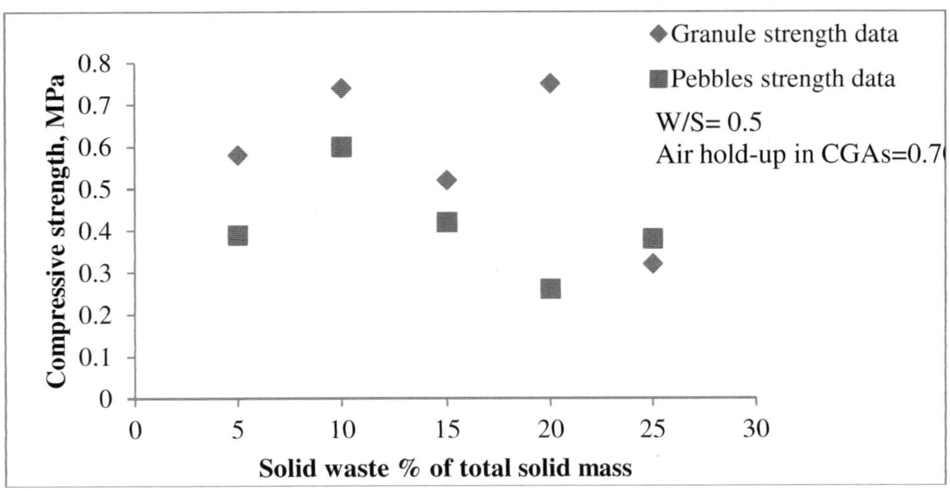

Figure 10. Compressive Strength vs. Solid Waste Percentage of Autoclaved Aerated concrete (AAC) Samples.

CONCLUSIONS

The results revealed that the industrial solid wastes such as fly ash, and aerated concrete, in the form of granules and pebbles, can be utilized using CGAs based technology. A wide range of strengths and densities were observed using varying composition and densities. It is, thereby, possible to utilize up to 25% of AAC solid waste with respect to total mass of solid. Moreover, there is a lot of scope for the improvement of compressive strength of the AAC solid waste sample in future, as well as utilization of other industrial solid wastes.

ACKNOWLEDGEMENT

The authors would like to thank and acknowledge foundation for innovation and technology transfer (FITT), India for their support in patenting the work and Biltech Building Elements Limited, Palwal, India for their help in getting solid waste and TSDP, DST for their financial support.

REFERENCES

Chavan, P G, and S B Patil. 2013. "Substantial Use of Solid Waste." *International Journal of Advanced Technology in Civil Engineering* 2(1): 61–63.

Erniati, M. Wihadi Tjaronge, Zulharnah, and Ulva Ria Irfan. 2015. "Porosity, Pore Size and Compressive Strength of Self Compacting Concrete Using Sea Water."

Jauregi, Paula, Geoffrey R. Mitchell, and Julie Varley. 2000. "Colloidal Gas Aphrons (CGA): Dispersion and Structural Features." *AIChE Journal* 46(1): 24–36.

Narayanan, N., and K. Ramamurthy. 2000. "Structure and Properties of Aerated Concrete: A Review." *Cement and Concrete Composites* 22(5): 321–29.

Sebba, F. 1987. *Foams and Biliquid Foams-Aphrons*. Wiley, Chichester, Ž. England.

Siddique, Rafat. 2004. "Performance Characteristics of High-Volume Class F Fly Ash Concrete." *Cement and Concrete Research* 34(3): 487–93.

Singh, S P, and S P Gupta, M Singh. 2013. "Fly Ash Production and Its Utilization in Different Countries." *Ultra Chemistry* 9(1): 156–60.

Stutzman, Paul E. 2001. "Scanning Electron Microscopy in Concrete Petrography." *Materials Science of Concrete Special Volume*: 59–72.

Zhao, Haitao, Qi Xiao, Donghui Huang, and Shiping Zhang. 2014. "Influence of Pore Structure on Compressive Strength of Cement Mortar." *TheScientificWorldJournal* 2014: 1–12.

Analysis of Municipal Solid Waste Producing Law and Internet Technology-based Enterprises in China

Jinyan Li, Haizhen Yang
(Tongji University, Shanghai, China)

ABSTRACT: Due to the large population and consequent production of municipal solid waste (MSW), the classification of waste is an increasing important part of environmental management in China. The effective classification helps to reduce the amount of MSW initially and the pressure of subsequent disposal. This paper presents the MSW producing law in China's urban area and investigates a number of information technology based enterprises engaged in MSW classification. The multiple regression analysis of the waste data and development level expressed by urban population, gross domestic product (GDP), disposable income per capita of urban residents, consumption level per capita of urban residents and consumer price index (CPI), revealed a positive relationship between them (R^2=0.9978). Furthermore, Chinese enterprises are exploring ways to waste classification and recycling with Internet innovation such as Computer Image, Sensor System, Android Application and so on. Some have obtained success by providing software and hardware services, some by developing specific and precise waste category classification. Due to the Internet practices, MSW management in cities becomes more intelligent than ever. The MSW producing law and waste classification practices provide essential references for decisions in structuring promotion of sustainable behavior and implementing local initiatives to get out of "junk dilemma".

INTRODUCTION

Municipal solid waste (MSW) is an increasing environmental challenge around the world, causing hazards to both inhabitants and society. Having a good command of MSW producing law and socioeconomic factors is beneficial to decrease the MSW initially. Literature on this issue reveals that the relationship between MSW generation and factors varies with respect to the developmental stage of an area. In developing countries, economic development contributes mainly to waste generation. The income and other five economic factors explained 88.8% of waste generation in Oyo, Nigeria (Afon & Okewole, 2007). In Brazil, a multicollinearity analysis resulted the income per capita (r_p = 0.607) as a major drive of waste generation(Vieira & Matheus, 2018). In India, upper-income class brought about more MSW than the lower income class(Khan, Kumar, & Samadder, 2016). On the contrary, waste generation in developed countries decreases as economic development increases. Production of MSW increased in Finland until 1990, declined to year 1997, increased to 2000 and then declined again while the economy increased continuously almost all the time(Sokka, Antikainen, & Kauppi, 2007). In New York City, MSW peaked in 1940, was at its lowest in the 1960s, and remained steady after the 1980s(Walsh, 2002). As for China, researchers have conducted many surveys and analysis on MSW generation factors in city or province level within a specific period. It is crucial to master the relationship between socio-economic factors and MSW generation on a national scale for policy making. Hence the focus is presented in this paper and the multiple regression analysis is chosen as the method.

Besides, with the rapid development of information technology, enterprises based on the Internet technology are playing significant roles in MSW reducing, reusing and recycling (3Rs). Innovative technologies such as Computer Image, Radio Frequency Identification (RFID), Sensor System, and Global System for Mobile Communication (GSM) / General Packet Radio Service (GPRS) are widely used to the whole management process of MSW, including collection, transportation, storage, recycling and disposal. J Bobulski et.al created a plastic waste database of objects images constituting the typical contents of municipal waste. By using methods of Computer Vision, plastic waste can be automatically selected on the sorting lines businesses for waste disposal (Bobulski & Piatkowski, 2017). M Arebeya et.al studies the application of advanced computer image processing techniques integrated with communication technologies such as RFID, GPS and GPRS with a camera. It aimed to solve the problem of MSW collection and automated bin level detection (Arebeya, Hannana, Hussaina, Zamanb, & Basric). By the means of advanced technologies, Chinese enterprises and cities are struggling their way out to an integrated and cost-effective MSW management. A data acquisition equipment and cloud platform are designed for Laogang Solid Waste Recycling Base, the biggest waste recycling center in Shanghai, for better monitoring and analysis of the energy internet of Laogang. And the cloud platform is based on the three-layer cloud technology: Infrastructure-as-a-Service (IaaS), Platform-as-a-Service (PaaS) and Software-as-a-Service (SaaS) (Feng et al., 2017; Mian, Zeng, Bin Nasry, & Al-Hamadani, 2017; Wang et al., 2018; Yi-Qi et al., 2017).

An attempt has been made to provide a comprehensive understanding of socioeconomic status on MSW generation and the software of Matlab was utilized in this study. And an overall picture of enterprises based on information technology in China was presented as well to offer reference of relieving the intractable problems.

MSW PRODUCING LAW

Population and MSW Generation. The data of MSW generation are collected from China Statistic Yearbook on Environment provided by National Bureau of Statistics of China and Ministry of Ecology and Environment of China, while the others are collected from National Data, organizations and papers.

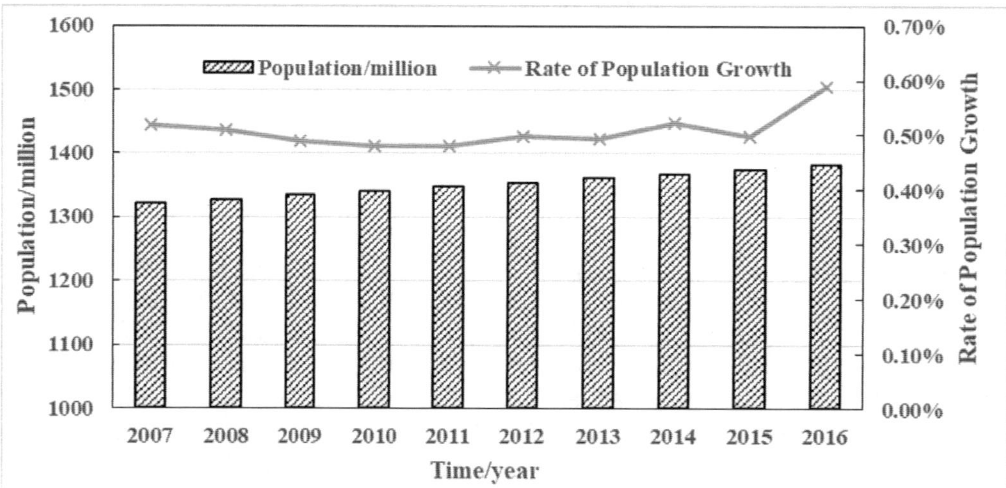

FIGURE 1. Population and Rate of Population Growth in China (2007–2016)

According to the data issued by the United Nations in 2017, China's population reached 1405 million while the world population reached 7500 million, of which 18.82% are Chinese. As a result of the government's great efforts, the growth rate of population in China is 5.19% and the average annual growth is only 0.51% in the past decade (2007-2016), as shown in Figure 1. China is composed of cities, counties and countries, with diverse MSW features and quantities. In 2016, the population of China's urban areas and counties was 631 million, accounting for 45.7% of the total population. The quantity of MSW reached 270 million tons totally, 200 million tons in urban and 70 million tons in countis respectively, which was 5% more than in 2015. In the past decade, MSW generation increased dramaticly from 222 million tons in

2007 to 270 million tons in 2016, as shown in Figure 2. The annual increase rate is 6–8% from 2007 to 2010 and 3–6% after 2010. Note that 2008 and 2013 both emerged slight garbage reduction.

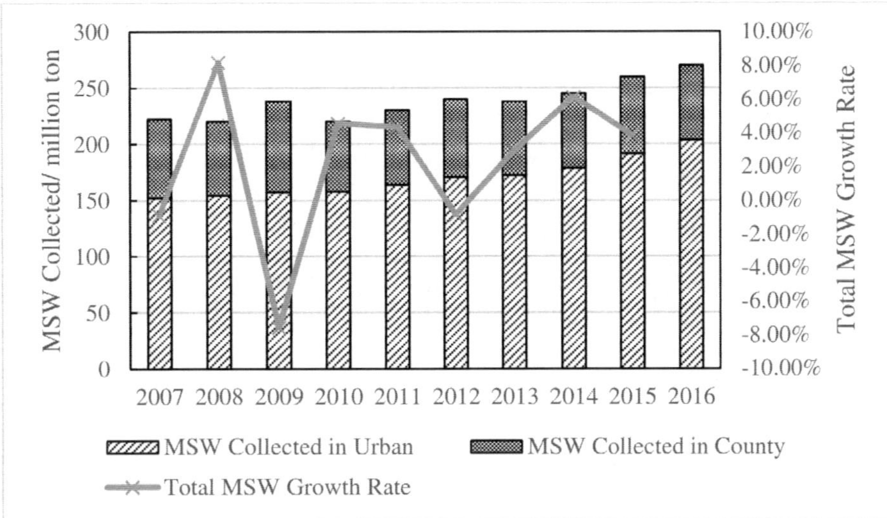

FIGURE 2. MSW Generation and Growth Rate in China (2007–2016)

China can be divided into seven areas: East China, South China, North China, Central China, Southwest China, Northeast China and Northwest China. The volumes of garbage disposal in these areas are as shown in Figure 3, with East China ranked the top followed by South China and Northwest China at the end of the rankings. Among the regions, Guangzhou Province has largest MSW generation with 23.91 million tons while Tibet ranked at the bottom with 0.46 million tons. Because of the geographical locations and population, volumes of garbage disposal vary greatly and are closely related to citys' size, function and economic condition. And it illustrates that the more prosperous a province is the higher MSW quantity in the province.

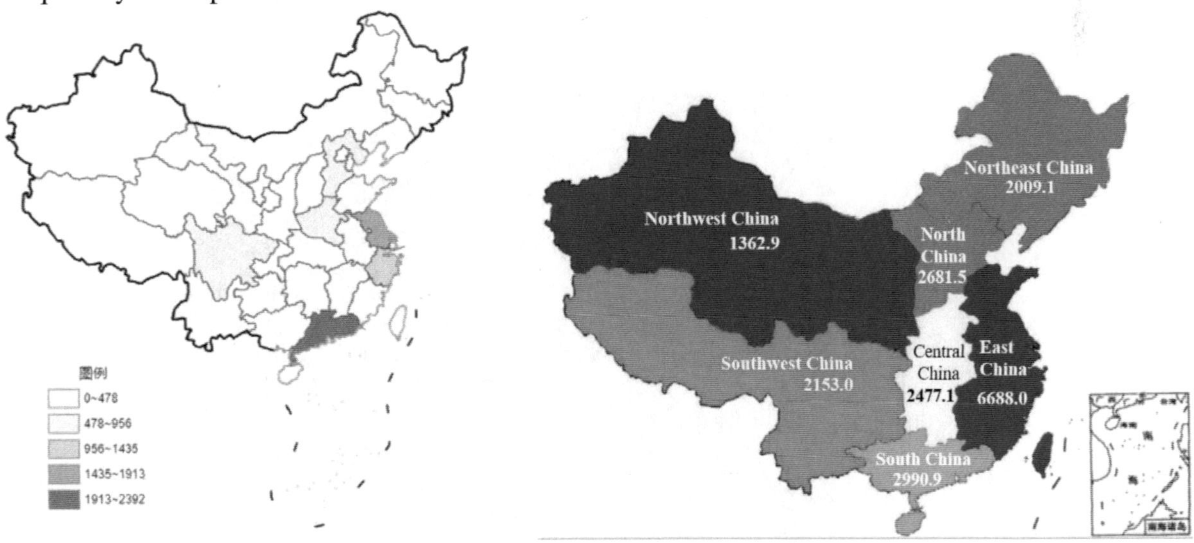

FIGURE 3. Distribution of MSW Generation in Province-Level and Area-Level in China (10^4 tons)

MSW Generation Factors. To explore the impacts on MSW generation, 5 socioeconomic factors were taken into consideration: urban population, gross domestic product (GDP), disposable income per capita of urban residents, consumption level per capita of urban residents and consumer price index (CPI), variables pertaining to: (a) the population; (b) the level of economic development; and (c) the living standards of urban residents. The data are shown in Table 1. The multiple regression analysis of MSW generation is expressed by the following regression equation (R^2=0.9978).

$$Y=-20498.3+30.21P-82.16G+115.33I+92.03C+116.93X$$

Where Y is the amount of MSW Collected in Urban in 10^{4} t/year
P is the Urban Population in 10^{6}
G is the GDP in 10^{11} yuan
I is the disposable income per capita of urban residents in 10^{2} yuan
C is the consumption level per capita of urban residents in 10^{2} yuan
And X is the CPI.
The residual analysis is shown in Figure 4.

TABLE 1. MSW Generation and Influencing Factors in China

Year	MSW Collected in Urban/10^{4} t	Urban population /10^{6}	GDP/ 10^{11} yuan	Disposable income per capita of urban residents/10^{2} yuan	Consumption level per capita of urban residents/10^{2} yuan	CPI
2007	15215	606.33	270.844	137.86	124.8	104.8
2008	15438	624.03	321.5005	157.81	140.61	105.9
2009	15734	645.12	348.4985	171.75	151.27	99.3
2010	15805	669.78	411.2652	191.09	171.04	103.3
2011	16395	690.79	484.7532	218.1	199.12	105.4
2012	17081	711.82	539.1165	245.65	218.61	102.6
2013	17239	731.11	590.4224	264.67	236.09	102.6
2014	17860	749.16	644.7911	288.44	254.24	102
2015	19142	771.16	686.4496	311.95	272.1	101.4
2016	20362	792.98	740.5987	336.16	292.95	102

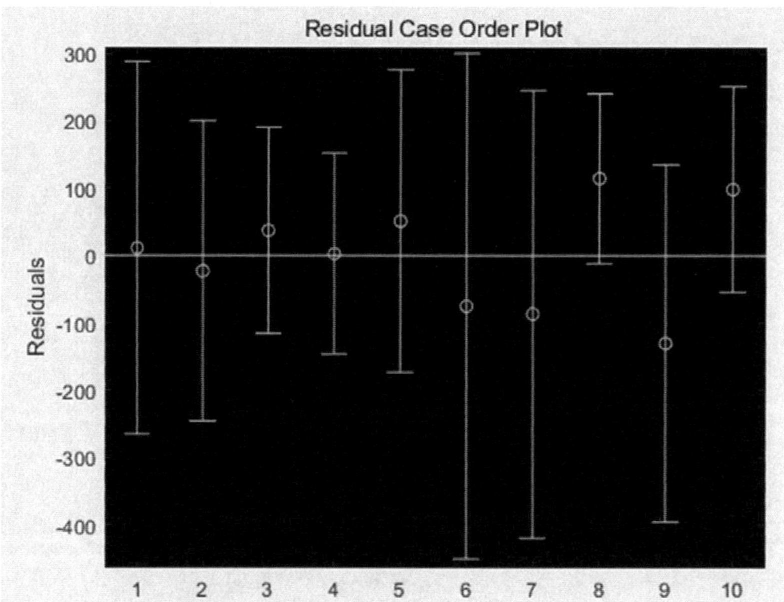

FIGURE 4. The Residual Analysis

INFORMATION TECHNOLOGY APPLICATION IN CHINA

IT Enterprises on MSW Management. With the booming development of Internet and big data, enterprises based on these up-to-date technologies appear strong momentum year by year. Commerce models like Online-to-Offline (O2O) emerge, which is a business strategy that draws potential customers from online channels to make purchases in physical stores. IT Enterprises in China starts with electronic waste recycling because of its high value. In 2011, the "Ai Hui Shou (Loving Recycling)", which focused on the recovery of electronic digital products, was born in Shanghai. In 2014, a number of "Internet plus" waste recycling platforms emerged, such as "Mei Yi Dian (More Beauty)", "Zai Sheng Huo (Re-life)", "Xian Dou Hui Shou (Idle Bean Recycling)" , "9 Shell" and so on. After the initial capital favor and the interim resplendence, many platforms can hardly maintain operation because of high operating costs and low recycling value, and some even declare a service termination. But there are still promising platforms, such as "Hui Shou Ge (Recycling Guys)" and "Ben Ge Ge (Clumsy Guys)". Characteristics of these platforms are as shown in Table 2.

TABLE 2. Characteristics of IT Enterprises on MSW Management in China

Platform	Services area	Foundation time	Business	Profit model	Customer	Remarks
Mei Yi Dian (More Beauty)	Beijing、Shanghai	2015	Provides solutions to MSW sorting, hardware and software products	Service based on Internet	Government and enterprises mainly	Less business
Zai Sheng Huo (Re-life)	Beijing、Shanghai	2014.5	Home waste recycling, convenience store commodity distribution	Daily necessities delivery	Person	Slow down in 2016 and shut off some base stations
Xian Dou Hui Shou (Idle Bean Recycling)	Beijing、Tianjin、Hebei	2015.7	Waste paper recycling mainly	Storage, logistics, and information systems built independently	Enterprises, merchants	Completed the B round 100 million yuan RMB financing in Sept. 2017
9 Shell	Hangzhou, Beijing, Shanghai, Wuhan	2015	All categories can be recycled and exchange of goods by credits	/	Person	Closed
Hui Shou Ge (Recycling Guys)	Wuhan, Tianjin	2015.7	7 kinds of waste including waste paper, waste plastics, waste batteries, large household appliances, small appliances, waste lamps, waste metal	Relys on the garbage disposal technology of the Ge Lin Mei (GEM) company	Person, merchants, enterprises	Works well
Ben Ge Ge (Clumsy Guys)	Beijing, Shanghai	2016.5	Waste paper mainly - packaging chain station - paper mill	Profits from sales to paper mill	Merchants	More packing stations and larger business scale

Among the Chinese enterprises, some provide both software and hardware services of garbage classification, such as "Mei Yi Dian (More Beauty)". Some access to the business of household electrical appliances maintenance and convenience store commodity distribution with the help of waste recycling, such as "Zai Sheng Huo (Re-life)". Others build up their own warehouse to refine and specialize recycling waste types.

IT on MSW Management in Cities. Efforts have been made to sort out wastes by the means of big data so far. As early as 2011, Shanghai launched a program "Green Account" to record the waste classification among residents. The "Green Account" offers residents credits to redeem gifts and motivate them to classify waste correctly. In recent years, the "Green Account" has gradually cooperated with Bank of China Shanghai Branch, Alipay, and Shanghai Municipal Investment (Group) Corporation. The big data statistics function makes it possible to facilitate the targeted promotion of unopened regions and analyze the correct rate of waste classification. The scope of the "Green Account" enable decision-makers to carry out further tracking and guidance on the waste classification in the community.

Jin Song Zhong Community located in Chaoyang District, Beijing, not only implements the garbage classification credit system, later in 2016, it also carried out facilities transformation of 378 garbage collection points in the pilot area. For the trash bins, the waterproof NFC electronic tags with mutual communication and calculation are embedded on them. For the garbage collecting and transporting vehicles, the Voltaire-mounted machines are installed to obtain the real -time position and working status with GPS. Managers can also learn about the status of daily garbage collection and transportation from the information system. Thus the data analysis software can evaluate the efficiency and provide the optimal transport mode, saving energy consumption and manpower.

In 2012, Hangzhou tried to use the "Coded Bag". By scanning the code on the trash bags, managers can trace the origin of the garbage, transparentize the responsibility for the garbage separation and urge residents to classify the garbage properly. The Xingan community in Hangzhou has also implemented the "QR Code Intelligent Garbage Bag Machine". It can collect the outgiving information of garbage bags automatically(Gao et al., 2017). By viewing the big data, management personnel can provide targeted attention and guidance to residents who are not actively receiving garbage bags, which helps to improve the enthusiasm and implementation of MSW classification.

TABLE 3. IT on MSW Management in Cities

Location	Measurement	IT Application
Shanghai	"Green Acount"	Alipay big data network platform statistics function
Beijing	NFC electronic tag, Voltaire-mounted vehicle	NFC mutual communication, GPS location information
Hangzhou	"Coded Bag"、"QR Code Intelligent Garbage Bag Machine"	Embedded information processing module, QR code identification technology, GPRS communication technology
Guangdong	"Qing Gong model"	Intelligent garbage bins recording account information

TABLE 4. IT on MSW Management

Information technology	Application
Computer Image	Waste identification for bin level detection/ sorting lines
RFID	Obtaining target objects information and relevant data through radio frequency signals
Sensor System	Volume reminder
GSM/GPRS	Information transmission to the required control room
Renesas Microcontroller	Communicate the sensor system with GSM system
Android Application	Monitor the information
HTML5	Garbage classification propaganda

As for universities, Guangdong Industry Polytechnic has formed a "Qing Gong model" for MSW classification. Intelligent classification trash bins can be seen everywhere on the campus. By swiping the student ID cards or user cards, students enable the bins to open automatically. Once the garbage is placed correctly, they can get the corresponding credits and can exchange the credits for daily necessities. The feedbacks will later be sent to the college personnel, who will instruct the students to develop good habits of conscious garbage sorting.

The information technologies applied in the cities are concluded in the Table 3.

IT on MSW Management. By the means of IT, including Computer Image, RFID, Sensor System, GSM and so on, the whole process of MSW management can be more effective and efficient. The applications of different information technologies are as shown in Table 4.

CONCLUSIONS

There is an increasing trend of MSW generation in China and the main influencing factors are urban population, GDP, disposable income per capita of urban residents, consumption level per capita of urban residents and CPI. The urgent situation calls for an integrated and innovative solution. Internet technology can arouse residents' waste recycling enthusiasm and lead to the intelligent environment-friendly management. In order to ensure the long-term effectiveness of garbage classification, the operating mode needs optimization combined with Internet.

ACKNOWLEDGEMENTS

Support for this research by Shanghai Administration on City Appearance and Environmental Sanitation is gratefully acknowledged.

REFERENCES

Afon, A. O., & Okewole, A. (2007). Estimating the quantity of solid waste generation in Oyo, Nigeria. *Waste Management & Research the Journal of the International Solid Wastes & Public Cleansing Association Iswa, 25*(4), 371-379.

Arebeya, M., Hannana, M. A., Hussaina, A., Zamanb, H. B., & Basric, H. Solid Waste Classification and Recognition for Intelligent Collection System.

Bobulski, J., & Piatkowski, J. (2017). *PET Waste Classification Method and Plastic Waste DataBase - WaDaBa.* Paper presented at the International Conference on Image Processing and Communications.

Feng, S. J., Gao, K. W., Chen, Y. X., Li, Y., Zhang, L. M., & Chen, H. X. (2017). Geotechnical properties of municipal solid waste at Laogang Landfill, China. *Waste Management, 63*, 354-365. doi:10.1016/j.wasman.2016.09.016

Gao, Y., Yan, Q., Xia, X. U., Yingga, A., Chen, H., & Zhao, Y. (2017). Investigation and Analysis of Municipal Solid Waste Classification Management of Hangzhou. *Environmental Science & Technology.*

Khan, D., Kumar, A., & Samadder, S. R. (2016). Impact of socioeconomic status on municipal solid waste generation rate. *Waste Management, 49*, 15-25.

Mian, M. M., Zeng, X. L., Bin Nasry, A. A., & Al-Hamadani, S. M. Z. F. (2017). Municipal solid waste management in China: a comparative analysis. *Journal of Material Cycles and Waste Management, 19*(3), 1127-1135.

Sokka, L., Antikainen, R., & Kauppi, P. E. (2007). Municipal solid waste production and composition in Finland—Changes in the period 1960–2002 and prospects until 2020. *Resources Conservation & Recycling, 50*(4), 475-488.

Vieira, V., & Matheus, D. R. (2018). The impact of socioeconomic factors on municipal solid waste generation in São Paulo, Brazil. *Waste Management & Research the Journal of the International Solid Wastes & Public Cleansing Association Iswa, 36*(1), 79.

Walsh, D. C. (2002). Urban residential refuse composition and generation rates for the 20th century. *Environmental Science & Technology, 36*(22), 4936-4942.

Wang, S., Nie, J., Zhang, D., Li, X., Tai, J., Yu, Z., . . . Xie, D. (2018). *Analysis of Laogang energy internet and construction of the cloud platform.*
Yi-Qi, L. U., Xie, D. A., Chen, S. P., Xue, X., Wang, H., & Jin, Q. H. (2017). Data Acquisition and Cloud Platform for the Energy Internet in Laogang Solid Waste Recycling Base. (epee).

Utilization of Phosphogypsum to Reduce Greenhouse Gases Emissions and to Improve the Conditions of Calcareous Soil on the Kingdom of Saudi Arabia

Saud S.. AL-Oud, Fahad N Al-Barakaha and Adel RA Usman
(College of Food Science and Agriculture, King Saud University, P.O. Box 2460, Riyadh 11451, Saudi Arabia)

In Saudi Arabia, there are some industrial plants have certain by-products that have adverse effects on the environment. Among these industrial plants, phosphate fertilizer industry which generates a huge amount of phosphogypsum (PG) as by-products. The aim of this study was to evaluate the effects of phosphogypsum (PG) on greenhouse gases (GHG). In addition, recycling of phosphogypsum to improve the conditions of calcareous, alkaline and saline soils was also studied The results indicated that the addition of PG to cow manure under anaerobic conditions increased the emission of methane (CH_4) and carbon dioxide (CO_2), but not for aerobic condition. The addition of PG to the industrial waste water increased both methane (CH_4) and carbon dioxide (CO_2), but deceased pH. Greenhouse experiments were conducted to evaluate the integrative role of PG (added at 0.0, 10, 30, and 50 g PG kg^{-1}) in the presence and absence of microbial inoculation (Azotobacter, phosphate solubilizing bacteria, and Pseudomonos) and cow organic manure (added at 5% w/w) on nutrients availability and maize growth in non-saline and saline sandy calcareous soils. In calcareous soil, results showed that applying PG decreased the soil pH by 0.08-0.11 units. Applying cow organic manure (COM) caused also a significant higher reduction in the soil pH. Results indicate there are no significant variations in the level of soil available P with increasing application rate of PG when applied alone to soil. However, the combined addition of PG with COM or PG with microbial inoculation (Incl) increased available P compared with the control soil. Inoculation with the Azotobacter, phosphate solubilizing bacteria, and Pseudomonos improved the nutrients uptake and growth of maize plants in both soils. However, the growth of plant amended with COM decreased, showing negative impacts on nutrients uptake. In saline soil, applying PG reduced soil pH and co-inoculation induced significant decreases in soil electrical conductivity EC_e values. Applying PG increased significantly soil available P. The co-inoculation along with PG effectively increased the soil available K. However, the soil available trace elements decreased significantly with PG. The maize grown on inoculated soils treated with PG showed significant higher plant dry weight (82.1-127.4% over control) and nutrients uptake than the control. It could be concluded that co-inoculation along with PG may be an important factor for alleviating negative effects of salinity on plant growth. It could be concluded that inoculation along with PG may be as alternative source for the chemical P fertilizer in sandy calcareous or saline soils, promoting plant growth. On the contrary, the use of animal organic manures with low quality to improve soil productivity is possibly to be unsuccessful.

Life Cycle Assessment of Bioplastics and Food Waste Disposal Methods

Shakira R. Hobbs[1], Tyler Harris[2], William J. Barr[3] and Amy E. Landis[2]
([1]University of Virginia, Charlottesville, VA, USA; [2]Colorado School of Mines, Golden, CO, USA;
[3]Arizona State University, Tempe, AZ, USA)

Many establishments (e.g. McDonald's in Sweden) are switching from petroleum-based plastics to plant-derived plastics, such as polylactic acid (PLA), to achieve waste diversions goals. Bio-based packaging materials such as PLA are also used for food to extend shelf life and serve as an oxygen and water barrier. Fruits and vegetables wrapped in bioplastics reduces the amount of food waste that is disposed of. However, PLA is often comingled with expired food when disposed of to landfills. There is an increase of bioplastic and food waste as well as demands for management strategies that responds to common concerns for environmental sustainability. The aim of the paper is to demonstrate the environmental consequences of landfilling food and PLA waste compared to other waste management options such as compost and anaerobic digestion. This study quantifies the environmental impact of five end-of-life waste management options for PLA and food waste:

1. anaerobic digestion (untreated),
2. anaerobic digestion (treated),
3. compost (untreated),
4. compost (treated), and
5. landfill.

using life cycle assessment (LCA). Treatment of bioplastic and food waste are used to accelerate degradation and environmental impacts are assessed along with controls that are not pretreated in anaerobic digestion and compost. Sending food waste and treated bioplastics to anaerobic digestion offers life-cycle and environmental benefits in the following impact categories: ecotoxicity (-186%), eutrophication (-19%), global warming potential (-11%) and human health non-carcinogenic (-178) for treated anaerobic digestion. In addition, life-cycle and environmental benefits are observed for untreated anaerobic digestion in impact category cumulative energy demand (-15%). Results from this study show that the landfill scenario had lower impacts than composting in all categories expect for ecotoxicity (-71%) and human heath non-carcinogenic (-63%) in the compost treated scenario. The use of digestate from anaerobic digestion and biosolids from composting as a fertilizer replaces the use of manufactured fertilizers and displaces material from going to landfill. The use of the digestate as an inoculum reduces the energy intensive processes used to stabilizes the digestate. The environmental assessment of this study highlights benefits and trade-offs of co-treating food waste and bioplastics.

Waste Management, Treatment and Disposal at Kuwait Oil Company (KOC)

Zainab Hussain, ***Kholood Yousef,*** Soud AL-Mutairy, Haitham Fouzy
(Kuwait Oil Company (KOC), P.O Box 9758, Ahmadi 61008, Kuwait)

Kuwait Oil Company (KOC), a subsidiary of Kuwait Petroleum Corporation (KPC), is involved in exploration, drilling and the production of oil and gas. The massive effort in producing this oil includes the employment of 9000 plus employees, involvement of thousands of contractors, operating thousands of wells, numerous gathering centers, gas compression booster stations, drilling rigs, crude oil bulk storage tank farms, marine transport loading facilities, and a significant number of transportation pipelines. The significant oil production related activities generating waste over large land area include drilling, production, maintenance, construction, materials procurement, as well as use of office and housing complexes.

The issue of waste management in KOC covers analyses of different types of waste streams (solid, liquid, and sludge), the activity upon waste, and a holistic view of the goals of waste management and approach Figure 1. The increasing volume of waste generated, which is exacerbated by a lack of proper waste management is a concern worldwide due to social, economic and environmental implications. KOC always ensures that the generated waste be managed effectively in compliance with applicable regulations of Kuwait Environment Public Authority (KEPA) and KOC HSE management system (HSEMS) procedures implemented for KOC facilities. In addition, KOC plays a dynamic role in facing the adverse effects of waste through the implementation of many projects in the regions of waste management and by focusing on the development of long-term waste management strategy covered all KOC activates. This study was conducted to present the Waste Management, Treatment and Disposal at KOC, In addition, in this paper we will discuss various processes used in KOC areas indicating how waste is being managed.

Utilization of Wood Waste for Container Media by Accelerated Composting

Hoda Bakhshizadeh[1], ***H.Borazjani***[2], R.C.Sloan[3], and S.S.Worthey[3]

1. Mississippi State University, Department of Sustainable Bioproducts. Current address: Soybean Plant Pathology Lab, Bayer Crop Science, Memphis, TN 38120, USA, hbzadeh_a@yahoo.com
2. Mississippi State University, Department of Sustainable Bioproducts, Box 9820, Mississippi State, MS 39762, hamid.borazjani@msstate.edu
3. Mississippi State University, Northeast Research & Extension Center, Box 9389, Mississippi State, MS 39762, USA.

ABSTRACT: The objectives of this study were: To determine the suitability of quick composted forest products wood wastes as ornamental plant media. In a 3-month accelerated composting study, wood wastes were amended with (20% & 40%) poultry litter, (1% & 2%) ammonium nitrate, or non-amended. Composting was run in fifteen 130- liter containers outdoor and samples were collected at day 90 for compost maturity (toxicity), carbon-to-nitrogen ratio and green house evaluations. The amendments containing 40% poultry litter showed significantly higher mass weight and vigor grading in transplanted zinnia and gardenia than other treatments and was comparable to commercial media. Overall, results indicated that the amendment of wood wastes with poultry litter could produce a comparable product to currently used commercial container media.

INTRODUCTION

The disposal of byproducts from forest product industries has become a problem throughout the southeastern United States. Hardwood bark (HB) and pine bark (PB) are both forest product industry byproducts. The U.S. census reveals in 1997 that there were 892 wood products industries in Mississippi (Borazjani et al., 2004). The manufacturers of wood products in Mississippi produce several million tons of wood waste and bark residues each year. Around 75 percent of these waste materials are used for energy and some to make other wood products, however, a thousand tons of unused residues still remain (Borazjani et al., 2004).

Thirty one percent of solid wood products generated in Mississippi annually are hardwood wastes in which bark residues make up around 31% (Garrard & Leightley, 2005). In 1994 in Mississippi, 7.5 million tons of wood wastes were generated in which 750,000 tons were hardwood bark (Alderman et al., 1999). Also, it is estimated that the amount of wood waste and bark residue increased in 2004 to 12.2 million tons from 376 wood products manufacturers in Mississippi (Borazjani et al., 2004).

There are many ways the forest products industry can dispose of their wastes. One of the most popular ways to dispose of this waste material is to burn them as an energy source for oriented strand board and plywood manufacturers, however, smaller manufacturers do not have enough space or facilities to burn the wood waste for energy. This wood waste may be transported elsewhere if it is economical. However, transportation is more expensive than the profit gained from the sale (Borazjani et al., 1997). Burning wood and bark residues can help the mills reduce the cost of using natural gas, but the manufacturers have to meet the current air pollution standards which are often costly (Sweet & Fetrow, 1975). Some companies use landfills for disposal of their wood wastes, but this method is relatively expensive and may cost as much as $6,000 per month for small companies (Borazjani et al., 1997). In addition, the protection of the ecosystem has recently become increasingly important and new laws have been approved which prevent dumping or burning of hardwood bark by wood manufacturers (Gartner et al., 1974).

The suitability of hardwood bark for use as a soil amendment is low because of its pH and the high level of tannin. Tannin is toxic for plants and causes plant growth inhibition (Hayden, 2005). One main reason for high marketability of PB is its great water holding capacity. The water holding capacity of HB is less than pine bark, but it increases gradually during a plant's growth cycle. Increasing water holding capacity can cause a decrease in air porosity which could increase the fungal populations in the compost (Jagdale et al., 2004). HB does not have a strong market in the United States because of these negative characteristics (Mason, 2004).Bosley (1969) began to look for a substitute for potting soil. He compared different species of hardwood bark which are produced in large quantities in the Midwest and mixed them with sand. When composted, these materials were great container media for ornamental plants. However, there were several problems such as pH and nitrogen requirements through the composting processes which were different when the species of hardwood bark was changed or mixed together (Bosley, 1969).

Hardwood bark (HB) can be used as soil amendment after it is completely composted. Hardwood bark contains high cellulose and low lignin; so it shrinks very fast and steals nitrogen from plants. Also, this defect cannot be balanced efficiently by adding additional nitrogen. However, well-aged, fine textured, composted hardwood bark can minimize this negative effect (Stoffella & Kahn, 2001). Except nitrogen which is an essential addition for composting HB (Bosley, 1969), iron sulphate should also be added after the initiation of the composting process since HB releases excessive quantities of manganese for some crops. As HB releases noticeable amounts of calcium, less lime is needed to be mixed into composting HB. Also, composted HB has shown better disease-suppressive properties than other kind of composted barks (Hoitink et al., 1997; Lanthier, 2009).

The objective of this study was: To determine the suitability of composted material for ornamental plant homeowner potting media.

MATERIALS AND METHOD

Compost Setup

Ground-up hard wood bark was provided by Penick Forest Products in Macon, Mississippi. Inorganic ammonium nitrate fertilizer and poultry litter was used as a nitrogen (N) source for this study. Dried poultry litter was collected from the Poultry Science Department at Mississippi State University campus from caged broilers and contained bedding materials such as sawdust, and manure. Ammonium nitrate was purchased from a local agriculture store. Samples were collected from hardwood bark and poultry litter and sent to Enviro-Labs, Inc. to determine the amount of Total Kjehldahl Nitrogen (TKN) (EPA Method 351.3), which includes Nitrate (NO3), Nitrite (NO2), and Ammonia (NH4). In addition, samples were tested for Phosphorous, and Potassium using EPA Method 6010 and toxicity (1/EC50) for the starting materials before mixing the treatments for this experiment. The results of these background tests are indicated in Table 1.

Table 1. 90-days Composted Treatments Pretest results.

Sample ID	Carbon mg/kg	Nitrogen mg/kg	C:N
C	52552	937	56.93
PL25	41352	2398	17.62
PL40	36235	3580	10.12
AMN1	44839	3766	12.85
AMN2	44930	2717	18.02

This study consists of five treatments that were replicated three times. The experiments were conducted on site and outdoors at the North Mississippi Research & Extension Center for horticultural research. Individual random samples from bark and poultry litter were taken to determine the overall moisture content and nutrient levels of each replicate within treatments. The treatments were as follows and weights given below are on dry-weight basis:

Control containing 11 Kg of ground up hardwood bark only
Hardwood bark plus 25 percent poultry litter 2.2Kg mixed with 8.8 Kg of ground up bark
Hardwood bark plus 40 percent poultry litter (4.5 Kg) mixed with 6.5Kg ground up bark
Hardwood bark plus ammonium nitrate (330gr) Containing 10.7 kg of ground-up bark with a onetime application of ammonium nitrate. Ammonium nitrate 3 percent plus water (110gr nitrogen each containers)
Hardwood bark (10.3 Kg) and another application of 330gr of ammonium nitrate or 110gr of nitrogen after 30 days again

Greenhouse Study

Zinnia Test. To determine the suitability and effect of composted materials as bedding media on ornamentals plants in the short term, transplanted Zinnias (*Zinnia elegans*) were grown for a period of 4 weeks on the day 90 composted substrates. The commercial potting media known as Metro Mix was used as commercial media to compare with composted treatments. *Zinnia elegans* requires growing conditions such as neutral pH, around 7, well drained bedding soil or compost, full sunny location, and warm weather (Rhyne, 2008).

A)

B)

Figure 1. A) Transplanted Zinnias in the greenhouse with random design. B) Transplanted gardenia bushes placed outdoors for a period of three months.

One transplanted Zinnia was planted in each pot and six pots were used for each replicate of each treatment. All pots were placed in greenhouse in same condition and light (Figure 1). All Zinnias planted in treatments and commercial media were fertilized with same regiment of commercially available fertilizer once at the beginning and irrigated regularly to keep pots moist. After four weeks Zinnias were visually graded for vigor, number of flowers/ buds and height. Finally, plants were harvested at the ground level for dry mass weight determination.

Gardenia Test. The Gardenia bushes (*Gardenia jasminoides*) were used and planted on all treatments to show the long term effect of finished compost on plants. Transplanted gardenia bushes were planted on day 90-composted substrate and their growth were compared with those planted on standard commercial composted pine bark. Six transplanted Gardenia bushes were planted in six pots for each replication and pots were placed outdoor randomly for a period of three months (Figure 1B).

The optimum conditions for growing gardenias are full sun to partial sun, well drained bedding media, and an acidic soil or compost (MacCubbin & Tasker, 2002). Plants were irrigated with tap water without adding any fertilizer. After ninety days, gardenias were visually graded for vigor and measured for width, and then harvested from ground level. Harvested plants were measured and weighed to determine which treatments produced the largest and heaviest plants.

Statistical Analysis. The weight reduction, nutrient concentration, pH analysis, germination and plant growth in compost, and greenhouse study results from this experiment were statistically analyzed to determine significant difference between treatments. A completely random design with three replications for each treatment was used. For comparing treatment mean differences at P=0.05 Tukey's multiple comparisons test was used. Data was processed by SPSS (Statistical Package for the Social Sciences) statistics software. Treatment descriptions, abbreviated ID, and the number of replications are listed in Table 2.

Table 2. Treatment Descriptions and Reference ID Used in the Experiment Design.

Treatments	Reference ID	Replication
Hardwood Bark	C	3
Hardwood Bark + 25% Poultry Litter	PL25	3
Hardwood Bark + 40% Poultry Litter	PL40	3
Hardwood Bark + 330g Ammonium Nitrate	AMN1	3
Hardwood Bark + 660g Ammonium Nitrate	AMN2	3

Note for Reference ID: C=Control, PL=Poultry Litter, AMN=Ammonium Nitrate. Letters illustrate amount of mentioned parameter.

RESULTS AND DISCUSSION

Results of Zinnia Test. The green house study of day 90 composted treatments with transplanted Zinnia (*Zinnia elegans*) for a period of 4-weeks showed increased vigor (height, bulb, and flower) comparable to commercial nursery container media for 40% poultry litter treatment (Table 3). The results of visual grading, length in centimeter, number of flowers and buds, and average weight per plant in grams are tabulated in Table 3. The Zinnia dry weights per plant grown in 90-day composted treatments are shown in Figure 3. The above ground dried weight of Zinnias for C, PL25, AMN1, AMN2 treatments were significantly less than the dried Zinnias weight planted in PL40 and commercial media for the day 90 . Yates and Rogers (1981) reported that plants grown in 180-day composted HB amended with inorganic nitrogen had less dry weight than plants grown in commercial media. Figure 4 shows the difference in size, color of leaves, and number of flowers and buds of plants grown in unamend treatments (C) and treatments amended with 40 % poultry litter (PL40). Also, the difference in appearance of zinnias planted in treatments amended with 330 and 660 gram ammonium nitrate compared to treatment PL40 is shown in Figure 4.

Table 3. Zinnias visual grading, number of flowers and buds, average of length per plant and average of dry weight per plant grown on day 90 compost samples.

Treatments	Ave. Grading per Treatments	Number of Flower and Bud	Ave. Dry Weight per Plant (g)	Ave. Length per Plant (cm)
C	*5*	*2*	*1.44*	*27.94*
PL25	6.16	2	1.7	33.02
PL40	9	18	4.8	58.2
AMN1	5.88	2	1.45	29.97
AMN2	6.55	11	1.65	33.02
Commercial	9.2	18	5.3	60.4

Note for Number of Plants: Each treatment was replicated three times and for greenhouse six pots were planted for each replication which means 18 plants per treatment.

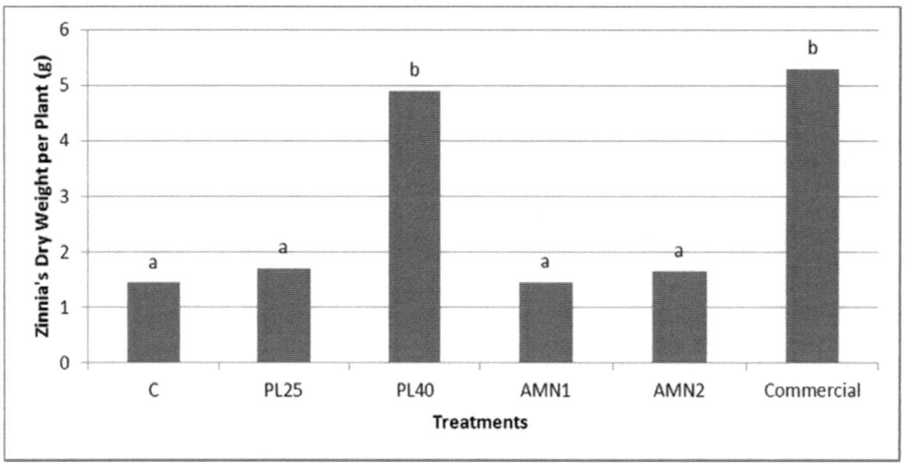

Figure 3. Zinnia Dry Weight Per Plant Planted in Day 90 Composted Treatments. Columns with different letters indicate a significant difference in zinnia dry weights at the P=0.05 probability level.

A) B) C)

Figure 4. A) Zinnias planted in 40 % poultry treatment on left side. B) Zinnias planted in treatments amended with 1% and 2% percent ammonium nitrate on right side. C) Zinnias planted in control treatments which was hardwood bark only (C) Vs. 40 % poultry litter Treatment (PL40).

Results of Gardenia Test. Results of dried above ground tissue mass weight of gardenias are summarized in Figure 5A. Plants grown in treatments amended with 40 % (PL40) showed significantly higher weight than unamend treatments (C). Unlike zinnias, gardenias showed comparable results in vigor to PL40 and commercial media for treatments amended with 330 and 660 gram ammonium nitrate. This was not surprising since gardenias favor low pH and the ammonia amended compost had a lower pH than other treatments (Maccabi & Tasker, 2002). The average lengths of gardenias per plants in centimeter are detailed in Figure 5B. The lengths of plants on PL40 were exactly the same as commercial media plants and were significantly higher than the C treatment. Figure 4C shows the plants in C and PL40 treatments and also shows the differences in appearance and size between gardenias planted on commercial media, C, and PL40. Overall results for gardenias showed that 90-day composted hardwood bark amended with 40 % poultry litter was comparable with commercial media.

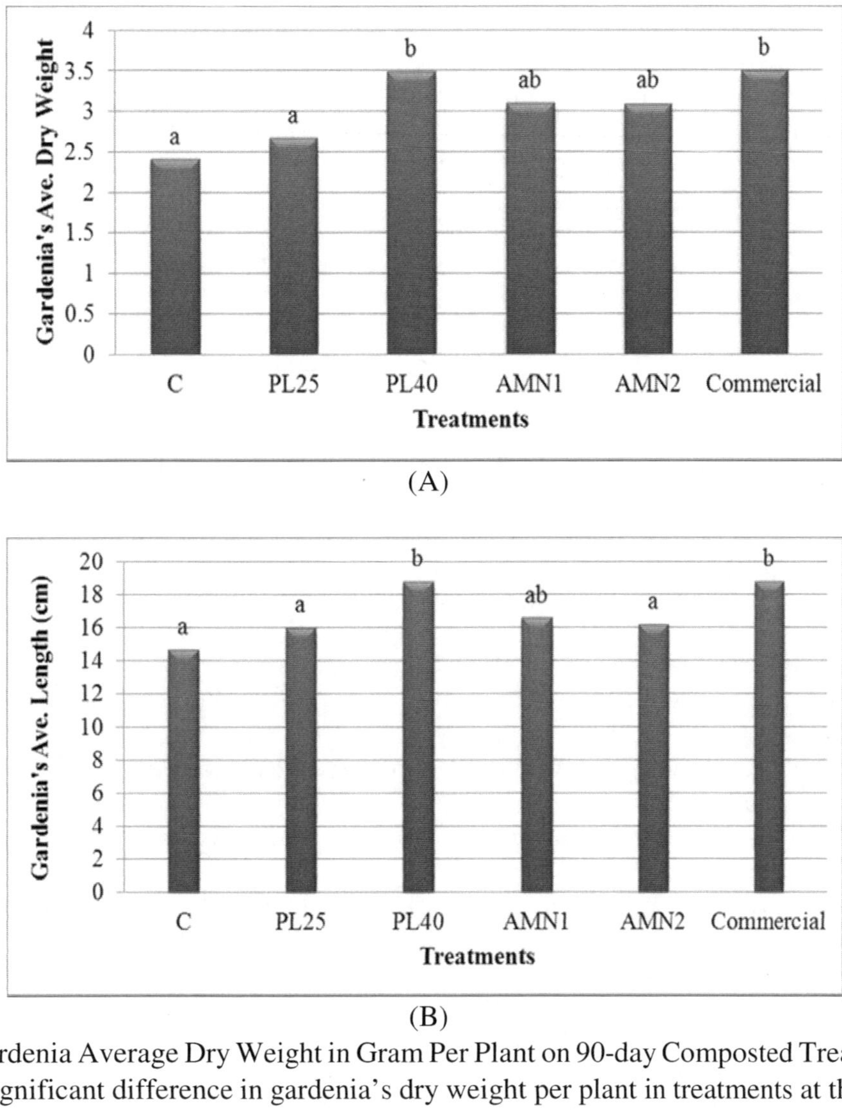

Figure 5. (A) Gardenia Average Dry Weight in Gram Per Plant on 90-day Composted Treatments. Different letters indicate significant difference in gardenia's dry weight per plant in treatments at the P=0.05 level of significance. (B) Gardenia average length in centimeter per plant planted on compost collected from day 90. Different letters above columns indicate significant difference between treatments in gardenia's length.

Figure 6. An Example of Difference in Appearance of Gardenias Planted in C on Right and PL40 on Left.

CONCLUSION

This study has shown that composting is a practical approach poultry and forest product industries to deal with HB and PL to make a valuable product in a relatively short period of time and reduce the disposal cost of HB and PL separately. The finished compost could be very popular among nurseries and farmers due to its organic characteristics, low cost and comparability to commercial media.

ACKNOWLEDGMENT

The article is approved for publication as Journal Article SB - 849 of Forest & Wildlife Research Center, Mississippi State University, Mississippi State, MS.

REFERENCES

Alderman Jr, D.R., Smith, R.L., and Reddy, V.S., 1999. Reviewed Articles-Assessing the Availability of Wood Residues and Residue Markets in Virginia. Forest Products Journal and Index, 49(4), pp.47-55.

Borazjani, H., Diehl, S.V., and Stewart, H.A., 1997. Production of Compost from Furniture Manufacturing Woodwastes. Forest Products Journal, 47(2), pp.47-48.

Borazjani, H., Diehl, S.V., Stewart, H.S., and Brasher, K., 2004. Composting Research Targets Forest Products and Poultry Industries. BioCycle-Journal of Composting and Recycling, 45(5), pp.42-43.

Bosley, R.W., 1969. Ground bark-a Container Growing Medium. ISHS Acta Horticulturae, 15, pp.17-20.

Garrard, A. and Leightley, L., 2005. Characterizing Wood Waste from Wood Products Companies in North Mississippi. Research Report. Mississippi State University: Forest and Wildlife Research Center.

Gartner, J.B., Still, S.M., and Klett, J.E., 1974. The Use of Bark Waste as a Substrate in Horticulture. ISHS Acta Horticulturae, 37, pp.2003-12.

Hayden, D., 2005. Soilless Substrate Management for Nursery Crops. Kentucky: University of Kentucky, College of Agriculture.

Hoitink, H.A.J. and Poole, H.A., 1977. Composted Bark Media for Control of Soil Borne Plant Pathogens. Ohio Florists' Association. Bulletin No. 567, Bulletin No. 567, pp.10-11.

Jagdale, G.B., Casey, L., Grewal, S., and Lindquist, R.K., 2004. Application Rate and Timing, Potting Medium, and Host Plant Effects on the Efficacy of Steinernema Feltiae Against the Fungus Gnat, Bradysia coprophila, in Floriculture. Biological Control, 29(2), pp.296-305.

Lanthier, M., 2009. Natural Suppression of Plant Diseases: a Research Update. A Research Update. North Okanagan Organics Association Workshops on Soil Biology.

MacCubbin, T. and Taske, G., 2002. Florida Gardener's Guide. Revised ed. Franklin, TN: Cool Springs Press.

Mason, J., 2004. Nursery Management. 2nd ed. Australia: Landlinks Press, pp.139-40.

Stoffella, P.J. and Kahn, B.A., 2001. Compost Utilization In Horticultural Cropping Systems. United States of America: CRC Press.

Sweet, H.R. and Fetrow, R.H., 1975. Ground-Water Pollution by Wood Waste Disposal. Groundwater, 13(2), pp.227-231.

Yates, N.L. and Rogers, M.N., 1981. Effect of Time, Temperature, and Nitrogen Source on the Composting of Hardwood bark for Use as a Plant Growing Medium. Journal of the American Society for Horticultural Science, 106(5), pp.589-93.

Sustainable Utilization for Lead Immobilization in Contaminated Soil by Recovered Struvite Supported Diatomite Complex: Potential, Mechanism, Efficiency and Risk Assessment

Huanping Jing, Xuejiang Wang
(Tongji University, Shanghai, China)

A recovered struvite supported diatomite complex (MAP@DIA), obtained from nutrient-rich wastewater, was applied to immobilize lead in simulated wastewater and contaminated contained soil. The influence of pH on the lead (II) adsorption process in aqueous solution was investigated, and the effectiveness of lead immobilization in contaminated soil was evaluated by determining the change in lead (II) available concentration and lead (II) fractions. In addition pH titration test and pH_{stat} leaching test were also investigated for certain risk assessments. Lead (II) adsorption capacity by MAP@DIA increased from 831.66 to 948.52 mg/g when the pH of the solution ranged from 3 to 7, due to the higher affinity of adsorbent surface to lead (II) and the enhanced ≡ Dia-FeOPbOH and ≡ Dia-SiOPbOH adsorption with increase of pH. The main mechanism of Pb immobilization by MAP@DIA could be contributed by the surface complexation and the dissolution of struvite followed by the precipitation of hydroxypyromorphite $Pb_{10}(PO_4)_6(OH)_2$, which was evidenced by XRD, SEM and EDS mapping. In the MAP@DIA -treated soils, lead (II) concentration was significantly reduced with the increase of incubation time and MAP@DIA application rates, as lead (II) species were converted from the nonresidual/potentially available phase to more inert/unavailable phase. The increased ANC and lower lead extraction yields in pH_{stat} leaching test in amended soil suggested higher buffering capacity against potential acidic stresses and delayed triggering of 'chemical time bombs'. This study revealed that the reutilization of MAP@DIA could be an effective amendment for lead (II) contaminated aqueous solution and soil.

Indigenization to Improve Access to Affordable Refuse Disposal Services in Non-Urban South African Households

Fannie Machete (Department of Environmental Sciences, Florida Park Science Campus, University of South Africa, City of Johannesburg, South Africa; Email: mtec.dr@gmail.com)

Municipal systems of waste management are not accessible in most historically disadvantaged households of the Republic of South Africa. This study presents indigenous systems of household waste management currently used in South Africa, despite being not a recognised and informal. The study also presents a comparison between the levels of waste management used by the entire population of South Africa between modern and indigenous knowledge, in waste disposal. This dependence in indigenous knowledge systems is especially common in nonurban provinces such as Limpopo, Mpumalanga, Northern Cape and Eastern Cape, with more than 50% of each province's total households having no access to municipal refuse collection services. This study used reliable data from Statistics South Africa. Quantitative data, in the form of interval data was analysed using descriptive statistical methods. The results confirm that many historically disadvantaged and communities, continue to use and rely on indigenous systems of waste disposal, 22 years after South Africa's democratic dispensation. Indigenous systems of waste management remain informal and not financed by government. Most of these communities that use indigenous knowledge and systems of waste management are classified as unserved communities, while they actually service themselves without government assistance. Thus, indigenous systems and practices are used as the last alternative for household waste disposal. Although indigenous knowledge systems continue to be practiced in most communities, such systems are not accepted or recognised by government as formal systems of services delivery. Thus, this paper recommends that IKS be formalised as one of the hybrid government alternative mechanism of services delivery in South Africa. Lastly that further research on IK as an alternative services delivery mechanism be undertaken and funded by government.

Decontamination of Liquid Effluents from Uranium Mine by Using Natural Zeolite

Ahmet Erdal OSMANLIOGLU
(Istanbul University-Cerrahpasa, Istanbul, Turkey)

ABSTRACT: Decontamination of liquid effluents generated during mining and processing of uranium operations became an important issue for the environment. Because of negatively impacts on aquatic ecosystems and drinking water resources these effluents should be recycled back to processing circuits for reuse or treated prior to disposal or release to the environment. Treatment and control of liquid effluents are required throughout the uranium production and reclamation cycle. This study is based on decontamination of liquid effluents from uranium mine by using natural zeolite as an economical and practical solution. In this study, contaminated surface waters were taken as liquid effluents from the outcrops of uranium mine deposit in the field. A filtration column was prepared by using natural zeolite matrix as filtration material in the laboratory. The characteristics of uranium liquid effluents depend on the composition of the ore and the type of processes used to extract the uranium. Climate conditions can also play a role in treatment and control of effluents. For this reason, decontamination factors were calculated according to temperature and effects of temperature on the decontamination factors for the natural zeolite filter were determined. Decontamination test results show that temperature is positively effective on the adsorption of the zeolite filter. Increasing temperature in the decontamination column was caused increasing of adsorption up to %15 for radioactivity.

INTRODUCTION

Several decontamination techniques have been applied for many years in radioactive waste management operations and other activities involving the treatment of radioactive liquids. Zeolite is used as ion exchange material in various stages of decontamination processes of nuclear industry. Such as; primary coolant water purification, treatment of primary effluents, treatment of fuel storage pond water, steam generator blow-down demineralization, liquid waste and drainage water treatments, boric acid purification for recycling, condensate polishing (for nuclear power plants with boiling water reactors) (Sun et al., 2012). Liquid radioactive waste was decontaminated and remained sludge or filter material should be solidified for transportation and disposal. Liquid effluents from uranium mine are decontaminated and the remaining material is immobilized by using solidification matrix. A wide range of ion exchange or filtration media is available for decontamination. Treatment media for decontamination varies from low cost naturally occurring inorganic materials too expensive synthetic organic and inorganic engineered to remove specific ion. The natural zeolite is one of these low-cost materials. The chemistry of most ion exchange media has been extensively studied for the nuclear industry and for other waste treatment applications (Li et al., 2012; Yasmen et al., 2011; Caijun et al., 2006; IAEA, 2003; Campbell et al., 1990). A wide range of materials is available for the ion exchange treatment of radioactive liquids (Hunce et al., 2012; Osmanlioglu, 2006). These materials are available in a variety of forms, have widely differing chemical and physical properties and can be naturally occurring material like zeolite. Ion exchange materials can be categorized according to their suitability for different applications. The type of material to be used is selected based on its ability to remove impurities and undesirable ions and to control pH. These materials are available in a variety of forms, have widely differing chemical and physical properties and can be naturally occurring or synthetic. To minimize disposal costs, in any process for the removal of radionuclide from a liquid waste it is important to minimize the volume of the secondary waste to be conditioned and disposed of in a final repository (El-Kamash, 2008; Calvo, 2009). The selection of an appropriate medium depends on the needs of the system. Generally, low-cost ion exchange media required large volumes. This leads to larger volumes of waste to treatment and disposal.

MATERIALS AND METHODS

The distribution coefficient (K_d) for each radionuclide and/or species present in the solution can be used directly to calculate the amount of ion exchange medium that is needed to achieve the desired decontamination for a certain amount of liquid in a batch process. Under dynamic conditions, in which the ion exchange medium is used in a packed bed or in a column, the distribution coefficient can be used to calculate the theoretical maximum capacity of the ion exchange bed. If the bed includes m kg of ion exchange medium, the total capacity of the bed V_{tot} can be calculated as:

$$V_{tot} = K_d \times m \tag{1}$$

Ion exchange processes can be implemented in a variety of ways, including in batches, in columns, in continuous loops and, as part of, or in combination with, membrane processes (Valsala, 2009). Each technique has features that make it more or less suitable for specific applications. In this study, ion exchange process has been applied by using column. A measured quantity of the ion exchange medium is mixed with the liquid waste in a suitable container. The amount of media required and the rate of exchange can be determined by laboratory tests or by calculation using the following equation:

$$K_d = (DF - 1) \times V/m \tag{2}$$

where;

K_d is the measured distribution coefficient,
DF is the required decontamination factor,
V is the volume of liquid to be purified,
m is the amount of the ion exchange medium needed to reach the required decontamination factor.

The natural zeolite used as a filtration media was obtained from Gordes region of western Turkey. The zeolite consists of clinoptilolite and k-feldspar. The cation exchange capacity of Gordes zeolite was determined about 70 meq/100 g. The chemical compositions were determined by inductively coupled plasma-atomic emission spectroscopy (ICP-ES). The results are presented in Table 1.

TABLE 1. Chemical compositions of natural zeolite (wt.%).

Zeolite	SiO_2	Al_2O_3	Fe_2O_3	MgO	Na_2O	K_2O	CaO
Gordes	63.74	11.80	1.66	1.18	0.51	2.46	1.86

RESULTS AND DISCUSSION

Zeolite works over a limited pH range (4–9) since they are soluble both in acidic and alkaline media. In general, a single ion exchange material is unlikely to be able to take up a particular radionuclide over the whole pH range. Effect of pH levels on the distribution coefficient (K_d) of the zeolites was determined to optimize the waste solution pH for the plant scale operations. Distribution coefficients at different pH levels were shown in Fig. 1.

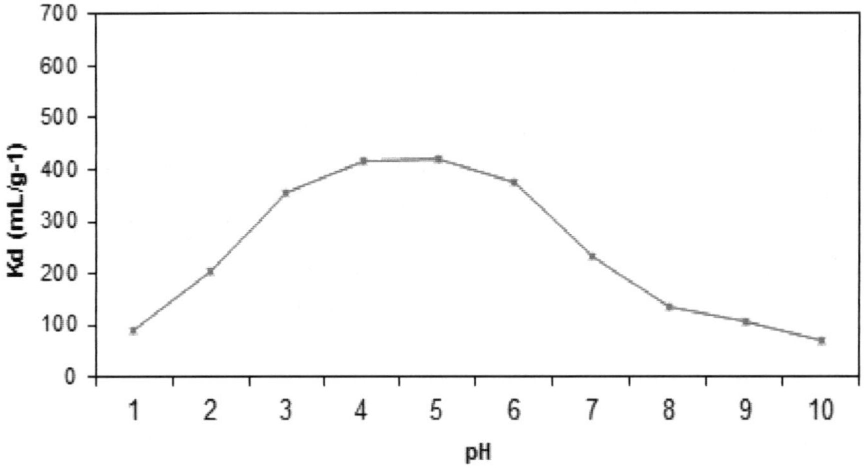

FIGURE 1. Distribution Coefficients (K_d) at different pH levels.

Filter media were prepared by using zeolite particles in size of −16 + 25 ASTM mesh in the column. Natural zeolite particles were loaded into decontamination column. The laboratory scale column was constructed of 0,5 L capacity. Column operation was a semi continuous process.

According to the test results, passed radioactivity from the column during the operation time is presented in Fig. 2. Decontamination factor (D) of the liquid was calculated by dividing initial radioactivity (A_i) to final radioactivity (A_f) of the liquid effluents. Decontamination factors were determined between 350-750 related to pH and temperature.

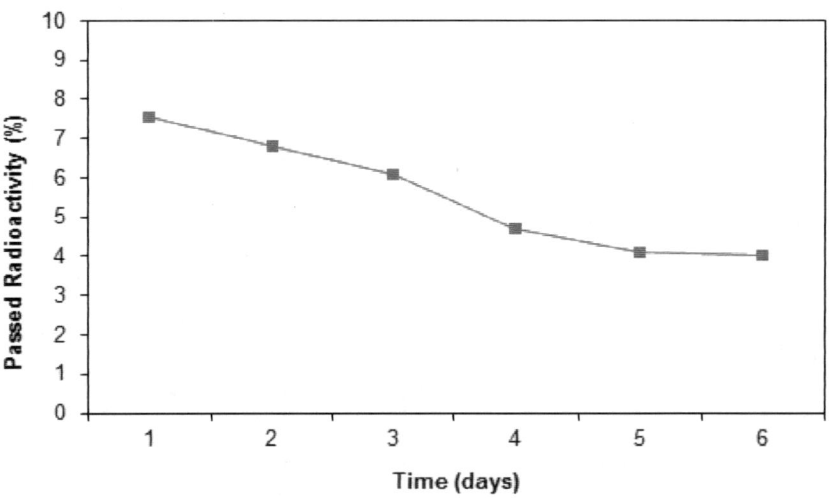

FIGURE 2. Passed Radioactivity versus time.

Effect of temperature on the radionuclide adsorption of the zeolite was determined to optimize the waste solution temperature for the plant scale operations. Purification percentages (%) at different temperatures were shown in Fig. 3.

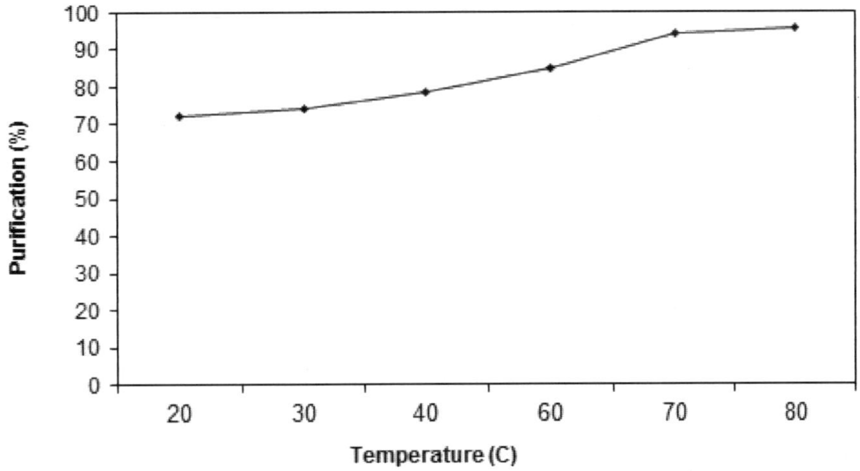

FIGURE 3. Purification level at different temperatures.

CONCLUSIONS

Decontamination of liquid effluents from uranium mine is an important issue for minimization of operational costs. After decontamination process filtration matrix remained as a concentrated solid waste. This radioactive waste usually stabilized by solidification process. Natural materials like zeolite can be used in decontamination process for liquid effluents. This method is cost effective and time-saving method with reasonable waste form product. Technically simple and less costly treatment methods are always attractive in uranium waste management applications. In this study, the results show that this decontamination method could be acceptable for most of the liquid effluents from uranium mines.

REFERENCES

Caijun, S., Fernández-Jiménez, A., 2006. Stabilization/solidification of hazardous and radioactive wastes with alkali-activated cement, Journal of Hazardous Materials 137 (3), 1656-1663.

Calvo, B., 2009. Continuous elimination of Pb, Cu, Zn, H and NH4 from acidic waters by the ionic exchange on natural zeolites. Journal of Hazardous Materials. 166, 619-627.

Campbell, D.O., Lee, D.D., Dillow, T.A., 1990. Low-Level Liquid Waste Decontamination by Ion Exchange. WM90. Proc. Waste Management Symposium, Vol.1. 389-398.

El-Kamash, A.M., 2008. Evaluation of Zeolite A for the sorptive removal of Cs and Sr ions from aqueous solutions using batch and fixed bed column operations. Journal of Hazardous Materials. 151, 432-445.

Hunce, S., Akgul, D., Demir, G., Mertoglu, B., 2012. Solidification/stabilization of landfill leachate concentrate using different aggregate materials. Waste Management 32 (7), 1394-1400.

IAEA, 2003. Combined Methods for Liquid Radioactive Waste Treatment. IAEA-TECDOC-1336. IAEA Publishing, Vienna (Austria).

Li, J., Wang, J., 2006. Advances in cement solidification technology for waste radioactive ion exchange resins: A review. Journal of Hazardous Materials 135 (1–3), 443-448.

Osmanlioglu, A.E., 2006. Treatment of radioactive liquid waste by sorption on natural zeolite in Turkey. Journal of Hazardous Materials.137, 332-335.

Sun, Q., Li, J., Wang, J., 2011. Solidification of borate radioactive resins using sulfoaluminate cement blending with zeolite. Nuclear Engineering and Design 241 (12), 5308-5315.

Valsala, T.P., 2009. Removal of radioactive cesium from low-level radioactive waste (LLW) streams using cobalt ferrocyanide impregnated organic anion exchanger. Journal of Hazardous Materials. 166, 1148-1153.

Yasmen, A.M., Maysoon, J.Z., 2011. Treatment of radioactive liquid waste (Co-60) by sorption on Zeolite Na-A prepared from Iraqi kaolin. Journal of Hazardous Materials. 196, 228-23.

A New Formula of Active Powder Concrete for Curing Radioactive Waste

Baosen Liu , Li'an Hou, Shuili Yu, and Yubing Ye
(Tongji University, Shanghai, China)

The safe disposal of radioactive waste is one of the restricting factors for the sustainable development of nuclear power. Previous studies had shown that the magnesium phosphate cement has certain advantage in disposing radioactive wastes containing Strontium and Caesium, Which existed stably in the form of $MgCsPO_4$ and $SrHPO_4$. Although it was ideal solidification carrier for long-term storage of radioactive waste, Wide application would bring phosphorus resources competition between agricultural fertilizer and cement production using large quantities of phosphate rock. In this study, we used activated Serpentine and Magnesite by high-energy ball milling to stabling Cs+ and Sr^{2+} in wastewater, and the removal rates were over 85%. Then, a new formula was proposes for the solidification of sediment produced from wastewater treatment. The Active powder, steel fiber, water reducer, etc. were added to the Portland concrete according to the specified ratio, and test of cemented waste form showed the performance indicators met 《Performance requirements for low and intermediate level radioactive waste form - Cemented waste form, 》 (GB 14569.1 -2011) , especially the compressive strength was more than 90 MPa. The innovation lied in the step-by-step implementation of chemical stabilization and physical package of radioactive waste, which greatly reduced the consumption of phosphorus resources by Reactive Powder Concrete and may be used in solidification of high-level radioactive waste.

Statistic Methods for Assessments of Risks and Damages at Nuclear Power Plants

Alexander Valyaev* (Nuclear Safety Institute of Russian Academy of Sciences, Moscow, Russia); Gurgen Aleksanyan (Yerevan State University, Yerevan, Republic of Armenia); Alexey Valyaev (Oklahoma State University, Stillwater, USA); Oleg Arkhipkin (Kazakhstan Institute of Oil and Gaz, Almaty, Kazakhstan)

Today NPP using for energy production is constantly increasing in the world with the growth of different threats, for example, caused by the natural and manmade catastrophes and directed terrorist attacks.

It is necessary to make possible correct assessments of corresponding risks levels in common case and for single separated NPP from moment of NPP projecting, building and especially during its exploitation. It demands to provide exclusive attention and special conditions under realization of complex integrated emergency NPP management.

We use our universal formula for assessment of the total vector of limited losses under NPP exploitation. The main problem is assessment of loss probability matrix elements [2]. If representative statistic data, obtained for long NPP exploitation period, are present, then some of its elements may be assessed by statistic methods. We have predicted the irradiation doses and corresponded risks for population under implementation of Russian Federal Program:" Development of Russian atomic energy industrial complex on 2007-2020 years at 10 Russian NPP that operated during some last decades. But early some types of NPP disasters were absent at researched NPP. Using of classic methods of expertise assessments is not correct in this case.

Some needed data may be obtained from primary virtual computer tests of concrete NPP with imitation of possible disasters. It allows to assess risk values and also to plan the actions for NPP operators and special services under serious NPP disasters or may be to prevent them at all.

The following important aspects and problems are under consideration in our communication: 1. The NPP researchers by our statistic methods; 2. Assessments of Risks and Possible Ecological and Economic Damages from Large-Scale Natural and Manmade Catastrophes in Ecology-Hazard Regions of Central Asia and the Caucasus. 3. Some particularizes, connected with the future development of nuclear energy industrial complex in Kazakhstan regions of Ust –Kamenogorsk city and NPP future construction near Lake Balshash and Kurchatov city are also presented in our communication.

Potentialities of Six Plant Species on Phytoremediation Attempts of Fuel Oil-Contaminated Soils

Matsodoum N. P. and Wanko N. A.
(University of Strasbourg, Strasbourg, France)
Djumyom Wafo G. V., Djocgoue P. F., Kengne Noumsi I. M.
(University of Yaoundé I, Yaoundé, Cameroon)

A field experiment investigating the phytoremediation potential of 6 plant species - Goosegrass (*Eleusine indica*), Bermuda grass (*Cynodon dactylon*), Sessile joyweed (*Alternanthera sessilis*), Benghal dayflower (*Commelinpa benghalensis*), Lovanga (*Cleome ciliata*) and Chinese violet (*Asystasia gangetica*) - on a soil contaminated with fuel oil (82.5 ml/kg of soil) have been conducted from March to August 2016. The experiments consider three modalities - Tn: unpolluted planted soils, To: unplanted polluted soils and Tp: polluted planted soil - randomized arranged. Only three (*E.* indica, *C. dactylon* and *A. sessilis*) of the six species survived while the others died 1 month after the beginning of experimentations. The relative growth indexes showed a strong similarity between the growth parameters of *E.* indica and *C. dactylon*, each on polluted and control soils, unlike *A. sessilis*. Total petroleum hydrocarbons (TPHs) removal efficiency were 82.56%, 80.69% and 77% on soil planted with *E. indica*, *C. dactylon* and *A. sessilis* respectively; and 57.25% on non-planted soil. According to the bioconcentration and translocation factors, *E. indica* and *A. sessilis* are involved on rhizodegradation and phytoextraction of hydrocarbons whereas *C. dactylon* is only involved into rhizodegradation. Overall, *E. indica* and *C. dactylon* out-yielded *A. sessilis* in the phytoremediation capacity of fuel oil-contaminated soils.

Study of Shear Strength of Root Reinforced Soil

Gloria James, Dr Manas Kumar Bhoi, Poornima Menon
(Department of Civil Engineering, Pandit Deendayal Petroleum University,
Ahmedabad, Gujarat, India)

ABSTRACT: The use of vegetation for soil stabilization is being done from ancient time. Vegetation can significantly increase the stability of sloping earth surface and bearing capacity of any type of foundation soil. The reinforcement depends upon the morphological characteristics of the tree root used and the tensile strength of the roots both in dry season and wet season. In this paper an experimental investigation has been conducted on the soil reinforced with plant roots in the laboratory using a standard CBR (California bearing ratio test) apparatus and triaxial test to study the stress strain behavior and the strength of the soil. The grass plant *Stipa capillata* is used to reinforce the soil. This experimental test program is conducted over a period of 4 to 5 months. The results obtained indicate that the strength decreases as the moisture content increases greater than the optimum moisture content. Through triaxial test it is found that root reinforced soil has a significant increase in cohesion factor compared to the non reinforced soil (87%).

Keywords: Root reinforced soil, Undrained Triaxial Compression test, California Bearing Ratio, Stress and strain relationship

INTRODUCTION

Soil stabilization is often important to provide a proper foundation for roadways, to prevent erosion and slope failure, and even to mitigate natural disasters such as landslides. The most widely used method seen for increasing the stabilization of the soil is by using vegetation and geofibers as reinforcement. Stabilization of slopes using plant vegetation is a long-practiced, effective, and environmentally friendly method that results in minimal disturbance to the subsurface soil.

Roots increase the tensile strength of the soil as roots have relatively high tensile strength and they bind with the soil particle and enhance the shear strength of the soil by forming a root matrix system. Although deep rooted trees have reinforcing effect up to the depths of 3 meters, the fine roots having 1-20 mm diameter also contribute to reinforcement as indicated in field studies of forested slopes conducted by O'Loughlin [11]. Grasses, legumes and small shrubs can have a reinforcing effect to depths of 0.75-1.5m. Various researchers have attempted studying shear strength of reinforced soil by various testing methods such as direct shear test & triaxial tests. Shear strength of a soil containing vegetation roots was studied by Ali and Osman[3] by using four different species of plant namely Vertiveria zizanoides, Leucaena leucocephala, Bixa orellana and Bauhinia purpurea. Experimental Studies on Soil Stabilization was done by Gopinath et al.[4] using Vetiver Root as Reinforcement. Soil strength Enhancements using Polymer-Infused Roots was studied by Miguelangel et al.(2011).

Stipa capillata is the most abundant form of grass found in Gujarat, India. There is an increase in strength of the surface soil if it is reinforced with it. In this paper CBR test and laboratory triaxial test of the soil samples reinforced with different quantity of roots are carried out in order to understand the stress-strain relationship with increase in root matrix for a given soil specimen.

METHODOLOGY

Materials.

(a) Empty Mould

(b) Mould with fully grown plant

Figure 1. Cylindrical Mould with Lower Cap Used for Growing the Plant Species

***Plastic Mould*:** 16 cylindrical plastic containers of diameter 15 mm and height 17 mm were used to prepare soil samples for growing the plant stipa capillata. These dimensions were selected (similar to CBR Mould) in order to perform CBR tests directly in the prepared sample mould so that the sample is in undisturbed state when performing the experiments. Identical soil samples and density were maintained in all the moulds.

Table 1. Physical & Mechanical Properties of Soil

Liquid limit	Specific gravity	Optimum moisture content	Coefficient of uniformity (Cu)	Coefficient of curvature (Cc)
4.16%	2.92	8.22%	5.012	1.58

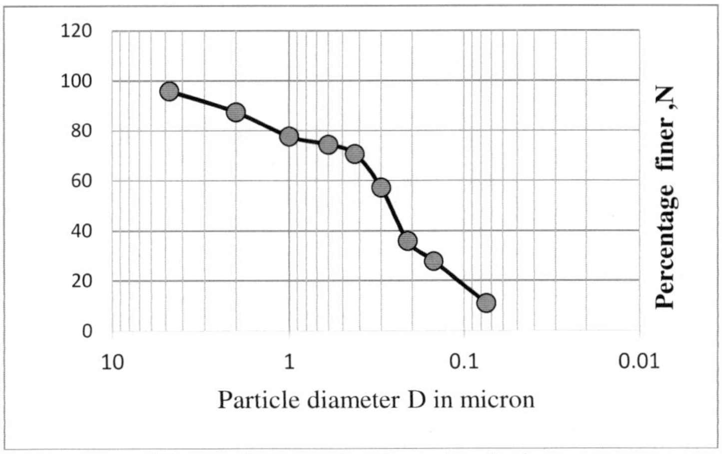

Figure 2. Particle Sieve Analysis

***Soil Material*:** Particle sieve analysis was done and the soil was classified as medium graded soil. Further sedimentation test was performed to determine the type of soil .The soil was classified as sandy silt soil.

***Plant Type*:** The plant species used is stipa capillata. It is a perennial bunchgrass species in the family Poaceae, native to Europe and Asia. It is available throughout the year in India. The diameter of the root ranges from 0.05-1mm. The time taken for the grass species to develop fully is 4-5 weeks. Soil was planted with this plant species and allowed to grow for 1- 2 months.

Test Methods. To determine the optimum moisture content (OMC) standard proctor test was conducted for the soil with roots separated. The laboratory tests are opted so that identical soil samples can be prepared & uniform saturation of samples can be carried out. 16 plastic cylindrical apparatus similar in size to the CBR test apparatus were prepared to study the variation in soil strength due to growth of root matrix system .Some of the soil samples were prepared by Procter equipment after taking out the roots out of the original reinforced soil sample and it is used as reference (i.e. no vegetation roots) soil sample to observe the improvement in engineering properties of reinforced soil. After determination of OMC, CBR test was done for unreinforced and root reinforced soil at the optimum moisture content. Further CBR test for reinforced soil was done under different moisture content with the same density maintained in all the samples to determine the effect of moisture content on CBR value.

Triaxial Tests was conducted to study the stress strain behavior of soil, and conducted for both unreinforced soil and reinforced soil (reinforced with different quantity of roots i.e. 0.2, 0.4 and 0.8 g per layer) sample with three layer roots. The quantity of roots was determined by the amount of roots present after 2, 3 and 5 (full grown condition) weeks respectively. The quantity of roots present was determined after preparing the sample using core cutter of the same internal dimensions as that of the triaxial specimen size. And the tests were carried out under the confining pressure of 50 Kpa, 100 Kpa and 150 Kpa.

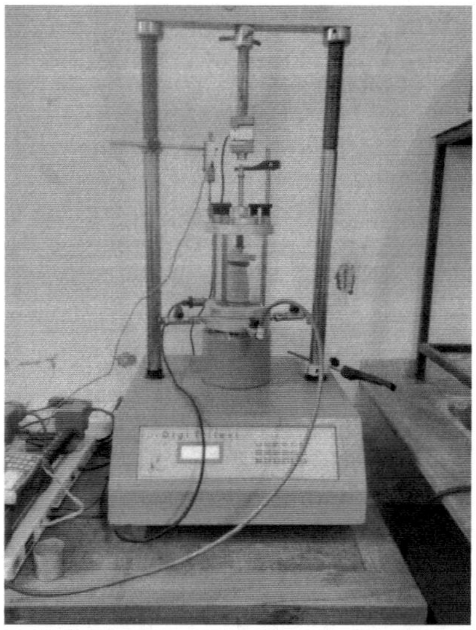

Figure 3. Triaxial Test on Reinforced Soil Specimen

RESULTS AND ANALYSIS

Procter Test Results. Optimum moisture content and maximum dry density of the unreinforced soil was found out using standard proctor test. The maximum dry density of the sample was observed at moisture content of 8.22% (Optimum Moisture Content).

CBR Test. CBR tests were conducted on the reinforced soil at different moisture content and the results are depicted in figure 5.

Further CBR test was conducted after 2,3 and 5 weeks and graph was plotted between CBR value and percentage of roots (wt /wt) as shown in figure 5. Here the 1 week time corresponds to 0.43% of roots by weight, 3 weeks time corresponds to 0.86 % of roots by weight and 5 weeks time corresponds to 1.735%

of roots by weight with respect to soil. Beyond this it is found that, there is no significant increase in CBR value with increase in number of weeks.

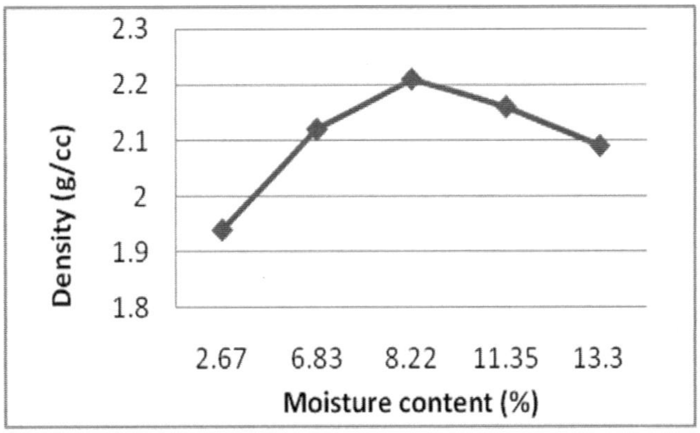

Figure 4. Optimum Moisture Content

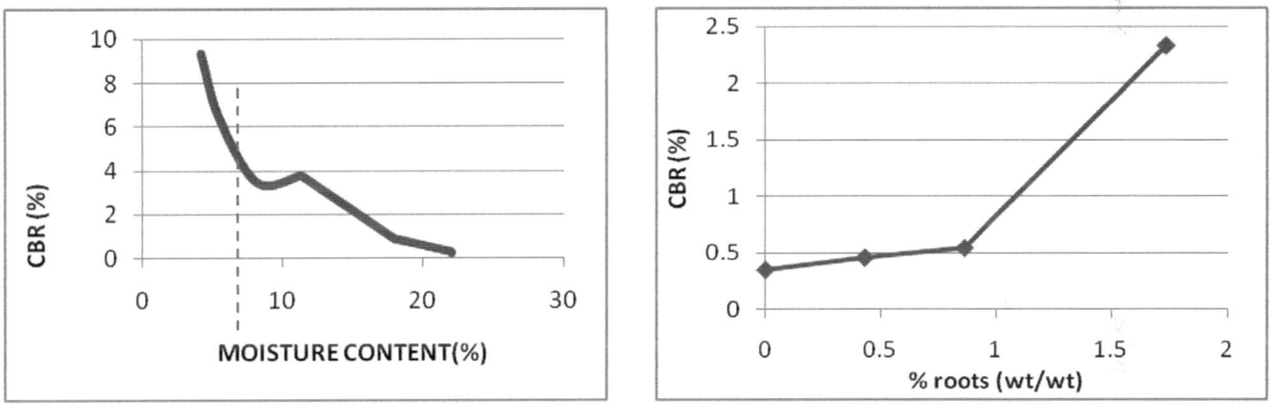

(a) CBR values at different moisture content

(b) CBR value w.r.t percentage of roots

Figure 5. CBR Variation with Moisture Content and Percentage of Roots

In general the CBR value increases as the root content increases. The reason being the presence of root matrix in soil mass which holds the soil particles together and it provides the resistance to the deformation caused by the CBR apparatus plunger penetration and that invariably led to increase in soil CBR value.

Triaxial Shear Test. Triaxial tests were done to study the stress strain behavior of un-reinforced and reinforced soil (under different quantity of roots). The results in figure 6 (a), (b), (c) shows the stress-strain relationship of soil specimens reinforced with roots under different confining pressure, plotted as principal stress difference against the axial strain at different confining pressures of 50, 100 and 150 kpa.

Under certain conditions, the strength increased up to certain quantity of roots and after that remained same or decreases. In the triaxial results, the strength increases up to 0.4g per layer and after that remained equal or decreased. The reduction in the soil internal angle of friction and cohesion as a result of increasing roots content beyond 0.4 g per layer or 0.75% could be attributed to the large quantum of roots matrix present in the soil. The excess roots decrease the contact between soil particles and with that decreases the frictional resistance developed to resist the deformation due to external forces.

As the shear strength is directly proportional to the friction factor and the cohesion of the soil, the decrease in cohesion of the soil particles due to increase in roots causes decrease in shear strength of the soil. Further Mohr's circle shows that there is 87% increase in cohesion of the soil when the root content are increased from 0 to 0.4 g per layer of soil in triaxial test specimen.

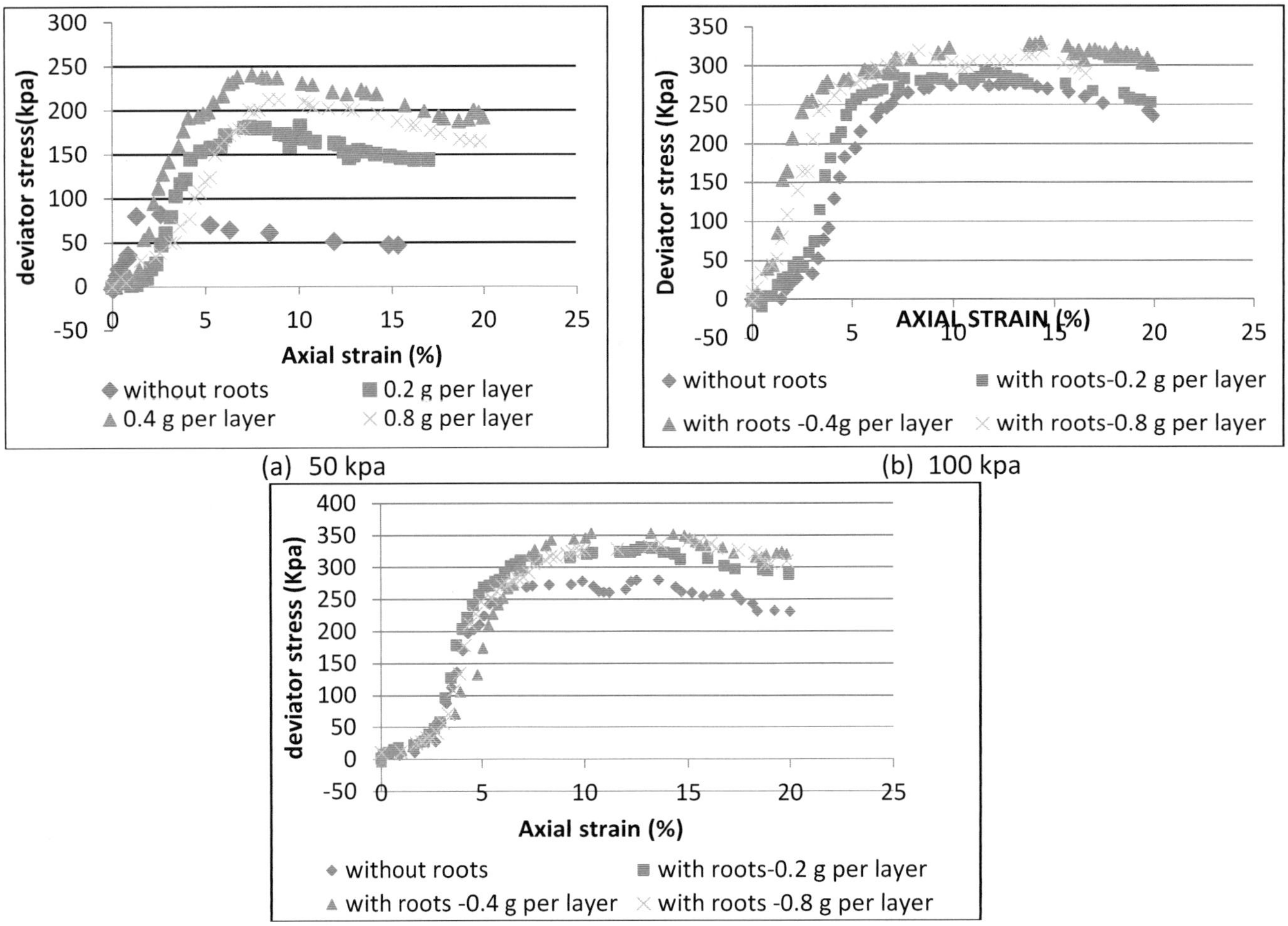

(a) 50 kpa

(b) 100 kpa

(c) 150 kpa

Figure 6. Stress Strain Variation Characteristics of Soil under Different Confining Pressure of 50kPa, 100kpa and 150kpa in Triaxial Test

CONCLUSIONS

Laboratory triaxial tests on soil specimens with different root quantity of stipa capillata plant species were carried out under the specific confining pressure. The test results indicate that the strength and capacity for resisting the deformation of soil reinforced with roots are better than those of unreinforced soil. Under the certain number of grassroots layers, the strength and capacity for resisting the deformation of soil reinforced with roots increase firstly and then reduce with the increasing of root quantity. In other words, there is an optimal quantity of the roots affecting the strength and capacity for resisting the deformation of reinforced soil. The strength of soil reinforced with roots had to do with the root quantity in the soil.

In case of CBR test the CBR value increases with the increase of root quantity upto 5 weeks and after that it remains constant (although not shown in graph) and in any case it is better than the unreinforced soil. The proposed natural reinforcement of soil using plant roots can be used to advantage in bearing capacity improvement and settlement reduction in the soil for design of pavements and other foundation soils.

REFERENCES

[1] Liu, k., Yang, X., Xie, X., Wu, C. & Liu, Y., (2011) *Laboratory Triaxial Test Study on Soil Reinforced with Roots of Manila Grass.*[online](updated 17 May 2011) Available at https://www.scientific.net/AMR.250-253.1366.

[2] Sauceda,M., Johnson, D., Huang, J., Sazzad ,Valerie M., & Appleford, M., (2013) .soil-Strength Enhancements from Polymer-Infused Roots. *Journal of Geotechnical and Geoenvironmental Engineering,* 140(2).

[3] Ali,F., & Osman, N., (2008).Shear strength of a soil containing vegetation roots. *Soils and Foundations* , 48(4): pp.587-596.

[4] R.Gopinath, G.P.Ganapathy, S.P.Saravanan, V.Vijayan, S.Muthukumar, M.Muthuseenivasan, (2015) .Experimental Studies on Soil Stabilization using Vetiver Root as Reinforcement. *International Journal of Applied Engineering Research,*10(53): pp.286-290.

[5] Aboshio, A., (2010). Comparative Study of the California Bearing Ratio Test and Undrained Triaxial Compressive strength Test for Lateritic Soils. *Journal of Engineering Technology*, 5(2).

[6] Tien, h., William, P., Swanston, D., (1978). Strength of tree roots and landslides on Prince of Wales Island: Alaska. *Canadian Geotechnical Journal*, 16(1): pp. 19-33.

[7] Helwany, S.,(2007). *Applied Soil Mechanics*. New Jersey: John Wiley & sons.

[8] Ayininuola, G., Oladotun, P., (2016). Geotechnical Properties of Coconut Coir Fiber Soil Mixture. *Journal of Civil Engineering Research,* 6(4):pp. 79-85.

[9] Butt, W., Gupta, K., Naik, H., Bhat, S., (2014). Soil Sub- grade Improvement Using Human Hair Fiber. *International Journal of Scientific & Engineering Research*, 5(12): pp. 977-981.

[10] Valerie M. & Appleford, M., (2014) .Durability of Polymer Infused Roots Used for Soil Stabilization. *Journal of Materials in Civil Engineering*, 26(8).

[11] O'Loughlin, C.L. (1984) "*Effectiveness of introduced forest vegetation for protecting against landslides and erosion in New Zealand's steeplands*", Paper presented to Symposium on effects of forest land use on erosion and slope stability, Honolulu, Hawaii.

ECOSYSTEM ASSESSMENT AND RESTORATION

Stress Factors to Fish Habitat in Urban Rivers

Yuta Yamauchi, Tetsuya Nakata, and Yutaka Sakakibara
(Waseda University, Shinjuku, Tokyo, Japan)

ABSTRACT: To assess crucial stress factors to life cycle of fish, a Life Cycle Risk Assessment (LCRA) taking into consideration of eight stress factors found in suburban and rural rivers was applied to urban rivers. Comparisons of predicted results by LCRA and observed results on surviving fishes suggested there are additional stress factors specific in urban rivers. Three additional factors, i.e. predation by large aquatic life-forms such as carp and red-eared slider turtles, the existence of estrogen (17alpha-ethinylestradiol; EE2), and washout by flooding were selected, and their effects to fish survivals were studied. As a result, it was found that predominant feeds of carp were bloodworms, waterweeds and crayfishes, while those of red-eared slider turtles were crabs, crayfishes and diving beetles, indicating the predation is not a crucial stress factor. Concentrations of EE2 were in the range of 0.06 to 1.5 (ng/L), suggesting EE2 is also not a crucial factor. However, it was considered significant fluctuations in flow rates in May and June may be an additional stress factor in urban rivers.

INTRODUCTION

In japan, a lot of river improvement works for flood control and water utilization had been done since high economic growth period to 1980s. These works were far from care to ecosystems. Biodiversity has been decreasing with increasing stress factors through exposures of artificial concrete structures, water pollutions by micropollutants, and invasions of alien species. In our former study (Aoki and Sakakibara, 2005), we proposed a Life Cycle Risk Assessment (LCRA) and showed LCRA is an effective tool to evaluate fish species surviving in urban and suburban rivers (Sakakibara and Nakada, 2008). Eight stress factors such as depletion of dissolved oxygen (DO), water temperature, and disappearance of habitats, shortage of depth in movement, disappearance of evacuation areas, disappearance of spawning areas, existence of obstacles, and shortage of food were taken into consideration in LCRA. In this study, LCRA was applied to urban rivers and additional stress factors specific in urban rivers were investigated.

LIFE CYCLE RISK ASSESSMENT

Life Cycle Risk Assessment (LCRA) is a method to assess crucial stress factors to life cycle of fish and to predict the possibility of fish inhabitation. In order that any organisms inhabit there and to keep their populations, continuous reproduction is essential. Life cycle of fish could be divided into 3 life-stages illustrated in FIGURE 1; fry fish, adult fish, and spawning. If certain fish cannot survive, it can be assumed that some life-stages are significantly affected by some stress factors. In LCRA, the following eight stress factors were considered; depletion of dissolved oxygen (DO), water temperature, disappearance of habitats, shortage of depth in movement, disappearance of evacuation areas, disappearance of spawning areas, existence of obstacles, and shortage of food. Threshold values were summarized in TABLE 1, which were referred to the literatures.

Comparisons of predicted results by LCRA and observed results are shown in TABLE 2. In this table, results shaded by green color mean disagreement that LCRA predicted 'survival' but fish survey showed 'no survival'. On the other hand, a result shaded by yellow color means disagreement that LCRA predicted 'no survival' but fish survey showed 'survival'. Most cases for disagreement were the case shaded by green color that LCRA predicted 'survival' but fish survey showed 'no survival'. Agreements were not shaded and concordance rate was 60%. In our former study, fish species in rivers flowing in suburban and rural areas could be reasonably well predicted by LCRA and resulted in 85% agreements (Sakakibara and Nakada, 2008). These results suggest there're additional stress factors specific in urban rivers.

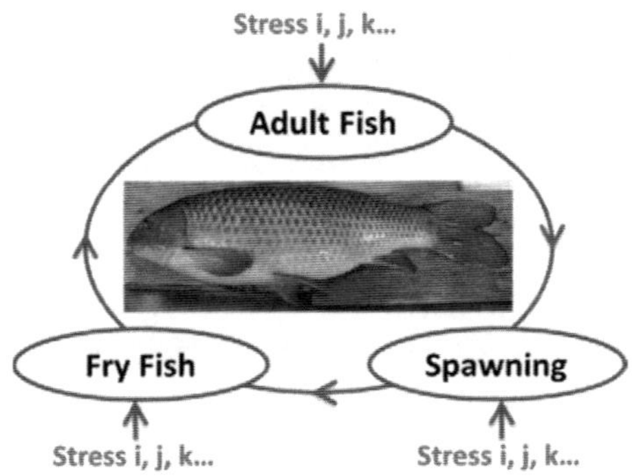

FIGURE 1. Illustration of LCRA. (Aoki and Sakakibara, 2005)

TABLE1. Threshold values for 8 stress factors (Sakakibara, 2013).

	Depletion of dissolved oxygen	Water temperature		Disappearance of habitats			Shortage of depth in movement			Disappearance of evacuation area	Disappearance of spawning area	non-existence of a food	Existence of obstructions
Life stages	All	Spawning	Fry fish, Adult fish	Spawning, Adult fish		Fry fish	Spawning	Fry fish	Adult fish	All	Spawning	All	Adult fish
	[%]	[℃]	[℃]	Rapids	Deeps	Flow rates [cm/s]	[cm/s]	[cm/s]	[cm/s]	※considering movement except fry fish	※considering movement	Feeding habitat	Bump [m]
Phoxinus lagowskii	80	29.8	29.8	$F_r>0.2$	$F_r<0.2$	5	10	5	10	Vegetation or stagnation	Gravel bottom	Omnivorous	1
Salvelinus leucomaenis	80	10	15	$F_r>0.2$	$F_r<0.2$	5	15	5	15	Vegetation or stagnation	Gravel bottom	Aquatic insect	-
Tribolodon Hakonensis	17.3	29.8	29.8	$F_r>0.2$	$F_r<0.2$	5	30	5	15	Vegetation or stagnation	Gravel bottom	Omnivorous	1
Zacco Platypus	19.1	32.7	32.7	$F_r>0.2$	$F_r<0.2$	5	10	5	10	Vegetation or stagnation	Gravel bottom	Omnivorous	1
Cottus pollux Gunther	80	27.3	27.3	$F_r>0.2$	$F_r<0.2$	5	20	5	10	Vegetation or stagnation	Gravel bottom	Aquatic insect	1
Pseudogobio esocinus	19.5	31.7	31.7	$F_r>0.2$	$F_r<0.2$	5	15	5	15	Vegetation or stagnation	Gravel bottom	Omnivorous	1
Cyprinus carpio	11.6	33	33	-	$F_r<0.2$	5	15	5	15	Vegetation or stagnation	Aquatic plants	Omnivorous	1
Cobitis biwae	50	29.8	29.8	$F_r>0.2$	$F_r<0.2$	5	15	5	15	Vegetation or stagnation	Aquatic plants	Omnivorous	1
Tanakia Lanceolata	50	33	33	-	$F_r<0.2$	5	15	5	15	Vegetation or stagnation	Bivalve	Omnivorous	1
Gnathopogon elongatus elongates	17.3	31.7	31.7	-	$F_r<0.2$	5	15	5	15	Vegetation or stagnation	Aquatic plants	Omnivorous	1
Misgurnus anguillicaudatus	20	33	33	-	$F_r<0.2$	5	15	5	15	Vegetation or stagnation	Aquatic plants	Omnivorous	1
Carassius auratus langsdorfii	7.3	33	33	-	$F_r<0.2$	5	15	5	15	Vegetation or stagnation	Aquatic plants	Omnivorous	1
Pungitius. sp	80	18	18	$F_r>0.2$	$F_r<0.2$	5	15	5	15	Vegetation or stagnation	Aquatic plants	Omnivorous	-
Pseudorasbora parva	20	32.7	32.7	-	$F_r<0.2$	5	15	5	15	Vegetation or stagnation	Stones, Aquatic plants	Omnivorous	1
Rhinogobius flumineus	50	32.3	32.3	$F_r>0.2$	$F_r<0.2$	5	20	5	10	Vegetation or stagnation	Gravel bottom	Omnivorous	-

TABLE 2. Comparisons of Predicted Results (Tokyo, 1999) by LCRA and Observed Results.

	Akebono Bridge	
object fish species	fish survey	LCRA
Phoxinus lagowskii	×	O
Salvelinus leucomaenis	×	×
Tribolodon Hakonensis	O	O
Zacco Platypus	O	O
Cottus pollux Gunther	×	O
Pseudogobio esocinus	×	O
Cyprinus carpio	O	O
Cobitis biwae	×	O
Tanakia Lanceolata	O	×
Gnathopogon elongatus elongates	O	O
Misgurnus anguillicaudatus	O	O
Carassius auratus langsdorfii	×	O
Pungitius. sp	×	×
Pseudorasbora parva	O	O
Rhinogobius flumineus	O	O
concordance rate (%)	60.0	

NEW STRESS FACTORS SPECIFIC IN URBAN RIVERS

As additional stress factors, predation by large aquatic life-forms such as carp and red-eared slider turtles, washout by flooding, and the existence of estrogen (17alpha-ethinylestradiol, EE2) were considered. Kanda river in Tokyo was selected in this study.

Predation by Carp and Red-eared Slider Turtles. Carp is omnivorous fish surviving even in polluted rivers. They have been planted to reduce chironomids in rivers. Red-eared slider turtles are carnivores native to the Mississippi valley. And they are imported as a pet. They have high ability to adapt a new environment and high fertility, so that they are increasing in natural environments. Carps and red-eared slider turtles were caught by fishing and dip net respectively. And we investigated what they ate by cutting their bodies.

Existence of EE2. EE2 is used as a main component of oral contraceptives. The use of oral contraceptives has been permitted since 1999 in Japan. Therefore, it was supposed that the amount of EE2 in environments has been increasing year by year. EE2 concentrations were measured by EE2 ELISA kit (microplate, Tokiwa Chemical Industries CO., LTD). Water samples were taken at 4 points.

Washout by Flooding. Urban rivers are generally covered with concrete and simplified and there are few evacuation areas. Therefore, once after rail falls, flow rate increases rapidly.

RESULTS AND DISCUSSION

Predation by Carp and Red-eared Slider Turtles. PICTURES 1 and 2 show the contents observed in intestines of carp and red-eared slider turtle, respectively. Predominant feeds of carp were bloodworms, waterweeds, and crayfishes, while those of red-eared sliders are crabs, crayfishes, and diving beetles. However, no fish was observed. Therefore, predation was not considered as a stress factor.

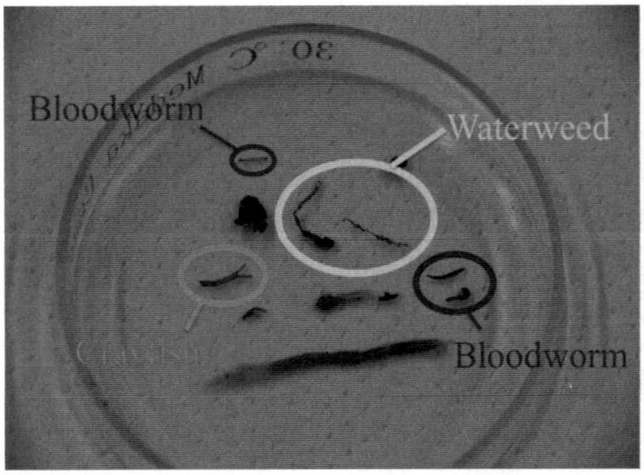

PICTURE 1. Feeds of Carp.

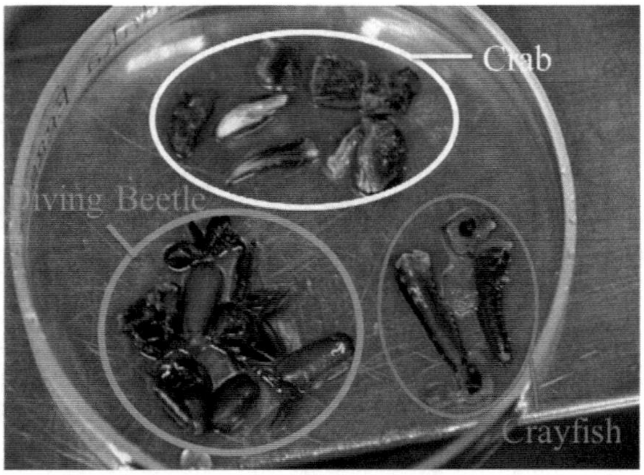

PICTURE 2. Feeds of Red Eared Slider.

TABLE 3. EE2 Concentrations at Kanda River

	upstream water	diversion aqueduct	Myosyoji river	Akebono bridge
EE2(ng/L)	0.193	1.510	0.088	0.060

TABLE 4. Critical values of Swimming Speeds (Suzuki, 1998)

Fish species	Fry fish [m/s]	Adult fish [m/s]
Phoxinus lagowskii	0.05	0.33
Salvelinus leucomaenis	0.05	0.66
Tribolodon Hakonensis	0.05	1.00
Zacco Platypus	0.05	0.38
Cottus pollux Gunther	0.05	0.38
Pseudogobio esocinus	0.05	0.50
Cyprinus carpio	0.05	1.50
Cobitis biwae	0.05	0.26
Tanakia Lanceolata	0.05	0.25
Gnathopogon elongatus elongates	0.05	0.25
Misgurnus anguillicaudatus	0.05	0.38
Carassius auratus langsdorfii	0.05	0.63
Pungitius. sp	0.05	0.09
Pseudorasbora parva	0.05	0.20
Rhinogobius flumineus	0.05	0.15

Existence of EE2. TABLE 3 shows the results of EE2 measurements. Concentrations of EE2 was in the range of 0.06-1.5 (ng/L). It was reported that reproductive behavior was decreased at 10 (ng/L) for Medaka

(Balch et al., 2004), and egg production and fertility rate were decreased at 10 (ng/L) for fathead minnow (Pawlowski et al., 2004). There were also some reports that EE2 have some negative effects on fish below 10 (ng/L), so that it was considered that EE2 affects fish survivals significantly above around 10 (ng/L). In addition, EE2 was not considered as a stress factor.

Washout by Flooding. Critical values of swimming speeds are summarized in TABLE 4. They were obtained from the literatures. And the range is 0.05-1.5 (m/s).

The flow rate of peak rainfall in 2014 was estimated at 2.24 (m/s) from the following expressions (Takasaki et al., 2008).

$$Q=4.564\times(H+0.236)^2 \quad (1)$$

$$Q=A\times v=B\times H\times v \quad (2)$$

Where A: Section Area〔km^2〕, B: River Width〔m〕, H: water Level〔m〕, Q: Flow Rate〔m^3〕 and v: Flow Velocity〔m/s〕 respectively.

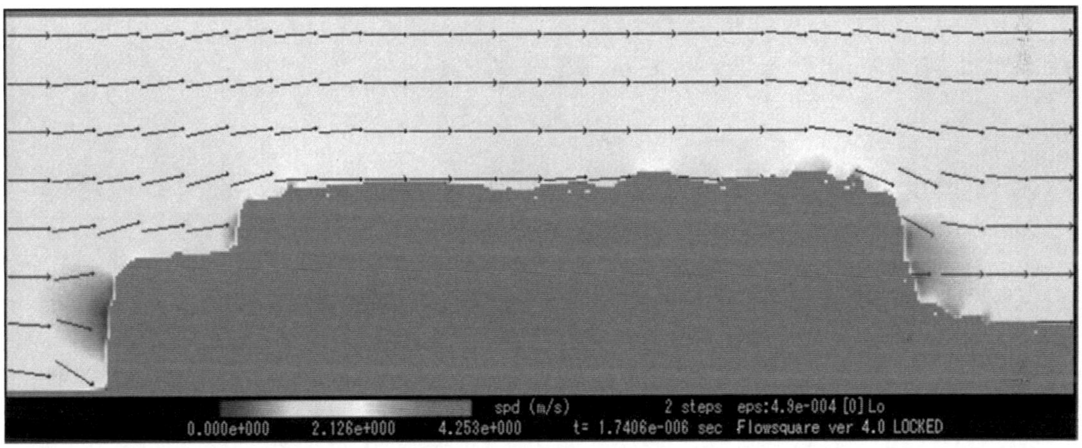

FIGURE 2. Numerical Calculation around Obstruction

From comparison of estimated flow rate and critical flow rates in TABLE 4, all fish would be washed out. However, if there are some obstructions in rivers, there may be a possibility that flow rate decreases below fish swimming speeds and therefore fish can survive in rivers.

FIGURE 2 is the numerical calculation results of flow rates around the obstruction under Akebono Bridge in Kanda river. Flowsquare ver. 4.0 was used in the calculation. As shown, it was found that there were some areas where flow rates were almost 0 (m/s). This means that flow rates were reduced significantly below swimming speeds (0.05-1.5m/s) in TABLE 4.

TABLE 5 shows the spawning seasons of fishes. In this TABLE, green color's arrows mean the fishes that LCRA predicted 'survival' but fish survey showed 'no survival'. Most fishes spawn eggs in May and June which are a rainy season in Japan. In this season, significant fluctuations in temperature and flow rate are observed, which may affect fish inhabitation. A further study will be needed to evaluate precisely the effects of fluctuations to fish life cycles.

TABLE 5. Spawning Seasons of Object Fish

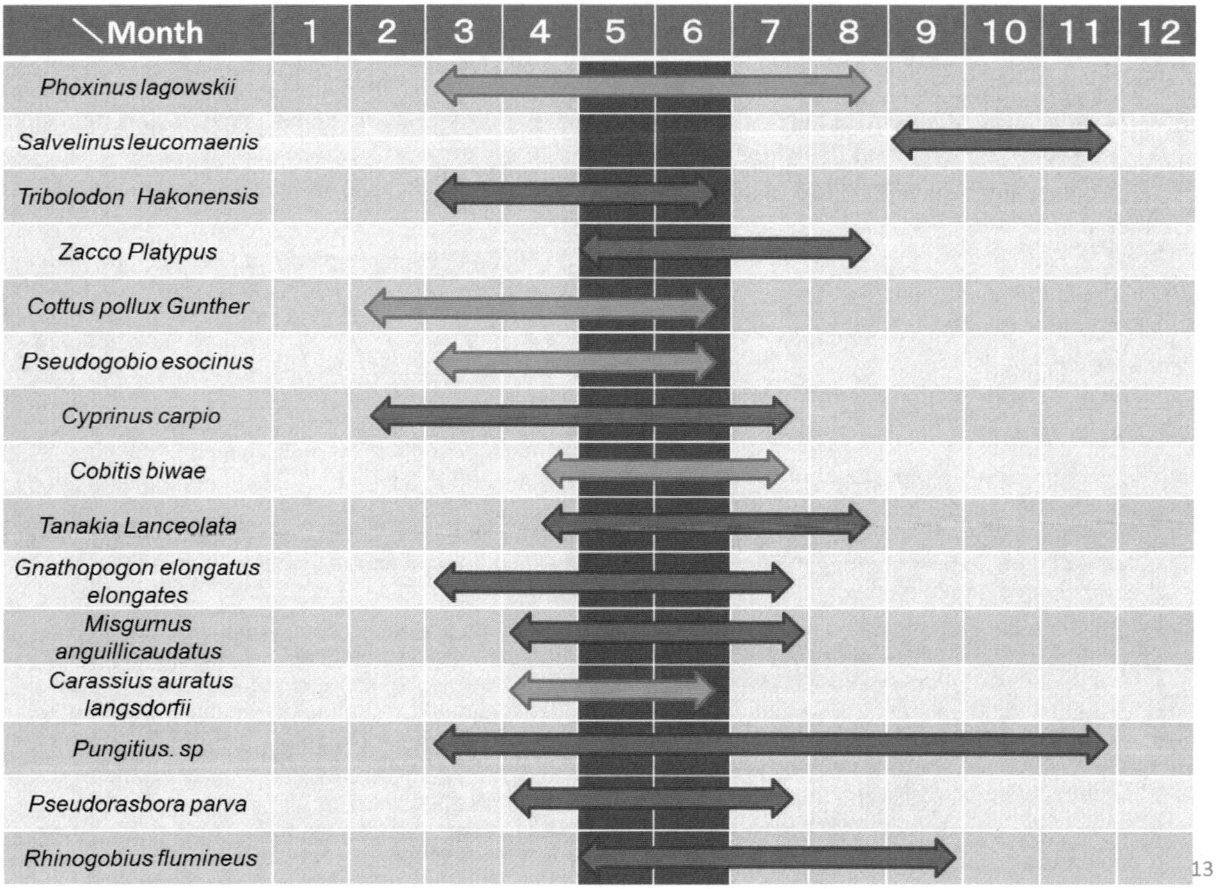

CONCLUSIONS

This study demonstrated that predations by carp and red-eared slider turtles were not a stress factor, in addition to EE2 in river. However, it was considered that fluctuations in flow rates in May and June may be a stress factor specific in urban rivers. A further study will be needed to evaluate the effects of fluctuations to fish life cycles.

REFERENCES

Aoki, T. and Y. Sakakibara 2005. "A life-cycle assessment for fishes in streams in suburban areas". *Proc. of 10th International Specialist Conference on Watershed and River Basin Management 2005, Calgary, Alberta, Canada.*

Balch, G.C., C.A., Mackenzie and C.D., Metcalfe 2004. "Alterations to gonadal development and reproductive success in Japanese medaka". *Environmental Toxicology & Chemistry,* 23(3): 782-791.

Pawlowski, S., R. van Aerle, C.R. Tyler and T. Braunbeck 2004. "Effects of 17alpha-ethinylestradiol in a fathead minnow (Pimephales promelas) gonadal recrudescence assay". *Ecotoxicol Environ Saf*, 57(3): 330-45.

Sakakibara, Y. and A. Nakada 2008. "Assessing crucial stress on life cycle of fish in suburban streams". *Water Science & Technology*, 58(3): 705-711

Sakakibara, Y. 2013. "4.3 Effects and evaluations of sewers that have effects on fish inhabitation in urban areas". *Ministry of Land, Infrastructure, Transport and Tourism, Water and Disaster Management Bureau, Sewer Department*, 158-166.

Suzuki, K. 1998. "Fish distribution and flow rate contribute to making rivers which are suitable fish". *Collection of Academic Papers Japan Society of Civil Engineers*, No.593/ Ⅱ -43, 21-29.

Takasaki, T., D. Sugihara and T. Iwaya 2008. "15.flow rate change and water characteristics of estuarine basin at small and medium rivers in urban areas". *Annual Report of Urban Civil Technical Centre*, 171-178.

Tokyo 1999. Internal data of Div. of green park of Dept. of civil and environmental engineering in Shinjuku Ward Office.

Study of Aquatic Macroinveretebrates as Bioindicators in Stream Ecosystems of Kashmir Himalaya

Sami Ullah Bhat, Inam Sabha, Sheikh Tajamul and Shabir Ahmad Rather
(School of Earth and Environmental Science, University of Kashmir Srinagar India-190006)

The valley of Kashmir is gifted with network of streams ecosystems which are world famous for its beauty and benefits. But today noticeable signs mostly in the form of deteriorating water quality and changing land use patterns have send the signals to policy makers for quick and continuous monitoring of these important stream ecosystems in the valley of Kashmir. The science of biological assessment and monitoring of aquatic ecosystems have been well-developed in many parts of the world namely Europe, Australia, United States and Great Britain but is limited in countries that are in their infancy of environmental protection. Since biological communities integrate the response to pollutants and human disturbance through their continuous exposure to the magnitude, duration, and frequency of these stressors/disturbances/pollutants, these indicators provide the most information on ecological condition. Different groups of macroinvertebrates are excellent indicators of human impacts, especially contamination. Most of them have quite narrow ecological requirements and are very useful as bioindicators in determining the characteristics of aquatic environments (Benetti and Garrido, 2010; Fernández-Díaz *et al.*, 2008; Pérez-Bilbao and Garrido, 2009). These stream ecosystems are usually characterised by peak (Spring-summer season) and lean periods (Autumn-winter seasons) which has a bearing on water chemistry and taxonomic structure of benthic macroinvertebrates. Water quality of streams in Kashmir valley varies considerably as the factors shaping the water quality differs ranging from pristine or nearly pristine conditions in contrast to most lowland streams being highly affected by urbanisation and changing land use patterns. Streams in the valley offer a large altitude gradient with streams originating on average from 4000m amsl (low order streams) and merging in the river Jhelum at about 1600m amsl (high order streams) . This altitudinal gradient is also accompanied by gradient in land-use changes from upstream to downstream. Macroinvertebrates were collected with the help of D-net, Rock pick method and Dredge. D-net and rock pick method was utilized for the upper and middle stream sites (cobbles or gravel) and for downstream sediment sites dredge was employed. The density and diversity patterns of stream ecosystem microhabitat such as riffles, pools, glides and run were analyzed. Standard methodology and taxonomic works were followed. McCafferty and Provonsha (1998); APHA (1989); Borror *et al.* (1976); Pennak (1978); Ward (1992); Engblom and Lingdell (1999) and Bouchard (2004).

The present study carried out is aimed to have a detailed information on taxonomic composition of aquatic macroinvertebrates collected at 52 sampling stations from 16 major stream ecosystems falling in ten districts of Kashmir valley covering 24 watersheds. A total of 56 taxa were reported and the phylum Arthopoda was found to be most dominant followed by Annelida and Mollusca. The dominant Class in phylum Arthopoda was Insecta, comprising of 50 taxa, Arachinida (2), Annelida (2), Crustacea (1) and Mollusca (1). Among Class Insecta, order Trichoptera (13) was most abundant followed by Ephemeroptera (11), Diptera (10), Plecoptera (5), Coleoptera (5) Odonata (2), Megloptera (2) and Hemiptera (2). The dominance of Class Insecta can be attributed to the suitable habitat provided by rocks and boulders at upper and middle stretches of streams. Moreover, a total of 4104 individuals were collected during the study. The highest number of taxa (19) were found at site 39 whereas only 1 taxa was recorded at site 34. Additionally a total 4104 individuals were collected with highest no of individuals being found at Site 38 (158) and the lowest at Site 14 (5) respectively. The majority of the taxa recorded indicate the water quality of the streams as good.

Carbon Sequestration on Application of Biochar Pellet Cooperated with Pig Manure Compost during Rice Cultivation

JoungDu Shin* · EunSuk Jang (Department of Climate Change & Agro-ecology, National Institute of Agricultural Sciences, Rural Development Administration, Republic of Korea)

This experiment was conducted to evaluate soil carbon sequestration and profit analysis for application of biochar pellet during rice cultivation in paddy. For nutrient leaching experiment, the biochar pellet was mixed with various ratio of pig manure compost. Each treatment was filled 5g of different ratios of biochar pellet in the glass column. The column was poured into deionized water and precipitated after a certain retention time. The leaching water was analyzed with NH_4-N, PO_4-P and K. For the field experiment, it was also consisted with four treatments as control, 100 % of pig compost pellet, biochar pellet (compost: biochar, 6:4), and slow release fertilizer in the rice paddy. The application amount of compost and biochar pellet was 2,500 kg ha^{-1}, and chemical fertilizer was applied at 85-45-57 kg ha^{-1} (N-P-K) based on recommendation rates of NIAS. All compost and biochar pellet, phosphorous and potassium were basal application. Nitrogen fertilizer was applied with 3 separations as basal, and 2 additional applications. For accumulation amount of NH_4-N, it was shown as order as pig manure compost> compost pellet> 2:8> 4:6> 8:2 ratios and its highest leaching amount was to be treatment of pig manure compost. Accumulated amount of PO_4-P and K were highest in the treatment of pig manure compost and lowest in biochar pellet (9:1). In the field experiment, carbon sequestration was 54.6T ha^{-1} relative to the control. Mitigation of CO_2-eq. emission as greenhouse gases was estimated to be at 537.3T ha^{-1} and its profit analysis was evaluated at $10,385(KAU) per hectare during rice cultivation. Rice yields in biochar pellet treatment were not significantly different with the control. Therefore, this technique for application of biochar pellet might be applied with Clean Development Mechanism (CDM).

Developing a Sustainable Eco-Fishing Village in Riverine Nigeria: A Review of Iworkiri Fishing Village Bonny Island, Nigeria

Ikechukwu Onyegiri. (Imo State University, Owerri, Imo State, Nigeria)
Ben Ugochukwu, Iwuagwu. (Abia State Polytechnic, Aba, Abia State, Nigeria)
Roland, Dan-Wachuku. (Imo State University, Owerri, Imo State, Nigeria)

Iworkiri fishing village is in Fanima community, a well-known community with outlying fishing settlements in Bonny Island Rivers State. Iworkiri is situated a few kilometers outside Bonny main town on the edge of the Atlantic Ocean. Fishing has been the most important activity around the Iworkiri coastal area for centuries and forms the biggest part of the Islands historical identity. Since fishing activities started at Iworkiri informal settlements, the fishing village has been faced with poor housing and environmental conditions. It still lacks good public infrastructure, planned spaces for public and other fishing activities. The existing fishermen Cabin which is the core of their everyday life are dilapidated Shacks. All these anomalies can be improved on by designing and constructing an ecological sustainable fishing village based on the existing settings, taking advantage of its untapped eco-friendly available building materials. For centuries, the people of Southern Nigeria have been using these eco-friendly materials in different forms as building material, both for support and as well as the house envelope. Studies show that eco-friendly building materials have proved to be a prized choice in the modern building construction as we search for sustainable building materials and that more active use of these materials would check the adverse effect of contemporary building materials and construction on our environment. This paper recommends the adoption of eco-friendly building materials for housing constructions that will improve the environmental and housing conditions in Iworkiri fishing village in particular and Bonny Island at large without adverse effects on the environment.

BIO-ASSESSMENT AND TOXICOLOGY

Are the 16 EPA PAHs in Need of Overhaul after 40 Years of Faithful Service?

Jan T. Andersson* and Christine Achten
(University of Muenster; Muenster, Germany; *E-mail: anderss@uni-muenster.de)

Due to their high carcinogenic/mutagenic potential and ubiquitous occurrence, polycyclic aromatic compounds (PAC) are among the most intensely studied environmental pollutants. Out of the thousands of representatives of this class of compounds, in most studies only 16 of them are selected, namely those that appear on the list that the Environmental Protection Agency (EPA) compiled in 1976 as "priority pollutants" for water quality studies. In the following decades this list of compounds has found many other uses, including monitoring hazards to human health, although it was not intended for such purposes by EPA.

The knowledge of PACs in the environment has grown enormously since the 1970s, and this is reflected in other selections of polycyclic aromatic hydrocarbons (PAH) that are now also used for regulatory purposes, i.e. in the U.S. Clean Water Act and in the 15+1 PAHs for food analysis of the European Commission. However, the original list is still used by many scientists as if it were a code of best practice.

The 16 compounds on the list are all hydrocarbons and contain from 2 to 6 rings. Some of the shortcomings of the list for adverse health effect estimations are related to this choice of compounds. Intense environmental studies have shown that other PACs show a stronger influence on human health and should therefore be preferentially analyzed if a truthful picture of the toxic properties of the sample is strived for.

Compound classes that show adverse effects on human health and are relevant in environmental studies but are absent from the EPA list, are larger PAHs, e.g. dibenzopyrenes, polycyclic aromatic heterocycles, especially those with sulfur or nitrogen as heteroatom, and alkylated PACs (for instance 5-methylchrysene with a toxicity equivalence factor equal to that of benzo[*a*]pyrene).

This realization has led to a call for establishing more suitable compilations of PACs for analysis for human health purposes. A fundamental question is whether one set of PACs would cover the needs for all kinds of samples and research interests or whether different ones are needed depending on the focus of the work. Representatives of the classes of compounds mentioned above should be included in such considerations. Other issues to be kept in mind are not only toxicity but also the frequency of occurrence, analytical practicability, if the compound may act as a proxy for other similar substances, etc.

Unraveling the Molecular Mechanism Behind UV-B Induced Skin Cancer

Rayala Suresh Kumar
(Department of Biotechnology, IIT Madras, Chennai, India)

Despite consistent exposure of human skin to solar UV-B irradiation, most of the epidermal keratinocytes does not transform, which is an indication of efficient DNA repair mechanism inherent capacity to either repair or restore the UV damaged cells. In this conference, i focus my talk on delineating the mechanistic role of p21 activated kinase (Pak1) in UV-B induced skin lesions. Molecular mechanistic studies revealed that Pak1 is activated in response to UV-B *via* epidermal growth factor receptor (EGFR) and cyclobutane pyrimidine dimers (CPD) pathways and both these membranous (EGFR) and nuclear (CPDs) events converge at Pak1 activation and contribute in a coordinated manner for yielding a complete response to UV-B. This is the first study that evaluates the mechanistic role of a signaling molecule - Pak1 in sun induced premalignant skin lesions.

Toxicity of Silver Nanoparticles on E. crypticus in Soils Applied with Biosolid: Effect of Transformation Ratio to Silver Sulfide Nanoparticles and Biosolid Application Rates

Emel Topuz and Ismail Koyuncu
(Istanbul Technical University, Maslak, Istanbul, Turkey)

Silver nanoparticles (AgNPs) are widely used in daily products since last decade. There are several studies concerning their release to the environment such as soil and aquatic media. Therefore, research is going on about their fate and effects in the environment. Recent studies showed that AgNPs are transformed to the silver sulphide nanoparticles (Ag_2SNP) in Waste Water Treatment Plants (WWTPs). Transformation of AgNPs are critical in terms of their toxicity since Ag_2S are known as less bioavailable compared to Ag. However, the transformation ratio is still speculative due to the variety of the conditions in WWTPs. AgNPs are mostly partitioned in the WWTP sludge which could be used as biosolid for agricultural purposes to increase the organic matter and nutrient content of the soil. This could lead to concern about the entrance of AgNP to the soil, however, biosolid application could reduce the toxicity due to the complexation of Ag with organic matter. Therefore, the effect of transformation rates of AgNPs and the application rate of biosolid to the soil should be investigated to understand the toxicity of AgNP and its transformation product, Ag_2SNP on soil organisms.

Polyvinylpyrrolidone coated AgNP (AgNP-PVP) and Ag_2SNP are provided from Nanografi, Turkey. $AgNO_3$ is tested as an ionic control for AgNP. *Enchytraeus crypticus* are kindly supported by VU Amsterdam University and cultured in our laboratory at 15 °C and 75% humidity. Animals are fed with oat meal mixture twice a week. Lufa 2.2 soil is used as soil medium. Biosolid is taken from sludge drying unit of a Waste Water Treatment Plant where anaerobic digestion and sludge dewatering is applied before drying. Survival and reproductive toxicity test are applied by following the revised method explained by Castro-Ferreira et al. Briefly, different concentrations of the pollutants are spiked to soil applied with biosolid and animals were introduced the next day. Test jars were kept at 15 °C and 75% humidity and 16:8 light:dark period are provided. Three different medium are tested as follows: only Lufa 2.2, Lufa 2.2:biosolid (500:3) and Lufa 2.2:biosolid (500:5). AgNP-PVP, Ag_2SNP and different transformation ratios (15, 50 and 60%) are tested. Therefore, lethal and/or reproductive toxicity are evaluated by considering the different transformation rates of AgNPs and different biosloid application rates in the agricultural applications.

It is expected that transformation of AgNPs to Ag_2SNP could decrease the toxicity due to the low bioavailability of Ag_2S. In addition, application of biosolid could lead to the decrease in toxicity since higher organic matter content in the soil can bind Ag and reduce the toxicity. As a conclusion, this study will show the possible threats to the soil organisms due to the presence of AgNPs and its transformation product, Ag_2SNP in the soil or biosolid.

Trans-Generational Immune Toxicity Effect of Benzo[A]Pyrene on Oryzias Melastigma

Xiaohan Yin and ***Kejian Wang***
(Xiamen University, Xiamen, Fujian, China)

The trans-generational toxicity effect of Benzo[a]pyrene (BaP) was investigated using marine medaka (*oryzias melastigma*). Adult marine medaka (F0) was exposed 0.5 μg/L BaP, after 14 and 28 days' exposure, the first generation of embryos were collected and raised without BaP exposure until juvenile fish (F1). F1 was used for the further experiments. Morphological observation showed some embryos developmental delay, deformity and even death. Moreover, the level of RNA methylation （m^6A） in F1 juvenile fish whose parents (F0) treated with BaP was significantly increased in comparison with control group, also the expression of *p53* and downstream genes markedly modulated. These results suggested that BaP exposure to adult marine fish might have a toxic effect on epigenetic modification and *p53* signaling pathway of offspring as well as cell cycle, leading to DNA damage and inhibiting the downstream DNA repair process, even though the F1 generation free of BaP. Subsequently, it might be involved in the event of inhibiting the downstream DNA repair process by abnormal methylation of RNA.

Persistent Physiological Concentration ROS Generated by BaP Exposure Blocking NF-κB Pathway in Medaka (*Oryzias melatigma*)

Qian Cui, and Kejian Wang
(State Key Laboratory of Marine Environmental Science, College of Ocean & Earth Science, Xiamen University, Xiamen, Fujian, China)

Benzo[a]pyrene (BaP) exposure can generate ROS, which might be involved in signaling pathway and plays a vital role in homeostasis. As reported, ROS could be a signal or a product of NF-κB pathway which is associated with many immune-related gene expressions. However, whether the BaP-induced ROS has any effect on NF-κB pathway is less studied. To elucidate the interaction between BaP exposure and the NF-κB pathway, a marine fish *Oryzias melatigma* was used in the study. Results showed that ROS generated by BaP exposure inhibited the activity of NF-κB pathway. Interestingly, the negative effect of ROS on NF-κB pathway was inconsistent with the observation on previous studies using the instantaneous ROS driven by H_2O_2 directly exposure, which is the conventional exposure way. Based on the modified luminol-NaOCl detection system, it was found that BaP exposure could generate sustained physiological concentration ROS, by which it was presumed that the persistent physiological concentration ROS generated by BaP exposure might have a different modulation on NF-κB pathway. These results were further confirmed by sustained sub-micromolar H_2O_2 system and western-blot detection.

In Vitro Study on the Joint Hepatoxicity upon Combined Exposure of Cadmium and BDE-209

Lixin Wang, Tong Wu, Lulu Zhang, Jiansheng Cui
(Hebei University of Science and Technology, Shijiazhuang, 050018, China)

Joint toxicity is an important issue during the risk assessment of environmental pollutants. Contamination of heavy metals and persistent organic pollutants (POPs) under the environmental and biological settings poses substantial health risk to humans. Although previous studies demonstrated the co-occurrence of cadmium and decabrominated diphenyl ether (BDE-209) in environmental mediums, food chains and even human body, their potentially joint toxicities remain elusive thus far. Our investigation here with respect to the hepatotoxicity in vitro clearly demonstrated that combined exposure of cadmium and BDE-209 aggravated the injuries in hepatocytes, which was evidenced by the additive effects on the induction of remarkable morphological alternations,LDH release, cell apoptosis and necrosis, impairment of mitochondrial activity and transmembrane potential. Enhanced ROS production was one of the mechanisms for cell apoptosis and death upon joint treatment. Additionally, more cadmium-treated cells underwent apoptosis than BDE-209-treated cells while more ROS was generated with BDE-209 treatment, indicating that other mechanisms might be involved in cadmium-induced apoptosis. Our results would be helpful for evaluating the joint-hepatotoxicity upon combined exposure of cadmium and BDE-209 as well as investigating the underlying mechanisms.

Cd-induced Phytotoxicity Reduces Lindane Accumulation in Rice Seedlings

Shidi Huang and G. Daniel Sheng
(Tongji University, Shanghai, China)

The present study was conducted to evaluate the phytotoxic effects of cadmium (Cd) pre-exposure on subsequent lindane (LDN) accumulations in rice seedlings. It was hypothesized that Cd-induced phytotoxicity could suppress the plant uptake of LDN. Rice seedlings (2-week old) were hydroponically pre-exposed to Cd^{2+} at the concentration of 0, 2, or 40 mg·L^{-1} for 3 days. They were subsequently exposed to LDN (2 mg·L^{-1}) for 5 days in 20%-strength Hoagland solution. The fatty acid profiles and free proline contents in rice seedlings were determined. Following the Cd^{2+} pre-exposure, varying degrees of growth inhibition and leaf etiolation of rice seedlings were observed, as compared to the control (without Cd^{2+}). An increase in Cd^{2+} concentration in the solution resulted in increased lipid peroxidation and water deficit in the rice seedlings, as evidenced by the reduction in double bond index (DBI, an index indicating lipid peroxidation and altered cell membrane fluidity) and the elevation in free proline contents (an index of plant water deficit) in both roots and aerial parts. Pre-exposure to much higher levels of Cd^{2+} brought about lower LDN uptake rate in roots and lower LDN accumulations in tissues. Meanwhile, both the DBI and the free proline contents of each group nearly leveled off. In summary, the subsequent LDN uptake by rice seedlings following Cd^{2+} exposure appeared to depend on the physiological status and cell membrane fluidity of the seedlings, which were determined by the severity of Cd-induced phytotoxicity.

Genotoxicity and Cytotoxicity of Nine Benzothiazoles: Development and Application of A High Content Screening in Vitro Micronucleus Test for Genotoxicity and Cytotoxicty Assessment

Mei Ma, ChaoHuang, Ye Yan, Na Li, Kaifeng Rao, Yiping Xu
(Research Center for Eco-Environmental Sciences, Chinese Academy of Sciences, Beijing, China)

Benzothiazole and benzothiazole derivatives (BTs), have been detected in various environmental matrices as well as in human beings, but insufficient available toxicological information on BTs makes assessing their health risks challenging.Recently, high content screening (HCS) has been proven to be a promising in vitro strategy for evaluating the toxicity and identifying the mechanisms of environmental pollutants.In the present study, a HCS in vitro micronucleus (MN) assay method based on OECD guideline 487 was developed and optimized for evaluating the cytotoxicity and genotoxicity of compounds.Then, this method was combined with SOS/umu test using Salmonella typhimurium TA1535/pSK1002 to comprehensively assess cyto-genotoxicity of nine BTs (benzothiazole [BT], 2-chlorobenzothiazole [CBT], 2-bromobenzothiazole [BrBT], 2-fluorobenzothiazole[FBT], 2-methylbenzothiazole [MeBT], 2-mercaptobenzothiazole [MBT], 2-aminobenzothiazole [ABT], 2-hydroxy-benzothiazole [OHBT] and 2-methythiobenzothiazole [MTBT]). Except for the cytotoxic effect of MBT on MGC-803 and A549, the other tested BTs showed more than 50% cytotoxicity at their highest concentrations in a dose-dependent manner, and their LC50s ranged from 19 (MBT in bacteria) to 270 mg·l–1 (CBT in A549). Activation and inactivation were observed for specific BTs after metabolism. On the other hand, no evidence of DNA-damaging effect was observed for BT, FBT and MBT in SOS/umu test, while MN was induced by ABT, OHBT, BrBT and MTBT in MGC-803, by MeBT in A549 and by CBT in both cells. Finally, through quantitative structure–activity relationship analysis, two structure alerts for chemical genotoxicity, including heterocyclic amine and hacceptor-path3-hacceptor are present in ABT and OHBT respectively. However, the underlying mechanisms still need further evaluation.

Nanoparticles as Emerging Contaminants of Concern: Ecotoxicity Effects on the Growth and Survival of Bacteria

Sidney Stokes, Lucas Ringo and Lynal Albert
(Tarleton State University, Member of the Texas A&M University System, Texas)

Nanotechnology has revolutionized technology and innovation in a wide range of fields including environmental engineering, biomedical engineering, textiles, food processing, cosmetics and pharmaceuticals. The recent surge in the use of engineered nanoparticles may pose a threat to living organisms and the environment thus causing them to be identified as emerging contaminants of concern. Conventional water treatment methods have not been designed to remove nanoparticles so after use they enter the environment. In this study, we have researched and identified the most widely used nanoparticles and are exploring their ecotoxicological effects on growth and survival of microorganisms. We are also examining the biological response of free swimming versus attached microorganisms to the presence of these nanoparticles. Growth kinetics and survival are both key parameters that are being examined experimentally. The results of this study can serve as a fundamental basis in the selection of nanoparticles for various applications and also in the design and engineering of nanoparticles. The findings will also help us improvise our treatment methods as and when necessary to use nanoparticles to our best advantage.

Multigeneration Effects of Pentachlorophenol and 2,2',4,4'-tetrabromodiphenyl ether on *Folsomia Candida*

Qian-qian Zhang and ***Min Qiao***
(Research Center for Eco-Environmental Sciences, Chinese Academy of Sciences, Beijing, China)

The multigeneration effects of pentachlorophenol (PCP) and 2,2',4,4'-tetrabromodiphenyl (BDE47) on the springtail *Folsomia candida* were evaluated. Multigenretion tests were performed according to two different methods. In the first method, the parental generation springtails (F0) were exposed to PCP or BDE47 for 28 days. The first filial generation (F1) springtails were transferred to unpolluted artificial soil for 28 days and reproduced the second filial generation (F2). In the second method, the F0 generation were exposed for 10 days and then transferred to unpolluted artificial soil to generate the F1 generation. The F1 generation were also transferred to unpolluted artificial soil for 28 days and reproduced the F2 generation. For PCP, significant effects were observed on F1 and F2 generation in the first method and F1 generation in the second method. This suggests that PCP influences the reproductive capacity of adult springtails and the hatching of eggs or the mortality of juveniles. For BDE47, significant effects were only observed on F1 generation in the first method, which shows that BDE47 affects egg hatching or juvenile survival rather than the reproductive capacity of adults. The affected endpoints of springtails can be inferred by the two methods. PCP and BDE47 do not influence completely the same endpoints.

Organotin Contamination in Commercial and Wild Marine Bivalves from China: Increasing Occurrence of Triphenyltin After the TBT Ban

Chunzhao Chen and Qinghui Huang
(Tongji University, Shanghai, P. R. China)
Wen Zhang
(Department of Civil and Environmental Engineering, New Jersey Institute of Technology, Newark, New Jersey, USA)

Organotin is a group of persistent and toxic organometallic compounds, widely distributed in coastal environment. This study investigated the levels, distribution and risk assessment of five common organotin compounds: tributyltin (TBT), dibutyltin (DBT), monobutyltin (MBT), triphenyltin (TPhT) and diphenyltin (DPhT) in 13 samples of commercial bivalves (oyster, mussel and clam) and 14 samples of wild oysters collected from the major China's coastline in 2014-2015. Organotin concentrations varied greatly in three types of commercial bivalves. Phenyltins were more dominant in oysters than in mussels and clams which had lower organotin levels. As for the wild oyster samples, TBT was the dominant species among butyltins with a concentration of 34.1±30.0 ng $Sn \cdot g^{-1}$ dw (dry weight). However, phenyltin levels in these samples varied spatially from 11.9-42.4 ng $Sn \cdot g^{-1}$ dw in the north China's coastline to 169-1977 ng $Sn \cdot g^{-1}$ dw in the south. This spatial variation could be attributed to the difference in TPhT applications (e.g., pesticides in mariculture and agriculture). TPhT in wild oysters from the south coastal zones accounted for 63%-81% of the total organotins and was significantly higher than TBT contamination of the whole study areas ($p<0.05$). Risk assessment indicated that there was no health impacts caused by commercial clams and mussels, while some oysters (both commercial and wild) with high TPhT contamination could pose significant health threats to the whole group (particularly children) as the risk quotient (RQ) was higher than 1. Organotin contamination still occurs in China's coastlines after the TBT ban, and TPhT in bivalves should be of increasing concern due to its higher human health risk for consuming TPhT-contaminated oysters.

Comparison of Inactivation of Disinfectant- and Antibiotic- Resistant Bacteria by Free Chlorine and UV

Jingyu Wang and Minghao Sui
(Tongji University, Shanghai, China)

The deactivation of disinfectant- and antibiotic- resistant bacteria by free chlorine and UV was compared in this work. *Mycobacterium fortuitum* (*M. fortuitum*), and *Mycobacterium mucogenicum* (*M. mucogenicum*) were chosen as the representative of rapidly and slowly growing disinfectant-resistant bacteria. *Enterococcus faecalis* (*E. faecalis*) and *Escherichia coli* (antibiotic-resistant *E. coli*) were selected as the representative antibiotic-resistant bacteria, respectively. Antimicrobial sensitivity of different disinfection methods was species-specific. *M. fortuitum* had a certain resistance to free chlorine and UV, while *M. mucogenicum* had a strong resistance to chlorine disinfection, a less tolerance to UV than *M. fortuitum*. Slowly growing strains were more resistant to chlorine than the more rapidly cells. *E. faecalis* and antibiotic-resistant *E. coli* were both susceptible to chlorine disinfection. Additionally, *E. faecalis* was resistant to UV disinfection, but antibiotic-resistant *E. coli* had little tolerance to UV. The Ct values of 3 log inactivation of *M. fortuitum* and *M. mucogenicum* were 8 and 67 mg•min/l in chlorination, and the two antibiotic-resistant strains were completely inactivated in less than one minute. Under UV irradiation, the Ct values for 3 log inactivation of the four organisms ranged from 4.7-7.3mJ/cm^2. According to fluorescence microscopy and flow cytometry, the attack on bacterial cell membrane was the first step in chlorine disinfection, while UV disinfection was not. For two strains of mycobacteria, their cell membrane damage was significantly slower than the other two kinds of antibiotic-resistant bacteria. And high hydrophobicity was also one of the reasons for chlorine resistance of disinfectant-resistant bacteria. Furthermore, the bactericidal property of sodium hypochlorite was stronger under acidic condition. Humic acid could affect the inactivation effect of chlorine and ultraviolet.

Existed to Acquire or Evolved to Destroy? -The story of Bacterial Phosphotriesterases

Dayananda Siddavattam
(Dept. of Animal Biology, School of Life Sciences, University of Hyderabad, Hyderabad-500 046, India)

Bacterial Phosphotriesterases (PTEs) have gained prominence due to their involvement in hydrolysis of triester linkage found in organophosphate (OP) insecticides and nerve agents. These PTEs have been classified into three distinct structural groups, viz organophosphate hydrolases (OPH), methyl parathion hydrolase (MPH) and Organophosphate Acid Anhydrolase (OPAA). Among these three PTEs, the OPAA is shown to be a dipeptidase (prolidase). However, quite independent views exist on the physiological role of other two PTEs. One of them proposes evolution of PTEs from their progenitors viz,, lactonases (OPH) and lactamases (MPH), in response to increased concentration of organophosphate insecticide residues in soil. The second view suggests their role in iron/phosphate acquisition and believes that OPs are fortuitous substrates for PTEs. The paper describes unique structural features of PTEs and draws logical conclusions on unique physiological role of bacterial PTEs.

Effects of Soil Moisture on Biodegradation of Atrazine by DN36

Yang Chen and G. Daniel Sheng
(Tongji University, Shanghai, China)

To better understand the pesticide degradation in soil over a wide range of climate conditions, the effects of soil moisture must be addressed. The soil water controls pesticide sorption in soil and hence influences pesticide degradation. The soil water also affects microbial activities toward pesticides. The degradation of atrazine by DN36 (a Nocardioides sp. strain) in a high OM soil with different moisture contents (50%, 75%, and 100%) was determined. We found that the degree of atrazine degradation over a period of 7 days was in the decreasing order of 100%>75%>50% (moisture content), indicating more rapid atrazine degradation in higher water content soil. The sorption isotherms of atrazine were linear in water-saturated soil but became nonlinear of the Langmuir type in the soil with moistures of 50%, 75% and 100%. These isotherms showed that the liquid phase atrazine concentration increased with increasing soil moisture. Microbial analysis also showed that the microbial quantity and activity increased with increasing soil moisture. It is therefore concluded that soil moisture influenced both atrazine sorption and DN36 in soil, thereby mediating atrazine degradation in the soil.

WETLANDS AND SEDIMENTS

Wetland Conservation and Protection

Kwang, Daniel Kwasi
(VIALE BALDASSARRE PERUZZI, NO 25 INT 3 CASTELVOLTURO, CASERTA, 81030, ITALY; kwasidaniel69@yahoo.com)

The most important thing to remember about managing to protect your wetland is to work with what you have. For many wetland property owners, the best way to manage the wetland for protection and maintenance of the functions it serves is a hands-off approach. The hands-off approach means that you are not actively changing the key components of the wetland to modify the functions it naturally provides. It means that you are conscious of the potential threats to your wetland and are actively managing those threats if you can protect a wetland and its surrounding uplands from potential threats, the wetland will take care of itself and provide a range of beneficial functions.

The simplest way to manage your wetland is not to engage in land use changes or hydrologic modifications. In addition, many landowners want to engage in low level enhancement activities or do more to ensure that their wetlands are protected for the long term. All too often our thoughts about what we can do to protect wetlands are limited to activities within the wetland. Many of the management activities that landowners can take to guard their wetlands can occur outside the wetland boundaries.

Waves from motorboats can cause erosion damage to riparian wetlands. Although wetlands serve to dampen wave energy, the waves from boats often generate much larger waves than the physical parameters of the water body would normally generate. One way to protect these areas is to establish "No Wake Zones." In other cases it may be appropriate to limit the speed or motor size on an entire lake. To find out how to establish a no wake zone near your wetland or a speed limit on your lake, contact the marine division of your county sheriff's office. Because of the damage to soils and vegetation, the use of ORVs such as dirt bikes, all-terrain vehicles, and even mountain bikes should not be permitted in wetlands.

In shoreline areas that are experiencing erosion, controlling erosion is very important. The first step in controlling erosion is to evaluate the nature and extent of the problem and determine if the problem is serious enough to warrant corrective action. There are two basic reasons for controlling erosion: to protect property and to protect the environment. If the erosion is largely the result of natural processes that do not harm the environment and that do not threaten property, then erosion control may not be necessary. Assuming that you determine the nature and extent of the erosion warrant corrective measures, the next step is to look at the alternative methods of control and select the one that is most appropriate.

Agricultural water scarcity in the predominantly rain fed agricultural system of sub-Saharan Africa (SSA) is more related to the variability of rainfall and excessive non-productive losses, than the total annual precipitation in the growing season. Less than 15% of the terrestrial precipitation takes the form of productive 'green' transpiration. Hence, rainwater harvesting and management (RWHM) technologies hold a significant potential for improving rainwater-use efficiency and sustaining rain fed agriculture in the region. This paper outlines the various RWHM techniques being practiced in SSA, and reviews recent research results on the performance of selected practices. So far, micro-catchment and *in situ* rainwater harvesting techniques are more common than rainwater irrigation techniques from macro-catchment systems. Depending on rainfall patterns and local soil characteristics, appropriate application of *in situ* and micro-catchment techniques could improve the soil water content of the rooting zone by up to 30%. Up to six fold crop yields have been obtained through combinations of rainwater harvesting and fertilizer use, as compared to traditional practices. Supplemental irrigation of rain fed agriculture through rainwater harvesting not only reduces the risk of total crop failure due to dry spells, but also substantially improves water and crop productivity. Depending on the type of crop and the seasonal rainfall pattern, the application of RWHM techniques makes net profits more possible, compared to the meager profit or net loss of existing systems. Implementation of rainwater harvesting may allow cereal-based smallholder farmers to shift to diversified crops, hence improving household food security, dietary status, and economic return. The much needed green revolution and adaptations to climate change in SSA should blend rainwater harvesting ideals with agronomic principles. More efforts are needed to improve the indigenous practices, and to disseminate best practices on a wider scale.

Highlights ► In sub-Saharan Africa, agricultural water scarcity is much linked to rainfall variability and non-productive losses. ► Micro-catchment and in situ rainwater harvesting techniques could improve the root zone soil water content by up to 30%. ► Integration of rainwater harvesting with soil amendments has shown spectacular water and crop productivity performances. ► Blending rainwater harvesting and agronomic principles is crucial for green revolution and climate-change adaptations.

Assessment of the Impact of Land Cover and Land Use Changes on the Water Quality of Sediments and Nutrients in Cahaba River Basin Using SWAT

Pooja P. Preetha, Ashraf Z. Al-Hamdan, and Michael D. Anderson
(Department of Civil and Environmental Engineering, University of Alabama in Huntsville, Alabama, USA)

Predictions of water quality constituents of sediments and nutrients in unmonitored small watersheds are limited because of the adversity in capturing the uniqueness of catchment hydrology on temporal and spatial scales. This study examines the impacts of land cover land use (LCLU) changes on the hydrologic response of the Cahaba river watershed in Alabama. An integrated approach of hydrological modeling and spatial-temporal analysis was applied to evaluate contributions of dominant LCLU patterns on changes in monthly runoff, sediment yields and nutrients loads of nitrogen and phosphorus. Hydrological modeling was conducted using Soil Water Assessment Tool (SWAT) using the LCLU data for the years 2006 and 2011 for the climate condition over the period 1980-2010. This study showed a higher effect of urban development on the monthly peak runoff and subsequent monthly peak sediment yields in September and October from 2007 to 2010. To evaluate the climate change impact on water quality, the analysis was also conducted for the 2006 and 2011 LCLU data over the periods 1980-2010 and 2010-2040. The annual average sediment yield in Cahaba watershed increased by about 250 tons due to the climate change effect. There was a reduction of 17% and 3% in the annual average loads of nitrogen and phosphorous, respectively, depicting the long term temporal effects of land cover land use and climate changes.

Assessment of Quality of Sediments and Nutrients with Land Use Land Cover Changes in Cahaba River Basin using SWAT

Pooja P. Preetha, Mohammad Z. Al-Hamdan and Michael D. Anderson
(University of Alabama in Huntsville, AL, USA)

Predictions of sediment and nutrient quality in small watersheds have been limited because of the adversity in capturing the uniqueness of catchment hydrology over large areas and different time periods. This paper discusses two land use land cover (LULC) classification on the Cahaba river watershed to address the impacts of land use and land cover changes on hydrology of the watershed. An integrated approach of hydrological modeling and spatial-temporal analysis was applied to evaluate contributions of dominant LULC patterns on changes in runoff, sediment yields and nutrient loads. Hydrological modeling was conducted using Soil Water Assessment Tool (SWAT) for each of the LULC maps in two time periods (1980-2010, 2010- 2040) in the Cahaba river watershed. Results of this study showed a higher effect of urban development on the peak runoff season and subsequent peak sediment yields in September and October from 2007 to 2010. The sediment yield in Cahaba watershed in a year would be 249,427 kg higher with LULC and climate change than when the sediment yield is affected by autonomous land-use changes only. 16.86% reduction in nitrogen loads and about 3% reduction in phosphorus loads from 2001 to 2035 depict the long term temporal effect of LULC with climate change. However, considerations on LULC transitions including construction, stream bank water body formation and erosion, effects of natural disasters etc. may give rise to more sediment and nutrient concentrations.

Enhanced Nitrogen Removal by the Integrated Constructed Wetlands with Artificial Aeration

Jizheng Pan
(Nanjing Institute of Geography and Limnology, Chinese Academy of Sciences, Nanjing, 210008, China)

The removal of organic pollutants, ammonia, and phosphorus from seriously polluted river water has been extensively explored using various traditional integrated constructed wetland systems. However, the removal efficiency of nitrogen was extremely low due to the limited oxygen concentration of these traditional systems. Herein, a novel pilot-scale integrated constructed wetland system (CWs), which consisted mainly of a biotic contacting oxidization pre-treatment (BCO), an artificial aeration vertical sub-surface flow CWs, and a horizontal subsurface flow CWs were applied to enhance the removal of total nitrogen (TN) under high hydraulic loading through artificial aeration. Results showed that, with the application of the novel CW systems, a total of 64.97% removal of TN was achieved with the average inflow of 11.39 mg/L and hydraulic load of 1.67 $m^3 \cdot m^{-2} \cdot d^{-1}$. Further analysis showed that the contributions of BCO, AVFCWs and HFCWs to TN removal were 13.86%, 15.97% and 35.36%, respectively. Besides the TN removal, the CW systems can also obtained about 75% of ammonia removal, within which artificial aeration vertical subsurface flow CWs unit occupied about 60% of contribution. Flowing by the application of this systems, the concentrations of ammonia and nitrate in the outflow decreased to 2.66 and 2.01 mg/L, respectively. The study indicated that the integrated constructed wetland system applied can be an efficiently potential technique for the removal of COD, NH_4-N and TP from some rural wastewater.

Hexabromocyclododecane (HBCDD) Biodegradation In Anaerobic Sediment Mesocosms

Irem Karahan, Hale Demirtepe, Ipek Imamoğlu
(Middle East Technical University, Ankara, Turkey)

Hexabromocyclododecane (HBCDD), a brominated flame retardant, is used in heat insulation materials, which are expanded polystyrene and extruded polystyrene, and in furniture, textile and other industries for incombustibility. HBCDD threatens human health and the environment due to its potential for biomagnification and bioaccumulative properties. It is a persistent organic pollutant (POP) regulated by the Stockholm Convention on POPs, and listed under Annex-A Elimination List of the Convention in 2013. Only in the last decade, studies are reported on HBCDD degradation in the literature, and they show that the degradation of HBCDD in soil and sediment samples is more favorable in anaerobic environments. There are currently a very limited number of studies in the literature on HBCDD degradation, and no mesocosm-scale study (to the best of our knowledge) on evaluation of biodegradation in HBCDD contaminated sediments. In this study, anaerobic biodegradation of HBCDD was investigated in two laboratory sediment mesocosms (2L total glass tank volume with 685 g sediment mass and 550mL overlying liquid volume), set up as natural attenuation and biostimulation sets. While natural attenuation contains only distilled water and HBCDD contaminated sediment, biostimulation set is prepared by adding organic medium containing electron donor (ethanol) and carbon source (sodium formate) to sediment contaminated with HBCDD. There are also 2 control mesocosms: contaminant control and sterile mesocosms. Contaminant control contains only distilled water and clean sediments (without adding HBCDD). Sterile set contains distilled water and HBCDD contaminated sediment, and it is prepared by autoclaving and poisoning with $HgCl_2$. All reactors were prepared as duplicate, completely under N_2:CO_2:H_2 environment in an anaerobic glovebox (PlasLabs 818GB/Exp) and incubated in the dark at 25°C. The mesocosms are operated for 49 days and sampling is done in a 7-day period. During sampling, triplicate sediment samples were taken from each mesocosm and extracted using an ultrasonic bath (USEPA Method No: 3550C). In the analysis of HBCDD, GC-MS was used for evaluation of total-HBCDD and LC-MSMS is used for evaluation of α, β, γ-HBCDD at different sampling times of operation. As a result, percent reduction in total-HBCDD, and α, β, γ-HBCDD as well as degradation rates were obtained for natural attenuation and biostimulation, followed by comparison with the limited literature available on HBCDD. In natural attenuation set, 92-96% reduction was observed in 42 days, while HBCDD was below detection in the last sampling day. In biostimulation mesocosms, 96-98% reduction was achieved in 35 days, after which HBCDD was below detection. As expected, HBCDD was never detected in the contaminant control set, indicating absence of any cross-contamination in laboratory procedures. However, degradation was observed in sterile sets, at 95% reduction of HBCDD in 49 days. Observation of degradation in sterile sets has been cited in the literature, and results of this study and those are comparatively discussed. Overall, this mesocosm study indicate HBCDD anaerobic biodegradation can be stimulated with the addition of organic rich medium containing an electron donor and carbon source. All degradation results, both terms of total-HBCDD and stereoisomer-specific-HBCDD are discussed with relevant data from the literature for the purpose of improving our understanding while remediation of sediments contaminated with this persistent organic pollutant.

GLOBAL WARMING
AND
CLIMATE CHANGE

Effects of Global Warming on Reproductive Functions, Heat Shock Protein Expression and Cellular Apoptosis of Eastern Oyster Gonad

Sarah Nash, and Md Saydur Rahman
(University of Texas Rio Grande Valley, Brownsville, TX, USA)

Global warming due to climate change is likely to intensify the heat/thermal stress in marine and coastal organisms, affecting their development, growth and reproductive performance. American oyster also called Eastern oyster (Crassostrea virginica, a native edible and commercially important marine oyster in the Eastern seaboard and Gulf of Mexico) is an excellent model species in response to global climate change. This marine species is a classic example of how global warming affects organism's normal reproductive functions. In this study, we tested gonadal development, heat shock protein expression and cellular apoptosis in gonad, and coelomic fluid (CF, an important body fluid that helps regulate important physiological functions) pH in American oyster under elevated sea water temperature. Oysters were placed in six different 20gallon aquariums with various temperatures under controlled laboratory conditions for one week. Two of these aquariums were at a controlled temperature (24oC), followed by two at medium temperature (28oC), and the remaining two at high temperature (32oC). Ten oysters from each aquarium were dissected and sampled for normal histological observations of gonadal functions, immunohistochemical analysis of heat shock protein expression, in situ Terminal deoxynucleotidyl Transferase (TdT) dUTP Nick-End Labeling (TUNEL) assay for goandal apoptosis, and also biochemical analysis for coelomic fluid. Oysters exposed to higher temperature showed an increase of heat shock protein expression in eggs of ovary and spermatogenic cells of testis, as well as an increase in cellular apoptosis in gonadal tissues. High temperature also significantly increased pH levels of the coelomic fluid in oyster. Collectively, these results suggest that high temperature has a negative impacts of gonad development and reproductive functions in American oyster.

A Socio-Epidemiological Study on Global Climate Change and Malaria Risk in Lombok, Indonesia: 2005-2014

Hisayoshi MITSUDA (Bukkyo University, Kyoto, Japan)

This study is focused on socio-epidemiological perspectives and future issues on malaria control program to fight against malaria outbreak resulted in global climate change. In order to examine the relationship between global climate change and malaria risk, the collaborative researches between Mitsuda, Bukkyo University and Mulyanto, School of Medicine, Mataram University have been executed in Lombok, Indonesia, the center of a malaria outbreak in 2005.

Firstly, we made a hypothetical scheme on the potential impact of global climate change, e.g., global warming on malaria outbreak in East Lombok, 2005 and carried out the Collecting Baseline Data of Epidemiological/Sociological Survey (CBDESS) in 2006 to explore how the social/environmental determinants expanded the outbreak in East Lombok. Secondary, along with field interviews with medical staffs, government officers, school teachers and religious leaders, etc., continuous malaria blood tests and socio-epidemiological questionnaire surveys on malaria knowledge and behavior have been conducted based on the sample of 2,000 respondents from over 100 malaria infectious villages in East Lombok until 2008. Finally, the fundamental determinants of the outbreak were elucidated by statistical analyses of 98 socio-environmental variables of CBDESS data as the four major social factors; poverty, education, gender discrimination and malaria risk behavior.

In response to the results, we launched "School Based Malaria Intervention (SBMI)" method from 2008 through 2010, a new environmental education program to fight against malaria, targeting the primary school children to play a role as community health messengers, "Malaria School Scout (MASCOT)." Approximately 800 children at 16 primary schools in East-North Lombok joined MASCOT program, which proved to be an effective socio-epidemiological approach to achieve a lower incidence of malaria in the outbreak communities.

In conclusion, we will address the future issues and political implications of malaria control by further integration of socio-epidemiological researches and environmental education program on global climate change and malaria risk.

CO_2 Absorption Studies using Amine Solvents with Fourier Transform Infrared Analysis

Funmilola Avoseh, Khalid Osman, Paramespri Naidoo and Deresh Ramjugernath
(Thermodynamics Research Unit, School of Engineering, University of KwaZulu-Natal, Durban 4041, South Africa)

The increasing global atmospheric temperature is of great concern and this has led to the development of technologies to reduce the emission of greenhouse gases into the atmosphere. Flue gas emissions from fossil fuel combustion are major sources of greenhouse gases. One of the ways to reduce the emission of CO_2 from flue gases is by post compustion capture process and this can be done by absorbing the gas into suitable chemical solvents before emmiting the gas into the atmosphere. Alkanolamines are promising solvents for this capture process. VLE of CO_2-alkanolamine systems are often represented by CO_2 loading and partial pressure of CO_2 without considering the liquid phase. The liquid phase of this system is a complex one comprising of 9 species. Online analysis of the process is important to monitor the concentrations of the liquid phase reacting and product species. Liquid phase analysis of CO_2-diethanolamine (DEA) solution was performed by attenuated total reflection Fourier transform infrared (ATR-FTIR) spectroscopy. The concentration of CO_2 captured by the chemical solvent was estimated using the spectroscopy information. Partial least square regression model was built based on the spectroscopy data.

METALS

Accumulation and Sources of Trace Metals in Roadside Soils in Shanghai, China: A Case Study of Two Urban/Rural Roads

Geng Yan and Ling Chen
(Tongji University, Shanghai, P. R. China)
Lingchen Mao
(University of Shanghai for Science and Technology, Shanghai, P.R. China)

The road traffic has become one of the main sources of urban pollution and could directly affect roadside soils. To understand the level of contamination and potential sources of trace metals in roadside soils of Shanghai, 10 trace metals (Sb, Cr, Co, Ni, Cu, Cd, Pb, Hg, Mn and Zn) from two urban/rural roads (Hutai Road and Wunign-Caoan Road) were analyzed in this study, which was the first study to investigate pollution level of Sb in roadside soils of Shanghai. Antimony, Ni, Cu, Cd, Pb, Hg and Zn concentrations were higher than that of soil background values of Shanghai, whereas accumulation of Cr, Co and Mn were minimal. Significantly higher Sb, Cd, Pb contents were found in samples from urban areas than those from suburban area, suggesting the impact from urbanization. The concentrations of Sb and Cd in older road (Hutai) were higher than that in younger road (Wunign-Caoan). According to the enrichment factors (EFs), 78.5% of Sb, Cu, Cd, Pb and Zn were in moderate or significant pollution, indicating considerable anthropogenic influence. In particular, recently introduced in automotive technology, accumulation of Sb has been recognized in 42.9% samples of both roads. Moreover, based on the results of BCR, Cd and Zn were largely in extractable, reducible and oxidable fractions, while other metals were mostly in residual fraction, indicating the high potential available of Cd and Zn. Multivariate statistical analysis revealed that Sb, Cu, Cd, Pb and Zn were mainly controlled by traffic activities (e.g. brake wear, tire wear, automobile exhaust) with high contamination levels found near traffic-intensive areas, including coach station, gas station, commercial center, expressway entrances and exits; Cr, Co, Ni and Mn derived primarily from soil parent materials; Hg was related to coal combustion. The accumulation of these traffic-derived metals causes potential negative impact to human health and ecological environment and should be concerned, especially the emerging trace elements like Sb.

Vertical distribution and Chemical Behavior of Uranium in the Tailings Material of Schneckenstein (Germany)

Taoufik Naamoun (University of Sfax, Faculty of Sciences, Department of Geology, Route de Soukra Km 3,5; P.B. 1171, Sfax, 3000, Tunisia; Email: naamoun_taoufik@yahoo.ca)
Broder Merkel (TU Bergakademie Freiberg,Institute of Hydrogeology, Gustav-Zeuner-Str. 12 . D-09596 Freiberg . Germany)

Abstract: The purpose of this study was to investigate the spatial distribution of uranium and its mobility. In this work, the total concentration of uranium was determined by means of gamma and alpha spectrometry. The study of uranium mobility in selected samples from different intervals was made by the application of a seven steps sequential extraction procedure. Also hydrochemical investigation was carried out. The distribution of uranium species and saturation indices were calculated by means of the hydrochemical model PHREEQC. Results showed that uranium content ranges from 185 Bq/kg in the heap material to 6193 Bq/kg in the processed material. Also the data analysis shows the high uranium contents in most analysed water samples. Moreover, the PHREEQC model indicates the change of the tailing sediments with depth from aerobic to post aerobic or anaerobic conditions. Moreover, it illustrates the presence of uranium mostly (90%) in its high soluble form $UO_2(CO_3)_3^{4-}$. This is in agreement with the results of the selective extraction procedure which proves the association of important amounts of its contents with the carbonate phase (until 24%). Furthermore, as between 30 and 80 % of the non residual uranium is in association with the nodular hydrogenous fraction, the decrease of the Eh values in the study areas enhance its solubility.

Accumulation of Metals and Hydrocarbons in Mangrove sediments of Arabian Gulf at the Eastern Province of Saudi Arabia

Md Iqram Uddin Al Amran, Omer Al Alhaiqi, Abideen Ojo Salawudeen and Mohamed Ali Qurban
(King Fahd University of Petroleum and Minerals, KSA)

Mangrove areas are an important ecosystem but are under the threat of anthropogenic contaminants. Contaminants like trace metal and petroleum hydrocarbon are considered to represent a potential risk to the marine ecosystem. This study evaluates the level of trace metals and PAHs deposited in the mangrove sediments of Arabian Gulf. Samples of sediment were collected by following the standard methods from the mangrove areas of Tarut and Rahima on the Gulf coast of Saudi Arabia. Sediment samples were analyzed for trace metals using ICP-OES while Hg was analyzed using NIPPON MA 3000. GC-MS and GC-FID were used for analyzing PAHs. The mean concentrations of As and Cd in the sediment samples from both Tarut and Rahima are equal. Concentrations of Cr, Cu, Hg, Pb, V, Ni, Al, Fe and Mn, except Zn and Ba, were found higher in Rahima than Tarut. The higher concentrations in Rahima were probably because of drilling mud chemicals from close oil fields while higher Zn and Ba in Tarut may be from agricultural and municipal discharges. Also, the mean of the total PAHs was higher for Rahima than Tarut. The same reason given for the higher trace metals may also be attributed for it. The contamination factor (CF), enrichment factor (EF) and pollution load index were used to assess the level of quality which suggest proper monitoring and management to keep unobjectionable level.

Potential of Mechanical Activation and Ultrasound Combination Towards Mineral Extraction: A Case Study of Low-Grade Rhodochrosite

Hongping He, Ning Duan
(Tongji University, Shanghai 200092, China)

The potential of ultrasound and traditional ball milling in cutting down the acid amount required for low-grade MnO (manganese ore as received) extraction was extensively investigated in this study. Results showed that ball milling and heating-up combination could obviously promote Mn dissolution kinetics and thereby achieve higher extraction efficiency under identical H_2SO_4 dosage. The positive role of ultrasound was found to be evident in the extraction of MnOA (manganese ore after ball milling), but negligible for MnO. Such a difference was correlated with the much different physicochemical properties of the liquid/solid systems that ultrasound involved in, as ball milling was confirmed to induce the dissociation of Mn-containing flocs and SiO_2 particles. High potential of ultrasound in weakening the agglomeration of ultra-fine MnO_A was also recommended. Semi-empirical Averami equation was found to well model the Mn extraction kinetics, which was much more diffusion controlled after ball milling based on the modeling. The modeling disclosed high capability of ball milling in decreasing the activation energy of Mn extraction (from 11.33 to 5.43 kJ/mol), whereas, ultrasound was effective in enhancing mass transfer or promoting pre-exponential factor A. Considering the potential in assisting efficient extraction via a cleaner way, ultrasound and ball milling combination provides a feasible alternative in mediating the hydrometallurgy of low-grade minerals.

Modeling

Probing the Water Quality Transformation of Underground Caverns in a Tropical Region Using Elcom-Caedym Model

Ming Chen, Jian Li, and ***Xiaosheng Qin***
(School of Civil and Environmental Engineering, Nanyang Technological University, Singapore)

ABSTRACT: This study aims to preliminarily model the water quality variations in an underground cavern environment using the coupled hydrodynamic and ecological model ELCOM-CAEDYM. The water quality transformation in terms of nitrogen and chlorophyll-*a* under both light and dark environment was analyzed. The ecological part of the model (i.e. CAEDYM) was calibrated and verified by experimental data. It was found that the chlorophyll-*a* concentration (phytoplankton biomass) decreased under the dark and light conditions with the decreasing rate being faster in a dark environment. The ammonia concentrations under both dark and light conditions increased rapidly first and then declined to an equilibrium state. The trend for the nitrite concentrations differed under these two conditions. Nitrite concentrations under the light condition only slightly fluctuated during the entire experiment, while they varied significantly under a dark environment. Nitrate concentrations generally increased with time except for the initial stage under both conditions. The modeling results demonstrated that ELCOM-CAEDYM could reasonably reproduce the experimental observations. Different scenarios for hypothetical design were also run for ascertaining the critical factors controlling the water quality in an underground cavern.

INTRODUCTION

Water shortage is a global problem today due to uneven distribution of water resources and fast economic and population growth. In 2006, it was reported that about one third of the world population (about two billions) was suffering from water scarcity (Oki and Kanae 2006). Unfortunately, the situation for water shortage is getting worse due to climate change and human-induced activities. In the past decades, Singapore has been developing at an accelerating rate, leading to an increasing water demand. Although Singapore receives considerable rainfall throughout the year, she has limited land to collect and store rainwater. The use of underground caverns has proved to be an effective technology to mitigate water resources shortage (Broch 2007; Zhao and Bergh-Christensen 1996). Due to enclosed and dark environment in underground caverns, water quality may deteriorate, rendering the water unsuitable for human usage. Thus, research is needed to investigate and predict how water quality changes during the storage in underground caverns.

Light is an important factor associated with many processes in an aquatic ecosystem. There is evidence that light controls the concentrations of dissolved inorganic phosphorus and dissolved total phosphorus at the water-sediment interface in Taihu, China (Jiang et al. 2008). It was reported that light intensity was related to nitrogen cycling rates in lake Maracaibo, Venezuela (Gardner et al. 1998). Many other researchers also investigated the light effects on various aspects of aquatic ecologic systems (Spears et al. 2008; Dalai et al. 2014; Bhuvaneshwari et al. 2015). Generally, an open environment allows oxygen to enter the water, while enclosed environment limits the normal air exchange between water and air interface. Although the underground reservoir may still be subject to intermittent episodes of inflows with certain nutrient and pathogenic loads, the absence of solar radiation would lead to severe limitations on algal dynamics, nutrient uptake and cycling. Previously, few efforts have been made in exploring water quality variations in underground caverns (Trolle et al. 2012). In this study, a preliminary modeling study is to be carried out for exploring the water quality changes in an underground cavern environment, with the aid of experimental test.

MATERIALS AND METHODS

Bench-scale Experiment

Singapore is a tropical country, with a total land area around 700 km^2 and population about 5 million. The annual rainfall is around 2,400 mm and mean daily temperature varies generally from 23-32°C (Gin et al. 2011). To better understand the mechanisms of the ecological processes in an underground environment in such a region and also to support verification of numerical modelling, a bench-scale experimental test was conducted. A local lake near Nanyang Technological University (NTU), Singapore, was selected for water sampling. The nutrification level in this lake is found higher than the general Singapore raw water and is useful for our experimental purpose. Two stainless-steel water tanks with one having light and the other without light were prepared to store the water samples. FIGURE 1a shows the dimension (40cm×30cm×20cm) and configuration of the water tanks. The tank with light input has its top open to sunlight to ensure phytoplankton photosynthesis, and the one without light was covered to mimic the underground environment. Both tanks have connection pipes to atmosphere and are not completely sealed, which is similar to real cavern condition. The experimental test started from September 9th, 2015 to October 27th, 2015. The analyzed parameters include dissolve oxygen (DO), pH, temperature, total carbon (TC), inorganic carbon (IC), ammonia, nitrate, nitrite and Chl-*a*. The experimental instruments used for the measurement of these parameters are: DO: YSI 5000 DO meter; pH: pH meter (Mettler toledo); Temperature: YSI 5000 DO meter; TC and IC: TOC Analyzer (Shimadzu); Ammonia: Ammonia TNTplus, ULR; Nitrate: Nitrate TNTplus vial test, LR; Nitrite: Nitrite TNTplus, LR; Chl-*a*: UV-vis Spectrophotometer (UVmini-1240, Shimadzu).

Ecological Modeling

The ELCOM-CAEDYM model was used to simulate the water quality deterioration in an underground cavern environment. The CAEDYM module was customized to reproduce the water quality changes in the experimental water tanks. The configuration and preprocessing of the model mainly include water-tank meshing, bathymetry processing, boundary/initial condition setup, and parameter estimation. The metrological factors were considered major driving forces influencing the water quality in the water tank as there was no inflow into the water tank and the water body was always static. The solar radiation and air temperature used in the modelling process was assumed following the Sine function as they varied diurnally. The air temperature was set to 29 °C considering no major changes for the studied period. The model used the actual size of the water tanks to study water quality changes with and without light input (see FIGURE 1a). The settling of the particle nutrients and phytoplankton cells may cause limited vertical distribution of concentration; in areal direction, it is assumed spatially uniform due to the small size of the system. The mesh size used in the ecological modeling was 0.02 m with grid numbers 20×15×10, which was deemed sufficiently dense for capturing the chemical/biological reactions in the tank.

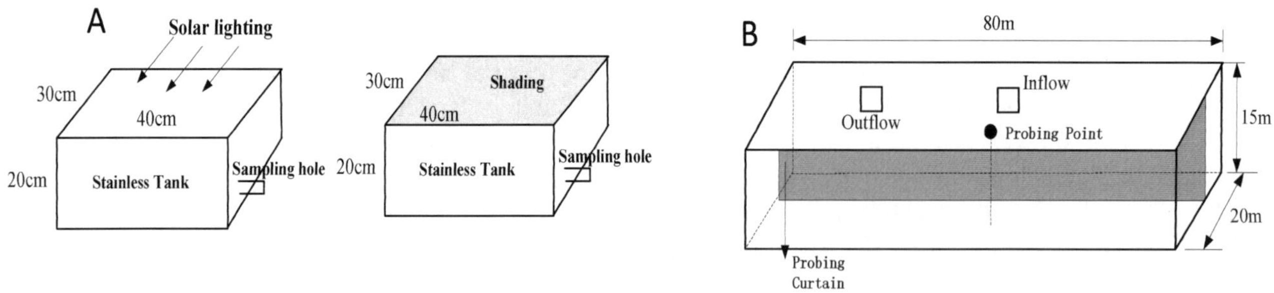

FIGURE 1. (A) Water Tank Configuration and (B) Hypothetical Underground Water Cavern.

A number of water quality variables including ammonia (NH_4), nitrate (NO_3), dissolved oxygen (DO), water temperature, pH, TC, and Chl-*a* concentration were measured. The phosphate (PO_4) concentration was below the detection limit of our equipment and its initial concentration in the model was provided from the related literature on Singapore reservoir water quality. The initial concentrations of NH_4, NO_3, DO, water temperature, pH, TC, and Chl-*a* were set based on the laboratory measurement data listed in TABLE 1. The initial concentrations of nutrients and Chl-*a* were higher in the light condition than those without light as some reactions may already have started before the first measurement on September 9th, 2015.

Modelling a Hypothetical Underground Water Cavern

The modelling of a hypothetical underground water cavern was carried out assuming frequent inflows into the cavern. This was to simulate the water input from surface catchment under high storm events. Also, the hydrodynamic cycling would affect the internal water quality changing process. As shown in FIGURE 1b, the modeling case was designed as a rectangular tank with dimensions of 80 m long, 20 m wide and 15 m high. The total volume of the modelling case is 24,000 m^3. One inflow point and one outflow point were setup in the middle of the left and right sections, respectively. The wall was assumed to be a concrete wall which means no light could penetrate into the tank. The size of the case was referred to two real projects including one pilot underground oil cavern built in Bukit Timah granite bedrock in Singapore (Zhao and Bergh-Christensen 1996) and one underground water storage cavern built in Trondheim, Norway (Broch 2007).

TABLE 1. The Initial Condition of Water Quality Variables in Modeling

Variables	Without Light	With Light	Unit	Value obtained from
DO	5.20	5.80	mg/L	Measured
Water temperature	28.0	28.0	°C	Measured
pH	8.03	8.06	-	Measured
NH_4	0.037	0.041	mg/L	Measured
NO_3	0.095	0.151	mg/L	Measured
PO4	0.01	0.01	mg/L	Reference (Low et al. 2010)
ON	1.10	1.10	mg/L	Reference (Low et al. 2010)
OP	0.05	0.05	mg/L	Reference (Low et al. 2010)
Chl-*a*	54.91	115.77	μg/L	Measured

The factors influencing the water quality changing process in the studied underground water storage tank include the inflow/outflow rates, the discharging frequency and duration, and the pollutant loads. The nutrients and Chl-*a* concentration of the inflow water will be studied through scenario analysis. But some other influencing factors such as the chemicals from the concrete wall, the suspended particulate matter and microorganisms were not considered in this study.

The initial conditions of DO, water temperature, nutrients and Chl-*a* concentrations were the same in all the scenarios of modeling studies, which aims to compare the effect of different boundary conditions on the underground water quality changing process. The total Chl-*a* concentration was the sum of the phytoplankton groups. Three kinds of algae including Cyanobacteria, Nodularia and Freshwater Diatoms were chosen according to the literature related to the reservoirs in Singapore (Low et al. 2010), and the ratio of the above algae was set as 2:2:1. The simulation period was set to 60 days for all scenarios.

In total, 4 cases were modelled to examine the effects of different boundary conditions on the water quality changing process in the hypothetical underground water storage cavern. Case-1 was used as a standard benchmark case and other cases will be used for comparison. All the boundary conditions used in the scenarios are listed in TABLE 2. The nutrient concentrations including both nitrogen and phosphorous increased in Case-2. The inflow Chl-*a* concentration was 10 times higher in Case-3, and the inflow frequency and duration all decreased by half in Case-4. It should be noted that the general Chl-*a* concentration in Singapore raw water is usually lower than 100 μg/L; we set it at a higher value for theoretical analysis only.

RESULTS AND DISCUSSION

Changes in Water Quality

Water sampling was performed continuously from these two water tanks for water quality analyses over about two months. It was found that DO concentrations exhibited a similar trend under the two types of conditions (FIGURE 2). Their mean DO concentrations were at similar levels (5.76 mg/L for no light and 5.78 mg/L for light conditions, respectively). We did not observe significantly different patterns for TC, IC and pH concentrations between water samples under light and dark conditions. However, the curves for Chl-*a* concentrations were notably different under these two conditions, although all of them showed a generally decreasing tendency. The declining trend in Chl-*a* concentration under the dark condition was faster than that under the light one. Temporal plots of ammonia, nitrate and nitrite were shown in FIGURE 2. The mean ammonia, nitrate and nitrite concentrations were 0.053, 0.540 and 0.036 mg/L for the dark condition, respectively, and 0.045, 0.515 and 0.022 mg/L for the light condition, respectively. There was a similar trend for the ammonia concentrations under the dark and light conditions. Both observed ammonia concentrations increased rapidly within the first week, and then quickly declined. After about 10 days, they would reach an almost equilibrium state. It is noted that, the trend for the nitrite concentrations was largely inconsistent under these two conditions. Nitrite concentrations under the light condition only slightly fluctuated during the entire experiment, while they varied significantly under the dark condition. Nitrate concentrations generally increased with time except for the initial stage under both conditions. But the increasing rate was obviously higher for the tank with dark condition.

TABLE 2. Scenarios Modeling Boundary Conditions

Variables	Case-1	Case-2	Case-3	Case-4	Unit
Discharge	5.0	5.0	5.0	5.0	m^3/s
Frequency	1	1	1	**2**	/10days
Duration	6	6	6	**3**	hours
Temperature	29.0	29.0	29.0	29.0	°C
DO	8.0	8.0	8.0	8.0	mg/L
NH_4	0.06	**0.20**	0.06	0.06	mg/L
NO_3	0.15	**1.50**	0.15	0.15	mg/L
PO_4	0.01	**0.02**	0.01	0.01	mg/L
ON	0.60	**3.0**	0.60	0.60	mg/L
OP	0.02	0.02	0.02	0.02	mg/L
Chl-*a*	91.5	91.5	**915**	91.5	μg/L

Modeling Results for Lab Water Tanks

CAEDYM model was used to simulate the changing process of water quality variables to repeat the lab test results. The variation of nitrogen nutrient components has been given a priority for analysis. The phosphate (PO_4) concentration was very low and limited the phytoplankton growth in the reservoir of Singapore (Gin et al. 2011), so it was not analyzed. The concentrations of ammonia would decrease gradually in the simulated period (30 days) because ammonia would transform into nitrate through the nitrification process and also the phytoplankton cells adsorption. From FIGURE 2a, the simulated ammonia concentration changed following a similar trend with observed one regardless of with light or without light.

From FIGURE 2b, the nitrate concentration would consistently increase from 0.1 to about 1.0 mg/L until the end of simulation period. The Chl-*a* concentration dropped from an initial high concentration to nearly zero due to the death of phytoplankton cells (FIGURE 2c).

Nevertheless, the total nitrogen in the water tank shows only a slight decrease which means the total mass of nitrogen was kept in equilibrium. The simulated nitrogen nutrients and Chl-*a* concentrations are in a good agreement with the observed data. It seems that the relatively weak solar lighting in the laboratory would not significantly affect the chemical reaction and phytoplankton dynamics; the adapted CAEDYM could well reproduce the ecological process in the dark environment. Generally, both the simulated results and measured data indicate that (i) the phytoplankton would die because of the limiting factors from phosphorous nutrients and organic carbon, (ii) the nitrate concentration, that are transformed from the initial organic nitrogen, would rise and keep at a high level, (iii) the transformation rates of ammonia and nitrate under the dark environment are somewhat higher than those under the light one, but the decay rate of Chl-*a* is somewhat lower under the light condition than under the dark one. The observed DO concentration decreased before the 10th day and then increased due to re-aeration from the atmosphere (FIGURE 2d). Interestingly, the trends for DO concentrations under light and dark conditions were very close, as shown in FIGURE 2d. The simulated DO concentration agrees with the observed data well to a large extent.

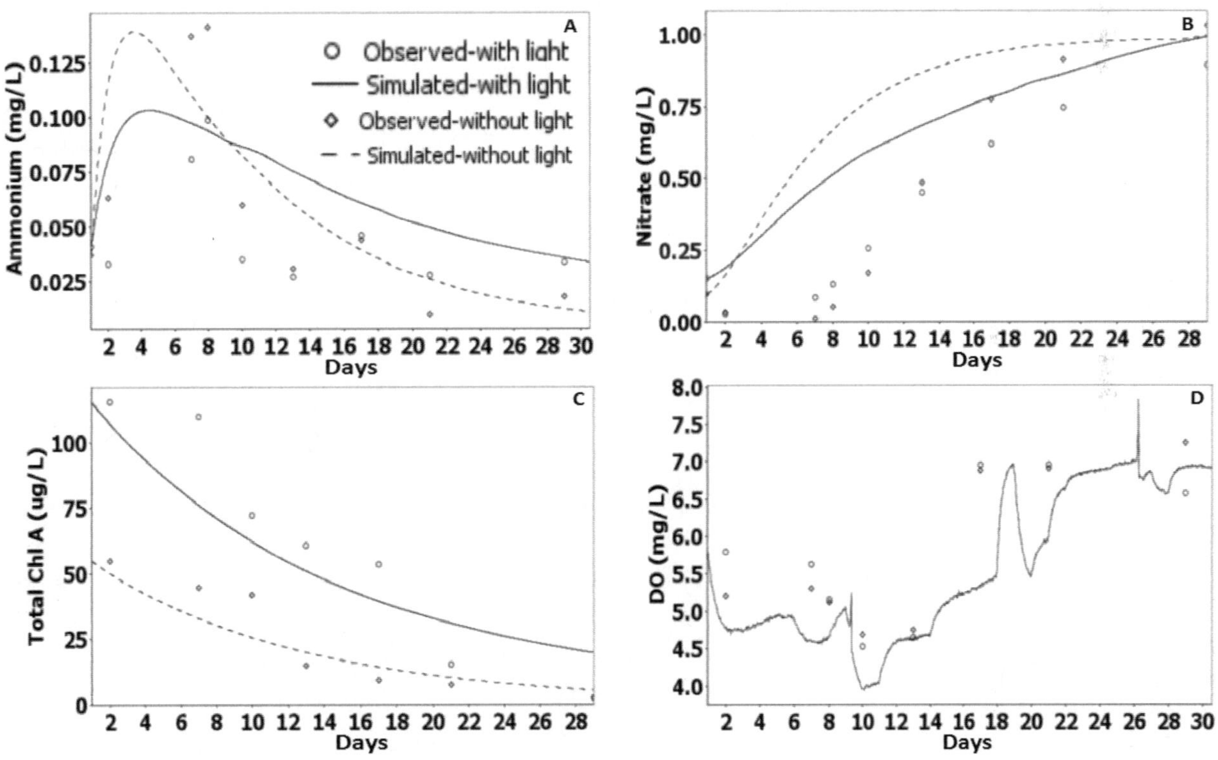

FIGURE 2. Dynamic Trend of the Measured Variables (a: Ammonia, b: Nitrate, c: Chl-a, and d: DO).

Modelling Results for the Hypothetical Underground Cavern

Based on the parameter setting for the lab model under no light condition, we conducted a modeling study for the hypothetical underground cavern. It should be noted that, there are multiple possibilities of results under different scenarios (e.g., DO concentration, inflow patterns, pollutant loadings, location of probing profile). Herein, only one scenario was presented (as shown in FIGURE 3). The NH_4 concentration variations were affected by the boundary conditions, as demonstrated from FIGURE 3a where the NH_4

concentrations of Cases 2, 3 and 4 were different from those of benchmark Case 1. The increase of NH_4 concentrations in the inflow water had a significant impact on those in the underground cavern. From FIGURE 3b, with a high Chl-*a* concentration, the frequency and duration of inflow had a minor effect on the nitrate concentrations. However, these nitrate concentrations would largely increase when the nitrate concentrations in the inflow water increase as reflected by Case 2. The Chl-*a* concentration would decrease gradually in all the simulation periods when no inflow was presented (FIGURE 3c). The presence of inflow would increase the Chl-*a* concentrations. The TN concentrations were higher in Case-2 and Case-3 with higher inflow nutrients and Chl-*a* concentrations than in Case-1 and Case-4, where the inflow boundary conditions related to N and Chl-*a* would affect the TN concentration more obviously than other factors (FIGURE 3d). This study only gives a projection of possible water quality changes under a hypothetical cavern case, with the main purpose of demonstrating the capability of ELCOM-CAEDY. The results may vary significantly under different settings of operating modes or physical characteristics of the cavern.

CONCLUSIONS

This study conducted a preliminary modeling study on the water quality variations in dark and light environment using the coupled hydrodynamic and ecological model (i.e. ELCOM-CAEDYM). The experimental data was used for model validation. Different varying trends for Chl-*a*, ammonia, nitrite, and nitrate concentrations were found in the experiment and they could be well reproduced by the applied model. Some scenarios for a larger scale system were run for ascertaining the critical factors controlling the water quality in an underground cavern environment.

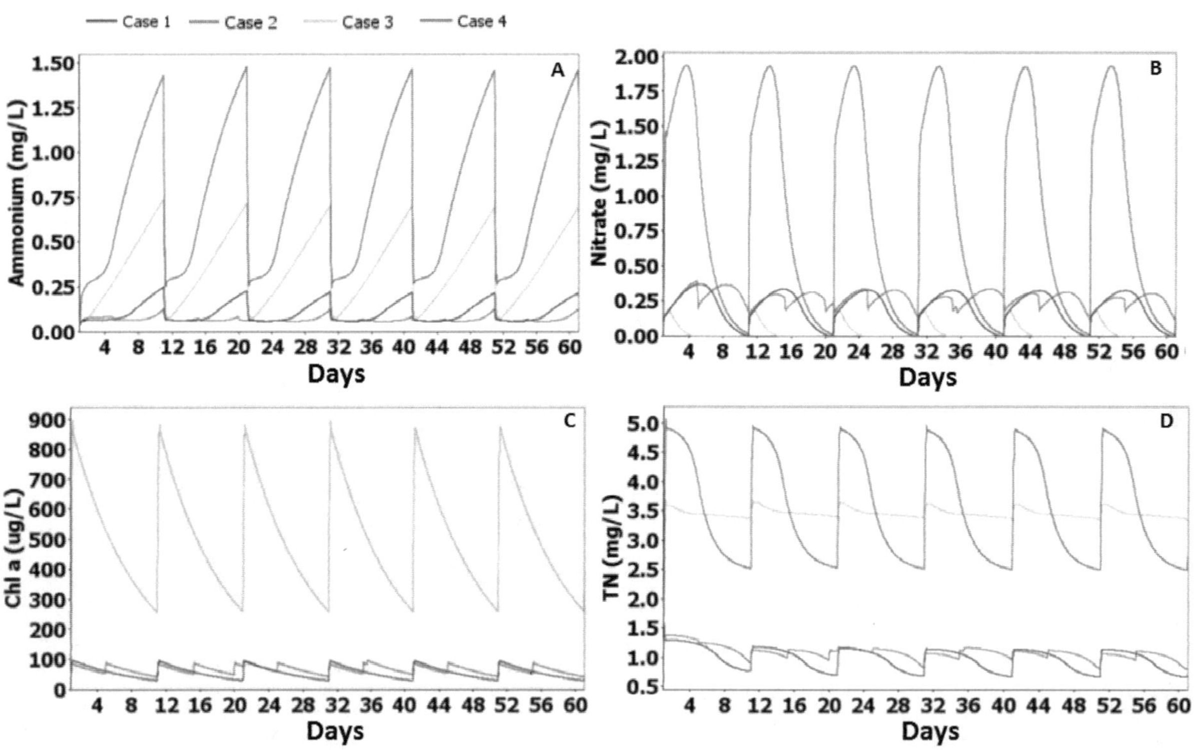

FIGURE 3. Variations of Water Quality Variables at The Probing Point.

ACKNOWLEDGEMENTS

This material is based on research/work supported by the Land and Liveability National Innovation Challenge under L2 NIC Award No. L2NICCFP1-2013-3 (WBS no.: M4061545.D63).

REFERENCES

Bhuvaneshwari, M., et al. 2015. "Cytotoxicity of ZnO NPs towards fresh water algae Scenedesmus obliquus at low exposure ..., visible and dark conditions." *Aquat Toxicol*, 162: 29-38.

Broch, E. 2007. "Use of the Underground in the City of Trondheim, Norway." 11th ACUUS Conference: Underground Space: Expanding the Frontiers, Sep. 10-13 2007, Athens - Greece.

Dalai, S., et al. 2014. "Studies on interfacial interactions of TiO2 nanoparticles with bacterial cells under light and dark conditions." *B Mater Sci*, 37(3): 371-381.

Gardner, W. S., et al. 1998. "Nitrogen cycling rates and light effects in tropical Lake Maracaibo, Venezuela." *Limnol Oceanogr*, 43(8): 1814-1825.

Gin, K. Y.-H., et al. 2011. "Comparison of nutrient limitation in freshwater and estuarine reservoirs in tropical urban Singapore." *J Environ Eng*, 137(10): 913-919.

Jiang, X., et al. 2008. "Effects of biological activity, light, temperature and oxygen on phosphorus release processes ... of Taihu Lake, China." *Water Res*, 42(8): 2251-2259.

Low, E., et al. 2010. "Top-down control of phytoplankton by zooplankton in tropical reservoirs in Singapore." *Raffles B Zool*, 58(2): 311-322.

Oki, T., and Kanae, S. 2006. "Global hydrological cycles and world water resources." *Science*, 313(5790): 1068-1072.

Sanudin, N., et al. 2014. "Feeding Activity and Growth Performance of Shrimp Post Larvae Litopenaeus vannamei Under Light and Dark Condition." *J Agr Sci-Cambridge*, 6(11): 103-109.

Spears, B. M., et al. 2008. "Effects of light on sediment nutrient flux and water column nutrient stoichiometry in a shallow lake." *Water Res*, 42(4): 977-986.

Trolle, D., et al. 2012. "A community-based framework for aquatic ecosystem models." *Hydrobiologia*, 683(1): 25-34.

Zhao, J., and Bergh-Christensen, J. 1996. "Construction and utilization of rock caverns in Singapore Part D: Two proposed cavern schemes." *Tunn Undergr Sp Tech*, 11(1): 85-91.

Numerical Analysis for a Series of Recent Catastrophic Floods (2011-2017): New Verification of the Groundwater and Tectonic Processes Possible Impact

T. Trifonova
(Lomonosov Moscow State University, Moscow, Russia)
S. Arakelian, D. Trifonov, S. Abrakhin
(Stoletovs Vladimir State University, Vladimir, Russia)
V. Koneshov, A. Nikolaev
(Schmidt Institute of Physics of the Earth, RAS, Moscow, Russia)

The principal item of present paper is to discuss the reasons of the existing often uncertainty and discrepancy for the flood/debris water balance estimation in the area under heavy rain.

The simplest and conventional model for the water balance estimation for disastrous floods shows the greater (up to 75%) water mass discharge observed during the events than it could be expected from the rainfall process only in the area under study. The fact gives us the founding to take into account the groundwater possible contribution to the events).

The key part of the concept is determined by impact of fractured bedrock in the water budget for the river basin as an unified 3D-system. We consider also a possible role of tectonic stresses in the Earth's crust in the groundwater basin functioning in dynamics. The reasons for that are, first, the pressure field variation in groundwater basin and, second, the modification of the crack-net itself by different factors, occurring both suddenly and/or smoothly (due to additional dynamic and static stresses and variation of physical/chemical properties of solid/rocks). The observable phenomena are a hydrology state variation, e.g. dramatic change of discharge parameters in different water channel and/or in wells, appearance/disappearance of new springs (both temporal and/or permanent), groundwater table change, changes in lifecycles of geysers and mud volcanoes.

We took into account both groundwater oscillations in the far/near field (especially for post seismic groundwater recession) and the data for triggered-hydrologic responses to selected earthquakes. Our statistical analysis over some selected catastrophic events – in Western Europe, USA, Russia, China, India (2011-2017) – showed that more significant impact on the groundwater exit on surface occurs for the earthquake hypocenter depth ~10 km when the magnitude value is about 5.0 ($\sim 10^{12}$ J) which may be associated with 7 points in earthquake epicenter on the land surface. The problem has been considered, in particular, for two real catastrophic events in USA: (1) the flood in Colorado (Sept. – Oct. 2013); can be identified with the earthquakes, first, in the Northern California on August 27, 2013 (magnitude – M 4.2), second, in North of Mexico on August 27, 2013 (M 4.3) and, third, in the East Texas on September 02, 2013 (M 4.5); (2) the flood in Texas (Houston/Dallas), April-May 2017; the earthquakes, first, in Nicaragua, April 18 (M4.5); second, near Mexico city, April 26 (M 5.3); third, in California Bay Area, April 26 (M 5.1).

The scale difference between groundwater response on the earthquake in both time and distance is sufficiently complicated and depends on many factors being not clear still: may vary in time from immediate reaction to hours and weeks, and in distance – up even to thousands kilometers.

We carried out some computer simulation for the subject. The most accurate correlation (Pierson coefficient K) can be estimated between two principal parameters: river-discharge and artesian water level in wells in some localized river-basin areas. We had K~0.97 but with some optimal day shift ~10-20 days.

The approach gives reasonable results to forecast the risks of some real water catastrophic events, including the conditions for a solitary destructive wave propagation which may occur over the land surface during such catastrophic events under trigger mechanism of the crack-net modification due adjustable earthquakes.

An Analytical Solution to the Problem of Ponded Ditch Drainage in a Stratified Soil

Gautam Barua and Subhadeep Chakrabarti
(Indian Institute of Technology Guwahati, Guwahati, Assam, India)

An analytical solution is worked out for predicting flow into a network of equally spaced ditch drains dug in a stratified soil overlying an impervious substratum, the drains being fed by a ponding head distribution introduced at the surface of the soil. The solution is new in the sense that apart from handling the usual complexities of the problem, it can also tackle situations with unequal water level heights in the drains. The steady state portion of the solution is general and is valid for all possible combinations of the flow parameters of the problem; however, so far as the transient part of the solution is concerned, it is valid only for situations where the directional conductivities and specific storage of the layers obey certain pre-defined relations among each other. Even with these restrictions, the transient solution is expected to be important as it is still quite extensive in nature with the capability of handling a wide variety of subsurface drainage situations. As a drainage system is often installed to have a control on waterlogging and salinity build up of irrigated lands, it is hoped that the generic drainage model proposed here would lead to a better design of subsurface drains in these fields than the existing solutions to the problem. Further, as subsurface drains are now also proving to be increasingly important in having a check on the release of the greenhouse gas methane from stagnated ponded paddy fields, the ponded drainage solution provided here is also expected to be quite useful in designing drains for controlling release of methane from such environments.

Improving Human Health in China through Alternative Energy

Melissa Scott[a], Robert Sander[b], Greg Nemet[ac], Jonathan Patz[ade]
([a]University of Wisconsin – Madison, Nelson Institute for Environmental Studies, Center for Sustainability and the Global Environment (SAGE), Madison, Wisconsin, USA;
[b]International Institute for Applied Systems Analysis, Laxenburg, Austria;
[c] University of Wisconsin – Madison, LaFollette School of Public Affairs, Madison, Wisconsin, USA
[d] University of Wisconsin – Madison, Global Health Institute, Madison, Wisconsin, USA;
[e]University of Wisconsin – Madison, Department of Population Health Sciences, Madison, Wisconsin, USA)

In this study we estimate the health benefits of more stringent alternative energy goals and the costs of reducing coal-fired power plant pollution in China in the study year 2030. One of our two overarching alternative energy goals was complete elimination of coal energy, supplemented by natural gas and renewables. The second was a policy scenario similar to the U.S. 2013 Climate Action Plan (CAP), which played a pivotal role leading up to the 2015 Paris Climate Agreement. We used the Greenhouse Gas and Air Pollution Interactions and Synergies (GAINS) model created by the International Institute for Applied Systems Analysis for our model simulations. We found that 350-416 million years of life can be saved if coal energy is completely replaced by alternative energy, and 111-133 million years of life can be saved if coal energy is replaced by alternatives in a CAP-like scenario. A CAP-like scenario using emission-controls in coal plants costs $11-18 per person. Reducing coal energy in China under a CAP-like scenario would free up $9.4-32 billion in the annual energy budget, to spend on alternatives. Assuming an average Chinese life expectancy of 76.1 years, these scenarios would save from 1.46 million to 5.47 million lives annually.

Numerical Simulation on Algae Bloom in Chaohu Lake, China

Xin Qian, Mei Li and Liuyan Yang
(State Key Laboratory of Pollution Control and Resources Reuse, School of the Environment, Nanjing University, China)

The Chaohu Lake Basin is located in the middle of the Anhui Province between the Yangtze River and the Huaihe River. It is among the five well-known freshwater lakes in China. The Chaohu Lake is the main water source of local industrial, agricultural, and living activities. It is also the main water body that integrates industrial and agricultural drainage with the domestic wastewater along Chaohu Lake. The water quality of several of the main inflowing rivers deteriorates continuously as a result of the excessive economic and population growth in the Chaohu Lake Basin, which in turn deteriorates the water quality of Chaohu Lake. The water environment in the Chaohu Lake Basin has undergone significant ecological stress in the past three decades. Since the 1970s, frequent "algal blooms" have been detected in Chaohu Lake, and this phenomenon is accompanied by water deterioration. As a result, the water quality is poorer than the national Class V standards and significantly affects drinking-water safety, fishery, and tourism. The water quality has not improved despite recent efforts in pollution control and lake management. The high levels of nonpoint source pollution in the basin are difficult to control in the long term although the loads of main point-source pollution in cities and towns are managed effectively in recent years. The water quality is difficult to improve because of the insufficient of knowledge regarding the formation, transfer, accumulation, and outbreak of the algae bloom in Chaohu Lake.

This research aims to forecast and provide early warning against algae bloom in Chaohu Lake and try to identify the hydrodynamics factors that influence the migration of nutrients and algae bloom in Chaohu Lake.In order to study the process of the algae bloom in the lake, a three dimensional numerical model was introduced to simulate the distribution of the water quality parameters and the area of algae bloom.

The field works were carried out to gather the sufficient data for calibration the coefficients and verification for the simulation results. The initial nutrients distribution data were collected from field data and distribution of algae were from MODIS data, the wind data are form the weather forecasting in the next days from the web page. In order to make reasonable early warning of distribution of algae bloom, the simulation scenarios are focus on mainly the weather conditions in the summer. The results show that the wind speed and the direction are the key factors for the algae bloom. The detailed wind forecasting is the most important factors for the early warning of the concentrations and areas of the algae bloom in the lake. The process for the early warning against algae bloom in Chaohu Lake has been recommended to the management bureau.

GIS, DATABASE, AND REMOTE SENSING

Geospatial Assessment of Selected Wheat Producing Areas of Punjab, Pakistan

Sheikh Saeed Ahmad
(Fatima Jinnah Women University, Rawalpindi, Pakistan)

Agriculture is the main pillar of Pakistan's economy. The impact assessment of shifting landuse practices on the vegetative health of agriculturally important areas is imperative for future sustainability of agriculture production. The present study examined the impact of shifting landuse practices on vegetative health in two major wheat producing areas of Punjab, Pakistan for the period 1999-2014. For this purpose Landsat dataset for the respective years were acquired and analyzed using ArcGIS 10.2, ERDAS Imagine 11 and ENVI 5.2. Shift in landuse practices was assessed via spatio-temporal change detection and vegetative health was assessed by NDVI. Therefore, this research can be used as a benchmark for sustainable landuse planning and supporting future sustainable agricultural production in the study area.

Groundwater Exploration and Geochemical Analysis of Its Sediments: A GIS Approach to Combat Climate Change for Sustainable Water Resource Management

S. N. Shashtri * (CEST, Fiji National University, Natabua); Salanieata Matai (Master Student - Sophia University, Tokyo); C.K.. Singh (TERI University, India) and S. Mukherjee (School of Environmental Sciences, JNU, India)

Information on the geological set up, lineament density and land use pattern and changes are required for groundwater exploration and rainwater harvesting. Multispectral and multi-temporal satellite data along with geological map has the potential to delineate sites for water resource management. In Fiji Archipelago, early Tertiary rocks are found only in Viti Levu where there is succession of volcanic rocks and their sedimentary derivatives with minor intercalations of carbonate rock. Exploration of groundwater in this particular zone at the boundary of basaltic and sedimentary basin is significant for the sustainable agriculture in the contemporary phase of the climate change. Apart from that the continuous recharge of the aquifer is taken into consideration for the suitable aquifer recharge from the heavy rainfall. The drilling site has been demarcated with help of multispectral satellite data (Landsat-8) and ancillary maps available. The sediment which came out at every 15feet has been analyzed under X-ray diffraction for heavy and some rare earth metal to ensure the water quality. Further GIS mapping of the groundwater potential zone has been demarcated with safety of salt water intrusion into the aquifer.

Investigation of the Spatial Relationships between Groundwater Quality and Hydro-geological Physical Controls using Geographically Weighted Regression Model

Jena S. and ***Panda R. K.***
(School of Infrastructure, Indian Institute of Technology Bhubaneswar, Odisha-752050, India)

The Hydro-geological Physical Controls like land use/cover (NDVI), soil characteristics (percentage sand, silt, clay and saturated hydraulic conductivity), topography (elevation, slope and ascent) and hydrology (recharge from rainfall and drainage density) have substantial impact on groundwater quality parameters. Understanding the spatial relationships between these physical controls and groundwater quality parameters are of paramount importance for effective and efficient planning for conservation and reversal of the resource. In the reported study, geographically weighted regression (GWR) and ordinary least squares (OLS) models have been used to develop spatial relationships between groundwater quantity parameters (pH, Electrical Conductivity, Turbidity, Alkalinity, Phenolphtalin Alkalinity, Nitrites, Total Hardness, Calcium Hardness, Ammonium, Chlorides and Iron) and Hydro-geological Physical Controls. The study investigated the spatial variability of the quality parameters using geostatistics and the dominant physical controls for the groundwater quality parameters were determined using GWR and OLS. The investigation was conducted for two hydrological years (2015-17) at 25 observation wells in the study groundwater basin in the mid-Mahandi river basin of eastern India. The groundwater quality data were also analysed for normality and spatial trend along with spatial autocorrelation to study the spatially varying clustering. Spatial statistics (Moran's I) showed a strong positive autocorrelation among the groundwater quality parameters within 2-km lag distance. Therefore, the groundwater quality parameters at unknown sites in the study area can be estimated within 2-km limiting distance using geostatistics. In the reported study, the corrected Akaike's Information Criterion (AICc) and coefficients of determination (R^2) were used to compare the relative measure of performance of the GWR and OLS models. Smaller values of AICc indicate better performance of the models; whereas, R^2 closer to 1 was considered as indicator for better goodness of fit. Moreover, higher R^2 values indicate the skill of the model to explain more variance in the dependent variable as a function of the independent variables. All the GWR models showed lower AICc values and higher local R^2 and adjusted R^2 values than the OLS model. This results suggests that, GWR is better than OLS for the study groundwater basin. The coefficients for the physical controls under hydrologic and soil classes in the regression model had higher values indicating their dominance in describing the spatial variability of the groundwater quality parameters. The recharge from rainfall was the most dominant physical control for all the quality parameters. The results of the study will be quite useful for the management of the groundwater quality at different locations of the study basin during different period of the year. The developed approach will be useful for similar locations.

Assessment of Soil Quality Using Decision Tree by Integration of Spatial Distribution of Soil Physio-Chemical Properties Using GIS

Mohamed A.E. AbdelRahman*, (National Authority for Remote Sensing and Space Sciences, Egypt); **Salah A. Tahoun** (Soil Science Department, Faculty of Agricultural Science, El-Zagazig University, Egypt)

The agricultural development of Egypt depends heavily on soil ecosystem goods and services. There is an urgent need to establish soil quality surveillance systems to sustain the rural self-supporting system, as well as the provision of reliable parameters to guide investments and monitor trends in soil status and impacts of interventions. Measurement of the soil quality indicators (SQI) is time-consuming and expensive Nevertheless assessment of soil quality is essential to monitor the sustainability of agricultural systems. The objective of the study is to assessing soil quality in the county using minimum appropriate data set of SQI. In this context, the case of Abo Hammad County of El-Sharkia Governorate offers an illustrative example. Abo Hammad County is a model for varieties of soil for whole country. The soils of the county occupy an area in the irregular fringe zone between the flood plain of the River Nile and the elevated plateau of the Eastern Desert. Consequently, soils of different textural classes, extending from clay to sand, are encountered. Irrigated agriculture has been practiced for more than a century, mostly without an efficient drainage system. As such, soluble salts have accumulated in soils, reaching high levels in localities. The simple additive SQI methodology (Amacher *et al.* 2007) was used to achieve the objective of this study. A series of minimum data set of information such physical (soil texture), chemical ($CaCO_3$ content, CEC, EC, ESP, and SAR) and biological (OM content) indicators is used for assessment of SQ. Analytical data were gathered from representative 53 soil samples covering most physiographic units of the county. Geostatistical approach was used to estimate the SQI and Arc GIS model builder also used to establish soil quality model (SQM) for this study. The results show that Universal Kriging (UK) has the best fit for the geostatistical methods than Inverse Distance Weight (IDW). Where the Cross Validation Results using R^2 were recorded best values for UK than IDW. Soils of the high soil quality occupy 35.4 % of the county; those of the intermediate quality occupy 38.0%, whereas those of the low soil quality occupy 26.6%. In the case of soil physiochemical properties, the largest part of the soils has medium to large values of these parameters. The largest part of the analyzed soil samples are characterized by low values in term of soil biological properties.

Evaluation Soil Organic Carbon in Some Areas in Nile Delta Using Remote Sensing and GIS Techniques

Mohamed Elsayed *a ; Mohamed A. E. AbdelRahman *a; Mohamed Abu-hashim*b ((a) Authority for Remote Sensing and Space Sciences (NARSS), Cairo, Egypt; (b) Soil Science Department, Faculty of Agriculture, Zagazig University, Egypt; salama55@mail.ru)

Soil organic carbon pool (SOCP) is affected by several factors particularly soil type, climate, topography, crop management, and anthropogenic factors. The study was carried out to clarify relationships between SOCP under different soil types and land-use changes east of Nile delta . Data of 20 soil profile and 150 surface soil samples were gathered, where the collected soil samples covered different soil types and land-uses. The results showed significant differences of SOCP among soils: loam and clay loams were rather similar. Clay soils were the most extensive and have mean SOCP of 4.08 ± 1.41 kg C m2 . The highest SOCP of 8 kg C m2 was in clay loam soil associated with bare soil, while the lowest of 3 kg C m2 in sandy clay loam soil associated with bare soil. Furthermore the overall average results of SOCP in cropland area showed 59.85 Mg C ha1 under different soils. Losing the arable lands to urbanization resulted in a decrease of 300.421 Gg C of SOCP. With the decrease in SOCP sequestrated within the soil surface, carbon dioxide would be emitted to the atmosphere. Land-use changes have marked impact on surface SOCP and C sequestration.

Apply Data Science to Rapid Modeling of Dual Fuel Technology for Life Cycle Assessment (LCA)

Fanxu Meng, Carolyn LaFleur, Asanga Wijesinghe and Richard Haut
(Houston Advanced Research Center, The Woodlands, TX, USA)

ABSTRACT: Dual Fuel diesel engines utilize natural gas and diesel fuel simultaneously, thereby reducing diesel fuel consumption and offering certain emissions reductions. Advances in dual fuel technology used in oilfield operations such as drilling and hydraulic fracturing offer opportunities to reduce fuel usage, costs and emissions. Emissions from dual fuel engines can vary greatly depending on how the engine is operated. Dual fuel engines can displace approximately 40% - 73% of typical diesel fuel consumption over an optimized range of operating conditions. Data science (e.g. machine learning) is applied to analyze field testing data and laboratory studies of dual fuel engines enabling rapid prediction of emissions, fuel consumption and other parameters to optimize operation of dual fuel engines. Machine learning such as regressions are used for rapid modeling. The average relative errors between predicted values and actual values mostly range from 3.9% to 10.4%. The model is integrated into life cycle assessment / analysis (LCA) to evaluate economic and environmental impacts for a range of operating conditions. This is an innovative application of data science to analyze dual fuel technology for the purpose of LCA.

INTRODUCTION

Data science is now one of the most important trends sweeping the oil and gas industry (Ball et al., 2017; Mishra & Lin, 2017). Oil and gas companies are increasingly using data science to support planning, operations and decision making (Mohaghegh et al., 2017) with advanced data analysis for added business value. Proof of concept practices include safety and risk (Caldwell & Hinton, 2015), real-time drilling (Cao et al., 2018; Kale et al., 2015), artificial lift (Pennel et al., 2018), reservoir characterization (Pankaj et al., 2018a), well completion (Pankaj et al., 2018b), production forecasting (Fulford et al., 2016) and management (Arciero & Ismail, 2017). A successful practice needs deep understanding of data science and the technology, and how to add business value between them (Dursun et al., 2014; Thajudeen, 2018). The Houston Advanced Research Center (HARC) has been providing leadership in dual fuel technology in field testing (Wijesinghe et al., 2018) and after-treatment system (ATS) (Meng et al., 2017). Dual fuel technology offers economic and environmental benefits because it utilizes widely available, low cost natural gas as the additional fuel for the diesel engine (Wei & Geng, 2016).

Regarding how to make decisions in adopting dual fuel technology, life cycle assessment (LCA) is a powerful tool. LCA can evaluate economic and environmental impacts for a range of operating conditions from a life cycle perspective. Life cycle inventory (LCI) is the basic unit to assembly for LCA. An LCI usually contains the function unit, input energy and material, and output materials and emissions, etc. For dual fuel technology, emissions are related to operational parameters such as engine loading, rpm, fuel consumption and ATS (if applicable). Data science is a powerful tool to estimate the uncertainty of life cycle inventory among different data sources (Xu et al., 2015). Identifying patterns with data science tools such as machine learning helps to estimate the uncertainty. Successful studies in chemical manufacturing (Cashman et al., 2016) and the oil and gas industry (Busby et al., 2017; Mishra & Lin, 2017) have been reported.

This paper presents an analysis of data concerning dual fuel technology. This proof of concept analysis is based on 114 observations of field test data (HARC, 2016; Johnson et al., 2017; Johnson et al., 2017; Wijesinghe et al., 2018). Each observation contains specific characteristics of an operation of a dual fuel engine and its emissions. After data cleaning and descriptive analytics, we identify several correlations between engine operation characteristics and emissions. We finally create a regression model to predict emissions of a dual fuel engine from its features. We also present the predicted results and their scores from 1,000 repeated random experiments. The scores include Mean Absolute Relative Error (MARE) and Root Mean Squared Error (RMSE). The average relative errors between predicted values and actual values mostly range from 3.9% to 10.4%. We come up the framework to use our model to calculate the life cycle inventory (LCI) of process of dual fuel technology.

METHODS

A dual fuel engine is a diesel engine that has been adapted to utilize natural gas fuel along with diesel fuel, co-combusted simultaneously. Using natural gas fuel reduces the amount of diesel fuel required. In most dual fuel engines, natural gas fuel in vapor phase is injected into the engine air intake. The gas fuel is blended with air and supplied to the engine. The engine can operate with the addition of gas fuel (dual fuel mode) or on diesel fuel alone (diesel-only mode).

TABLE 1. Feature List

Feature	Unit	Data Type	Calculation and Method
Source	N/A	String	Which field test is the observation data from
Activity	N/A	Category	HF-hydraulic fracturing; DR-drilling
Engine_make	N/A	String	The manufacturer of the engine
Engine_model	N/A	String	The model of the engine
Rated_speed_rpm	RPM	Numeric	The rated speed of the engine
Rated_power_kw	kW	Numeric	Power output at 100% loading
Power_kw	kW	Numeric	Actual power output during the test
NG_LHV_btupercf	Btu/cf	Numeric	lower heating value of NG used during the test
Fuel_consump_DLEperkwh	DLE/kWh	Numeric	Total fuel consumption in diesel liter equivalent (DLE)
Diesel_consump_DLEperkwh	DLE/kWh	Numeric	Total diesel consumption in diesel liter equivalent (DLE)
NG_consump_DLEperkwh	DLE/kWh	Numeric	Total NG consumption in diesel liter equivalent (DLE), including converted and loss
Engine_load	1	Numeric	Engine loading during the test
Fuel_efficiency_ZECE	1	Numeric	power out / (NG power in - CH_4 loss + diesel power in)
Diesel_disp	1	Numeric	1-DF diesel rate / DO diesel rate
Substitution_ratio_corrected	1	Numeric	(NG power in - CH_4 loss) / Total Fuel in
Substitution_ratio_industry	1	Numeric	NG power in / Total Fuel in
Methane_loss	1	Numeric	Methane out / NG in
ATS	N/A	Category	After-treatment system, Y- after DOC; N- before DOC or DOC is now applied during test
Emission_GHG	CO_2e kg/kWh	Numeric	GHG emission includes CO_2, CH_4 (GWP=25) and N_2O (GWP=298)
Emission_NMHC_NOx	g/kWh	Numeric	NMHC emissions + NOx emissions
Emission_CO	g/kWh	Numeric	CO emission

Researchers may conduct field testing based on protocols designed to capture specific information, but a typical test design is described here (Meng et al., 2017; Wijesinghe et al., 2018). Briefly, the tests of engine emissions and performance were conducted for the diesel baseline and natural gas-diesel dual fuel operation. Natural gas and diesel flow rates in parallel along with the collection of engine control unit data were continuously measured. Simultaneously, emissions were measured by the Fourier-transform infrared spectroscopy (FTIR) gas analyzer. Usually, data at different engine loads and rpms were automatically recorded. We harmonized the calculations among difference sources and summarize significant features as shown in Table 1.

We split 80% of the data to train the model and used the rest 20% to evaluate the performance, which is considered one experiment. We used Mean Absolute Relative Error (MARE) and Root Mean Squared Error (RMSE) to show the relative (percentage) and absolute differences between the predict values and actual test values. We conducted 1,000 random experiments and showed the average of MARE and RMSE. We mainly discussed the relative error in our result section, though we calculated both MARE and RMSE.

$$MARE = \frac{\sum \left| \frac{Predicted\ value\ - Test\ value}{Test\ value} \right|}{Number\ of\ test\ value}$$

$$RMSE = \sqrt{\frac{\sum (Predicted\ value\ - Test\ value)^2}{Number\ of\ test\ value}}$$

We predicted fuel efficiency, diesel displacement, natural gas (NG) substitution ratio, emissions of greenhouse gas (GHG), non-methane hydrocarbon (NMHC) plus nitrogen oxides (NOx) and carbon monoxide (CO). GHG is the sum of CO2, CH4 and N2O, after weighted by their Global warming potentials (GWPs). Notably, if some features could be directly used to calculate the predicted value, these features were not all included in the prediction of that value. For example, the fuel efficiency could be directly calculated by the method of dividing power output by fuel input, so fuel input and power output were not both included to predict the fuel efficiency.

RESULTS

We conducted the initial exploration of the data beginning with some summary and descriptive. Summary statistics for mean, minimum, maximum, median and standard deviation were calculated for numeric columns as shown in Table 2.

TABLE 2. Summary Statistics

Feature	Unit	Count	Mean	STD	Min	25%	50%	75%	Max
Rated_speed_rpm	RPM	114	1514	297	1200	1200	1500	1800	1950
Rated_power_kw	kW	114	1277	220	1050	1101	1133	1382	1678
Power_kw	kW	114	766	272	210	590	704	951	1423
NG_LHV_btupercf	Btu/cf	114	979	39	943	943	982	1031	1055
Fuel_consump_DLEperkwh	DLE/kWh	73	0.418	0.044	0.305	0.397	0.422	0.445	0.551
Diesel_consump_DLEperkwh	DLE/kWh	73	0.125	0.036	0.090	0.101	0.113	0.138	0.243
NG_consump_DLEperkwh	DLE/kWh	73	0.292	0.055	0.168	0.264	0.301	0.330	0.423
Engine_load	1	114	0.56	0.15	0.20	0.45	0.55	0.65	0.85
Fuel_efficiency_ZECE	1	114	0.266	0.021	0.202	0.254	0.268	0.275	0.331
Diesel_disp	1	73	0.604	0.096	0.400	0.540	0.637	0.688	0.728
Substitution_ratio_corrected	1	73	0.547	0.058	0.386	0.526	0.562	0.585	0.660
Substitution_ratio_industry	1	73	0.698	0.090	0.489	0.663	0.731	0.765	0.790
Methane_loss	1	73	0.213	0.062	0.060	0.186	0.226	0.257	0.309
Emission_GHG	CO_2e kg/kWh	73	1.880	0.453	0.837	1.641	1.956	2.176	3.082
Emission_NMHC_NOx	g/kWh	73	7.763	3.450	2.395	5.496	6.596	9.612	17.549
Emission_CO	g/kWh	73	12.417	12.200	0.050	0.197	13.916	25.384	32.298

We explored how the activity (hydraulic fracturing or drilling) could affect the fuel efficiency, diesel displacement and substitution ratio of dual fuel engines in Figure 1. When the engines do the hydraulic fracturing tasks they slightly reduce their fuel efficiency but have higher diesel displacement and substitution ratio. These findings of diesel displacement and substitution ratio indicate that dual fuel engines will use relatively more natural gas when they conduct hydraulic fracturing than drilling. This explains the slightly lower fuel efficiency of the dual fuel engines for hydraulic fracturing. The engines certainly have methane loss out of the cylinders and consequently reduce the overall fuel efficiency.

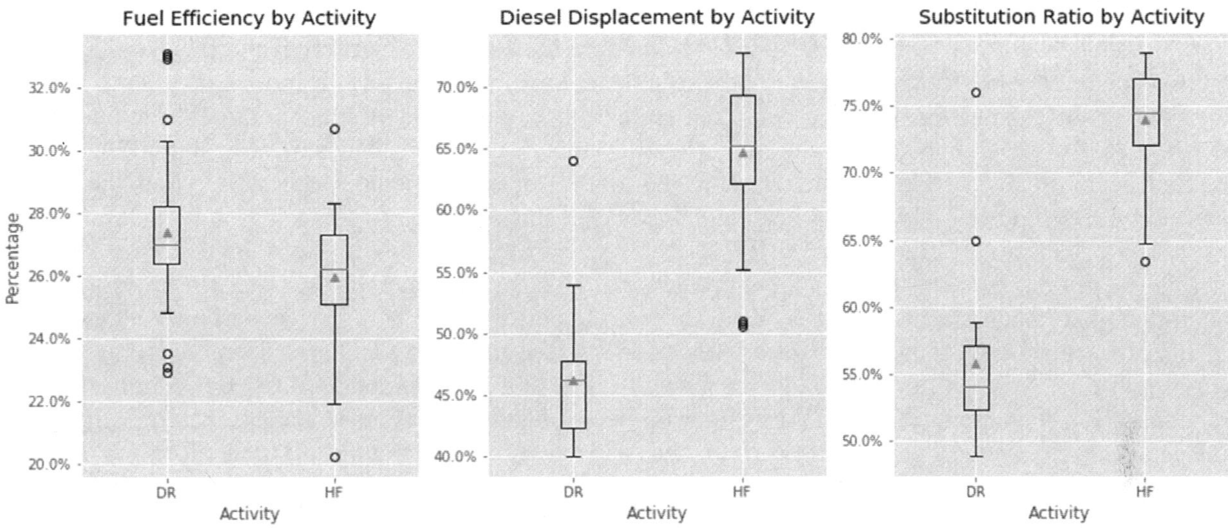

FIGURE 1. Efficiency and Substitution Ratio by Activity

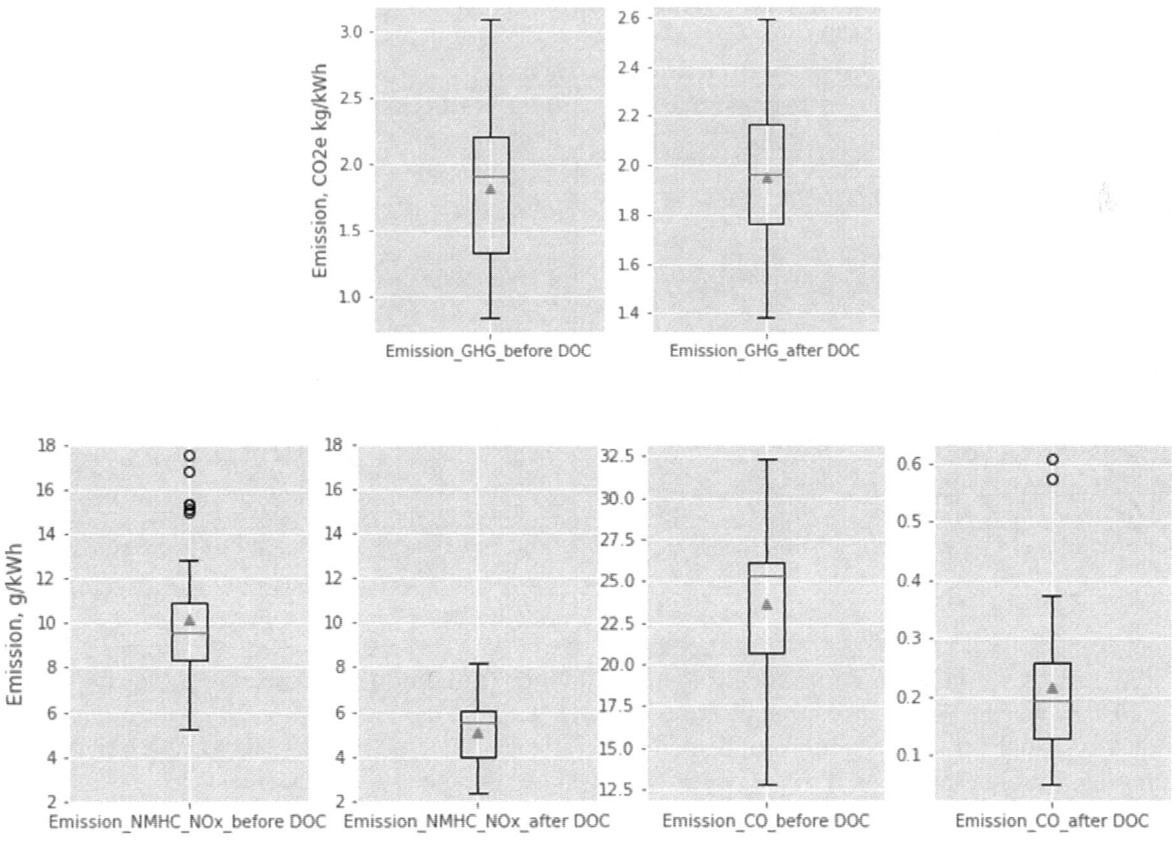

FIGURE 2. The Effects of ATS to Emissions

We explored how after-treatment system (ATS) could affect the emissions from the dual fuel engine. The ATS here mainly refers to diesel oxidation catalyst (DOC). ATS does not significantly affect GHG emission, but will largely reduce NMHC + NOx and CO emissions (Figure 2). This is because NMHC and CO will be oxidized to CO_2 by DOC. However, since the produced CO_2 takes much less portion than the CO_2 in the exhaust, the GHG emissions will not change much.

Table 3 shows a 4.3% MARE of predicted fuel efficiency that we could obtain from our machine learning model. The range of MAREs from 1,000 random tests is approximately from 2.3% to 6.5%. We need to include more observations and filter the trouble makers in high-error predictions in order to narrow down the distribution and have a more confident prediction. The RMSEs of predicted fuel efficiency vary between 0.75% to 2.5%, considering the fuel efficiency shown in Table 2 has a mean value of 26.6% and a median value of 26.8%. The features used in the model for fuel efficiency are activity, engine make, model, RPM, rated power, power, NG heating value and engine loading.

Table 3 shows the prediction MARE by machine learning mostly vary from 3.9 to 10.4% with the selection of optimized features. Notably, prediction of CO with ATS has a 21.5% MARE. It was challenging to predict CO emission after ATS, because the performance of ATS was hard to evaluate. That is, if the efficiency of ATS increases from 98.5% to 99%, the remaining CO will decrease from 1.5% to 1%. A slight fluctuation of the ATS performance will result in a 50% reduction in CO emission. The ATS performance is tightly related to the features such as operation temperature, the catalysts materials, life time of the catalysts, etc. Including those features can help with better understanding and prediction. It is inferred that this is the case in the future when we evaluate emissions which ATS can effectively reduce. Interestingly, we can predict CO emissions without ATS with a 4.8% MARE.

Predicting NMHC+NOx is also a challenge. The components are too complicated for the instruments to collect and measure. And many of these components are intermittent and are tightly related to operation from time to time, even if our analysis is based on stationary scenarios.

TABLE 3. Prediction Performance Summary

Parameters	Unit	Representative Value*	Best Prediction MARE**	Best Prediction RMSE***
Fuel efficiency	1	0.266	4.3%	0.015
Diesel displacement	1	0.604	7.2%	0.055
Substitution ratio	1	0.698	4.8%	0.044
GHG emission	CO_2e kg/kWh	1.880	3.9%	0.095
NMHC+NOx w/o ATS	g/kWh	10.194	10.4%	1.15
NMHC+NOx w/ ATS	g/kWh	5.123	10.4%	0.61
CO w/o ATS	g/kWh	23.65	4.8%	1.36
CO w/ ATS	g/kWh	0.217	21.4%	0.063

*Mean value in Summary Statistics; **MARE: Mean Absolute Relative Error; ***RMSE: Root Mean Squared Error

We came up with the framework (Figure 3) to show how the LCA researchers can utilize data science to estimate the emissions output of life cycle inventory for a dual fuel process. We can predict the emissions by collecting or assuming the operation parameters such as activity, RPM, loading and fuel consumption. In some scenarios, the researchers can even predict the fuel displacement.

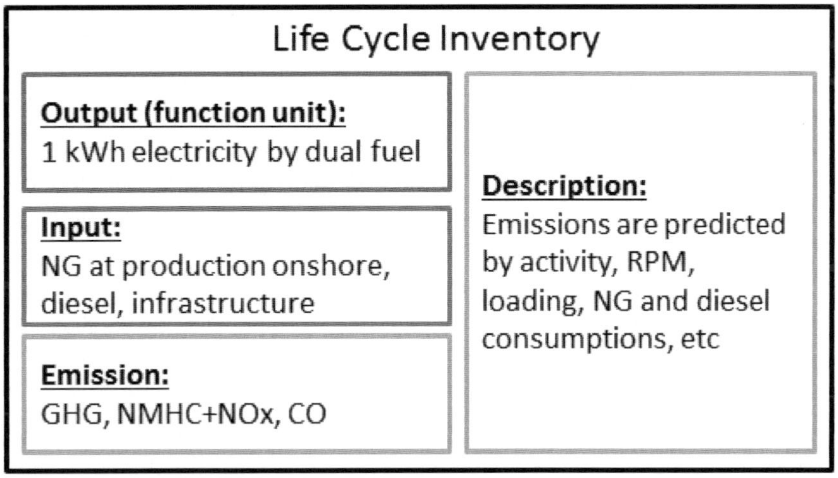

FIGURE 3. Apply Data Science to Life Cycle Inventory of Dual Fuel Technology

We came up with the framework (Figure 3) to show how the LCA researchers can utilize data science to estimate the emissions output of life cycle inventory for a dual fuel process. We can predict the emissions by collecting or assuming the operation parameters such as activity, RPM, loading and fuel consumption. In some scenarios, the researchers can even predict the fuel displacement.

CONCLUSION

This proof of concept practice demonstrates the powerful ability and potentiality to predict both value and uncertainty of life cycle inventory of dual fuel technology. The average relative errors between predicted values and actual values mostly range from 3.9% to 10.4% based on available 114 observations. The predicted error increases involving unknown factors, such as the performance of ATS. Collecting various data sets from more sources and optimizing feature selection can further improve the results. We expect to apply data science for confident decisions on adopting dual fuel technology and many other innovative technologies. It requires the joint efforts from industrial and academic partners to be able to complete the full cycle research.

ACKNOWLEDGEMENTS

The authors would like to thank the Program Activity Funding from the Houston Advanced Research Center (HARC) and Environmentally Friendly Drilling (EFD) Program.

REFERENCE

Arciero, B. V., & Ismail, H. 2017. *Applying Data Management and Visualization to OPEX Reduction and Improved Asset Integrity Programs*. SPE Annual Technical Conference and Exhibition, San Antonio, Texas, USA.

Ball, K., Arbus, T., Odi, U., & Sneed, J. 2017. *The Rise of the Machines, Analytics, and the Digital Oilfield: Artificial Intelligence in the Age of Machine Learning and Cognitive Analytics*. SPE/AAPG/SEG Unconventional Resources Technology Conference, Austin, Texas, USA.

Busby, D., Pivot, F., & Tadjer, A. 2017. *Use of Data Analytics to Improve Well Placement Optimization Under Uncertainty*. Abu Dhabi International Petroleum Exhibition & Conference, Abu Dhabi, UAE.

Caldwell, B., & Hinton, J. 2015. *Data Drilling: Changing the Way the Oil and Gas Industry Manages Safety and Risk*. SPE E&P Health, Safety, Security and Environmental Conference-Americas, Denver, Colorado, USA.

Cao, D., Loesel, C., & Paranji, S. 2018. *Rapid Development of Real-Time Drilling Analytics System*. IADC/SPE Drilling Conference and Exhibition, Fort Worth, Texas, USA.

Cashman, S. A., Meyer, D. E., Edelen, A. N., Ingwersen, W. W., Abraham, J. P., Barrett, W. M., . . . Smith, R. L. 2016. Mining Available Data from the United States Environmental Protection Agency to Support Rapid Life Cycle Inventory Modeling of Chemical Manufacturing. *Environmental Science \& Technology, 50*(17), 9013--9025.

Houston Advanced Research Center (HARC). 2016. Coastal Impacts Technology Program - Final Report.

Dursun, S., Duman, K., Tuna, T., Abbas, M., & Ding, J. 2014. *A Workflow for Intelligent Data-driven Analytics Software Development in Oil and Gas Industry*. Paper presented at the SPE Annual Technical Conference and Exhibition, Amsterdam, The Netherlands.

Fulford, D. S., Bowie, B., Berry, M. E., Bowen, B., & Turk, D. W. 2016. Machine Learning as a Reliable Technology for Evaluating Time/Rate Performance of Unconventional Wells. SPE Annual Technical Conference and Exhibition, 28-30 September, Houston, Texas, USA

Johnson, D., Heltzel, R., Nix, A., Clark, N., & Darzi, M. 2017. Regulated Gaseous Emissions from In-use High Horsepower Drilling and Hydraulic Fracturing Engines. *Journal of Pollution Effects \& Control, 5*(2), 187.

Johnson, D. R., Heltzel, R., Nix, A. C., Clark, N., & Darzi, M. 2017. Greenhouse gas emissions and fuel efficiency of in-use high horsepower diesel, dual fuel, and natural gas engines for unconventional well development. *Applied Energy, 206*, 739--750.

Kale, A., Zhang, D., David, A., Heuermann-Kuehn, L., & Fanini, O. 2015. *Methodology for Optimizing Operational Performance and Life Management of Drilling Systems Using Real Time Data and Predictive Analytics*. SPE Digital Energy Conference and Exhibition, The Woodlands, Texas, USA.

Meng, F., Wijesinghe, A., Colvin, J., Lafleur, C., & Haut, R. 2017. *Conversion of Exhaust Gases from Dual-Fuel (Natural Gas-Diesel) Engine under Ni-Co-Cu/ZSM-5 Catalysts*. SAE Technical Papers.

Mishra, S., & Lin, L. 2017. *Application of Data Analytics for Production Optimization in Unconventional Reservoirs: A Critical Review*. SPE/AAPG/SEG Unconventional Resources Technology Conference, Austin, Texas, USA.

Mohaghegh, S. D., Gaskari, R., & Maysami, M. 2017. *Shale Analytics: Making Production and Operational Decisions Based on Facts: A Case Study in Marcellus Shale*. SPE Hydraulic Fracturing Technology Conference and Exhibition, The Woodlands, Texas, USA.

Pankaj, P., Geetan, S., MacDonald, R., Shukla, P., Sharma, A., Menasria, S., & Judd, T. 2018a. *Need for Speed: Data Analytics Coupled to Reservoir Characterization Fast Tracks Well Completion Optimization*. SPE Canada Unconventional Resources Conference, Calgary, Alberta, Canada.

Pankaj, P., Geetan, S., MacDonald, R., Shukla, P., Sharma, A., Menasria, S., Judd, T. 2018b. *Application of Data Science and Machine Learning for Well Completion Optimization*. Offshore Technology Conference, Houston, Texas, USA.

Pennel, M., Hsiung, J., & Putcha, V. B. 2018. *Detecting Failures and Optimizing Performance in Artificial Lift Using Machine Learning Models*. SPE Western Regional Meeting, Garden Grove, California, USA.

Thajudeen, S. S. 2018. *Advanced Analytics Solutions: Towards a Data Driven Organisation*. Offshore Technology Conference Asia, Kuala Lumpur, Malaysia.

Wei, L., & Geng, P. 2016. A review on natural gas/diesel dual fuel combustion, emissions and performance. *Fuel Processing Technology, 142*, 264--278.

Wijesinghe, A., LaFleur, C., Meng, F., Colvin, J., & Haut, R. 2018. *Fuel Economy and Emission Characteristics of a High Horsepower Natural Gas/Diesel Dual-Fuel Engine in Oil & Gas Operations*. IADC/SPE Drilling Conference and Exhibition, Fort Worth, Texas, USA.

Xu, M., Cai, H., & Liang, S. 2015. Big Data and Industrial Ecology. *Journal of Industrial Ecology, 19*(2, SI), 205--210.

ENVIRONMENTAL ANALYSIS AND MEASUREMENTS

Rapid Identification and Determination of Trace Polycyclic Aromatic Hydrocarbons in Water by Nylon 6 Nanofiber Membrane Enrichment/ Solid-Phase Surface Spectrofluorimetry

Zhi-gang Wang, Jian Chen
(College of Environmental Science and Engineering, Yangzhou University, Yangzhou, Jiangsu, China 225127)

Using nylon 6 nanofiber membrane prepared by self-made electrospinning device as absorbing material, a method for rapid determination of trace polycyclic aromatic hydrocarbons (PAHs) in water by solid-phase surface spectrofluorimetry has been established. Nylon 6 nanofiber membranes with a diameter of 5cm were used for the filtration of three kinds of PAHs (phenanthrene, pyrene, fluoranthene) solution, then the membranes were moved to the variable angle powder sample pool after air-drying, and three dimensional excitation-emission solid-phase surface fluorescence spectra of PAHs which was absorbed on the membrane was determined using fluorescence spectrophotometer. According to the three dimensional excitation-emission solid-phase surface fluorescence spectra, the excitation/emission wavelength pair with the maximum fluorescence intensity was found out and the linear relationship between fluorescence intensity and the initial concentration of each PAHs in water were investigated. The results show that: the excitation/emission wavelength pairs with the maximum fluorescence intensity located at Ex_{255nm}/Em_{368nm}, Ex_{340nm}/Em_{376nm} and Ex_{290nm}/Em_{437nm} for phenanthrene, pyrene, and fluoranthene respectively. In the situation of 500mL sample solution for suction filtration, the calibration curves between the fluorescence intensity and the initial concentration for three kinds of PAHs were listed as follow: y=9432.4x+261.1 with linear concentration range of 5～500 ng/mL for phenanthrene; y=753480x+805.51 with linear concentration range of 0.2～10 ng/mL for pyrene; and y= 9946.06x+603.48 with linear concentration range of 10～400 ng/mL for fluoranthene. The detection limits were 0.973ng/mL for phenanthrene, 0.0162 ng/mL for pyrene and 0.0896ng/mL for fluoranthene. For phenanthrene standard solution of 100 ng/mL, pyrene standard solution of 10 ng/mL and fluoranthene standard solution of 50 ng/mL, the relative standard deviations(RSD) for seven time determinations were 7.0%, 2.6% and 5.1% respectively, and relative errors of mean values were 1%、2% and -0.2% respectively. For tap water containing 50 ng/mL, 200 ng/mL and 500 ng/mL phenanthrene, the average recoveries were 87.2%～98.2%; for tap water containing 0.5 ng/mL, 2 ng/mL and 10 ng/mL pyrene, the average recoveries were 101%～120% and for tap water containing 10 ng/mL, 50 ng/mL and 100 ng/mL fluoranthene, the average recoveries were 85.8%～92.3%. Therefore, the method has the advantages of simplicity, economy, high sensitivity, which is suitable for the rapid identification and detection of trace polycyclic aromatic hydrocarbons in water.

On-Line Monitoring of Heavy Metals in Industrial Wasterwater by LIBS Technique

Nanjing Zhao, Yao Jia, Li Fang, Mingjun Ma and Deshuo Meng
(Anhui Institute of Optics and Fine Mechanics, Chinese Academy of Sciences, Hefei, China)

Heavy metal pollution from industrial wastewater is an important source. Understanding of the emission status timely and control of pollution sources are the key to current governance. A method of heavy metals determination in industrial wastewater based on laser induced breakdown spectroscopy (LIBS) technique was studied，the on-line system with graphite enrichment automatically and plasma spatial confinement detection were developed and field demonstrated.

The main difficulties of LIBS in quantitative analysis include stability, repeatability and limit of detection (LOD), especially in liquids. Here, the graphite enrichment method is studied to solve the problems of water splashing, shorter life of plasma and weaker spectrum signal, and the hemispherical confinement device of laser plasma is designed for further enhance the spectral emission intensity and duration of laser plasma. The experimental results show that the stability and repeatability of the measurement spectrum were improved, the LOD of different elements (e.g. Pb, Ni, Cd, Cr, Zn and Cu) was reduced with the increase of liquid quantity and can be reached several ppb.

The automatic enrichment method suitable for continuous online work based on graphite substrate was also studied, including graphite substrate transfer mode, liquid filling and drying, and the automatic enrichment device was designed. An on-line monitoring system of heavy metals determination for industrial wastewater with automatic sample enrichment and fast spectral measurement was designed, realizing on-line automatic measurement of heavy metals in industrial wastewater, and a demonstration operation was carried out in a copper smelting enterprise in Tongling, China.

Influence of Physical Properties and Constitutions of Soil on the Laser Induced Fluorescence Spectra of PAHs in Soil

Deshuo Meng, Nanjing Zhao*, Yao Huang, Zhaolu Zuo, Mingjun Ma, Xiang Wang, Wenqing Liu and Jianguo Liu
(Anhui Institute of Optics and Fine Mechanics, Chinese Academy of Sciences, Hefei, China)

The PAHs in soil have brought a huge environmental risk. The laser induced fluorescence can realize the rapid and on-site detection of organic pollutants, while the physical and chemical properties of soil have a great impact on the LIF spectrum of the soil, which hasn't been revealed. Therefore, this paper studied the effect of soil solid and liquid components, soil moisture, porosity, etc. on the spectrum of PAHs in soil. The results showed that soil porosity, particle size did not affect the stability of soil fluorescence spectrum intensity. But the soil moisture had a much greater influence on soil fluorescence spectrum intensity. When the soil moisture was low, the fluorescence intensity was positively correlated to the humidity. The primary minerals and secondary minerals in the soil did not affect the soil fluorescence, but the type and concentration of humus in the soil had a great influence on the fluorescence intensity of the pollutants. The extreme difference analysis of orthogonal experiment showed that, the effect of humic acid is greater than that of fulvic acid. In addition, the contents of heavy metal elements in soil did not affect the spectral intensity of thorium. The results show that the effect of soil physical and chemical properties cannot be ignored when using laser induced fluorescence technology to detect organic pollutants in soil. This paper provides an important reference for the LIF measurement of PAHs in soil.

Occurrences, Sources, and Transport of Organophosphorus Flame Retardants in the Waters of Fildes Peninsula, Antarctica

Kaifeng Rao, Xiaozhong Gao, Yiping Xu, and Mei Ma
(Research Center for Eco-Environmental Sciences, Chinese Academy of Sciences, Beijing, China)

As a pristine continent, Antarctica provides a good opportunity to study the spatial transport and temporal accumulation of persistent toxic substances (PTS) and the impacts of anthropogenic activities, both of which have given rise to ongoing public concern. In this research, an approach of coupling aquatic time-integrated passive sampling with chemical analysis and bioassays was used to assess organophosphorus flame retardants (PFRs) pollution in the Antarctic. Passive samplers (triolein-embedded cellulose acetate membranes, TECAMs) were deployed in lake and marine waters of Fildes Peninsula, Antarctica, for an average of 23 days. Chemical analyses of eight hydrophobic PFRs were conducted using extracts obtained by passive sampling. In total, six PFRs (Tris(1,3-dichloroisopropyl)phosphate, tris(phenyl) phosphate, 2-ethylhexyl diphenyl phosphate, tris(2-ethylhexyl)phosphate, tri-m-cresyl phosphate, tri-p-cresyl phosphate) were quantified in Antarctic waters with the total free dissolved concentrations ranged from 0.6 ng/L to 36.5 ng/L. The sources and transport pathways of PFRs and PAEs in Fildes Peninsula were studied, and multiple local sources (wastewater, air traffic, research stations, and animal feces) for different PFRs were proposed. Furthermore, a comparison among a variety of Antarctic water sampling methods revealed that passive sampling can be a tool for aquatic time-integrated investigations in polar regions.

Water Biosensor Challenge to Address Toxicity of Water

Eunice Varughese, Sid Hunter, Denice Shaw, Jay Garland
(U.S. Environmental Protection Agency, Office of Research & Development)

An ongoing concern for water treatment systems and resource managers is the need to monitor for the presence of increasing number of pollutants from agricultural, municipal, and industrial outfalls that are present in U.S. source waters. The associated environmental compounds can include pesticides, heavy metals, personal care products, natural toxins such as cyanobacterial toxins, and a host of other organic and inorganic chemical pollutants and their transformation products. Current methods for detecting and identifying many of these contaminants are expensive, time-consuming, and require the use of specialized laboratories. Moreover, multiple methods are needed to detect the variety of contaminants of interest. In addition, water monitoring can be even more complex in the presence of new or emerging contaminants. A chemically "agnostic" approach to water quality testing could allow for detection of multiple contaminants that are biologically active and trigger specific toxicity or adverse health outcome pathways. Biosensors, (i.e. those sensors which take advantage of biological phenomenon that are altered in the presence of contaminants), can potentially provide faster detection as well as portability, continuous monitoring, and/or detection in complex matrices using minimal sample preparation. Over the past years, the EPA has used a challenge approach that uses crowd-sourcing to find solutions to difficult problems. In particular, the Water Biosensor Challenge has the goal of producing design solutions for a biologically-based effects monitor/biosensor capable of responding to multiple environmental contaminant exposures that result in toxicity or adverse health effects when host organisms are exposed. This poster will describe the challenge process as well as provide information related to the challenge of designing a Water Biosensor to analyze the toxicity of water.

Design of Chlorophyll Fluorescence Detection System based on Pseudo-random Sequence Modulation

Chaoyi Shi, Xianhe Gao, Jun Lu, and Yanhong Gu
(Hefei University, Hefei, Anhui, China)

A fluorescence detection system based on pseudo-random sequence modulation was designed for the detection of weak chlorophyll fluorescence of phytoplankton. Based on the analysis of the noise characteristics of fluorescence detection circuit, an I/V conversion circuit employing T-type feedback network and a secondary amplification circuit were designed. A fluorescence signal demodulation circuit was designed using AD630, according to the pseudo-random sequence demodulation principle. The system was applied on the chlorophyll fluorescence measurement of *chlorella pyrenoidosa*. The results indicated that the signal-to-noise ratio reached 22dB for 10 μg/L chlorophyll concentration, and the linear correlation coefficient between the output voltage of the detection system and the chlorophyll concentration was 0.995 in a chlorophyll concentration range of 5~30 μg/L.

The Influence of Urban Ecology on the Native Flora

Mohamed A. El-Sheikh[1] and *Amro M. El-Sheikh[2*]*
([1]King Saud University, College of Science, Botany & Microbiology, Riyadh, Saudi Arabia);
([2*]Tanta University, Faculty of Engineering, electronic & Communication Department, Tanta, Egypt)

The construction and use of tracks, roads, canals, railways and air ports have involved many changes, some of them are direct and others are indirect. Direct influences include the destruction of the existing habitats and the provision of new ones, which have special characteristics. These have provided more or less continuous stretches of open habitats extending for hundreds of miles and forming a nation-wide network, with opportunities for rapid colonization and spread. The advantage taken of these networks by a few alien species is well known but no serious attempts have been made to study their effect on "native" plants. The indirect influence of transport is too big as the opening up of new facilities inevitably promotes economic developments of great importance to botanists. Just as the construction of the railways which opened up the Prairie Provinces in Canada, so transport improvements have been followed by agricultural and commercial exploitation of these areas. These changes in agriculture carried, many foreign seeds, about the country to grow in new places. The transport of building materials has facilitated the destruction of vegetation by the construction of towns and factories in some places with materials dug out of the ground in others.

SOCIETY AND THE ENVIRONMENT

Environmental and Social Sustainability

Sukhmander Singh
(Santa Clara University, Santa Clara, CA, USA)

Sustainable development should include both environmentally and socially sustainable solutions. This is so because environment, society and technology are interlocking parameters. Society's economic system influences the production systems that are appropriate to that society. The production system in turn impacts the ecosystem. It is now well recognized that the production system with an ever increasing per capita consumption can lead to the depletion of resources as well as to the pollution of the ecosystem; and hence the importance of sustainable solutions. Since the production is often motivated by profit, which can be exploited by the rich, even in the use of a sustainable solution, there is a need for a sustainable solution to be both environmentally and socially just. A model is presented for addressing sustainability in society. It is argued that both environmentally and socially sustainable solutions can be achieved by incorporating locally available renewable or recycling material with the use of local labor. Such an approach would help set up employment base for local people who otherwise may be denied such an opportunity should a large scale system/entrepreneurship is allowed to employ an approach of mass production versus production by masses. In some poor countries where poverty, hunger and poor sanitation exist, environmentally acceptable, economically accessible and socially sustainable solutions must be found to bring about a developed and socially just society. These solutions ought to be simple, inexpensive and environmentally safe. Examples of such solutions are presented in the paper. It is recommended that affordable technologies based on the concept of sustainability be developed to help marginal communities. The paper points out that there's a need for interference between technology and society while developing technology based sustainable solutions.

Migration, Climate Change and Livelihood Crisis in Fiji: A Historical and Holistic Perspective

DMumtaz Alam and Dr. S. N. Shashtri
(CHE and CEST, Fiji National University, Natabua Campus, Fiji)

Migration of the people in the history of Archipelago islands is not a new thing in terms of time and space. The study area mainly considered the Viti Levu Island of Fiji($17^0 48^/$South and 178^0East) which is the biggest among the other islands. Recent migration tendency has been coupled with climate change and sea level rise in the low lying islands has been in question and in consideration globally. In the present paper study has been conducted with questionnaire distributed to the people who have migrated due to the various reasons. This questionnaire considered the factors from the period of the early colonization to the contemporary scenario. The factors which were put in consideration where sea-level rise and its implications in islands like Tuvalu and Kiribati which is manifested by the Ciguatoxin and salt water intrusion in the aquifers., better education and livelihood earnings, political instability in some of Islands like Papua New Guinea, Fiji and Solomon. Another factor which was considered is the natural disaster like Tsunami and Cyclones (Cyclone Pam 2015 in Vanuatu,) but found it out to be the temporary relocation. Natural resources available in the Fiji are exhausting due to the overburden and change in the land use pattern. Samoan Pathway has been proven effective in combating this crisis due to climate change and sea level rise with the sustainable development. The study has concluded that holistic development of the bigger islands which are not at the threat of sea level rise monitored by the high resolution satellite sensors should be focused. Institutional development and unemployment has to be checked simultaneously to stop the migrating tendency of the islanders and climate change and sea level rise has affected the low lying islands badly from where relocation is indispensible.

Water Environment Protection in the Farmhouse Tourism Village—A Case Study on Shiling Village, Anji County

Ancheng Luo, Yanping Sun, Zhiwei Liang
(Zhejiang University, Hangzhou, P R China)

A case study was conducted in Shiling Village Baofu Town Anji County, a typical mountainous village where the main economic source for the residents is from farmhouse tourism. According to the statistics in 2014, the annual tourists of Shiling Village had amounted to 15,000 trips. The business income of farmhouse tourism had reached 3,500,000 yuan, accounting for more than 80% of the total income. Proportionally according to the farmhouse business types, the average annual income of each bed was about 15,900 yuan. The water environment impact from farmhouse tourism in Shiling Village was also assignable. During the two month's tourist season in July and August, the emission of COD amounted to 5136kg, while the ammonia emission upped to 400kg. The monitoring carried out in October 2014 showed that the river water quality did not meet the standard level, indicating that pollutants discharged from farmhouse tourism to the river exceeded the capacity of the river. Based on the analysis of present situation of the local water quality and the contribution of farmhouse tourism to the local economy and water pollutant emission, countermeasures for environment protection was proposed to provide guidance and technical support for the local economic development and environment protection. Hopefully, these results might also provide a demonstration for other similar situations.

Access to Environmental Benefits: Interest Driven or Need Driven?

Ajay Karki
(Ministry of Forests and Soil Conservation, Department of Forests, Manang, Nepal)

This study is based on the data from 45 community-forestry user groups including 1111 households, from three mid-hill districts, namely Kaski, Gorkha and Baglung of western Nepal. Basic objective of this research is to know the relation of environment (forest) with the people and how different group of people access the environmental resources. Being specific, the aim of the study is to determine the level of access to different forest products (firewood, fodder and timber) among different groups of users. The influence of different variables influencing on access to forest products is also investigated. The users are differentiated based on social (caste), political (executive committee membership, political elite) and socioeconomic categories (wellbeing ranks). Pairwise test was done to understand, if there is any significant difference between these groups. Logistic regression was run to know the relationship of different independent variables on access to forest products. Statistical analysis reveals very few statistically significant differences in access to benefits between households when grouped on the basis of caste group, which is followed by wellbeing groups. This statistical difference is found in the case of less lucrative benefits like firewood and fodder. However, political status and connectedness, namely membership of executive committee is significantly and positively associated with more lucrative benefits namely timber. Female-headed households are always negatively (and significantly in some benefits) associated with all benefits. Thus, this study indicates that there is a need in paradigm shift of studies, priorities, policies from caste and wealth based analysis to political status notably presence in Executive Committee and connectedness to the decision-making bodies.

SUBSTAINABLE DEVELOPMENT AND ENVIRONMENTAL MANAGEMENT

Demonstration of Enhanced Net Anthropogenic Nitrogen Inputs (Nani) Model for Estimating Commodity-Specific Agricultural Nitrogen Flows

Mikaela Algren, Amy Landis, and Pragnya Eranki
(Colorado School of Mines, Golden, CO, USA)
Chris Costello
(University of Missouri, Columbia, MO, USA)

ABSTRACT: To ensure that food and renewable fuel needs can be met into the future, new protocols for management of nitrogen (N), especially with regard to its use in agricultural fertilizer, must be developed. Application of fertilizer N in excess of plant consumption has a wide range of environmental impacts, such as soil and stream acidification, air pollution, contribution to global warming through greenhouse gas (GHG) emissions, drinking water pollution and eutrophication in water bodies. The Net Anthropogenic N Inputs (NANI) model developed for the United States (US) at Cornell University was modified in 2015 to track commodity-specific N inputs from 2002 and has now been modified to consider the growing role of the ethanol coproducts in animal feed in order to quickly generate national life cycle (per unit) metrics. The model's potential for application in life cycle assessment (LCA) studies and ecofootprinting are demonstrated in this work as NANI is used to examine the growth of the ethanol industry and the corresponding shifts in per unit fertilizer and fixation N inputs to corn, ethanol, and ethanol feed coproducts in 1997, 2002, 2007, and 2012, corresponding with pre- and post-corn ethanol market development.

INTRODUCTION

From 1997 to 2012, the area planted with corn in the U.S increased by over 50%, and the fraction of U.S. corn used in ethanol production increased from 2 to 37%. Data from the United States Department of Agriculture (USDA) Economic Research Service (ERS), shows that ethanol production in the United States has increased tenfold since 1999 (USDA 2018), with corn ethanol supplying 88-96% of total ethanol production from 2011 to 2017 (USDA 2018). U.S. corn crops are fertilized with N at an average rate of 16,252 kg per square km (USDA 2018). However, about 30-50% of this fertilizer N is lost to the environment (USDA 2018). Increased rates of reactive nitrogen (N) and phosphorus (P) releases to the environment associated with agriculture have significantly disrupted environmental N and P balances (Steffen et al. 2015). N_2O, a potent greenhouse gas and ozone-depleting substance, is released as reactive forms of N are converted back to non-reactive N (Ravishankara et al. 2009, Elgood et al. 2010). Increased concentrations of N and P in waterbodies can lead to eutrophication and hypoxia, damaging oxygen sensitive ecosystems, and nitrate (NO_3^-) that enters groundwater can pollute drinking water sources (Fan et al. 1996, Costello 2009).

The two major ethanol production processes are dry milling and wet milling. In dry milling, which is responsible for over 90% of U.S. corn ethanol production (USDA 2018). the grain mass remaining after starch fermentation becomes distillers grains and solubles (DGS). DGS is increasingly used in concentrated livestock feeds. Wet milling also produces two concentrated feed coproducts, corn gluten feed (CGF) and corn gluten meal (CGM). For each liter of corn ethanol, which requires 2.69 kg corn grain (Pimentel et al. 2005), approximately 0.73 kg of DGS or 0.59 kg CGF and 0.10 kg of CGM is produced (Bremer et al. 2010, O'Brien et al. 2010).

Life cycle assessment (LCA) has been the primary tool applied to compare the environmental performances of ethanol and gasoline (Heath et al. 2009); in fact, LCA was specified in the Energy Independence and Security Act of 2007 as the tool required to ensure that specified greenhouse gas reductions occurring with the substitution of renewable fuels for gasoline are met (2 H.R.6 § 202(A)).

Accounting for crop production shifts that occur as ethanol feed coproducts offset production of other feed crops is critical in demonstrating reduced GHG impacts of corn-based ethanol when compared to those of gasoline (Farrell et al. 2006, Bremer et al. 2010). In LCA, when more than one useful product results from a production system, the impacts are allocated between the coproducts, in this case ethanol and feed coproducts, i.e., DGS or CGF and CGM. The increase in production of ethanol and corresponding ethanol feed coproducts also impacts livestock feed mixes, and, therefore, N inputs required to produce animal food commodities like beef. This could result in a reduction in the per unit impacts associated with an animal food product, while the overall system experiences an increase in reactive nitrogen inputs. In this work, we calculate fertilizer and fixation nutrient inputs to agricultural commodities to track changes over time in N inputs per unit of beef, corn, ethanol, and ethanol feed coproducts. The phrase "N inputs," in this work will always refer to lifecycle inputs of N via fertilizer application or N fixation. For example, N input to livestock is greater than N consumed, because not all N fixed or applied in fertilizer is taken up by the crops consumed by livestock.

The 2002 version of the Net Anthropogenic Nitrogen Inputs (NANI) model was transferred from an Excel Visual Basic format to a series of matrices that are manipulated within MATLAB by Costello et al. (2015). The NANI and Net Anthropogenic Phosphorus Inputs (NAPI) Toolboxes (Swaney 2018) calculate human nutrient inputs to U.S counties as the sum of atmospheric N deposition, fertilizer N and P application, N fixation, and net imports of food and feed N and P. Further, NANI was modified to link crop production, including ethanol and ethanol feed coproducts, to livestock production and calculate crop-specific fertilizer N inputs (Costello et al. 2015). Both of these changes enable tracking of commodity-specific nutrient inputs. Additional improvements are made herein to easily load data for additional study years, and an analysis is completed to track the N inputs to corn, ethanol, and beef production per unit of product in years 1997, 2002, 2007, and 2012, from the beginning of the ethanol boom to the most recently updated agricultural year available for the NANI toolbox.

METHODS

The NANI toolbox allocates commodity crops (**TABLE 1**) and their N inputs on a fractional basis to humans and animals (**TABLE 2**), tracks agricultural and non-agricultural fertilizer N application at the U.S. county scale, and applies average fixation rates for N fixing crops to determine the reactive N input to the land. For this study of commodity-specific impacts, modifications to the NANI toolbox were made to allocate specific crops and feed products to specific animals according to animal N consumption requirements, forage to feed ratios, and crop quantities allocated to animals in each data year. This allocation method, implemented previously for 2002 data only, is described in Costello et al. (2015) and enables commodity-specific nutrient input calculations. The modified NANI toolbox also uses average fertilizer application rates for each crop to enable crop-specific quantification of N inputs to the land.

Ethanol feed coproducts, which have contributed with increasing significance to the U.S.'s livestock feed mix, have been incorporated into the modified NANI model by creating three new "crop" categories for DGS, CGF and CGM (17-19 in **TABLE 1**). The crop and feed mix was adjusted by first subtracting from corn grain the mass of "corn for fuel ethanol," as reported in the USDA ERS feed yearbooks (USDA 2018) to produce values for corn grain with ethanol use removed (1 in **TABLE 1**). DGS, CGF, and CGM quantities were then calculated using **Eqs. (1)-(3)**:

$$T_{\mathrm{DGS,y}} = T_{\mathrm{corn,y}} \times P_{\mathrm{dm,y}} \times R_{\mathrm{DGS/corn}} \quad (1)$$
$$T_{\mathrm{CGF,y}} = T_{\mathrm{corn,y}} \times P_{\mathrm{wm,y}} \times R_{\mathrm{CGF/corn}} \quad (2)$$
$$T_{\mathrm{CGM,y}} = T_{\mathrm{corn,y}} \times P_{\mathrm{wm,y}} \times R_{\mathrm{CGM/corn}} \quad (3)$$

where y is the year, T is the total corn grain mass, DGS, CGF, or CGM produced, dependent on the subscript, and P is a proportion of corn utilized to produce ethanol allocated to dry milling (*dm*) or wet milling (*wm*). R is the proportion of DGS, CGF, or CGM produced per mass of corn processed, also dependent on the subscript.

TABLE 1. Production Quantities for the 20 U.S. Crop Commodities Tracked in the NANI Model. Production data comes from the USDA ERS (USDA 2018). Values for crops and ethanol coproducts are in million metric tons. Ethanol values are in billion liters. The commodities for which per unit impacts are examined here are grayed.

		1997	2002	2007	2012
0	Corn grain (total)	218	219	323	262
1	Corn grain (ethanol use removed)	207	198	255	160
2	Corn silage	80.2	88.1	94.6	103
3	Wheat	57.5	41.1	52.0	57.0
4	Oats	2.20	1.59	1.30	0.95
5	Barley	7.32	4.67	4.51	4.68
6	Sorghum grain	14.2	8.47	12.3	6.71
7	Sorghum silage	4.21	3.54	5.23	5.12
8	Potatoes	20.9	20.5	20.2	21.1
9	Rye	0.17	0.18	0.17	0.17
10	Alfalfa hay	59.8	62.2	59.3	44.9
11	Other hay	66.7	75.7	78.1	65.8
12	Soybeans	68.2	73.7	70.3	79.7
13	Cropland pasture	58.4	54.9	32.4	11.6
14	Noncropland pasture	193	193	198	200
15	Rice	8.27	9.54	9.01	9.08
16	Peanuts	1.53	1.42	1.68	3.02
17	Corn gluten feed (CGF)	1.27	1.72	1.34	0.35
18	Corn gluten meal (CGM)	0.23	0.31	0.24	0.06
19	Distillers grains and solubles (DGS)	1.34	3.57	17.5	28.2
20	Ethanol	5.55	8.10	24.7	50.0

TABLE 2. Populations For the 19 U.S. Livestock Types Tracked in the NANI Model and Total Edible Beef Mass. Data comes from the USDA ERS (USDA 2018). Livestock populations are in millions, and edible mass of beef is in billion kg. The groups that contribute to beef production in the model are grayed.

		1997	2002	2007	2012
0	Beef (Edible Mass)	12.2	12.1	12.0	10.9
1	Fattened Cattle	10.9	11.3	11.0	10.2
2	Milk Cows	9.09	9.10	9.26	9.25
3	Hogs For Breeding	6.83	6.17	6.29	5.71
4	Hogs For Slaughter	54.1	61.8	69.5	67.5
5	Chicken Layers	313	334	349	350
6	Breeding Turkeys	5.50	4.65	5.36	5.04
7	Chicken Pullets	80.7	85.3	86.9	92.7
8	Chicken Broilers	1120	1410	1510	1430
9	Slaughter Turkeys	125	113	123	117
10	Beef Breeding Herd	35.7	35.0	34.4	30.3
11	Beef Calves	11.5	11.4	11.2	9.97
12	Dairy Calves	2.43	2.43	2.47	2.47
13	Beef Heifers	2.09	2.03	1.99	1.76
14	Dairy Heifers	0.75	0.75	0.76	0.76
15	Beef Stockers	26.8	26.3	27.0	25.5
16	Dairy Stockers	0.00	0.00	0.00	0.00
17	Sheep	7.80	6.32	5.80	5.34
18	Horses	2.42	3.64	4.02	3.61
19	Goats	0.96	0.59	0.54	0.57

TABLE 3. Proportion of Total Corn Production Used in Wet Milling and Dry Milling. For each year, y, the proportion of total corn to wet milling is $P_{wm,y}$ and the proportion to dry milling is $P_{dm,y}$. These proportions were calculated using corn disappearance and DDGS production data from USDA ERS feed grains yearbook tables (USDA 2018).

	1997	*2002*	*2007*	*2012*
$P_{wm,y}$	0.0264	0.0358	0.0188	0.0060
$P_{dm,y}$	0.0219	0.0584	0.1935	0.3838

TABLE 4. Proportions of Ethanol Feed Coproducts Produced Per Mass of Corn Processed. Proportions (R) are shown for corn processed through either dry milling (which produces DGS) or wet milling (which produces CGF and CGM).

RDGS/corn	*RCGF/corn*	*RCGM/corn*
0.27[1]	0.22[2]	0.04[2]

Notes: [1]Bremer et al. (2010), [2]O'Brien et al. 2010

TABLE 5. Output Mass-Based Allocation Factors for Coproducts of Dry and Wet Milling. These allocation factors are provided by Shapouri et al. (2002) and used to distribute N inputs to ethanol production between ethanol and ethanol feed coproducts.

	Ethanol	*Feed coproducts*
Dry milling	0.51	0.49
Wet milling	0.52	0.48

The proportions used in these calculations are shown in **TABLE 3** and **TABLE 4**. In order to allocate corn fertilizer, land use, and GHG impacts between ethanol and its feed coproducts, allocation by mass was used (Shapouri et al. 2002) with the allocation fractions listed in **TABLE 5**.

RESULTS AND DISCUSSION

Per unit N inputs to corn grain and ethanol feed coproducts (CGF, CGM, and DGS) are unchanging from 1997 to 2012. Notably, fertilizer N inputs (the model attributes no fixation N inputs to corn products) per kg of product are significantly higher for ethanol feed coproducts than for corn (**FIGURE 1A and B**). However, higher protein and, therefore, N content in ethanol feed coproducts results in lower fertilizer N inputs per kg protein for these products as compared to corn. Since the model allocates feed products to livestock with the constraint that livestock N requirements are met, reducing average fertilizer and fixation N input per N content of feed, would result in lower N inputs to livestock.

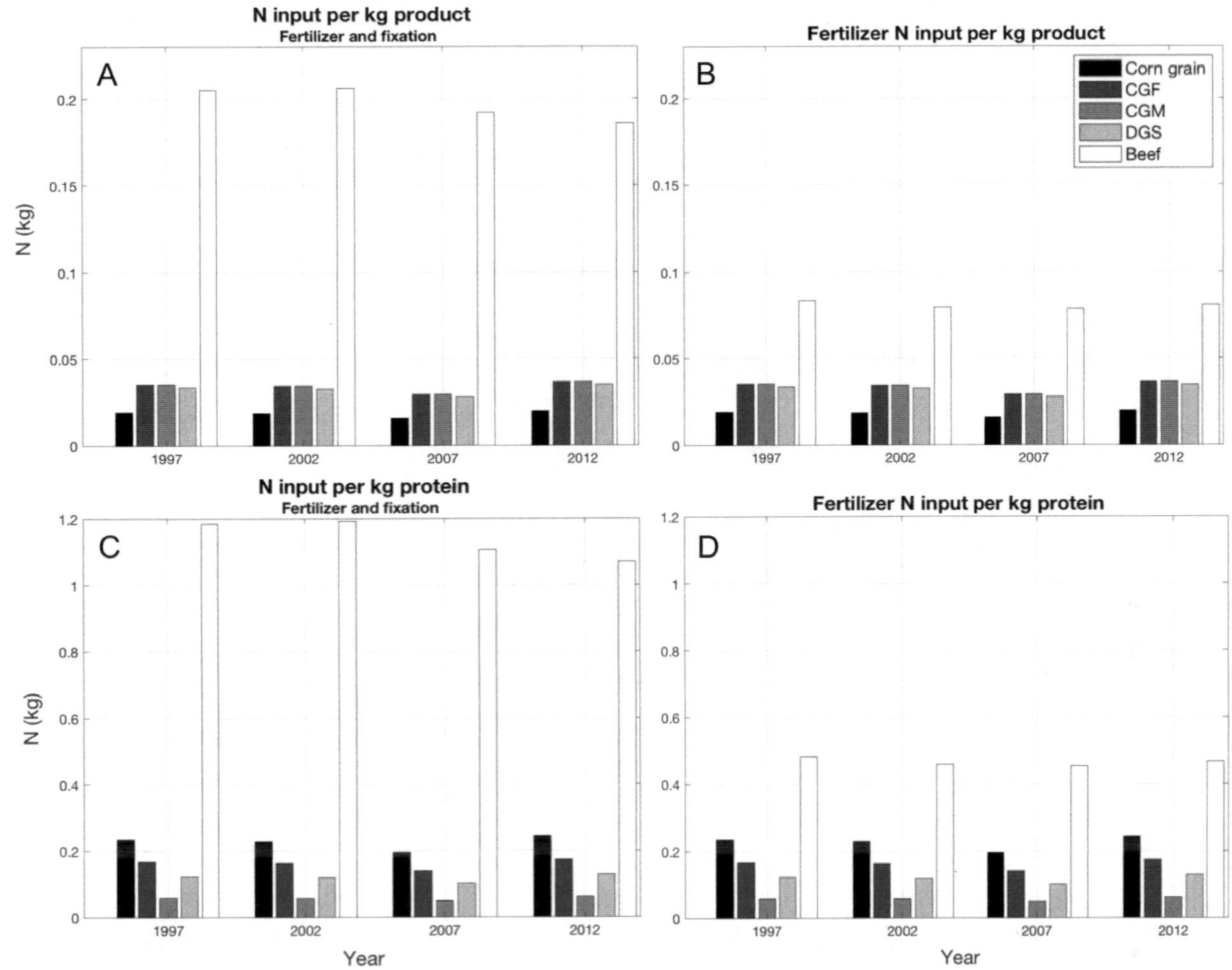

FIGURE 1. Total and Fertilizer N Inputs to Production of Corn, Ethanol Feed Coproducts, and Beef. The total inputs (A and C) and fertilizer inputs (B and D) are each normalized to kg product (A and B) and kg protein (C and D) from 1997 to 2012. CGF = corn gluten feed, CGM = corn gluten meal, and DGS = distillers grains and solubles. Note that the y-axis scale of A and B is different from that of B and C.

From 2002 to 2012, per unit N inputs to beef production do decrease (**FIGURE 1**). With total beef production decreasing (**TABLE 2**), this will contribute to a decrease in total N inputs to beef. However, despite the decrease in N input per kg protein in ethanol feed coproducts, specifically DGS, we are not seeing a fertilizer input decrease for beef (**FIGURE 1B and D**). In this feed allocation model based on produced quantities of crops, N requirements of livestock, and livestock grain to forage consumption ratios, DGS is not replacing corn in the feed mix for beef (**FIGURE 2**).

In **FIGURE 2**, we see shifts in the modeled feed mix that occur with the inclusion of ethanol feed coproducts in cattle diets. From 2002 to 2012, model results show a significant decrease in the role of N fixing crops in beef cattle diets, whereas the role of crops that require the most significant fertilizer N inputs remains roughly the same. Noting the draught- reduced corn yield in 2012 (Mallya et al. 2013), we hypothesize that replacement of cropland pasture for cornfield drives these modeled changes in the beef feed mix. It may also be that DGS directly replaces soybeans previously produced for livestock. This trend, however, is difficult to discern from 2007 and 2012 data alone, especially considering the effects of draught on crop yields in 2012.

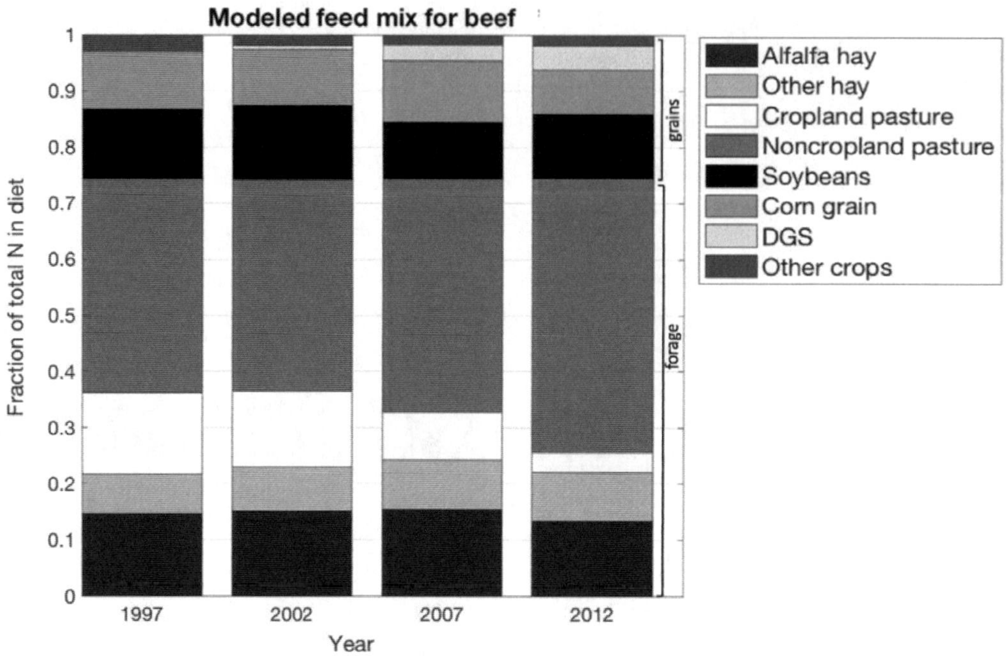

FIGURE 2. Modeled Feed Products Meeting Beef N Requirements in Each Year. The model's feed allocation process distributes crops to livestock based on livestock-specific grain to forage ratios and N consumption requirements, and crop quantities produced for animals in each data year. The top three sections—other crops (which are all feed grains), DGS, and corn grain—are responsible for nearly all of the fertilizer N inputs to beef. The bottom three sections, cropland pasture, other hay, and alfalfa hay, together with soybeans, are responsible for all of the fixation N inputs.

CONCLUSIONS

This research leads to further questions about how to improve the modeled livestock feed mixes to accurately represent livestock diets and allocation of N inputs to between ethanol and feed coproducts. Fixed grain to forage ratios make the model unable to capture changes that may be occurring in the proportions of forage and grain used to meet livestock N requirements. Additionally, the model does not include contributions from the dairy industry, which may be significant; according to a publication from the Ohio State University Extension, beef from dairy cows made up 17.9% of U.S. beef production in 2002 and 22.7% in 2016 (Geiser 2017). The results of impact allocation between coproducts can be highly sensitive the allocation method used. The allocation method used here is based on the output masses of ethanol and feed coproducts, producing a nearly 50-50 allocation (**TABLE 5**). Other allocation methods, such as those based on replacement or market values, attribute over 70% of impacts to ethanol (Shapouri et al. 2002).

It will also be important to determine the potential impact of these reduced beef commodity impacts on the N loading problem in watersheds. Decreasing N inputs to a major commodity like beef do not necessarily lead to decreased watershed-scale N inputs. Planted area of corn has increased from 318,000 square kilometers in 1997 to 389,000 square kilometers in 2012 (USDA 2018). The corresponding fertilizer N inputs, although some are allocated to the renewable fuel industry instead of the agricultural industry, will still contribute to the N pollution problem.

This work showed that the NANI model modified herein can be used to produce commodity-specific impacts for the US agricultural system. In this particular case study, per unit N inputs were examined for corn grain, ethanol, ethanol feed coproducts, and beef. Development of commodity-specific capacity for the NAPI model is in progress.

ACKNOWLEDGEMENTS

This research was supported by a United States Department of Agriculture (USDA) National Institute for Food and Agriculture (NIFA) grant, award number 2017-67019-26449. Special thanks to Dr. Jay Devkota at Lawrence Berkeley National Lab (LBNL) for guidance and support.

REFERENCES

2 H.R.6 § 202(A). C. (110th). Energy Independence and Security Act of 2007.

Bremer, V. R., A. J. Liska, T. J. Klopfenstein, G. E. Erickson, H. S. Yang, D. T. Walters and K. G. Cassman (2010). "Emissions Savings in the Corn-Ethanol Life Cycle from Feeding Coproducts to Livestock " *Journal of Environmental Quality* **39**(2): 472-482.

Costello, C., W. M. Griffin, A. E. Landis and H. S. Matthews (2009). "Impact of Biofuel Crop Production on the Formation of Hypoxia in the Gulf of Mexico." *Environmental Science & Technology* **43**: 7985-7991.

Costello, C., X. Xue and R. W. Howarth (2015). "Comparison of production-phase environmental impact metrics derived at the farm-and national-scale for United States agricultural commodities." *Environmental Research Letters* **10**(11): 114004.

Elgood, Z., W. D. Robertson, S. L. Schiff and R. Elgood (2010). "Nitrate removal and greenhouse gas production in a stream-bed denitrifying bioreactor." *Ecological Engineering* **36**(11): 1575-1580.

Fan, A. M. and V. E. Steinberg (1996). "Health Implications of Nitrate and Nitrite in Drinking Water: An Update on Methemoglobinemia Occurrence and Reproductive and Developmental Toxicity." *Regulatory Toxicology and Pharmacology* **23**(1): 35-43.

Farrell, A. E., R. J. Plevin, B. T. Turner, A. D. Jones, M. Hare and D. M. Kammen (2006). "Ethanol Can Contribute to Energy and Environmental Goals." *Science* **311**(5760): 506.

Geiser, J. a. B. B. (2017). Dairy Cattle Impact on Beef Supplies. Ohio Beef Cattle Letter, Ohio State University Extension Beef Team.

Heath, G. A., D. D. Hsu, D. Inman, A. Aden and M. K. Mann (2009). *Life Cycle Assessment of the Energy Independence and Security Act of 2007: Ethanol—Global Warming Potential and Environmental Emissions*. ASME 2009 3rd International Conference on Energy Sustainability collocated with the Heat Transfer and InterPACK09 Conferences, American Society of Mechanical Engineers.

Mallya, G., L. Zhao, X. Song, D. Niyogi and R. Govindaraju (2013). "2012 Midwest drought in the United States." *Journal of Hydrologic Engineering* **18**(7): 737-745.

O'Brien, D. and M. Woolverton (2010). "Updated trends in US wet and dry corn milling production." *AgMRC Renewable Energy Newsletter.*

Pimentel, D. and T. W. Patzek (2005). "Ethanol Production Using Corn, Switchgrass, and Wood; Biodiesel Production Using Soybean and Sunflower." *Natural Resources Research* **14**(1): 65-76. Ravishankara, A., J. S. Daniel and R. W. Portmann (2009). "Nitrous oxide (N2O): the dominant ozone- depleting substance emitted in the 21st century." *Science* **326**(5949): 123-125.

Shapouri, H., J. A. Duffield and M. Wang (2002). The energy balance of corn ethanol: an update, EERE Publication and Product Library.

Steffen, W., K. Richardson, J. Rockström, S. E. Cornell, I. Fetzer, E. M. Bennett, R. Biggs, S. R. Carpenter, W. De Vries and C. A. De Wit (2015). "Planetary boundaries: Guiding human development on a changing planet." *Science* **347**(6223): 1259855.

Swaney, D. P., Hong, B., and Howarth, R.W. . (2018). "NANI/NAPI Calculator Toolbox Version 3.1 Documentation." from http://www.eeb.cornell.edu/biogeo/nanc/nani/NANI.

USDA (2018). Economic Research Service (ERS) database. USDA (2018). National Agricultural Statistics Service (NASS).

RF Radiation Potential Environmental Health Impact on Riggers in Telecommunication Industry in Ghana

Lydia Michelle Tawiah (Madina Polyclinic Ghana Health Service, Accra GHANA)
Owusu Akyaw Kwarteng (Vodafone Ghana Telecommunication Company LtD ACCRA GHANA)

The telecommunications industry continues to grow rapidly. The increasingly common high density of antennas at a site, low antenna mounting heights and the frequent need for personnel to work near the vicinity of these antennas too often results in personnel being exposed to RF fields.

The environmental risks for exposure to significant levels of RF radiation are higher for tower climbers than they are for perhaps any other occupation. While the RF field levels on many towers such as those that contain only cellular antennas are not very high, the RF field levels near FM radio and television antennas are extremely high.

The methods used include direct measurements around towers close to settlements using hand held RF monitor, asking riggers in a questionnaire about their awareness and training provided them and how much they are aware of possible danger as well general perception on environmental effects

RF values recorded by RF monitors were compared with acceptable limits from WHO standards and ICNIRP. Much as telecommunication is desirable for national development, it seems Ghanaians are not ready to coexist with telecom infrastructure, without which the required quality mobile network service cannot be possible.

Public concern about the environmental health impact of Radio Frequency Radiation (RFR) from telecom towers, vis-a-vis the need for quality telecom service and possible challenges of co-location, presented a difficulty for the locals to deal with.

Results found was that the 85% of total exposures levels are low- typically 5,000 times lower than what could be harmful to humans. For the 15% with medium to high level exposures the effective use of personal protective equipment such as radiation suites in the FM radio and television antennas has truly protected the worker. The implications can be complicated because the research did not confirm the fears of residence and riggers.

The Challenges to Sustainable Development in Rural Regions of the Kyrgyz Republic

Kasymbekova L. and Matsui K.
(University of Tsukuba, Tsukuba, Japan)

ABSTRACT: Agriculture has played an important role in developing the economy of the Kyrgyz Republic. Recently the Republic has actively promoted regional agribusiness development but faced challenges, including climate change consequences and financial limitation. The objective of this paper is to examine these challenges and to identify the role public administrations can play in securing sustainable agricultural developments in rural regions. This examination is based on the questionnaire survey among 82 farmers in two rural administrative regions. We chose these regions mainly because of their importance in developing agriculture. We wanted to find out what prevent these farmers from successfully developing the regional economy today and in the future. We asked them questions related to the impact of climate change to agriculture, their willingness to start non-agricultural businesses, and their needs of public intervention. We also asked what these farmers expected from the public administration. The results show that the main challenges largely differ by regions though most farmers identified lack of marketing and machinery as well as impacts of climate change as formidable obstacles to advancing their businesses. This paper also discusses what these farmers perceived about their future survival.

INTRODUCTION

In the Kyrgyz Republic, one third of the population is under the poverty line and 70% of the poorest families live in rural areas (WFP, 2018), where agriculture is the main economic activity. The situation has become worsened partly due to climate change, including prolonged drought events. In response to climate change the Kyrgyz Government has developed agricultural adaptation programs. However, these programs have not clearly benefited farmers partly because planners of these programs did not sufficiently understand the real needs of rural farmers.

In order to better understand perceptions of rural farmers regarding their challenges like climate change and what they expect from the public administration to support, we conducted the questionnaire survey in rural areas. We chose Issyk-Kul and Jalal-Abad oblasts as our study areas as these two local regions are blessed with agricultural and some regional development potentials. These areas also appeared to have suffered from poverty and climate change impacts. In the following discussion, we first introduce our study areas briefly. Then we explain how we conducted our survey before detailing about our findings from the survey.

STUDY AREAS

Issyk-Kul oblast is situated in northeastern part of the Kyrgyz Republic. The population is 464,000 (2015) in 11 km^2 of the area. It takes its name from Lake Issyk-Kul ("warm lake"), the second largest saline lake in the world. The Lake never freezes despite its location in the high altitude along the Tian Shan Mountains (Wikipedia, 2018a). The area attract more than one million tourists every year. It has about 30% of poverty level. Issyk-Kul is the second largest industrial region in the country (CRP, 2017). Agriculture here plays an important role in the development of the national economy as 19% of nation's farms exist within this oblast (2016) (Table 1) (NSC 2018a). These farms produce 11% of the total agricultural production of the country. They provide 10% of crops and 12% of animal products to the market (Table 2) (NSC, 2017).

TABLE 1. Number of Farms by Area, 2016

	Units	%
Country (total)	312,922	100
Jalal-Abad oblast	120,916	39
Issyk-Kul oblast	60,149	19

Jalal-Abad oblast is situated in the southern part of the country. Its population (2016) is 1.1 million in 34 km^2 of the area. The Sary-Chelek State Biosphere Reserve, UNESCO Biosphere Reserve, is situated in this oblast (Wikipedia, 2018b). Sary Chelek ("yellow bucket") Lake has been tourist destination. The oblast has the highest poverty level (45%) in the nation and the third contributor to the total industrial production (10%, 2015) (NSC, 2017). Jalal-Abad is one of key contributors to the agriculture production of the country. There are about 120,000 farms in the oblast, consisting of 39% of nation's total farms. Many farmers here conduct mixed farming (crop and animal production) (Table 3) (NSC, 2018b).

TABLE 2. Contribution of Oblasts to Total Agriculture Production, 2016.

	Agro-production (mil. $)	Share in nation's total (%)
Jalal-Abad	542	19
Issyk-Kul	311	11

TABLE 3. Agricultural Products in Issyk-Kul and Jalal-Abad, 2016.

	Crops, mil. $	Share in national production (%)	Animal product (mil. $)	Share in nation's animal production (%)
Jalal-Abad	283	20	259	19
Issyk-Kul	148	10	163	12

MATERIALS AND METHODS

The questionnaire survey was conducted in April and May 2018 with permission from the Kyrgyz Ministry of Agriculture. We randomly selected 82 farmers in two oblasts (44 persons from Issyk-Kul oblast, 38 persons from Jalal-Abad oblast). The main goal of this questionnaire was to better understand their perceptions about sustainably developing rural regions so that future policies and programs can be better designed to support farmers. The questionnaire consisted of four sections. The first section is devoted to collecting socio-economic characteristics of the respondents, such as age, gender, education, farming experience, annual income and household information. The second section was designed to understand farmers' perceptions about challenges in developing the regional economy. They were asked to identify the degree of challenges in operating and starting agribusiness. They assessed activities of governmental organizations, investors and non-governmental organizations that support agriculture. The third section attempted to find out how farmers see the attractiveness of agriculture. The last section focused on climate change impacts. For example, the respondents evaluated the condition of pastures in the past five years.

In processing the collected data, we consulted some staff members in the Ministry of Agriculture, Food Industry, and Melioration to better interpret. Also, the lead author, Lira Kasymbekova, had two years of experience in the Ministry of Agriculture, and we used her experience to better understand farmers' perceptions behind the data.

RESULTS AND DISCUSSION

The results of the first part of the questionnaire show that most of the 82 respondents were male farmers (78%). The average age was 52 years old. The average period of their experience in agricultural activities was more than 20 years. In Issyk-Kul, about 52% of the respondents received university education. In Jalal-Abad, 40% received university education (Table 4). These show that our respondents generally have many years of agricultural experience with relatively high education background. Those in Issyk-Kul are relatively better connected to Bishkek, the capital, where there are more higher education opportunities.

The level of education in urban and rural areas in the Kyrgyz Republic shows that the majority of rural people receive up to high school education. Considering the national average of university graduates among rural people, the education level of the respondents in both Issyk-Kul and Jalal-Abad is considerably high (Table 5) (NSC, 2014).

TABLE 4. Completed Highest Education of the Respondents

Oblasts	High school	Vocational school	University
Issyk-Kul	16%	32%	52%
Jalal-Abad	26%	34 %	40%

TABLE 5. Level of Highest Education in Urban and Rural Areas
(Percentage of the total economic active population at 15 years old and more in 2013)

	High school	Vocational school	University
Urban	46.4%	21.5%	32.1%
Rural	71.7%	14.4%	13.9%

The questionnaire result shows that mixed activity of farming and livestock grazing was most common, consisting of 59% of the respondents. Mixed agriculture is more preferable for agribusiness because farmers can mitigate the risks of loses by diversifying income sources (Graph 1).

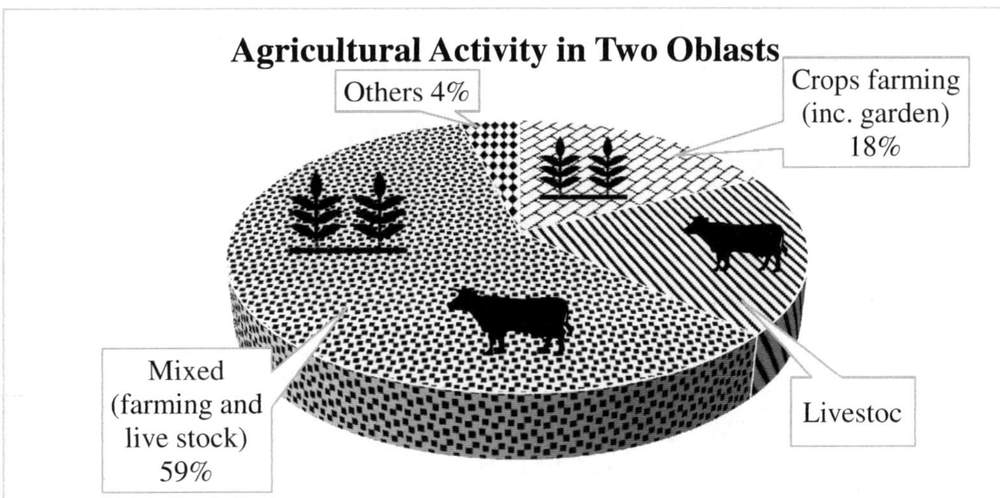

GRAPH 1. Agricultural Activity in Two Oblasts

In the Kyrgyz Republic, GDP per capita is about US$3,700 (2017). In rural areas, this amount can generally sustain one farm household. So, we asked if the annual income of the respondents is above or below this amount. In Issyk-Kul oblast more than 70% earned more than US$3,700, meaning that they could cover their agricultural expenses and sustain their livelihood. On the other hand, in Jalal-Abad, less than 40% of the respondents had the average annual income (Graph 2).

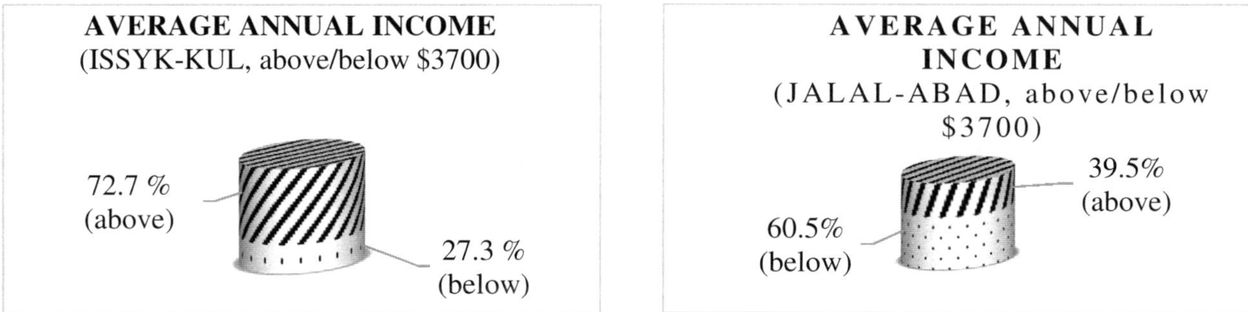

GRAPH 2. Average Annual Income in Issyk-Kul and Jalal-Abad Oblast

These figures somewhat correspond with national statistical records. According to the National Statistic Committee, in 2016, the level of poverty in Issyk-Kul oblast was 24.7%, slightly below the average national rate whereas more than 30% of people in Jalal-Abad was poor (Table 6) (NSK, 2018c).

TABLE 6. Territorial Poverty Level (in Percentage of Number of Population), 2016

	Kyrgyz Republic	Issyk-Kul	Jalal-Abad
Poverty level	25.4%	24.7%	32.2%

In the second part of the questionnaire, we wanted to find out how farmers find the level of challenges to advancing their agricultural businesses in their regions. We identified eight major challenges and asked the respondents to identify the level of each challenge by high importance, medium importance, and least importance.

The result shows that in Issyk-Kul more than 70% of the respondents found marketing most challenging to advance their businesses. This is partly because, in remote regions of the Kyrgyz Republic, the access to the large market is limited due to cost of transportation, lack of storage facilities, logistic centers, and insufficient information about market demand. Even though many farmers produced high quality crops in this region, lack of branding, attractive packaging or geographical information through marketing prevented them from fetching full potential prices. For example, many farmers produce their crops without using agrochemicals, but no organic certification is available at major markets in Bishkek and other big cities.

Another major challenge was the availability of technologies. The high cost of agricultural machines and their parts, for example, have inhibited agricultural producers from modernizing their agricultural practices. Currently about 20% of farmers purchase agricultural machines by themselves (InfoStream, 2018). In 2006, there were 0.3 tractors per one farm, but it dropped to 0.08 in 2015. Farmers have requested government financial support for having a better access to tractors and harvesting machines with no vail. Most farmers, therefore, have mainly applied simple methods, which often resulted in decreasing soil fertility and yield. They do not have technological solutions to the spread of pests, crop diseases, and weeds (Ukibaeva and Kocherbaeva, 2016) (Graphs 3, 4). Other than marketing and technology, we found that about 60% of the respondents had experienced water shortage such as drought.

In Jalal-Abad oblast, more than 73% of the respondents found that the lack of financial support as the most inhibiting factor for their agricultural businesses. They additionally needed fertilizer, pesticide, agricultural machinery, among others. This region experienced severe drought lately, partly due to climate change, and about 60% of the respondents suffered from water shortage.

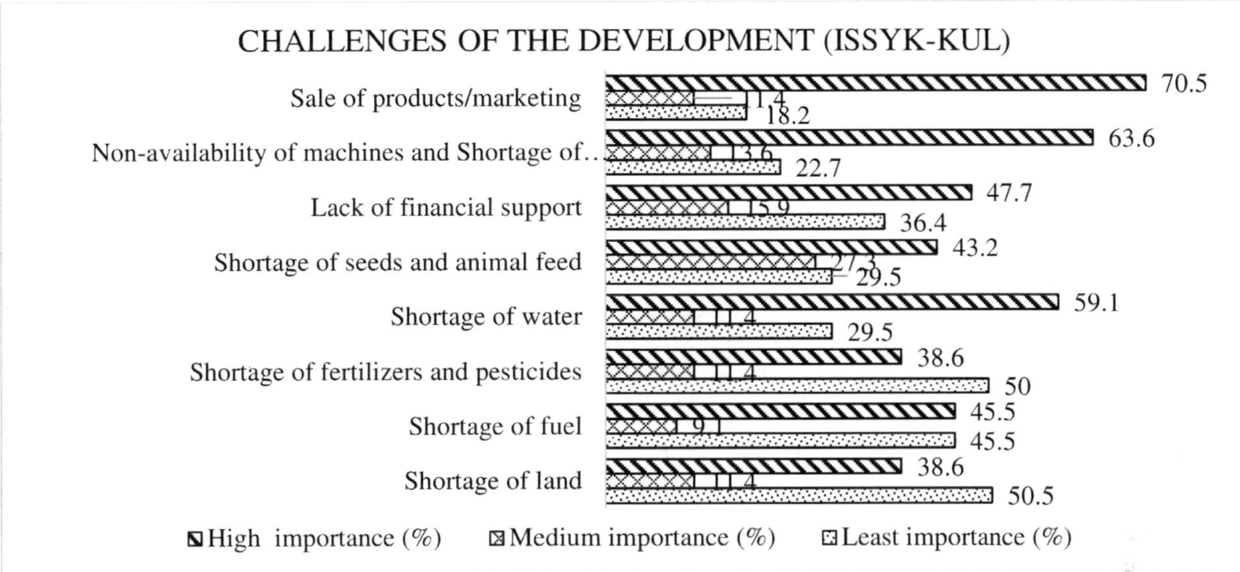

GRAPH 3. The Challenges of the Development in Issyk-Kul Oblast

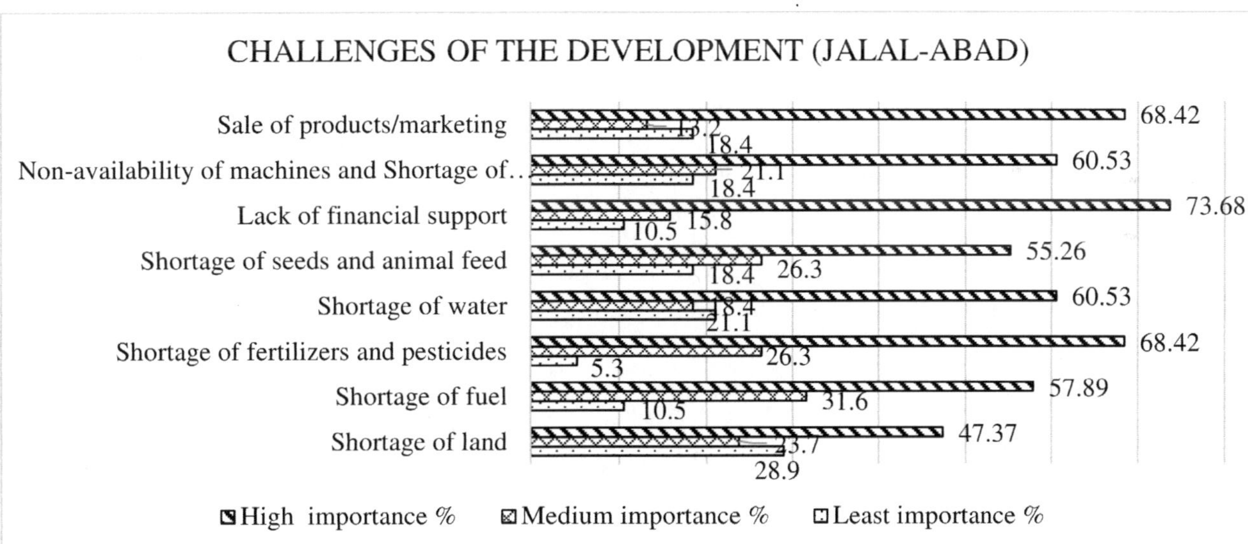

GRAPH 4. The Challenges of the Development in Issyk- Kul Oblast

Past studies often point out about the correlation between land size and agricultural productivity (Kyei and Matsui, 2018; Oyewo and Fabiyi, 2008; and Singh, 1992). Kyei and Matsui examined the efficiency of two rice producer groups in Kassena-Nankana Municipality of the Upper East Region of Ghana. The first farmer group practiced irrigation, and the second one engaged in rain-fed agriculture. The results showed that land size, age, labor, education and years of farming experience significantly affected the efficiency of farmers (Kofi and Matsui, 2018). Singh (1992) found a correlation between farm size and productivity by examining farmers who participated in an intensive agricultural district program (1960-1961) at Aligarh district, India. The results revealed that farms with larger land sizes tend to increase production than those with smaller land sizes. A Nigerian study also showed a positive and significant relationship between farm size and maize output (Oyewo and Fabiyi, 2008).

However, our respondents in both Issyk-Kul and Jalal-Abad did not find this particularly important. This means that most farmers were not interested in expanding their farms. In Issyk-Kul, the reason can be lack of labor and machinery. In Jalal-Abad, the reason can be attributed to lack of financial support to buy land, fertilizer, and others. Also, farmers in these two oblasts as well as other regions tend to focus on domestic needs. As Table 7 shows, agricultural exports from the Kyrgyz Republic are on the declining trend. In particular, livestock production, which requires large tracts of land, declined sharply since 2014 (Kaktakto, 2018).

TABLE 7. Export of Agricultural Products (Thousand US Dollars)

	2014	2015	2016
Livestock products and animals	225,647.9	71,605.9	27,032.3
Crops	170,101.3	101,624.8	126,587.6
Total	395,749.2	173,230.7	153,620

In the third section of our questionnaire, reflecting on difficulties and the declining trend of agricultural export, we asked how agricultural businesses still attract them today and in the future. In spite of many challenges, more than 70% of the respondents in both oblasts were still interested in agribusiness and did not want to switch to non-agricultural businesses. Some of them, especially in Issyk-Kul, have received substantial financial and technical support from government, international/donor organizations, or investors. In Jalal-Abad oblast, 68% of the respondents emphasized government support as other funding opportunities were not readily available. However, the farmers of Issyk-Kul oblast highlighted their own contribution (27%) and governmental authorities (28%).

The final section focused on climate change impacts. In Issyk-Kul oblast, 75% of the respondents had experienced negative impacts of climate change on agriculture whereas 29% of the respondents in Jalal-Abad oblast had experienced climate change impact. Those farmers who recognized climate change impacts in two oblasts highlighted the rising temperature and drought.

According to the Center of Climate Change in the Kyrgyz Republic, the national tendency of rising annual temperature is similar to global tendency (UNEP, 2016). The rising temperature is recognized as consequences of climate change (IPCC, 2013), which affects agricultural sustainability. So, to meet climate change challenges target governmental entities adopted sectorial programs in agriculture, forest, and health sectors (Governmental Resolution, 2013).

CONCLUSIONS

This study examined rural farmers' perceptions about agricultural businesses for the future of two agricultural regions in the Kyrgyz Republic. We found different regional needs and trends among farmers. This means that a unilateral national policy cannot properly help solve rural development issues. Issyk-Kul farmers needed more marketing support and advanced machinery to partly deal with labor shortage. They were also well-informed about the availability of agricultural machineries so that technological support quite likely will show some good results. The region is relatively successful in sustaining agricultural businesses within the country. In Jalal-Abad, as a quarter of the population faces poverty, farmers need not only technological support but also financial support and fertilizer/pesticide provisions. They expect that the Kyrgyz government provide more support to their businesses.

We also found that climate change has already affected agriculture. Considering more than 20 years of farming experience among our respondents, it is safe to conclude this much. We also found that climate change, especially water shortage, affects farmers in the Issyk-Kul region.

ACKNOWLEDGEMENTS

We are grateful for generous support we received for this study from the Ministry of Agriculture, Food Industry and Melioration of the Kyrgyz Republic and the farmers of Issyk-Kul and Jalal-Abad oblasts is gratefully acknowledged.

REFERENCES

CRP. 2017. *Conception of regional policy for 2018-2022*. Governmental Resolution № 194 March 31, 2017. The Government of the Kyrgyz Republic. (in Russian).

Government Resolution. 2013. *Priority directions for adaptation to climate change in the Kyrgyz Republic till 2017*. №549 February 10, 2013. The Government of the Kyrgyz Republic. (in Russian).

InfoStream. 2018. *Provision of agricultural machinery is 22%*. . Information center "ELVISTI". Accessed on May 25, 2018. (in Russian). http://uaport.net/news/kg/t/1405/15/5011895

IPCC. 2013. *Climate Change 2013: The Physical Science Basis*. Intergovernmental Panel on Climate Change. Switzerland.

Kaktakto. 2018. *Kyrgyzstan in EAEU. Forecast and Reality*. Accessed on 23 May 2018. https://kaktakto.com/analitika/kyrgyzstan-v-eaes-prognozy-i-realnost/

Kyei, K., and K. Matsui. 2018. "Efficiency Analysis of Rice Farmers in the Upper East Region of Ghana." *IAFOR International Conference on Sustainability, Energy and Environment in Conference Proceedings*, pp. 1-8. Hawaii, USA. http://25qt511nswfi49iayd31ch80-wpengine.netdna-ssl.com/wp-content/uploads/papers/iicseehawaii2018/IICSEEHawaii2018_38924.pdf

NSC. 2018a. *Brief Statistic Digest 2015-2017*. National Statistic Committee. Bishkek. The Government of the Kyrgyz Republic. (in Russian).

NSC. 2018b. *Gross agriculture production by territory*. National Statistic Committee. Bishkek. (in Russian).

NSC. 2017. *Kyrgyzstan and Regions*. National Statistic Committee. Bishkek. (in Russian).

NSC. 2014. *Education and science in the Kyrgyz Republic*. Statistical digest. National Statistic Committee. Bishkek. (in Russian).

NSK. 2018. *Level of poverty in the Kyrgyz Republic*. National Statistic Committee. Bishkek. (in Russian). http://www.stat.kg/ru/publications/uroven-bednosti-v-kyrgyzskoj-respublike/

Oyewo, L. O., and Y. L. Fabiyi. 2008. *Productivity of Maize Farmers in Surulere Local Government Area of Oyo State*. International Journal of Agricultural Economics and Rural Development. Volume 1(2).

Singh, A. 1992. *Farm Size and Agricultural Productivity in Aligarh district*. In: Mohammed, N. (Ed), Anthropogenic Dimensions in Agriculture. Concept Pub, New Delhi.

Ukibaeva G., and A. Kocherbaeva 2016. *Analysis of modern condition of Kyrgyzstan agricultural industry and perspectives of its development*. International research journal № 12 (56). Part 5 December. Yekaterinburg. P. 206-209. (in Russian).

UNEP. 2016. *Climate change: the world, Central Asia and Kyrgyzstan in facts and figures*. United Nations Environment Programme. Bishkek.

WFP. 2018. *Kyrgyzstan*. World Food Programme. Accessed on 24 May 2018. http://www1.wfp.org/countries/kyrgyzstan

Wikipedia.2018a. *Issyk-Kul Region*. Accessed on 24 May 2018. https://en.wikipedia.org/wiki/Issyk-Kul_Region

Wikipedia.2018b. *Jalal-Abad Region*. Accessed on 24 May 2018. https://en.wikipedia.org/wiki/Jalal-Abad.

Engineering Approach Used in Hydroponics Designs and Automation

Riley Horner, Elizabeth Alvizo, Michelle Mata, and Bahram Asiabanpour
(Ingram School of Engineering, Texas State University, San Marcos Texas, USA)

ABSTRACT: In this research the engineering design approach is used to design and implement a controlled hydroponic system within a shipping container that is independent of utilities. This system should include as little human interaction as possible. A multitude of aspects go into creating a successful hydroponic system. The portion this study will be addressing is finding the most efficient automated way to deliver nutrients to plants. To efficiently and successfully make this system the engineering design process was directly followed. First, constraints were identified based on previous research and predetermined factors such as the shipping container dimensions and previously purchased metal racks. Next, multiple concepts were generated, and variables of these concepts were tested. After comparing all alternatives, a sustainable and automated design was implemented on one rack. This system was then used to determine other vital variables within the overall hydroponic system. The engineering design process has allowed for a systematic way to improve and expand upon these variables.

INTRODUCTION

The engineering design process has been around for centuries and is applied every day in the engineering world. "The engineering design is defined as component, or process to meet desired needs. It is a decision making process (often iterative) in which the basic sciences, mathematics, and engineering sciences are applied to convert resources optimally to meet a stated objective. Among the fundamental elements of the design process are the establishment of objectives and criteria, synthesis, analysis, construction, testing and evaluation." (Page 1)(Tayal, 2013) The goal of this research is to implement this process to a topic that it has never been applied to: hydroponics.

Hydroponic farming techniques have the potential to revolutionize the agricultural industry. This has not occurred yet due to a multitude of common problems that occur in these systems. These problems include but are not limited to high start-up cost, slow profit return, excessive energy usage, time consuming maintenance methods, and the need for a constant environment to grow healthy crops. To combat these problems both the engineering design process and a Failure Mode and Effects Analysis (FMEA) chart will be exercised to design and implement a hydroponic system. This process and analysis will allow the construction of a controlled environment within a shipping container, needing minimal human interaction to be upkept and is void of common problems.

To begin applying this process to the focus of this research, the nutrition delivery system, common hydroponic practices were explored to better identify and organize the needs for the system. Following this, many preliminary designs were created that complied with the constraints. These constraints included predetermined factors such as the size of the shipping container, a five degree slant the container is at for drainage purposes, pre-purchased metal racks, and the basic needs for plant life. Next, promising alternatives were developed, tested and compared. To develop these, methods such as water jet cutting, drilling, plastic welding, and additive manufacturing techniques were used. Things that needed to be determined through these comparisons were the pump, trays for the plants, drainage valves and their placement, and the number of reservoirs.

Once these variables were resolved a system was implemented into the shipping container. With this system aspects such as, number of plants per shelf allowed, how long and often the pump from the reservoir will be on, the lighting system, and how much nutrient solution will be needed could be settled.

This process and analysis allow a systematic way for not only these variables to be determined but also future additions to be incorporated. Expansion on the system will include nutrient concentration, water compensation, natural and artificial light, ambient control, as well as insect control.

METHOD

There is a multitude of slightly varying engineering design processes out there. For the purposes of this research the process shown in Figure 1 was utilized since it contains all the necessary steps for designing an automated hydroponic system.

Ask. Following this process, we started by identifying the need for an automated hydroponic system. This system would need to have the following constraints: efficiently grow plant without the need for soil, use Food and Drug Administration (FDA) approved materials, include as little human interaction as possible, be independent of utilities, avoid traditional problems found in hydroponics and fit inside an insulated shipping container.

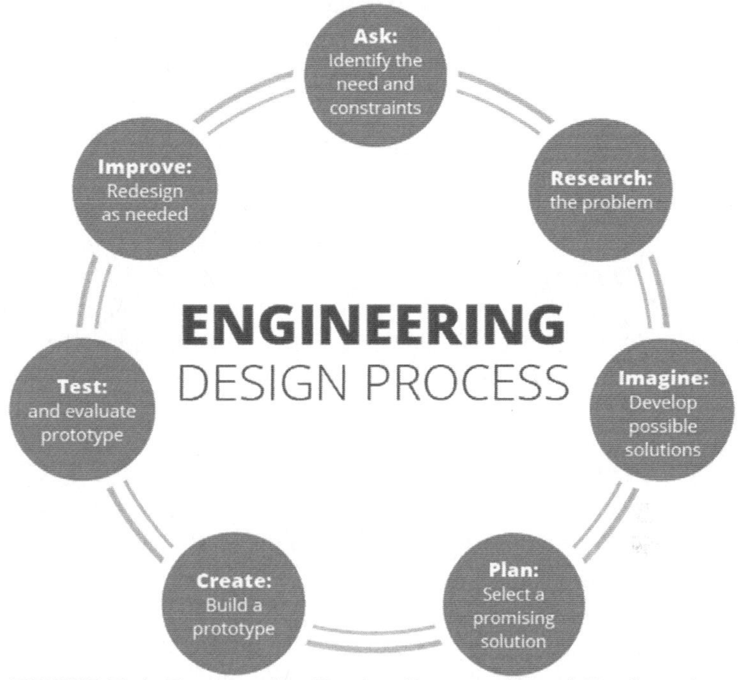

FIGURE 1. Engineering Design Process (TeachEngineering, nd)

Research. In this step multiple hydroponic systems, the problems they encountered, and how those problems have been solved were investigated. The first traditional hydroponic system studied was ebb and flow also known as flood and drain. In this system, trays are flooded in a nutrient solution for five to ten minutes then drain though an overflow tube. Plants are held in a substrate, generally perlite, rockwool, or clay pebbles. This is demonstrated in Figure 2.

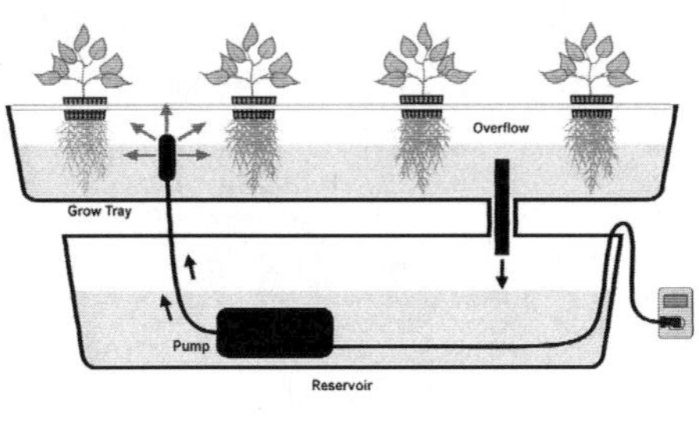

FIGURE 2. Ebb and flow (Ebb and Flow, nd)

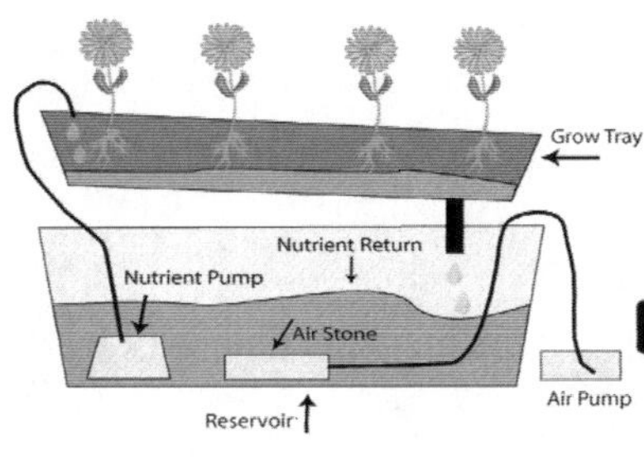

FIGURE 3. NFT (Hanna Instruments, 2018)

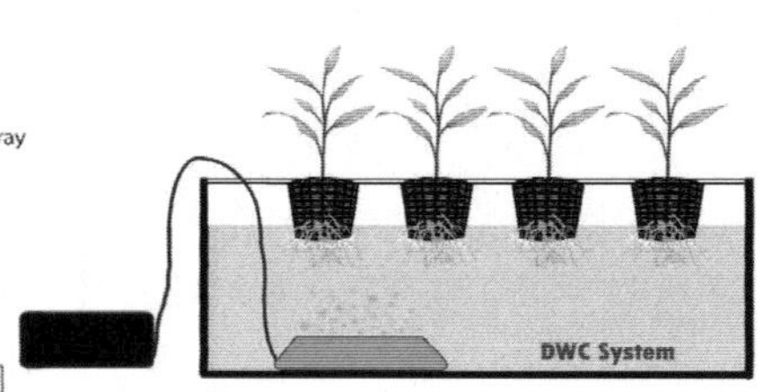

FIGURE 4. Deep water culture (Home & Garden, 2016)

In a nutrient film technique (NFT) system the nutrient solution continuously flows over the plant roots. Plants are held in a channel that maintains moisture in the top of the roots and allows the bottom of the roots to be in direct contact with the nutrient solution. This is demonstrated in Figure 3.

Deep water culture also known as a float system is where plants are held by a buoyant material and float top a heavily aerated nutrient solution. This is demonstrated in Figure 4. (Sheikh, 2006)

While there are additional hybrid variations of hydroponic growing systems out there these are the most efficient and easily implemented.

Imagine. Multiple designs were drawn up based not only on these traditional hydroponic systems but also a standard insulated shipping container, 90 in. high X 90 in. width X 24 in. depth metal rack with adjustable shelves and predetermined coco coir substrate. All designs will include a total of four racks filling half of the shipping container. The initial designs included a four-rack system, two-rack system, and one-rack system.

In the four-rack system, all four racks are connected and to an external reservoir that feeds the top tray. The top tray then utilizes a valve to release the nutrient solution to the trays below it eventually feeding back into the reservoir. These valves would be places on alternating sides of the tray to help ensure aeration and drainage. The plants would be suspended in substrate filled cups at the top of each tray allowing the roots to be submerged when the tray is filled.

The only difference between the four-rack system, two-rack system, and one-rack system is how the reservoir is connected. The two-rack system has one reservoir feeding two racks while the one-rack system has one reservoir feeing one rack.

Plan. To determine not only which of these designs would work best but also the materials that would be used aspects of each design were tested. Beginning with the trays, it became quickly apparent that traditional hydroponic trays would not work since the dimension were not compatible with the previously purchased racks. Other possibilities such as getting trays custom made, making the trays our self, and other commercially available tray options were explored. The chosen alternative would need to be made of a FDA approved material, have a water tight seal, not warp when filled with water, and efficiently utilize the space on the racks. Many companies were contacted pertaining to the customization of hydroponic trays. The companies willing/able to do it gave a quote far outside the budget. To make the trays our self, different materials and processes were considered. High density polyethylene (HDPE) was decided on as the material since it is FDA approved and will not damage due to water over time. Joining this material can be quite difficult, therefore two processes were chosen. First the HDPE slabs were aligned and screwed together, then they were fused together using a plastic welder. This process took about 3 hours to complete and a water tight seal was not achieved. Plastic welding the tray a second time may have given this seal but would still not have been ideal since the materials were costly and the process was extremely time consuming. After investigating commercially available trays and bins intended for different uses, trays traditionally used as house hold storage containers were found. With these trays the design would have to be slightly altered to incorporate two trays per shelf, due to their dimensions. Since this alternative was extremely inexpensive, and complied with all the constraints they were decided as the best option for the system.

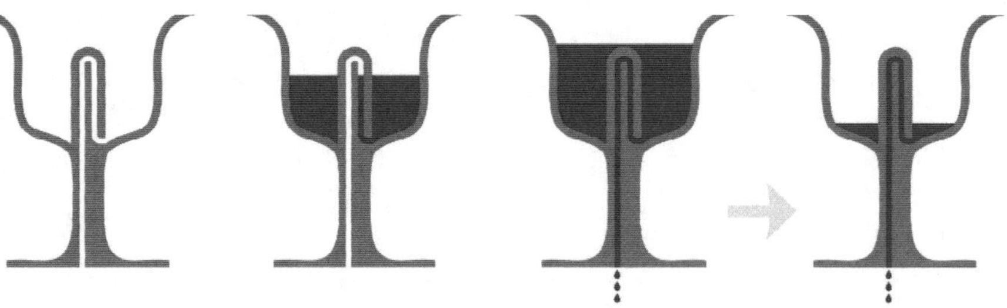

FIGURE 5. Pythagoras Cup or Siphon Pump Explanation (Kohlstedt, nd)

Next the type of valve that would release water from one tray to the next was explored. Initially automated valves that would control how long water would stay in each tray were considered. These were not ideal since electricity is needed, which counters our goal of being independent of utilities. Hence other options were sought out.

An experiment was conducted to compare traditional valves and a Pythagoras cup, commercially known as a siphon pump, was conducted. An explanation of how a Pythagoras or siphon pump works is given in Figure 5. For this experiment manual ball valves were used instead of electric valves. Multiple prototype siphon pumps were created out of PVC pipe. These prototypes varied in diameter for the inner and outer pipes as well as the size of hole cut into the bottom of the outer pipe. This experiment proved that the same results could be reached using either a ball valve or a siphon pump. Since the siphon pump does not need electricity to function it was chosen for the system.

Once these aspects were decided upon reservoirs that fit the dimensional need of the racks were sought out. A 100-gallon bin was decided on due to it being inexpensive and made of a FDA approved material. After some preliminary test of the system it became quickly apparent that a one rack system would be necessary since the chosen reservoir could only hold enough water to feed one rack.

Create. To implement this system eight trays had holes cut in the bottom that fit the siphon pumps. These holes were formed using saw drill bits and made in the back left or right corner of the tray. Two trays with siphon pumps were placed on each shelf. The reservoir containing the pump was placed on the floor beneath the rack. The hose connected to the pump was attached and guided along the side of the rack to the top trays where it was attached to a device that distributed the water between the two trays as seen in Figure 6.

FIGURE 6. Hydroponic system prototype

Test. To test this design, the reservoir was filled and the pump was turned on for 2 hours. Data was taken on how much water was needed to fill each tray, how long it took each tray to fill, and the rate of water flow from the pump as well as from tray to tray.

Improve. Improvement will be a continuous process with this system. Updates to the design will need to be made when major problems are encountered or when a better solution is found for a specific portion. To systematically decide which problem or improvements are done first a Failure Modes and Effects Analysis (FMEA) chart will be utilized. The use of this chart is best described by Kim K.O. and Zuo, M. J. in "General model for the risk priority number in failure mode and effects analysis" as "assuring the safety and reliability of a system, engineers identify all independent failures that may occur during a given stage of the system life cycle, and prepare actions to reduce the occurrences of failures or mitigate the severity of their consequences. Failure mode and effects analysis (FMEA) is a structured method used for fully understanding failure modes and their consequences in a given stage of the system life cycle." (page 321)(Kim, K. and Zuo, M., 2018) A visual representation of this process can be seen in Figure 7.

RESULTS

Utilizing this process has allowed the creation of a semi-automated hydroponic system with a clear plan on optimizing and fully automating the system. This technique minimized major downfalls in initial designs, ensured the best design was chosen, and created a systematic future improvement method. Currently this system can run water continuously for at least 2 hours without supervision. This means other variable like nutrition solution, lights, and plants can be added.

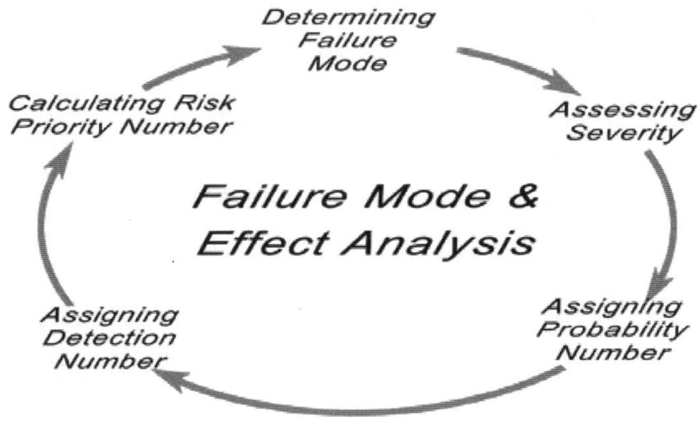

FIGURE 7. FMEA Process (Shmula, 2018)

DISSCUSION AND CONCLUSION

Based off the data collected it was determined that the height of the inner tube in the siphon pump needed to be 3.5 inches from the base of the tray to allow enough water to collect in the tray and reach the plant roots. It was also discovered that there is a minimum rate of flow needed to ensure the siphon pumps drain correctly. To ensure this rate of flow is met, a minimum amount of water is need in the reservoir to submerge the pump and maintain a steady intake of water.

A big risk factor encountered based on the FMEA chart was the potential for the system to over flow and spill out into the shipping container. To prevent this possibility, drainage tubes were implemented into the tops of each tray as a backup if the siphon pump were to fail.

Reflective paint has also added to the bottoms of each tray to decrease the amount of light loss throughout the system as well as protect the roots from damaging rays.

Preliminary data proves this system can be built upon and grow as further research is developed. This arrangement successfully decreased the amount of electricity needed to transport water throughout the system. Whether it decreases how much energy is needed to run the whole system from seedling to harvest will be determined once multiple crops have been successfully harvested.

FIGURE 8: LED lights implemented into hydroponic system

FUTURE WORK

This project is far from done. Red and blue Light Emitting Diode or LED lights (as seen in Figure 8) and a variety of leafy greens have been chosen and will be implemented and tested once purchased. In addition to this, different growing practices will be explored, such as deep-water culture for vertical growing. Beyond this, Natural lighting solutions and water conservation techniques are currently being explored by other teams on this project to later be implemented into this system.

REFERENCES

"Deep Water Culture (DWC) Hydroponics System at Home." *Home & Garden*, www.lintangsore.net/2016/06/deep-water-culture-dwc-hydroponics.html.

"Ebb and Flow." *Dosatron*, www.dosatronusa.com/portfolio-item/ebb-and-flow/.

Kim, Kyungmee O., and Ming J. Zuo. "General model for the risk priority number in failure mode and effects analysis." *Reliability Engineering & System Safety* 169 (2018): 321-329.

Kohlstedt, Kurt. "Pythagorean Cup: Practical Joke Chalice Overflows with Ancient Greek Humor." *99% Invisible*, 99percentinvisible.org/article/pythagorean-cup-practical-joke-chalice-overflows-ancient-greek-humor/.

"N.F.T. Growing System." *PH Meters, Photometers, Titrators, Controllers - Hanna Instruments*, Hanna Instruments, 2018, hannainst.com/hydroponics-nft-growing-system.

Sheikh, B. A. "Hydroponics: key to sustain agriculture in water stressed and urban environment." *Pak. J. Agric., Agril. Eng., Vet. Sci* 22 (2006): 53-57.

Shmula. "5 Common FMEA Mistakes to Avoid |." *Shmula*, 6 Jan. 2018, www.shmula.com/5-common-fmea-mistakes-to-avoid/24817/.

Tayal, S. P. "Engineering Design Process." *International Journal of Computer Science and Communication Engineering*, no. NCRAET-2013, 2013, pp. 1-5., doi:JCSE.SI040113

"TeachEngineering." *Www.teachengineering.org*, www.teachengineering.org/k12engineering/designprocess. The source of this material is the Teach Engineering digital library collection at www.TeachEngineering.org.

Aqueous Phosphorus Removal and Recovery by Pervious Concrete

Vincent D. Hwang
(University of Texas, Austin, TX, USA)
Yaileen M. Acevedo Badillo, Anaira Román Santiago, Carla M. Moreno Cortés, Rafael A. Terán Rondón, and ***Sangchul S. Hwang***
(University of Puerto Rico, Mayagüez, PR, USA)

Pervious concrete pavement is one of the Best Management Practices for stormwater runoff management. In this study, pervious concrete was made of seawater as the mixing water and contained fly ash as a partial cement replacement to enhance sustainable development in green pervious concrete production. The developed pervious concrete was used to reduce and recover phosphorus, a non-renewable resource, in water to improve resource recovery and environmental protection. Recovered amorphous phosphates were transformed to a slow-release alginate-phosphate fertilizer and were applied for the growth of common bean.

Bayesian Network Analysis of Reforestation Decisions by Local Communities in Vietnam

Thi Mai Anh Tran, Hai Dinh Le, Dongwook W. Ko
(College of Science and Technology, Kookmin University, Seoul 02707, South Korea)

Forest is important to protect head watershed in Cao Phong district of Vietnam, where a large number of household characteristics, socio-economic, biophysical, and institutional and farm characteristics can directly or indirectly influence the plantation decision of local people. The purpose of this study is to utilize Bayesian Network (BN) to identify 1) the main drivers that affect landholders 'planted forest area, 2) how the key drivers affect among themselves, and 3) what discourages tree planting. To identify the significant drivers, we surveyed 100 households and used bivariate analysis to build a structured system of the associations among variables, which was validated by stepwise linear regression. By using the significant drivers, we created a Bayesian network to predict scenarios with different household's perception regarding the planted forest area, and to provide suggestions for increasing area of forest. The results revealed five significant factors that influence farmers' agreement in planted forest area. Eighty-nine percent of the people planted trees on their own lands to secure or claim their land area. Decisions not to plant trees were due to the difficulties in protecting properties from illegal logging (59%) and lack of resources (58%). This paper is the first study exploring the key factors influencing the planted forest area in Vietnam, and our results can assist design efficient forestry programs in other comparable areas.

Maximizing Biochar and Biooil Yield from Pyrolysis of Albizia: An Invasive Plant Evaluation

Hafiz Ahmad* and Brandon Madden (Florida State University, Panama City, FL, USA)
Jordan Mayers (AECOM, Tallahassee, FL, USA)

In recent years, production of pyrolytic byproducts (biochar, biooil, biogas) is becoming very popular due to their versatile application. These byproducts are used directly or with additional processing for soil amendment, water purification, energy storage, fuel, medical usage, important chemicals production etc. But the production yields of the byproducts depend on the pyrolysis operating conditions (e.g. temperature, N_2 purging rate, temperature ramp rate etc.). In this study, biochars and biooils were produced from the biomass, Albizia (chips), an invasive plant by pyrolysis process. Various temperatures (850 °F, 1000 °F, 1100 °F, 1200 °F) with a temperature ramp rate of 6 °F/minute at a N_2 purging rate of 4-7 ft^3/hr were used to produce these pyrolytic byproducts. Models were developed to relate the product yield (biochars and biooils) and pyrolysis temperature for process optimization. The study suggested that the pyrolysis temperature of 850 °F will give maximum biochar (31.4%) and biooil (31.8%) yield. To utilize this outcome for commercial biochar and biooil production, Albizia pellets (same biomass but different form) were pyrolyzed at this optimum pyrolysis condition (850 °F). More biochar (32.9%) and biooil (50.8%) yield was observed with a gross calorific value 12,728 lb/BTU and 7,642 lb/BTU respectively.

An Appraisal on the Potential of Ecosan (Ecoloical Sanitation) to Serve the Three Tiers of Water-Food-Energy Nexus

Bilsen BELER-BAYKAL
(Istanbul Technical University, Department of Environmental Engineering, Istanbul, TURKEY)

The recent domestic wastewater management approach ECOSAN (ECOlogical SANitation) claims that wastewater is not a waste to be discarded but a source to be revaluated. It is a concept which has its roots in source control in the form of separation of domestic wastewater streams at the very point that they are produced and processing of each stream as a source or a "raw material" for further use without mixing. As such, ECOSAN contributes to environmental sustainability both through resources conservation and pollution control by suggesting the use of a "waste" (wastewater) as a source.

Within this context, domestic wastewater is segregated as grey, yellow and brown water where grey water refers to all domestic wastewater with the exception of toilet wastewater, while yellow water is separately collected human urine and brown water is mainly source separated human feces possible with flush water and toilet paper.

Results show that all of those streams have a potential to contribute as sources to all three tiers of the Food-Water-Energy (FWE) Nexus which focuses on ensuring food, water and energy security for the increasing global population as well as the interdependence among the three systems.

There are several pieces of work in the literature showing that grey water with its milder polluting potential is a meaningful alternative water source, which may be returned to almost any point of the water cycle after proper treatment, including toilet flushing, landscape and urban irrigation, industrial use as well as providing water for agriculture. Of primary importance is its contribution to water savings through toilet flushing which can spear 25-30% of pristine water currently consumed in daily domestic use which could otherwise be used for more worthwhile purposes.

With its rich nutrient (N, K, P) content, yellow water is a valuable source of fertilizers, either through direct application after storage and probably dilution, or through indirect use where source separated urine is used as a "raw material" for production of fertilizers. It has been shown that marketable fertilizers like P-rich struvite and N-rich ammonium sulfate can be produced through processing urine as well as nutrient enriched zeolites which are rich both in N and P. The global potential from human excreta is estimated to be equivalent to about 35% of the global fertilizer demand and plant nutrients in one person's urine are enough to produce one loaf of bread on a daily basis, so benefits can be vast.

Organic material rich brown water can be processed to produce compost which is a soil conditioner used in agriculture or it may be anaerobically digested to produce biogas to be used as energy.

In addition to individual benefits of each stream, the entire ECOSAN streams may be helpful simultaneously in securing food production through the use of fertilizers from yellow water, water from grey water and soil conditioner from brown water, all of which aid agricultural yields.

In conclusion, domestic "waste"water produced by mankind may be valorized with the ECOSAN concept to help ensure food, water and energy security and to contribute to the welfare of mankind.

Ecotourism and Sustainable Development: Evidence from Lope National Park, Gabon

Cornelius FRIMPONG
(Department of Biological and Environmental Sciences University of Stirling, Stirling, Stirlingshire, FK9 4LA, Scotland, UK)

Ecotourism is widely applied to the conservation of tropical forest resources as a means of raising funds for conservation and to support social and economic development of local communities. With these three objectives, ecotourism has been embraced as a sustainable management approach. On this premise, this study has produced a new schematic diagram to guide ecotourism conceptualization. Consequently, using the case of Lope National Park, Gabon, this study engaged a purposive sample of 20 conservation scientists and Park staff (Gabonese and non-Gabonese) to analyse how the three sustainability objectives are achieved and how this is influenced by respondents' nationality. Data analysis and hypotheses testing were conducted through MANOVA, Wilcoxon signed rank test and paired-sample t-test in SPSS Version 21 (IBM). The study found that ecotourism is perceived to fulfil all three objectives of sustainability. However, statistical analysis showed significant difference in the distribution of ecotourism revenue between Gabonese and non-Gabonese [$F\,(3, 16) = 5.66$, $p = .008$, partial $\eta 2 = .515$]. Follow-up ANOVA conducted revealed that, whereas Gabonese distributed revenue nearly equally across the three pillars, non-Gabonese placed overemphasis on conservation [$F\,(1, 1.21) = 17.056$, $p = .001$, partial $\eta 2 = .487$]. Consequently, ecotourism management at Lope was found to fall short of meeting international standards. The study concluded with recommendations and a suggestion for further research to broaden this scope by recruiting a larger sample of various stakeholders.

A Computable General Equilibrium Model for Liberia with Carbon Taxes and Exemptions

Presley K. Wesseh, Jr, Boqiang Lin
(Department of Technical Economics and Management, Xiamen University, Xiamen 361005, Fujian, China)

Small developing countries and their larger counterparts alike, face, among others, three major challenges in their drive for development, that is, economic growth and employment, stable energy supply, and environmental sustainability. While ensuring stable energy supply appears to provide a good support for economic growth and employment, quite often, these benefits come at the compromise of environmental standards. In order to avoid adverse output and energy shocks and to maintain support for environmental objectives, imposing environmental taxes with exemptions for energy- and labor-intensive sectors considered to be strategically critical for economic development could provide opportunities. The problem, however, is that, exemptions can be environmentally costly and this is exacerbated by the fact that the thin financial resources of many small developing countries cannot afford them the luxury of meeting their renewable energy investment requirement needs.

The goal of this paper, therefore, is to study various scenarios for a unilateral carbon tax design in a small developing economy and the implications of exemptions. For this purpose, our empirical analysis has sought to consider the following questions: (i) do exemptions raise the costs of carbon taxes? (ii) If yes, then by what magnitude? Which is better for saving output and employment: exemptions or direct wage subsidies? (iii) How do exemptions affect sectoral demand and supply?

To address these questions, this paper develops a static computable general equilibrium (CGE) model for a small developing country calibrated to the most recently available and consistent energy and economic data for Liberia. Our results suggest that exempting some polluting industries from carbon taxes would generate welfare gains. However, this compromise was found to be environmental costly. Hence, the general insights of these results point to the need for exploring other options for saving employment and output (e.g. direct subsidies) as opposed to environmental tax exemptions.

RENEWABLE ENERGY DEVELOPMENT

Feasibility Comparison of Passive and Active Optical Fiber Daylighting Systems Utilizing PMMA Optical Fiber

Richard Ramirez, Katherine Casey, Juan Silva Febres, Bahram Asiabanpour
(Ingram School of Engineering, Texas State University, San Marcos, TX, USA)

ABSTRACT: Increasing residential and commercial energy consumption has drawn attention to the reduction of energy consumption in buildings where artificial lighting is utilized. Adopting novel daylighting methods have shown promising results in reducing energy consumption and cost. Research has shown that daylighting is an efficient method of delivering light for indoor applications. In this research, the performance of both active and passive optical fiber daylighting (OFD) systems utilizing PMMA optical fiber is compared to determine their effectiveness for utilizing natural light in indoor applications.

INTRODUCTION

In 2010, the U.S. Department of Energy reported an energy consumption of 40% for commercial and residential buildings (Buildings Energy Data Book, 2010). Electricity and heating accounted for 72% of greenhouse gas (GHG) emissions with a 3.19% annual growth rate according to the Intergovernmental Panel on Climate Change (Brucker, T. et. al., 2014). An estimated 40% of energy used in the United States is taken up by residential and commercial buildings, 40-50% of which is from artificial lighting (Ullah, I. & Shin, S. 2015). Ullah and Shin also stated that up to 50-80% of a building's energy usage due to artificial lighting can be reduced when efficient daylighting is used. An optical fiber daylighting system is proposed as a solution to the environmental, energy, and health issues associated with artificial light. Optical fiber daylighting systems collect sunlight, concentrate that light into fiber optic cables, and then transmit it deep into the interior of a building through luminaires.

Active vs. Passive Photovoltaic Systems. When a dual-axis tracker and fixed position photovoltaic (PV) system were compared, up to 34.6% increased daily power was observed in the dual-axis tracking system (Maatallah, T., El Alimi, S., Nassrallah, S.B., 2011). In addition, this article found that in the past 10 years, the photovoltaic industry has seen a 30% increase in the development rate on average. This is indicative of the increased support of renewable energy and the willingness to spend large upfront costs for long-term environmental and utility cost benefits. Testing in Germany with a latitude of 48.58 degrees yielded a power density of 1149.02 kWh/m^2 for a fixed tilt system of a 40-degree angle (Helwa et. al., 2000). The researchers found that a two-axis tracker had a 30% increase in annual power density compared to the fixed-tilt system. Cammarata (2014), moved away from universal full-mobility parallel robots exhibiting six degrees of freedom (6 DOF) in favor of robots with lower DOFs for an increase in workspace. The researcher also noted that parallel kinematic machines (PKMs) also exhibit precise orienting applicable to solar applications. The U-2$\underline{P}$US system (Latitude of 37° 30′27′′68 N & Longitude of 15° 4′27′′12 E) had a 35.1% increase of power production over a 30-degree fixed photovoltaic (PV) panel, and an overall increase of 17.2% accounting for actuation power at a maximum tracking error of 1-degree. Another study found an 82% increase in efficiency for active PV panels when compared to passive PV panels. This study also noted an economic analysis with a 6-year breakeven point for a 9-panel configuration (Asiabanpour, B. et al., 2017).

Active vs. Passive Daylighting. Garcia-Hansen (2006) studied the impact of daylighting systems in commercial buildings. The researcher suggested different ways to implement these systems for indoor applications. Garcia-Hansen described two different light collection systems. She found that a light pipe, although cheaper, requires more space availability in a building and has more limitations on the angle of acceptance. On the other hand, she found that active systems collect the maximum daylight available at any time because it is constantly tracking the location of the sun, however, this system is more expensive and requires more maintenance (Garcia-Hansen, V.R. 2006). Seung Jin Oh and his colleagues (2012) demonstrated that an active daylighting system was more efficient for altitudes below 50° and a passive daylighting system provided more daylight for solar altitudes above 50°. They used a parabolic dish concentrator for their active system and a light tube for their passive system. When active optical fiber daylighting systems were compared, André and Schade (2002) determined that the 1-axis system was not the most efficient for sunlight collection due to limitation of angle precision. On the other, they found the 2-axis system had higher accuracy when obtaining light concentration into the fiber.

Passive Optical Fiber Daylighting. A commercially available tubular daylighting system trade name VELUX was modified to validate the hypothesis that the new design could be adapted for optical fiber daylighting (Courture, P., et. al. 2011). Two modifications were made; the primary change was to approximate linear Fresnel lens or light guide ridges based on pertinent solar azimuth charts and secondarily the researchers proposed increasing the surface area by redesigning the dome with revolved half circle protrusions. The primary modification showed a favorable factor of intensity in the summer from 14° to 61° elevation, while the winter measurements showed favorable amounts at the early morning and mid-day.

Fiber Optic Cables. Optical fibers are an important component of the optical fiber daylighting system and can be 76% of the total cost of an OFD system (Vu, N., Pham, T., & Shin, S. 2016). Plastic optical fiber (POF) is the most cost-effective solution but has been shown to have a 10-15% loss in efficiency. The attenuation of light through the fiber optic cable is also a concern when the goal is to transmit daylight to multiple levels in a building. Studies have shown that a 20-meter plastic optical fiber (POF) is the maximum practical length for an OFD system (Vu, N., Pham, T., & Shin, S. 2016).
The motivation of this study is to investigate the economic feasibility of two types of optical fiber daylighting systems: active and passive.

MATERIAL AND METHODS

Optical fibers with different diameters were used in this study dictated by the designs of their respective system. The passive system was designed in house and outfitted with a 20mm PMMA optical fiber. A commercialized optical fiber daylighting system was characteristic of an active system with 3mm PMMA optical fiber and was orientated manually at the necessary time increments to record photosynthetic photon flux density (PPFD).

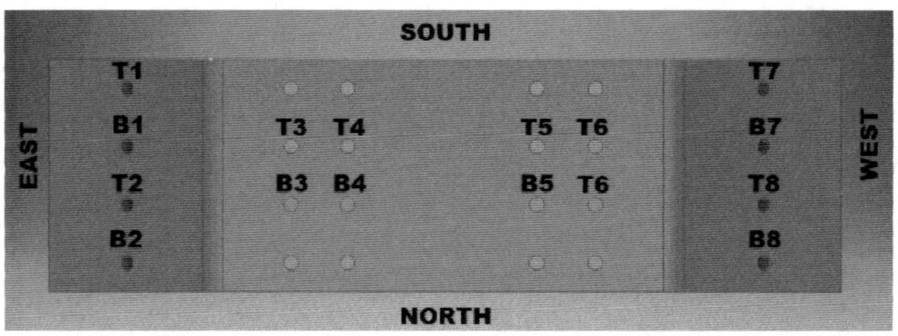

FIGURE 1. Map to Locations of Optical Fibers on Fixed Plate

The passive system was constructed to obtain performance data on large diameter optical fiber with PMMA core. Data collection was done on April 11th, 2018 between 10:30 am and 3:30 pm at the following global coordinates: Latitude 29.94° N and Longitude 98.00° W. Figure 3 is the relevant sun chart for the location and date. A Sekonic Spectromaster C-7000 was employed for the radiation data collection. The map provided in Figure 1 and the associated diffuser coupling locations inside a storage container are depicted in Figure 2. The "T" labeled areas are locations where the UV lens (or filters) were employed, whereas the "B" labeled areas represent locations with no UV lens and the optical fiber was exposed to the sun. The passive system study utilized Green.L 30mm multi-coated UV filters for protection and were positioned as described above. The eastern and western flanges are 41° from the middle plane of the end effector. This was done to increase the amount of incident radiation on the flanges from the sun's orientation in the early and late hours of the day. The end effector was positioned at an elevation of 51° and was directed toward the south.

FIGURE 2. Diffusers Labeled for Mapping to Corresponding Locations on Fixed Plate in Figure 1 (**Asiabanpour, B.**, Estrada, A., Ramirez, R., & Downey, M., 2018)

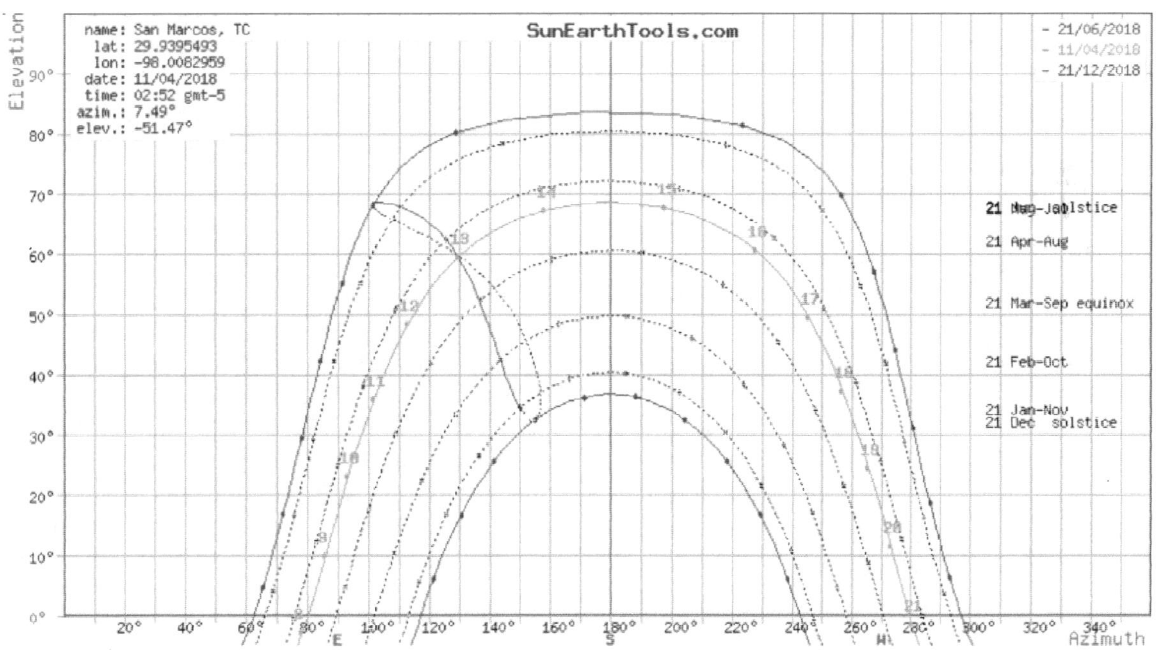

FIGURE 3. Sun Chart for Freeman Ranch in San Marcos, TX on April 11th

A cardboard box was used to isolate each diffuser from neighboring radiation sources when recording data with the Sekonic Spectromaster in all cases in a insulated storage container.

RESULTS AND DISCUSSION

The data that was collected by the fixed end effector is partition into columns organized by the planes and the presence of UV lens. Therefore, in Table 6A the column "T1-T2 Avg" represents the average between T1 and T2 with positioned UV filters. This nomenclature is carried out in the tables and charts below ("B1 and B2 Avg" represents no UV lens). Table 1 give values in photosynthetic photon flux density (PPFD) a unit used for measuring the responsive radiation according to photosynthesis using the quantum mechanical model. The general trend in each graph is the gradual decline in the radiation output on the eastern orientated flange and transpires through the adjacent flange, both intersecting near the solar noon. The PPFD of the center plane on the fixed end effector shows greater uniformity throughout the day, but only exceeds the exterior curves representing the flanged areas as they approach the intersection.

TABLE 1. Average Values of PPFD vs Time
A: Data of UV Lens Utilization; B: Data of No UV Lens Utilization

A

	PPFD (μmol/m^-2/s^-1)		
Time	T1 & T2 Avg	T3-T6 Avg	T7-T8 Avg
10:30 AM	195.4	22.9	6.4
11:00 AM	132.9	30.7	7.4
11:30 AM	307.8	83.9	7.3
12:00 PM	78.4	87.4	5.4
12:30 PM	131.0	64.0	10.6
1:00 PM	59.6	39.8	9.3
1:30 PM	53.0	142.5	19.5
2:00 PM	14.4	141.4	58.3
2:30 PM	12.2	113.8	112.2
3:00 PM	10.6	95.1	182.9
3:30 PM	11.1	63.6	174.1

B

	PPFD (μmol/m^-2/s^-1)		
Time	B1 & B2 Avg	B3-B6 Avg	B7-B8 Avg
10:30 AM	316.15	15.4	7.9
11:00 AM	165.7	21.5	5.2
11:30 AM	310.9	54	10.65
12:00 PM	76.95	89.375	4.95
12:30 PM	136.85	95.05	3.75
1:00 PM	132.45	117.575	15.4
1:30 PM	92.1	191.1	52.9
2:00 PM	31.1	131.025	138.9
2:30 PM	19.4	158.725	178.4
3:00 PM	14.45	108.8	112.25
3:30 PM	12.35	101.825	383.05

The graphs in Figure 4 shows a comparison of UV lens to no UV lens was analyzed in three zones divided by opposed flanges. In Figure 5, at time stamps 11:00 am and 12:00 pm, an inconsistency can be seen and a general downtrend as the sun moves toward the middle of the panel. When the sun moves over the middle portion of the panel the PPFD value increases until solar noon at 1:33 pm depicted in Figure 6. The radiation slowing decreases on the middle panel, but high values of radiation are recorded on the western flange shown in Figure 7. Another questionable drop in PPFD in the no UV lens scheme was recorded at 3:00 pm on Figure 7. Figure 7 showed a general increase in PPFD when the sun was aligning orthogonal to the western flange. The radiation incident on the middle panel was expected to be higher than the flanged portions. The lower values of the middle portion may be due to misaligned elevation. The end effector was 51° in elevation and was not aligned to elevation maxima, this may have contributed to the low PPFD recorded.

Figure 8 is a three dimensional plot comparing the passive system to the optical fiber daylighting (OFD) system. In Figure 8 the acronyms WF, MB, and EF signify western flange, middle base, and eastern flange respectively. The amount of radiation transmitted through the optical fibers was substantial in the case of the active system compared to the fixed system. The passive system utilized larger diameter optical fibers, but the active system still out performed it. This was expected since numerous reports indicated active systems typically outperform passive systems by approximately 30%.

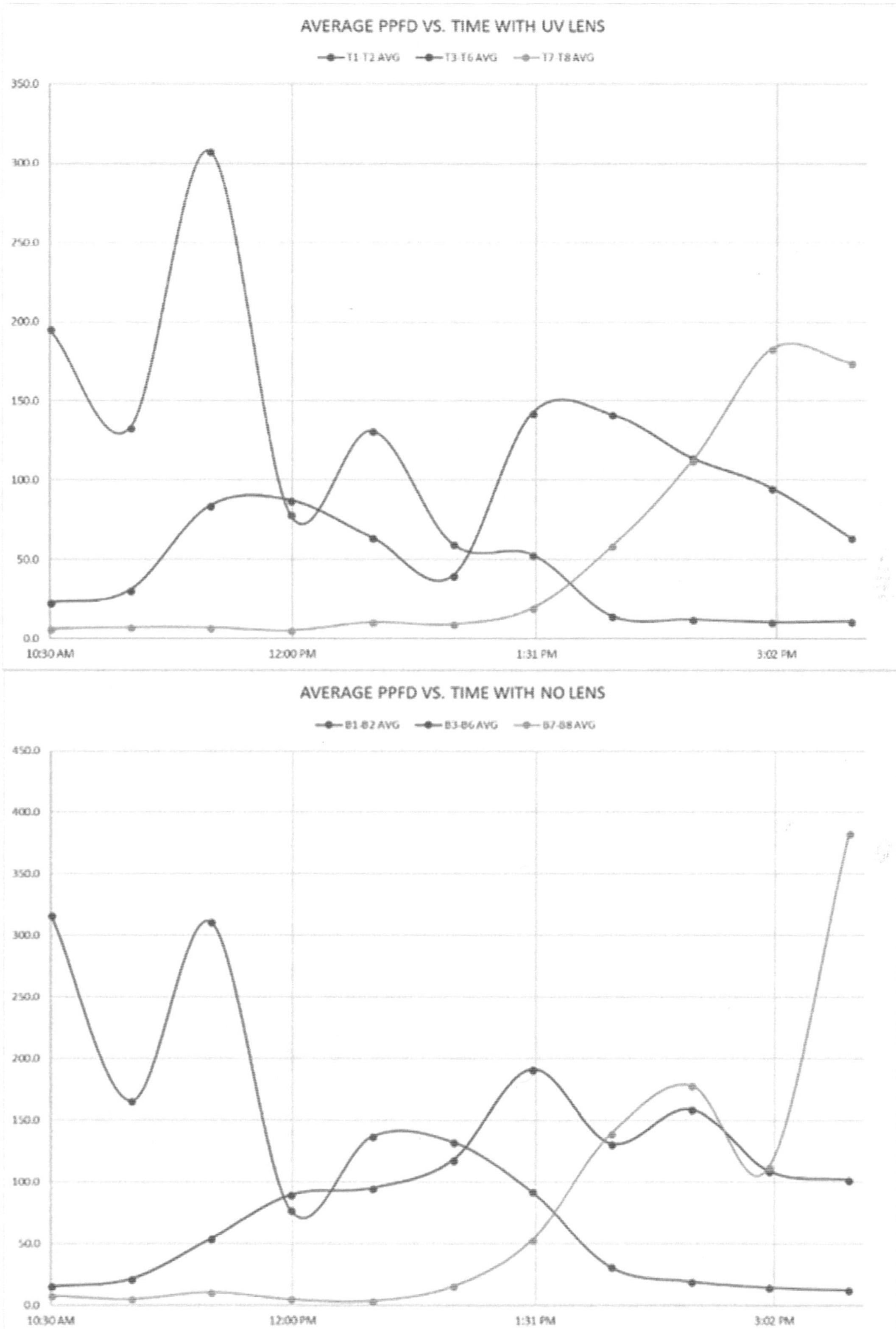

FIGURE 4. Average PPFD vs. Time. A: Data of UV Lens Utilization; B: Data Of No UV Lens Utilization

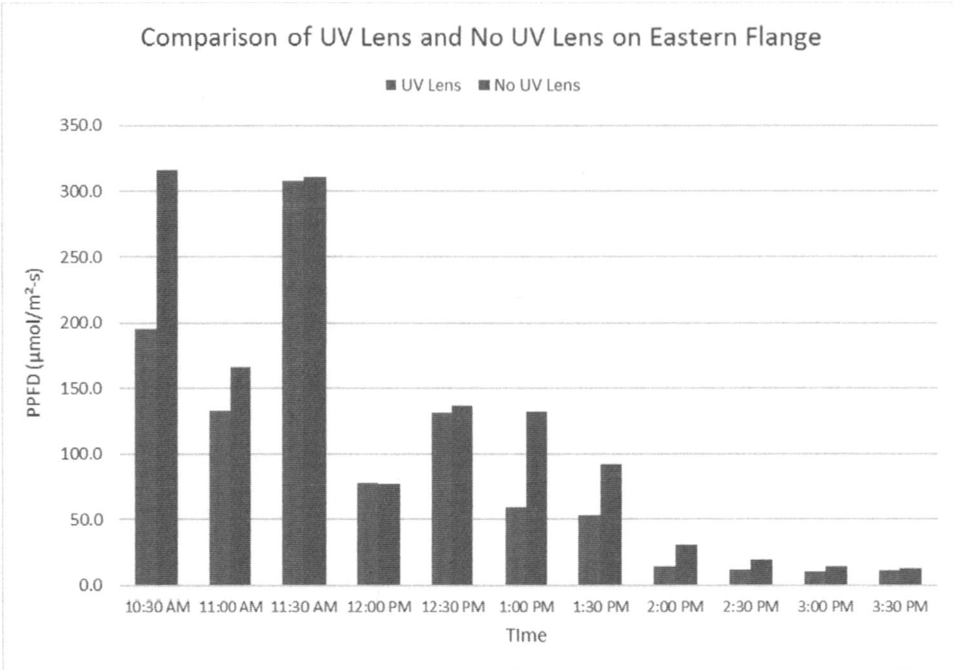

FIGURE 5. Comparing UV and No UV Lens on Eastern Flange

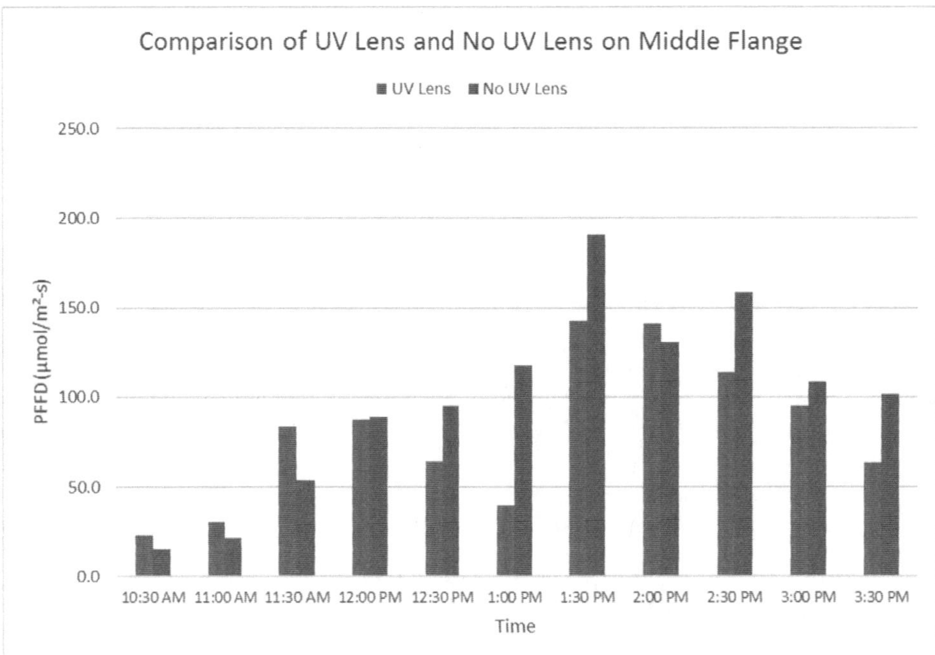

FIGURE 6. Comparing UV and No UV Lens on Middle Flange

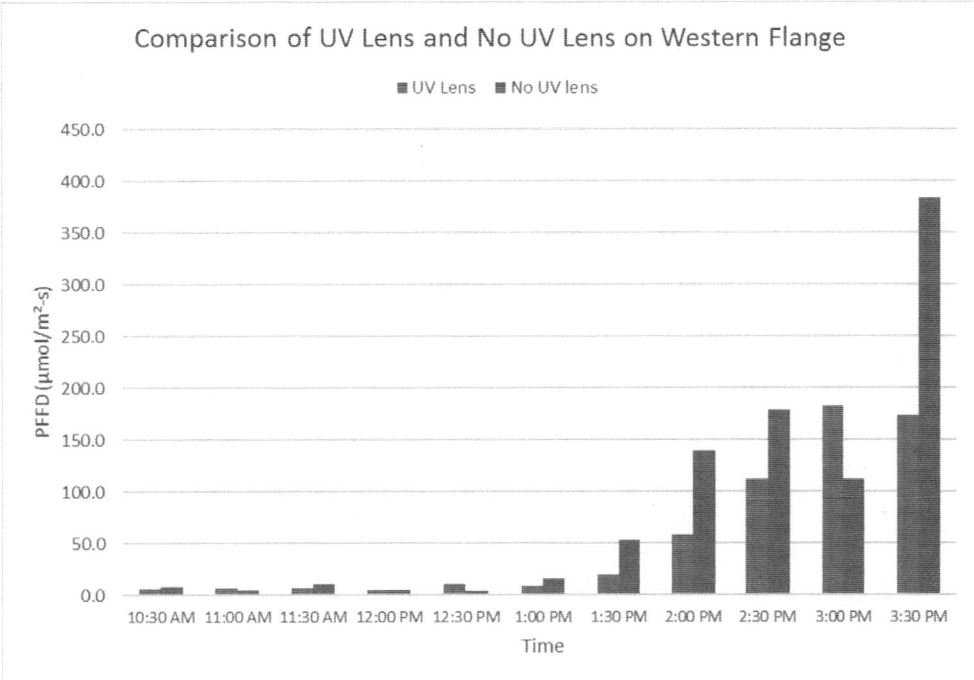

FIGURE 7. Comparing UV and No UV Lens on Western Flange

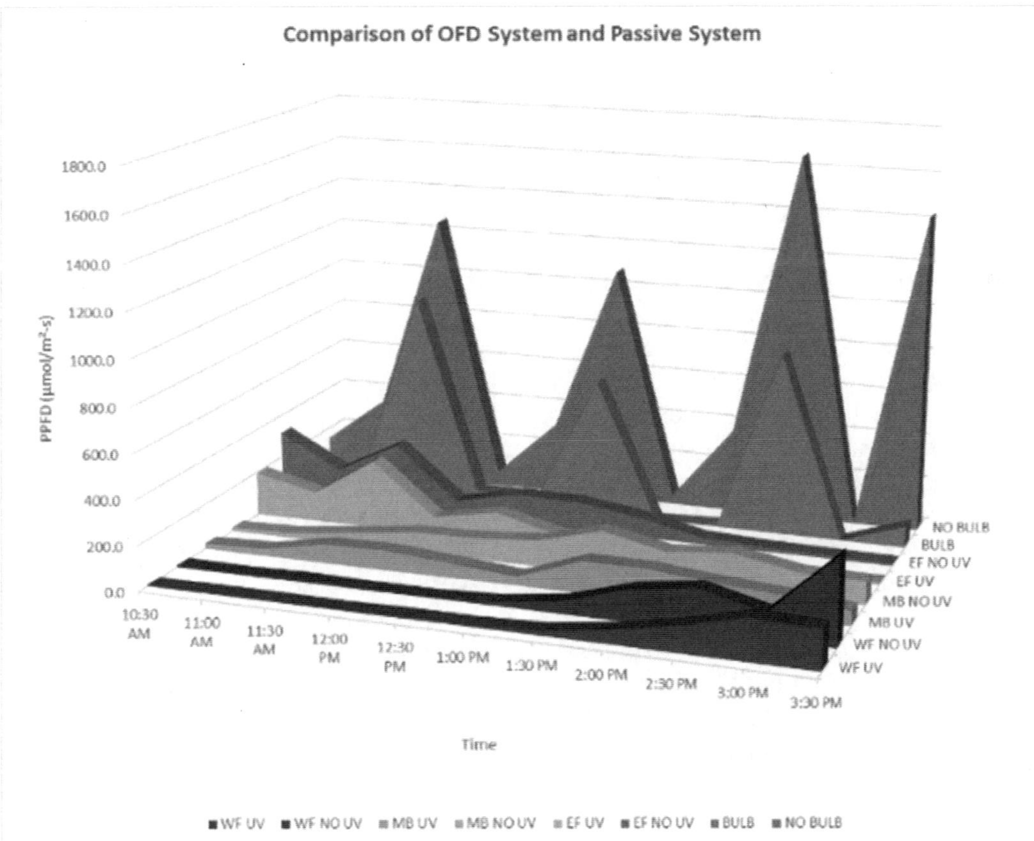

FIGURE 8. Three-Dimensional Plot of All Data Collected on April 11th, 2018

CONCLUSIONS

The active system in general demonstrated better illumination performance during the day but showed inconsistencies which may be contributed to the manual alignment throughout the data collection process. The fixed passive system was successful in providing consistent light albeit at lower PPFD values. The passive system provided a low cost robust design composed of no electrical components and supported large core optical fibers. One major drawback in the passive system is the inability to simultaneously illuminate with comparable levels since the optical fibers are divided in zones of opposed planes. An investigation on the spectral changes throughout both optical fibers will be performed to evaluate the quality of light at different lengths. A chart or table of this information will enable a quick guide to determine optimal length for specialized applications such as the cultivation of indoor crops leading to an accurate cost estimate of optical fiber.

ACKNOWLEDGEMENTS

This work was completed with funding from the US Department of Agriculture (Grant # 2016-3842225540) and the Department of Education-MSEIP program (Grant #P120A140065). The authors would like to thank the USDA, Department of Education, and Texas State University for providing funding and access to both infrastructure and laboratories. The sponsors are not responsible for the content and accuracy of this article. The authors declare that there is no conflict of interest regarding the publication of this paper.

REFERENCES

André, Erik, "Daylighting by optical fiber". 2002. Master's Thesis, *Lulea University of Technology*.

Asiabanpour, B., Almusaied, Z., Aslan, S., Mitchell, M., Leake, E., Lee, H., & ... Bland, A. (2017). "Fixed versus sun tracking solar panels: an economic analysis". *Clean Technologies & Environmental Policy*, 19(4), 1195-1203. doi:10.1007/s10098-016-1292-y

Asiabanpour, B., Estrada, A., Ramirez, R., & Downey, M. (In Press). Optimizing Natural Light Distribution for Indoor Plant Growth Using PMMA Optical Fiber: Simulation and Empirical Study. *Journal of Renewable Energy*.

Bruckner T., I.A. Bashmakov, Y. Mulugetta, H. Chum, A. de la Vega Navarro, J. Edmonds, A. Faaij, B. Fungtammasan, A. Garg, E. Hertwich, D. Honnery, D. Infield, M. Kainuma, S. Khennas, S. Kim, H.B. Nimir, K. Riahi, N. Strachan, R. Wiser, and X. Zhang, 2014: "Energy Systems. In: Climate Change 2014: Mitigation of Climate Change". Contribution of Working Group III to the Fifth Assessment Report of the Intergovernmental Panel on Climate. *Cambridge University Press*, Cambridge, United Kingdom and New York, NY, USA.

Couture, P., Nabbus, H., Al-Azzawi, A., & Havelock, M. (2011). "Improving passive solar collector for fiber optic lighting". *2011 IEEE Electrical Power And Energy Conference*, EPEC 2011, (2011 IEEE Electrical Power and Energy Conference, EPEC 2011), 68-73. doi:10.1109/EPEC.2011.6070255

Cammarata, A. (2015). "Optimized design of a large-workspace 2-DOF parallel robot for solar tracking systems". *Mechanism And Machine Theory*, 83175-186. doi:10.1016/j.mechmachtheory.2014.09.012

Garcia-Hansen, V. R. (2006). "Innovative daylighting systems for deep-plan commercial buildings" (Doctoral dissertation, *Queensland University of Technology*).

Helwa, N. H., Bahgat, A. G., Shafee, A. E., & Shenawy, E. E. (2000). "Maximum Collectable Solar Energy by Different Solar Tracking Systems". *Energy Sources*, 22(1), 23-34. doi:10.1080/00908310050014180

Maatallah, T., El Alimi, S., & Nassrallah, S. B. (2011). "Performance modeling and investigation of fixed, single and dual-axis tracking photovoltaic panel in Monastir city", Tunisia. *Renewable And Sustainable Energy Reviews*, 154053-4066. doi:10.1016/j.rser.2011.07.037

Oh, S. J., Chun, W., Riffat, S. B., Jeon, Y. I., Dutton, S., & Han, H. J. (2013). "Computational analysis on the enhancement of daylight penetration into dimly lit spaces: Light tube vs. fiber optic dish concentrator". *Building And Environment*, 59261-274. doi:10.1016/j.buildenv.2012.08.025

Ullah, I., & Whang, A. J. W. (2015). "Development of optical fiber-based daylighting system and its comparison". *Energies*, 8(7), 7185-7201.

Vu, N., Pham, T., & Shin, S. (2016). "Modified optical fiber daylighting system with sunlight transportation in free space". *Optics Express*, 24(26), A1528-A1545. doi:10.1364/OE.24.0A1528

Renewable Energy Storage & Utilization for Cooling and Desalination: Energy & Environmental Impact Analysis

Muhammad Wakil Shahzad, Muhammad Burhan and Kim Choon Ng (Water Desalination and Reuse Centre, King Abdullah University of Science & Technology, Thuwal, 23955-6900, Saudi Arabia; Muhammad.shahzad@kaust.edu.sa)

The inevitable escalation in economic development have serious implications on energy and environment nexus. The International Energy Outlook 2016 (IEO2016) predicted that the Non Organization for Economic Cooperation and Development (non-OECD) countries will lead with 71% rise in energy demand in contrast with only 18% in developed countries from 2012-2040. In GCC countries, about 50% of primary energy is consumed for cogeneration based power and desalination plants. In the past, many studies were focused on renewable energies based desalination and cooling processes to accommodate 5 fold increase in demand by 2050 but they were not commercialized due to intermittent nature of renewable energy such as solar and wind. We proposed highly efficient energy storage material, Magnesium oxide (MgO), system integrated with innovative hybrid desalination and cooling cycle for future sustainable water & cooling supplies. The condensation of $Mg(OH)_2$ dehydration vapor during day operation with concentrated solar energy and exothermic hydration of MgO at night can produce 24 hour thermal energy without any interruption. Combined system mathematical model was developed and simulation was conducted in System Advisory Model (SAM) and FORTRAN. It was showed that, $Mg(OH)_2$ dehydration vapor condensation produce $120^{O}C$ and MgO hydration exothermic reaction produce $140^{O}C$ heat during day and night operation respectively correspond to energy storage of 81kJ/mol and 41kJ/mol. The produced energy can be utilized to operate desalination and cooling cycle to reduce CO_2 emission and to achieve COP21 goal.

Hydrogen Production from Spirulina Algae by Atmospheric Pressure Microwave Plasma

Sumarlin SHANGDIAR, Ken-Lin Chang, Yu-Chieh Lin, Feng-Chih Chou, Yi-Hsing Hsiao, ***Yuan-Chung Lin****** (Institute of Environmental Engineering, National Sun Yat-Sen University, Kaohsiung 804, Taiwan)

This study investigates hydrogen production from dry spirulina algae under atmospheric-pressure microwave plasma reactor with downstream feeding of the biomass from resonator cavity. Various methodologies are adopted to describe and quantify the chemical composition of biomass and the biomass-produced syngas. By analyzing the sample it was found that the organic compound present contains mainly cellulose and the sum of H_2, CO_2, and CO volume fractions at microwave power of 800, 900, and 1000 are 87.09, 87.52, and 85.22 % with hydrogen conversion rates of 24.31, 32.88, and 42.26 respectively. With the increase in microwave power the hydrogen gas yield accounted to be as twice as that at lower power. The concentration of produced hydrogen accounts for about 30.80%, 33.20%, and 37.58% at microwave power level of 800, 900 and 1000. It was also observed that no methane was produced in this study since most of the methane produced from microwave plasma conversion has reacted with CO_2 and produced CO and H_2, hence the study shows that the concentration of CO_2 dropped as the power decreased.

Role of Microwave Irradiation in Enhancing the Saccharification Rate of Water Hyacinth

Sumarlin SHANGDIAR, Shang-Cyuan Chen, Yu-Chieh Lin, Feng-Chih Chou, Che-An Cho, Yuan-Chung Lin (Institute of Environmental Engineering, National Sun Yat-Sen University, Kaohsiung 804, Taiwan)

Extraction of sugar from free floating water hyacinth (*E. crassipes*) provides a long term solution with lesser cultivation and preparation cost. Sugar generated from aquatic plant biomass with high cellulose and hemicellulose content is considered as a dynamic progression towards the advancement in bio products. It has been renowned that monosaccharides and polysaccharides structures of water hyacinth holds different types of sugars and starch with high cellulose (20 %), hemicellulose (48%) and low lignin (3.5%) content. Water hyacinth can be considered as a possible feedstock for fermentable sugar production by applying different techniques and procedures in biomass treatment. Henceforth, this study investigates the possibility of enhancing the saccharification rate of sugar from water hyacinth (Eichhornia crassipes) by microwave heating method. Using of microwave irradiation successfully helps in loosen the cellulose and deteriorates the lignin from lignocellulosic biomass [Guo et al., 2012, Xia et al., 2013, Zhu et al., 2006]. In Microwave heating, the heat is transmitted to the solution via radiation. Hence, the solution is heated directly without the need for heat transfer through the vessel. This mode of action increases the rate of heating and reduces energy loss during thermal conduction [Kim et al., 2010]. Therefore, Microwave heating has a potential to increase the temperature and cellulosic structure of biomass with the expansion of water molecules. Accordingly, the crystallinity of the cellulose decreases and the hydrolysis efficiency increases. In addition, the factors affecting the hydrolytic saccharification rate of lignocellulose from water hyacinth (*Eichhornia crassipes*) was also studied by taking into consideration different parameters such as reaction temperature, hydrolysis time, solvent volume and dilute sulfuric acid concentration. The method of full factorial experimental design was adopted to explore different parameters that influence the conversion rates of biomass fibers into fermentable sugar. The structural changes of the fibers after microwave treatment with dilute sulfuric acid was observed on the residue by using X- ray Diffractometer (XRD), Fourier Transform Infrared Spectrometry (FTIR) and Environmental Scanning Electron Micrograph (ESEM) analysis. The results obtained illustrates that hydrolysis time reduces to approximately 40% and effectively improve the rate of saccharification to 13.94 % with optimal sugar concentration of 4650 mg/L.

Harvesting Microalgae from Urban Sewage for Biodiesel Production by Magnetic Flocculation Using Nano-Fe_3O_4 Coated with Polyethyleneimine

Yuxi Liu, Hongyi Chen, ***Wenbiao Jin**** and Peng Zhao
(Harbin Institute of Technology, Shenzhen, China)

As one of the most promising new energy, microalgal biodiesel has been widely studied worldwide. However, the conventional microalgae harvesting procedures of high-cost and low-efficiency restrict the development of microalgae biodiesel production. In this study, a novel magnetic flocculation method by using Fe_3O_4@PEI (nano-Fe_3O_4 coated with polyethyleneimine) was developed to harvest microalgae from urban sewage for biodiesel production, while urban sewage treatment was simultaneously realized. During the experiments, the effects of the magnetite nanoparticles' properties and dosage, microalgae cultivation and growth phase, sewage characteristics, and stirring condition were investigated. Based on a series of optimization experiments, the highest microalgae harvesting efficiency of 97% was finally achieved under the Fe_3O_4@PEI dosage of 20 mg/L when microalgae were at their logarithmic growth phase. The growth phase and morphology of microalgae significantly affect the algal harvesting efficiency. Meanwhile, it was observed that the sewage pH adjustment was not needed during the magnetic flocculation of microalgae by using Fe_3O_4@PEI, thus greatly reducing the cost of chemical addition. It has been demonstrated that the magnetic flocculation using nano-Fe_3O_4 coated with polyethyleneimine is an effective way for microalgae harvest from urban sewage, being worthy of further full-scale study for industrialization.